# Atlantis Studies in Differential Equations

Volume 4

## Aims and Scope of the Series

The "Atlantis Studies in Differential Equations" publishes monographs in the area of differential equations, written by leading experts in the field and useful for both students and researchers. Books with a theoretical nature will be published alongside books emphasizing applications.

For more information on this series and our other book series, please visit our website at: www.atlantis-press.com/publications/books

AMSTERDAM – PARIS – BEIJING
ATLANTIS PRESS
Atlantis Press
29, avenue Laumière
75019 Paris, France

More information about this series at www.atlantis-press.com

Stanislav Antontsev · Sergey Shmarev

# Evolution PDEs with Nonstandard Growth Conditions

## Existence, Uniqueness, Localization, Blow-up

ATLANTIS PRESS

Stanislav Antontsev
Center for Mathematics and Fundamental Applications
University of Lisbon
Lisbon
Portugal

Sergey Shmarev
Department of Mathematics
University of Oviedo
Oviedo, Asturias
Spain

ISSN 2214-6253 ISSN 2214-6261 (electronic)
Atlantis Studies in Differential Equations
ISBN 978-94-6239-111-6 ISBN 978-94-6239-112-3 (eBook)
DOI 10.2991/978-94-6239-112-3

Library of Congress Control Number: 2015935210



Printed on acid-free paper

Stanislav Antontsev · Sergey Shmarev

# Evolution PDEs with Nonstandard Growth Conditions

## Existence, Uniqueness, Localization, Blow-up

ATLANTIS PRESS

Stanislav Antontsev
Center for Mathematics and Fundamental Applications
University of Lisbon
Lisbon
Portugal

Sergey Shmarev
Department of Mathematics
University of Oviedo
Oviedo, Asturias
Spain

ISSN 2214-6253 ISSN 2214-6261 (electronic)
Atlantis Studies in Differential Equations
ISBN 978-94-6239-111-6 ISBN 978-94-6239-112-3 (eBook)
DOI 10.2991/978-94-6239-112-3

Library of Congress Control Number: 2015935210



Printed on acid-free paper

*This work is dedicated to our families,*
*Tamara, Nikolay, Stanislav and*
*Elena, Dmitry, Andrey,*
*to whom we owe so much.*

# Preface

This monograph is a contribution to the theory of second order quasilinear parabolic and hyperbolic equations with the nonlinear structure that may change from one point to another in the problem domain. In the past decade, there was an impetuous growth of interest in the study of such equations, which appear in a natural way in the mathematical modeling of various real-world phenomena and give rise to challenging mathematical problems. The aim of this work is to give an account of the known results on existence, uniqueness, and qualitative properties of solutions.

The parabolic equations studied below can be conventionally divided into several groups. Chaps. 2 and 3 are devoted to study the generalized porous medium equation

$$u_t = \operatorname{div}\left(|u|^{m(x,t)}\nabla u\right) + f(x,t) \tag{1}$$

with a given exponent $m(x,t) > -1$ and its generalizations, such as equations with lower order terms or anisotropic equations. We establish conditions of existence and uniqueness of weak solutions and show that for definite ranges of the exponent $m(x,t)$ the solutions exhibit properties typical for the solutions of equations with constant $m$, those of the finite speed of propagation and extinction in finite time. The former means the following: if the support of the initial data is compact, then the support of the solution remains compact for all time but may expand in space with finite speed. The latter property means that the solution corresponding to a nonzero initial datum may extinct in a finite time.

Chapters 4–6 concern the homogeneous Dirichlet problem for the nonlinear degenerate parabolic equations

$$u_t - \operatorname{div}\mathscr{A}(x,t,u,\nabla u) + \mathscr{B}(x,t,u) = 0 \tag{2}$$

with the function $\mathscr{A} \equiv (\mathscr{A}_1, \ldots, \mathscr{A}_n)$ whose components are of the form

$$\mathscr{A}_i(x,t,r,\xi) = a_i(x,t,r)|\xi_i|^{p_i(x,t)-2}\xi_i \quad \forall r \in \mathbb{R}, \xi \in \mathbb{R}^n, \tag{3}$$

where $p_i(x,t) \in (1,\infty)$ are given measurable functions and the coefficients $a_i$ are Carathéodory functions, $a_i \in [a_0, a_1]$ with positive constants $a_0, a_1$. The function $\mathscr{B}$ is assumed to satisfy the growth condition $|\mathscr{B}(x,t,r)| \leq d|r|^{\lambda} + f(x,t)$ with constants $d \geq 0$, $\lambda > 1$ and a given function $f$. Special attention is paid to the model case when

$$\mathscr{B}(x,t,r)r = c(x,t)|r|^{\sigma(x,t)} - f(z)r \quad \forall r \in \mathbb{R} \tag{4}$$

with a continuous exponent $\sigma(x,t) \in (1,\infty)$ and a given coefficient $c(x,t)$. Most of the results remain true if the operators $\mathscr{A}_i$ are substituted by the Leray-Lions operators with variable coercivity and growth conditions.

The assumptions on the functions $\mathscr{A}_i$ show that in the case of variable exponents $p_i$ there is a gap between the coercivity and growth conditions; for this reason such equations are often termed PDEs with nonstandard growth. We extend this name to PDEs of the type (1), to PDEs with anisotropic but possibly constant nonlinearity and to equations with lower order terms of variable growth. All these equations share the same important property: they are not scaling-invariant, which makes inapplicable many of the traditional methods and requires new approaches to the study of their solvability and the analysis of the qualitative properties of solutions.

The main results of Chap. 4 are the theorems of existence and uniqueness of weak solutions. The natural analytic framework for the study is furnished by the Lebesgue and Sobolev spaces with variable exponents which are introduced in Chap. 1. This chapter collects all the information about the properties of these spaces used throughout the text. The constructed weak solutions are the so-called energy solutions which can be taken for the test-function in the corresponding integral identity. Chapter 4 contains results on the dependence of the regularity of the weak energy solutions on the regularity of the data and the nonlinearity exponents $p_i$ and $\sigma$, and on the unique solvability of the Cauchy problem for the evolution $p(x)$-Laplace equation. In the final section we provide a review of other results and approaches to the study of parabolic equations of the type (2).

The properties of space localization are studied in Chap. 5. The study is confined to the energy solutions of Eq. (2) with the lower order terms of the form (4) with $c(x,t) \geq 0$. We show that there are two different mechanisms which can give rise to the property of finite speed of propagation. The first one is due to a suitable balance between the diffusion and absorption part of the equation. This is possible only if $c(x,t) \geq c_0 > 0$ and is expressed in terms of conditions on the variable rates of growth of the functions $\mathscr{A}_i$ and $\mathscr{B}$. The other is caused by the anisotropy of the diffusion operator and works even in the case when $c(x,t) \equiv 0$. It turns that if the anisotropic Eq. (2) combines the directions of slow and fast diffusion, $p_i > 2$ or

$p_i \in (1,2)$ respectively, the disturbances from the data run only a finite or even zero distance in the direction of the slowest diffusion.

Chapter 6 is devoted to the study of the large time behavior and the phenomenon of vanishing in a finite time for energy weak solutions of the homogeneous Dirichlet problem for Eq. (2). It is assumed that the lower order term $\mathscr{B}$ satisfies (4) with $c(x,t) \geq 0$. The study is based on the analysis of behavior of the local energy functions, which satisfy nonlinear ordinary differential equations. The effect of total vanishing in finite time is provided by a suitable relation between the exponents $p_i$, $\sigma$ in conditions (3), (4). Besides, the same effect is possible in several situations specific for equations with the nonstandard growth conditions. It turns out that a solution may vanish in a finite time even if the equation eventually transforms into the linear heat equation, which does not admit localized solutions. Extinction in a finite time is possible also in the case when the coefficient $c$ is allowed to vanish on a set of zero measure. Another effect is due to anisotropy: the equations of anisotropic diffusion admit solutions localized simultaneously in space and time.

In Chap. 7 we derive conditions of nonexistence of global in time bounded solutions to Eq. (2). We prove that under certain conditions on the data every energy solution becomes infinite in a finite time: there exists a finite $t^*$ such that $\|u(\cdot,t)\|_{\infty,\Omega} \to \infty$ as $t \to t^*$. The following two versions of Eq. (2) are considered: the semilinear equation with $\mathscr{A}(x,t,\nabla u) = \nabla u$ and $\mathscr{B}$ satisfying condition (4) with superlinear growth and $c(x,t) < 0$, and the quasilinear equation with the exponents $p_i$ and $\sigma$ independent of $t$. The results are extended to nonlocal equations, equations that become linear as $t$ grows to infinity, equations with nonnegative but not strictly positive coefficient $c(x,t)$.

Parabolic equations with double degeneracy are studied in Chaps. 8–10. Chapter 8 is devoted to study the homogeneous Dirichlet problem for the isotropic equations

$$u_t = \operatorname{div}\left(a|u|^{\alpha(x,t)}|\nabla u|^{p(x,t)-2}\nabla u\right) + f(x,t).$$

It is shown that under suitable conditions on the regularity of $p$ and $\alpha$ this problem has a weak energy solution. Under additional restrictions on the data, the uniqueness and comparison theorems are proven for the solutions that possess better regularity and satisfy the inclusion $\partial_t u \in L^1(Q)$. It is shown that this class is nonempty and every weak solution falls into it, provided that the initial data meet additional regularity assumptions.

Anisotropic doubly degenerate equations are studied in Chaps. 9, 10. We consider the homogeneous Dirichlet problem for the equation

$$\partial_t \Psi(x,t,u) = \operatorname{div}\mathscr{A}(x,t,\nabla u) + c|u|^{\sigma(x,t)-2}u + f \tag{5}$$

with $\Psi = |u|^{m(x,t)-2}u$, given exponents $m$, $\sigma$ and the anisotropic operator $\mathscr{A}$ that satisfies conditions (3). If the exponents $p_i$ and $m$ were constant and the operator $\mathscr{A}$ were isotropic, this equation would be formally equivalent to the equation studied in

Chap. 7, but in the case of Eq. (5) such a reduction is impossible and an independent analysis is required. Chapter 9 is devoted to prove the existence of strong solutions of Eq. (5) which possess the extra regularity property: $m|u|^{m-2}u_t^2 \in L^1(Q)$. Sufficient conditions for local and global in time existence of bounded strong solutions are proven and the energy relations are derived. Unlike the case of weak solutions to Eq. (2) the proof of existence of strong solutions to Eq. (5) requires certain monotonicity properties of the exponents $p_i$, $\sigma$, $m$ and the coefficients $a_i$ and $c(x,t)$.

The energy relations derived in Chap. 9 are used in Chap. 10 to study the phenomenon of extinction in a finite time, the large time behavior and the possibility of a finite time blow-up for strong solutions of Eq. (5).

In Chaps. 11, 12 we present results of the study of quasilinear and semilinear hyperbolic equations with nonstandard growth conditions. The homogeneous Dirichlet problem for the equation

$$u_t = \operatorname{div}\mathscr{A}(x,t,\nabla u) + \varepsilon\Delta u_t + c|u|^{\sigma(x,t)-2}u + f$$

with the isotropic operator $\mathscr{A}$ of the form (3) is studied in Chap. 11. Two different cases are considered: the equation with the damping term, $\varepsilon > 0$, and the equation with $\varepsilon = 0$. It is shown that for every $\varepsilon > 0$ the problem admits global or local in time weak solutions, provided that the exponents of nonlinearity satisfy certain regularity assumptions. The solutions of the damped problem may blow-up in a finite time and the blow-up moment $t^*$ admits the two-sided bounds, which depend on the problem data but are independent of $\varepsilon$. If $\varepsilon = 0$, the question of existence of weak solutions is left open. Nonetheless, it is proven that in this case the problem admits the weaker solution in the sense of Young measure, which is obtained as the limit of the sequence of weak solutions to the damped problems as $\varepsilon \to 0$. Stronger results are obtained in Chap. 12 for the semilinear hyperbolic equation

$$u_{tt} = \Delta u + c(x,t)|u|^{\sigma(x,t)-2}u + f.$$

We derive sufficient conditions for local in time existence of weak and strong solutions and prove uniqueness of a strong solution. It is shown that in the case of superlinear growth, that is, if $\sigma(x,t) > 2$, every nonnegative strong solution blows-up in a finite time. The nonexistence result is extended to equations with nonlocal lower order terms, to equations which transform into the linear wave equation as $t \to \infty$ and to the case when the coefficient $c(x,t)$ is not separated away from zero in the problem domain.

The bulk of the presented material is constituted by the original results of the authors obtained in course of the past 10 years. The other pertinent results scattered in the literature are reviewed in each chapter. Although the selection and presentation of the supplementary material always reflect the interests of the authors, we expect that the reader will find them quite complete.

The authors acknowledge the support of several institutions received at the final stage of the work: the Research Grants CAPES-PVE-88887.059583/2014-00 and

88881.0303888/2013-01 (Brazil), FCT-PTDC/MAT/110613/2009 and FCT-POCI/MAT/61576/2004 (Portugal), MTM2011-26119, MTM2010-18427 and MTM 2008-06208 (Spain).

It is a pleasure to express here our most sincere gratitude to our colleagues and friends whose collaboration, advice, and discussions have contributed to the preparation of the monograph. In no special order, we would like to give our thanks to M. Chipot, B. Andreianov, J.F. Rodrigues, H. Borges de Oliveira, G. Vallet, J. Simsen, M. Stefanello Simsen, J. Ferreira, J.I. Díaz, J.L. Vazquez, V. Zhikov, S. Pastukhova.

Lisbon, Portugal
Oviedo, Spain
January 2015

Stanislav Antontsev
Sergey Shmarev

# Contents


**1 The Function Spaces** 1

1.1 Preliminaries 1

1.2 Lebesgue Spaces with Variable Exponents 4

1.2.1 Definition and Basic Properties 4

1.2.2 Equivalent Norm and Completeness of $L^{p(\cdot)}(\Omega)$ 12

1.2.3 Dense Sets in $L^{p(\cdot)}(\Omega)$ 16

1.3 Sobolev Spaces with Variable Exponents 18

1.4 Anisotropic Spaces of Functions Depending on $x$ and $t$ 22

1.4.1 Approximation by Smooth Functions 23

1.4.2 A Subspace of W (Q) 27

1.5 Formulas of Integration by Parts 29

1.6 Embedding Theorems in Anisotropic Spaces 34

1.7 Interpolation Inequalities 40

**2 A Porous Medium Equation with Variable Nonlinearity** 45

2.1 Introduction 45

2.2 Model Equation: Assumptions and Results 46

2.3 Regularization 47

2.4 A Priori Estimates 48

2.5 Passage to the Limit 54

2.6 Uniqueness of Weak Solutions 56

2.7 Equations with Lower-Order Terms 60

2.8 Equations with Anisotropic Nonlinearity 69

2.9 Stationary Solutions 70

2.10 Remarks 79

**3 Localization of Solutions of the Generalized Porous Medium Equation** 81

3.1 Notation and Assumptions 82

3.2 Finite Speed of Propagation 83

3.3 Formula of Integration by Parts: The Energy Relations . . . . . 84
3.3.1 Ordinary Differential Inequality for the Energy Function . . . . . 88
3.3.2 Analysis of the Differential Inequality . . . . . 90
3.4 The Waiting Time Effect . . . . . 91
3.5 Asymptotic Behavior for Large $t$ . . . . . 93
3.6 Vanishing in a Finite Time . . . . . 96
3.7 A Limit Case in the Finite Time Vanishing: $\gamma^+ = 0$ . . . . . 97
3.8 Nonlinearity of Other Type: An Example . . . . . 100
3.8.1 Asymptotic Behavior for Large $t$ . . . . . 102
3.8.2 Finite Speed of Propagation . . . . . 103
3.9 Bibliographic Remarks . . . . . 104

**4 Anisotropic Equations with Variable Growth and Coercivity Conditions** . . . . . 107
4.1 Introduction . . . . . 107
4.2 Existence Theorems . . . . . 110
4.2.1 Galerkin's Approximations . . . . . 112
4.2.2 A Priory Estimates . . . . . 112
4.2.3 Compactness and Passage to the Limit . . . . . 116
4.3 $L^\infty$ estimates . . . . . 122
4.3.1 Global Estimates . . . . . 122
4.3.2 Global Existence via Boundedness . . . . . 126
4.3.3 Local Existence via Boundedness . . . . . 129
4.4 Uniqueness Theorems . . . . . 130
4.5 Global Regularity for a Class of Model Equations . . . . . 136
4.6 Singular Perturbation of the $p(x,t)$-Laplace Operator . . . . . 140
4.7 The Cauchy Problem . . . . . 144
4.7.1 Steklov's Means . . . . . 146
4.7.2 Problems in Bounded Cylinders . . . . . 150
4.7.3 Unbounded Domain . . . . . 157
4.7.4 Boundedness and Uniqueness of Weak Solutions . . . . . 159
4.8 Stationary Solutions . . . . . 160
4.8.1 A Model Equation . . . . . 162
4.8.2 The General Case . . . . . 163
4.8.3 Bounded Weak Solutions . . . . . 167
4.8.4 Systems of Elliptic Equations . . . . . 169
4.9 Remarks . . . . . 171
4.9.1 Renormalized and Entropy Solutions . . . . . 176
4.9.2 Equation with Measurable Exponent $p(x,t)$ . . . . . 178
4.9.3 On the Regularity of Weak Solutions . . . . . 181

**5 Space Localization of Energy Solutions** . . . . . . . . . . . . . . . . . . . . 185
5.1 Introduction . . . . . . . . . . . . . . . . . . . . . . . . . . . . . . . . . . . . 185
5.2 Localization via Diffusion-Absorption Balance . . . . . . . . . . . 187
5.2.1 The Basic Energy Relation . . . . . . . . . . . . . . . . . . 188
5.2.2 Finite Speed of Propagation of Disturbances . . . . . . 189
5.2.3 The Waiting Time Property . . . . . . . . . . . . . . . . . 197
5.3 Localization Caused by Anisotropic Diffusion . . . . . . . . . . . 198
5.3.1 Differential Inequality for the Energy Function . . . . 201
5.3.2 Directional Stable Localization . . . . . . . . . . . . . . . 204
5.3.3 Infinite Waiting Time . . . . . . . . . . . . . . . . . . . . . 205
5.4 Localized Stationary Solutions . . . . . . . . . . . . . . . . . . . . . 208
5.5 Remarks . . . . . . . . . . . . . . . . . . . . . . . . . . . . . . . . . . . . 213
5.5.1 Equations of General Form. . . . . . . . . . . . . . . . . . 214
5.5.2 On the Cauchy Problem. . . . . . . . . . . . . . . . . . . . 215
5.5.3 Localized Unbounded Solutions: An Example. . . . . 216

**6 Extinction in a Finite Time and the Large Time Behavior** . . . . . . 219
6.1 Introduction . . . . . . . . . . . . . . . . . . . . . . . . . . . . . . . . . . . . 219
6.2 The Energy Relations . . . . . . . . . . . . . . . . . . . . . . . . . . . . 221
6.2.1 Estimate on the Total Energy . . . . . . . . . . . . . . . . 222
6.3 Differential Inequality for the Energy Function . . . . . . . . . . . 224
6.4 Vanishing in a Finite Time. . . . . . . . . . . . . . . . . . . . . . . . . 227
6.4.1 Sufficient Conditions for the Finite Time Extinction . . . . . . . . . . . . . . . . . . . . . . . . . 227
6.4.2 On the Balance Between Slow and Fast Diffusion. . . . . . . . . . . . . . . . . . . . . . . . . . 229
6.5 Vanishing at a Prescribed Moment . . . . . . . . . . . . . . . . . . . 231
6.6 Large Time Behavior. . . . . . . . . . . . . . . . . . . . . . . . . . . . . 234
6.7 Limit Cases . . . . . . . . . . . . . . . . . . . . . . . . . . . . . . . . . . . 236
6.8 Remarks . . . . . . . . . . . . . . . . . . . . . . . . . . . . . . . . . . . . 238

**7 Blow-up in Equations with Variable Nonlinearity**. . . . . . . . . . . . . 241
7.1 Semilinear Equation with Variable Source . . . . . . . . . . . . . . 242
7.1.1 The Differential Inequality . . . . . . . . . . . . . . . . . . 243
7.1.2 Analysis of the Differential Inequality . . . . . . . . . . 245
7.2 Equations with Vanishing Sources. . . . . . . . . . . . . . . . . . . . 247
7.3 Various Extensions . . . . . . . . . . . . . . . . . . . . . . . . . . . . . . 250
7.3.1 Eventually Linear Equations . . . . . . . . . . . . . . . . . 250
7.3.2 Equations with Vanishing Sources. . . . . . . . . . . . . 251
7.3.3 Regional Blow-up . . . . . . . . . . . . . . . . . . . . . . . . 251
7.3.4 Equations with Nonlocal Sources . . . . . . . . . . . . . 252
7.3.5 Semilinear Equations of General Form. . . . . . . . . . 254

7.4 Equations with Anisotropic $p(x)$-Growth . . . . . . . . . . . . . . . 255
7.4.1 The Energy Relations . . . . . . . . . . . . . . . . . . . . . . 257
7.4.2 Second-Order Differential Inequalities . . . . . . . . . . 259
7.5 Remarks . . . . . . . . . . . . . . . . . . . . . . . . . . . . . . . 261
7.5.1 Higher-Order Equations . . . . . . . . . . . . . . . . . . . . 262
7.5.2 $p(x,t)$-Curl Systems Arising in Electromagnetism . . . . . . . . . . . . . . . . . . . . . . 263
7.5.3 Navier-Stokes Type Equations with Variable Structure . . . . . . . . . . . . . . . . . . . . . . . 264

**8 Equations with Double Isotropic Nonlinearity** . . . . . . . . . . . . . . . 267
8.1 Assumptions and Results . . . . . . . . . . . . . . . . . . . . . . . . . 268
8.2 Regularized Problems . . . . . . . . . . . . . . . . . . . . . . . . . . . 270
8.2.1 A Priori Estimates . . . . . . . . . . . . . . . . . . . . . . . 271
8.3 Existence of Weak Solution . . . . . . . . . . . . . . . . . . . . . . . 274
8.3.1 Step 1: $J_1^{(\varepsilon)} \to 0$ as $\varepsilon \to 0$. . . . . . . . . . . . . . . . . . 275
8.3.2 Step 2: $J_2^{(\varepsilon)} \to F(v,v)$ as $\varepsilon \to 0$. . . . . . . . . . . . . . . 276
8.4 Comparison Principle and Uniqueness for Strong Solutions . . . . . . . . . . . . . . . . . . . . . . . . . . . 278
8.4.1 Proof of Theorem 8.2: The case $a \equiv 1$. . . . . . . . . . 279
8.4.2 Proof of Theorem 8.3: The case $\alpha = 0$. . . . . . . . . . 284
8.5 Existence of Solutions $u \in \mathscr{V}(Q)$: $L^1$-Estimate for $\partial_t \Phi(z,v)$ . . . . . . . . . . . . . . . . . . . . . . . . . . . 286

**9 Strong Solutions of Doubly Nonlinear Anisotropic Equations** . . . . 291
9.1 Existence of Strong Solutions. . . . . . . . . . . . . . . . . . . . . . . 292
9.2 Regularized Problem . . . . . . . . . . . . . . . . . . . . . . . . . . . 295
9.2.1 Galerkin's Approximations. . . . . . . . . . . . . . . . . . 296
9.2.2 A Priori Estimates. . . . . . . . . . . . . . . . . . . . . . . 297
9.3 The Energy Inequality . . . . . . . . . . . . . . . . . . . . . . . . . . 299
9.4 Weak Solution of Regularized Problem . . . . . . . . . . . . . . . 305
9.5 Bounded Solutions of Regularized Problem . . . . . . . . . . . . . 309
9.6 Proof of the Existence Theorem . . . . . . . . . . . . . . . . . . . . 314
9.7 The Energy Relations for Strong Solutions. . . . . . . . . . . . . . 316

**10 Anisotropic Equations with Double Nonlinearity: Blow-up and Vanishing** . . . . . . . . . . . . . . . . . . . . . . . . . . . . . . 319
10.1 Strong Solutions . . . . . . . . . . . . . . . . . . . . . . . . . . . . . . 321
10.2 Sufficient Conditions of the Finite Time Blow-up . . . . . . . . . 322
10.3 Embedding Inequalities in Anisotropic Spaces . . . . . . . . . . . 327
10.4 Ordinary Differential Inequalities for the Energy Function . . . 331

10.5 Extinction in a Finite Time . . . . . . . . . . . . . . . . . . . . . . . . . 335
10.5.1 The Regular Case . . . . . . . . . . . . . . . . . . . . . . . . . 335
10.5.2 The Limit Case . . . . . . . . . . . . . . . . . . . . . . . . . . 337
10.5.3 Equations with Critical Growth . . . . . . . . . . . . . . . 339
10.6 Large Time Behavior . . . . . . . . . . . . . . . . . . . . . . . . . . . 340

**11 Wave Equation with $p(x,t)$-Laplacian** . . . . . . . . . . . . . . . . . . . . . 343
11.1 Local and Global Existence of Weak Solutions . . . . . . . . . . . 345
11.1.1 Galerkin's Approximations . . . . . . . . . . . . . . . . . . 347
11.1.2 Compactness and Passage to the Limit . . . . . . . . . . 354
11.2 Nonexistence of Global Solutions: Finite Time Blow up . . . . 359
11.3 Young Measure Solutions . . . . . . . . . . . . . . . . . . . . . . . . 364
11.3.1 Passage to the Limit as $\varepsilon \to 0$ . . . . . . . . . . . . . . . 365
11.3.2 Lower Estimates for $\|u^{\varepsilon}\|_{\sigma,\Omega}$ Blow up . . . . . . . . . . 367

**12 Semilinear Hyperbolic Equations** . . . . . . . . . . . . . . . . . . . . . . . . 369
12.1 Weak and Strong Solutions . . . . . . . . . . . . . . . . . . . . . . . 370
12.2 Local Existence Theorems . . . . . . . . . . . . . . . . . . . . . . . . 371
12.3 Nonexistence of Global Nonnegative Solutions . . . . . . . . . . . 381
12.3.1 Ordinary Differential Inequalities . . . . . . . . . . . . . 381
12.3.2 Blow-up . . . . . . . . . . . . . . . . . . . . . . . . . . . . . . 385
12.3.3 Regional Blow-up . . . . . . . . . . . . . . . . . . . . . . . . 388
12.4 Semilinear Equations of General Form . . . . . . . . . . . . . . . . 389

**References** . . . . . . . . . . . . . . . . . . . . . . . . . . . . . . . . . . . . . . . . 393

**Index** . . . . . . . . . . . . . . . . . . . . . . . . . . . . . . . . . . . . . . . . . . . 407

# Chapter 1
# The Function Spaces

This chapter is devoted to a brief exposition of the theory of function spaces that provide the analytic framework for the study of PDEs with variable nonlinearity. These are the Lebesgue and Sobolev spaces with variable exponents, which can be regarded as special cases of Orlicz's spaces [215], or semi-modular spaces [211–213]. The theory of such spaces is very interesting in itself, this challenging subject has been developing very rapidly in the last decades. A detailed discussion of the theory of Lebesgue and Sobolev spaces with variable exponents is beyond the scope of the monograph, for this reason we confine ourselves to outlining a minimal set of properties needed in the further proceeding.

Our presentation follows the papers [49, 127, 128, 184, 275]. A deep and thorough insight into the theory of variable exponent Lebesgue and Sobolev spaces, as well as the discussion of the available bibliography on this issue, can be found in the monographs [105] and [115].

## 1.1 Preliminaries

Prior to defining the spaces with variable exponents, let us introduce several notation and recall a few well-known facts most heavily used throughout the text.

Unless specially indicated, $\Omega$ always stands for a bounded simple-connected domain in $\mathbb{R}^n$ with Lipschitz-continuous boundary $\partial\Omega$. The notation

$$Q = \{(x, t) : x \in \Omega,\ t \in (0, T)\}$$

is used for a generic cylinder in $\mathbb{R}^n \times (0, \infty)$ with the base $\Omega$ and an arbitrary finite height $T < \infty$. The lateral boundary of $Q$ is denoted by $\Gamma = \partial\Omega \times (0, T)$. If the



S. Antontsev and S. Shmarev, *Evolution PDEs with Nonstandard Growth Conditions*, Atlantis Studies in Differential Equations 4,
DOI 10.2991/978-94-6239-112-3_1

value of $T$ is important for the proceeding, we use the notation $Q_T$ and $\Gamma_T$. For the sake of brevity, for the points of the cylinder $Q = \Omega \times (0, T)$ we often use the notation $z = (x, t)$.

Given a function $\phi(x, t) \in C^0(\overline{Q})$, by convention we denote

$$\begin{aligned} \phi^+(t) &= \sup_{\Omega} \phi(x, t), \qquad & \phi^-(t) &= \inf_{\Omega} \phi(x, t), \\ \phi^+ &= \sup_{Q} \phi(x, t), \qquad & \phi^- &= \inf_{Q} \phi(x, t). \end{aligned} \tag{1.1}$$

The Banach space $L^q(\Omega)$ with constant $q \in (1, \infty)$ is the set of measurable functions with bounded norm

$$\|u\|_{q,\Omega} = \left(\int_{\Omega} |u|^q \, dx\right)^{\frac{1}{q}}.$$

By $L^\infty(\Omega)$ we denote the set

$$L^\infty(\Omega) = \left\{ u : \Omega \mapsto \mathbb{R} \;\middle|\; \begin{array}{l} u \text{ is measurable, } |u| \le M \text{ a.e. in } \Omega \\ \text{for some finite constant } M \end{array} \right\}.$$

The norm of $L^\infty(\Omega)$ is defined by

$$\|u\|_{\infty,\Omega} = \inf\{M > 0 : \; |u| \le M \text{ a.e. in } \Omega\}.$$

The space $L^2(\Omega)$ is a Hilbert space with the inner product

$$(u, v)_{2,\Omega} = \int_{\Omega} uv \, dx.$$

$W^{k,q}(\Omega)$, $k \ge 1$, $1 \le q \le \infty$, is the space of $k$ times weakly differentiable functions with bounded norm

$$\|u\|_{W^{k,q}(\Omega)} = \left(\int_{\Omega} \sum_{0 \le |\alpha| \le k} |D^\alpha u|^q \, dx\right)^{\frac{1}{q}},$$

where $\alpha$ is the multi-index, $\alpha = (\alpha_1, \dots, \alpha_n)$ with integer $\alpha_i \ge 0$, $|\alpha| = \sum_{i=1}^{n} \alpha_i$.

The space $W_0^{k,p}(\Omega)$ is the closure of $C_0^\infty(\Omega)$ (the set of smooth functions with compact support in $\Omega$) in the norm of $W^{k,p}(\Omega)$. An equivalent norm of $W_0^{1,p}(\Omega)$ is given by

$$\|u\|_{W_0^{1,p}(\Omega)} = \|\nabla u\|_{p,\Omega}.$$

In the case $p = 2$ we denote $H^k(\Omega) = W^{k,2}(\Omega)$ and $H_0^k(\Omega) = W_0^{k,2}(\Omega)$. The inner product in the Hilbert space $H_0^k(\Omega)$, $k \geq 1$, is defined by

$$(u, v)_{H_0^k(\Omega)} = \sum_{0 \leq |\alpha| \leq k} (D^\alpha u, D^\alpha v)_{2,\Omega}.$$

We denote by $L^p(0, T; L^q(\Omega))$, $p, q \geq 1$, the space of measurable functions with the bounded norm

$$\|u\|_{L^q(0,T;L^p(\Omega))} = \left(\int_0^T \|u(\cdot, t)\|_{p,\Omega}^q \, dt\right)^{\frac{1}{q}}.$$

In the case $p = q$ we write $\|u\|_{L^p(0,T;L^p(\Omega))} = \|u\|_{p,Q}$. The notation $L^q(0, T; W_0^{1,p}(\Omega))$ stands for the space of functions which are measurable, weakly differentiable in $x$, and bounded in the norm

$$\|u\|_{L^q(0,T;W_0^{1,p}(\Omega))} = \left(\int_0^T \|\nabla u(\cdot, t)\|_{p,\Omega}^q \, dt\right)^{\frac{1}{q}}.$$

For every $f \in L^p(\Omega)$, $g \in L^{p'}(\Omega)$ with $1 \leq p \leq \infty$ and $p' = \dfrac{p}{p-1}$ *the Hölder inequality* holds:

$$\int_\Omega |f||g| \, dx \leq \|f\|_{p,\Omega} \|g\|_{p',\Omega}.$$

For all nonnegative real numbers $x, y$ and every $p \in (1, \infty)$ *the Young inequality* holds:

$$xy \leq \frac{1}{p} x^p + \frac{1}{p'} y^{p'}, \quad p' = \frac{p}{p-1}.$$

We will often use this inequality in the following form: for every $\varepsilon > 0$, $p \in (1, \infty)$ and real nonnegative $x, y$

$$xy = \left((\varepsilon p)^{\frac{1}{p}} x (\varepsilon p)^{-\frac{1}{p}} y\right) \leq \varepsilon x^p + \frac{(p-1)\varepsilon^{-\frac{1}{p-1}}}{p^{p'}} y^{p'}.$$

*The Gronwall inequality.* Let a nonnegative function $v(t) : [0, T) \mapsto \mathbb{R}$ satisfy the differential inequality

$$v'(t) \leq a(t)v(t) + f(t) \quad \text{a.e. in } (0, T), \qquad v(0) = v_0,$$

with given $a(t), f(t) \in L^1(0,T)$. Then for a.a. $t \in (0,T)$

$$v(t) \le v_0 \exp\left(\int_0^t a(\tau)\,d\tau\right) + \int_0^t f(\tau) \exp\left(\int_\tau^t a(s)\,ds\right) d\tau.$$

A function $f(z,r) : Q \times \mathbb{R} \mapsto \mathbb{R}$ is called *Carathéodory function* if $f(z,r)$ is measurable in $Q$ for every $r \in \mathbb{R}$ and continuous with respect to $r$ for a.e. $z \in Q$.

The notation $C$ is reserved for constants which can be explicitly calculated, or estimated, but whose precise value is unimportant and may change from line to line.

## 1.2 Lebesgue Spaces with Variable Exponents

### *1.2.1 Definition and Basic Properties*

Let

$$\begin{cases} \Omega \subset \mathbb{R}^n \text{ be a bounded open domain} \\ \text{with the Lipschitz-continuous boundary } \partial\Omega \end{cases} \tag{1.2}$$

and

$$p : \Omega \mapsto (1, \infty) \text{ be a measurable function.} \tag{1.3}$$

On the set of all functions $f : \Omega \mapsto \mathbb{R}$ we define the functional

$$A_{p(\cdot)}(f) = \int_\Omega |f(x)|^{p(x)}\,dx < \infty$$

and introduce the set

$$L^{p(\cdot)}(\Omega) = \left\{f \text{ measurable on } \Omega : A_{p(\cdot)}(f) < \infty\right\},$$

which is a linear space. It is easy to check that

(1) $A_{p(\cdot)}(f) \ge 0$ for every $f$,
(2) $A_{p(\cdot)}(f) = 0$ if and only if $f = 0$,
(3) $A_{p(\cdot)}(f) = A_{p(\cdot)}(-f)$ for every $f$,
(4) $A_{p(\cdot)}(f)$ is convex.

Every functional that meets properties (1)–(4) is called *convex modular*.

The modular $A_{p(\cdot)}(f)$ possesses also the properties

(5) if $|f(x)| \ge |g(x)|$ for a.e. $x \in \Omega$ and if $A_{p(\cdot)}(f) < \infty$, then $A_{p(\cdot)}(f) \ge A_{p(\cdot)}(g)$, the inequality is strict if $|f| \not\equiv |g|$,
(6) if $0 < A_{p(\cdot)}(f) < \infty$, then the function $\lambda \mapsto A_{p(\cdot)}(f/\lambda)$ is continuous and decreasing on the interval $[1, \infty)$

Property (5) is obvious. To prove property (6) we notice that for a.e. $x \in \Omega$ $|f(x)/\lambda|^{p(x)}$ is monotone decreasing as a function of $\lambda > 0$, and that $A_{p(\cdot)}(f/\lambda)$ is monotone by virtue of (5). Let $0 < A_{p(\cdot)}(f/\lambda) < \infty$ and $\lambda_k \searrow \lambda$. Then $|f(x)/\lambda_k| \nearrow |f(x)/\lambda|$ and the continuity of $A_{p(\cdot)}(f/\lambda)$ follows from the monotone convergence theorem.

Let us introduce the functional

$$\|f\|_{p(\cdot),\Omega} \equiv \|f\|_{L^{p(\cdot)}(\Omega)} = \inf\left\{\lambda > 0:\ A_{p(\cdot)}\left(\frac{f}{\lambda}\right) \leq 1\right\}. \tag{1.4}$$

Whenever it doesn't cause a confusion, we use the shorthand $\|f\|_{p(\cdot),\Omega} = \|f\|_p$.

**Proposition 1.1** *If* $0 < \|f\|_p < \infty$, *then* $A_p(f/\|f\|_p) \leq 1$.

*Proof* Take a sequence $\{\gamma_k\}$ such that $\gamma_k \searrow \|f\|_p$. By virtue of the definition of $\|f\|_p$, property (6) and the Fatou Lemma

$$A_p(f/\|f\|_p) \leq \liminf_{k\to\infty} A_p(f/\gamma_k) \leq 1.$$

□

**Corollary 1.1** *If* $0 < \|f\|_p \leq 1$, *then* $A_p(f) \leq 1$.

*Proof*

$$A_p(f) = \int_\Omega |f(x)|^{p(x)}\,dx \leq \int_\Omega \frac{|f(x)|^{p(x)}}{\|f\|_p^{p(x)}}\,dx = A_p(f/\|f\|_p) \leq 1.$$

□

**Proposition 1.2** (Luxemburg's norm of $L^{p(\cdot)}(\Omega)$) *The functional*

$$\|\cdot\|_{p(\cdot),\Omega} : L^{p(\cdot)}(\Omega) \mapsto [0,\infty)$$

*defines a norm of the space* $L^{p(\cdot)}(\Omega)$.

*Proof* We have to check that

1. $\|f\|_{p(\cdot),\Omega} \geq 0$ for every $f \in L^{p(x)}(\Omega)$;
2. $\|f\|_{p(\cdot),\Omega} = 0 \quad \Leftrightarrow \quad f = 0$ a.e. in $\Omega$;
3. $\forall\, \mu \in \mathbb{R},\ f \in L^{p(x)}(\Omega)\quad \|\mu f\|_{p(\cdot),\Omega} = |\mu|\|f\|_{p(\cdot),\Omega}$;
4. $\forall\, f,\ g \in L^{p(x)}(\Omega)\quad \|f+g\|_{p(\cdot),\Omega} \leq \|f\|_{p(\cdot),\Omega} + \|g\|_{p(\cdot),\Omega}$.

1. The first property is obvious.

2. Let us denote

$$\mathscr{I}(f) = \left\{\lambda > 0:\ A_{p(\cdot)}(f/\lambda) \leq 1\right\}.$$

It is clear that for $f = 0$ we have $\mathscr{I}(0) = (0, \infty)$. If $\|f\|_{p(\cdot),\Omega} \neq 0$, it follows from property (6) of the modular $A_{p(\cdot)}(\cdot)$ and Proposition 1.1 that $\mathscr{I}(f) = [\lambda, \infty)$ with $\lambda = \|f\|_p$. By (1.4)

$$\|f\|_p = \inf\ \mathscr{I}(f),$$

which is why for $f = 0$ we have $\|f\|_p = \inf \mathscr{I}(0) = 0$. Let us assume, for contradiction, that there exists $f \in L^{p(x)}(\Omega)$ such that $\|f\|_{p(\cdot),\Omega} = 0$, but $f \neq 0$ on a set of nonzero measure in $\Omega$. This means that

$$\int_\Omega |f(x)|^{p(x)}\, dx \geq \varepsilon > 0$$

for some $\varepsilon > 0$. Since $\inf \mathscr{I}(f) = 0$, there is a sequence $\{\lambda_k\} \subset \mathscr{I}(f) \cap (0, 1)$ such that $\lambda_k \to 0$ and $A_{p(\cdot)}(f/\lambda_k) \leq 1$, which is impossible because

$$1 \geq A_p(f/\lambda_k) = \int_\Omega |f/\lambda_k|^{p(x)}\, dx \geq \frac{1}{\lambda_k}\int_\Omega |f|^{p(x)}\, dx \geq \frac{\varepsilon}{\lambda_k} \to \infty \quad \text{as } k \to \infty.$$

3. For every constant $\mu \in \mathbb{R}$

$$\begin{aligned}
\|\mu f\|_p &= \inf\left\{\lambda > 0 : A_p(\mu f/\lambda)\right\} \\
&= \inf\left\{\lambda > 0 : \int_\Omega |\mu|^{p(x)} \left|\frac{f(x)}{\lambda}\right|^{p(x)} dx \leq 1\right\} \\
&= \inf\left\{\lambda > 0 : \int_\Omega \left|\frac{f(x)}{\lambda/|\mu|}\right|^{p(x)} dx \leq 1\right\} \\
&= \inf\left\{|\mu|\delta > 0 : \int_\Omega \left|\frac{f(x)}{\delta}\right|^{p(x)} dx \leq 1\right\} \\
&= |\mu| \inf\left\{\delta > 0 : A_p(\mu f/\delta) \leq 1\right\} \\
&= |\mu|\|f\|_{p(\cdot),\Omega}.
\end{aligned}$$

4. Let us introduce the set

$$\mathscr{M} = \left\{f \in L^{p(x)}(\Omega) :\ A_p(f) \leq 1\right\}$$

and notice that $\mathscr{I}(f) = \{\lambda > 0 :\ f/\lambda \in \mathscr{M}\}$. Since $A_p(\cdot)$ is convex, the set $\mathscr{M}$ is also convex. For every $f,\ g \in L^{p(x)}(\Omega)$ and an arbitrary $\varepsilon > 0$

$$\frac{f}{\varepsilon + \|f\|_p},\ \frac{g}{\varepsilon + \|g\|_p} \in \mathscr{M},$$

whence

$$\frac{\theta f}{\varepsilon + \|f\|_p} + \frac{(1-\theta)g}{\varepsilon + \|g\|_p} \in \mathscr{M} \qquad \forall \theta \in (0, 1)$$

due to the convexity of $\mathscr{M}$. Let us choose $\theta$ in the special way:

$$\frac{\theta}{\varepsilon + \|f\|_p} = \frac{1-\theta}{\varepsilon + \|g\|_p} \quad \Leftrightarrow \quad \frac{\theta f}{\varepsilon + \|f\|_p} + \frac{(1-\theta)g}{\varepsilon + \|g\|_p} = \frac{f+g}{\|f\|_p + \|g\|_p + 2\varepsilon} \in \mathscr{M}.$$

It follows that for every $\varepsilon > 0$

$$\|f\|_p + \|g\|_p + 2\varepsilon \in \mathscr{I}(f+g),$$

that is,

$$\|f+g\|_p \leq \|f\|_p + \|g\|_p + 2\varepsilon.$$

Since $\varepsilon$ is arbitrary, the assertion follows. □

It follows from Proposition 1.2 that the set $L^{p(\cdot)}(\Omega)$ equipped with the Luxemburg norm $\|\cdot\|_{p(\cdot),\Omega}$ is a normed linear space. In the special case $p = const$ the modular $A_p(\cdot)$ generates the Banach space $L^p(\Omega)$ with the norm $\|f\|_{p,\Omega} = A_p^{\frac{1}{p}}(f)$, which is the classical Lebesgue space $\left(L^p(\Omega)| \ \|\cdot\|_{p,\Omega}\right)$.

The space $L^{p(\cdot)}(\Omega)$ is a special case of the Orlicz-Musielak spaces $L^M(\Omega)$. $L^M(\Omega)$ is constituted by all functions $f$ on $\Omega$ such that

$$\int_\Omega M(x, \lambda f(x))dx < \infty \quad \text{for some } \lambda > 0,$$

where $M : \Omega \times \mathbb{R} \mapsto [0, \infty)$ is a nonnegative measurable function, lower semicontinuous, convex, even for a.e. $x \in \Omega$, and satisfying the condition $\lim_{u\to 0} M(x,u) = M(x,0) = 0$. For the Lebesgue-Orlicz spaces $L^{p(\cdot)}(\Omega) = L^M(\Omega)$ with $M(x,u) = |u|^{p(x)}$.

Let us denote

$$p^- = \operatorname{ess\,inf}_\Omega p(x), \qquad p^+ = \operatorname{ess\,sup}_\Omega p(x).$$

**Proposition 1.3** *If $p^+ < \infty$, then for every $f$ with $0 < \|f\|_{p(\cdot),\Omega} < \infty$*

$$A_{p(\cdot)}\left(f/\|f\|_{p(\cdot),\Omega}\right) = 1.$$

*Proof* By virtue of Proposition 1.1 we only have to check that the inequality $A_{p(\cdot)}\left(f/\|f\|_{p(\cdot),\Omega}\right) < 1$ is impossible. For every $0 < \lambda \leq \|f\|_{p(\cdot),\Omega}$

$$\begin{aligned} A_{p(\cdot)}(f/\lambda) &= \int_\Omega \frac{|f|^{p(x)}}{\lambda^{p(x)}}\,dx \\ &= \int_\Omega \left(\frac{|f|}{\|f\|_{p(\cdot),\Omega}}\right)^{p(x)} \left(\frac{\|f\|_{p(\cdot),\Omega}}{\lambda}\right)^{p(x)} dx \\ &\leq \left(\frac{\|f\|_{p(\cdot),\Omega}}{\lambda}\right)^{p^+} A_p(f/\|f\|_{p(\cdot),\Omega}). \end{aligned}$$

If $A_{p(\cdot)}(f/\|f\|_{p(\cdot),\Omega}) < 1$, we may choose $\lambda < \|f\|_{p(\cdot),\Omega}$ such that $A_{p(\cdot)}(f/\lambda) \le 1$, which contradicts the definition of the norm $\|f\|_{p(\cdot),\Omega}$. □

**Corollary 1.2** *For $p^+ < \infty$ properties* (2) *and* (4) *of the modular $A_{p(\cdot)}(f)$ together with* Proposition 1.3 *yield:*

$$\text{if } \|f\|_{p(\cdot),\Omega} \le 1, \text{ then } A_{p(\cdot)}(f) \le \|f\|_{p(\cdot),\Omega}.$$

Let us now assume that

$$\forall \text{ a.e. } x \in \Omega \qquad p(x) \in [p^-, p^+] \subset (1, \infty). \tag{1.5}$$

The relation between the norm of $L^{p(\cdot)}(\Omega)$ and the modular $A_{p(\cdot)}(\cdot)$ is given in the following assertion.

**Lemma 1.1** *Let conditions* (1.3) *and* (1.5) *be fulfilled. Then for every $f \in L^{p(\cdot)}(\Omega)$*

$$\min\left\{\|f\|^{p^-}_{p(\cdot),\Omega}, \|f\|^{p^+}_{p(\cdot),\Omega}\right\} \le A_{p(\cdot)}(f) \le \max\left\{\|f\|^{p^-}_{p(\cdot),\Omega}, \|f\|^{p^+}_{p(\cdot),\Omega}\right\}. \tag{1.6}$$

*Proof* Let us assume first that $\mu = \|f\|_{p(\cdot),\Omega} \neq 0$ and consider the function $h(x) = f(x)/\mu$. By Proposition 1.3 $h \in L^{p(\cdot)}(\Omega)$ with $\|h\|_{p(\cdot),\Omega} = 1$. On the other hand,

$$1 = \|h\|_{p(\cdot),\Omega} \le A_{p(\cdot)}(h) \le \begin{cases} \mu^{-p^+} A_{p(\cdot)}(f) & \text{if } \mu \le 1, \\ \mu^{-p^-} A_{p(\cdot)}(f) & \text{if } \mu > 1, \end{cases}$$

whence $A_{p(\cdot)}(f) \ge \min\left\{\mu^{p^+}, \mu^{p^-}\right\}$. Further,

$$A_{p(\cdot)}(f) = \int_\Omega |f(x)|^{p(x)}\,dx = \int_\Omega \mu^{p(x)}\,|h(x)|^{p(x)}\,dx \le \begin{cases} \mu^{p^-} & \text{if } \mu \le 1, \\ \mu^{p^+} & \text{if } \mu > 1, \end{cases}$$

which yields $A_{p(\cdot)}(f) \le \max\left\{\mu^{p^-}, \mu^{p^+}\right\}$. If $\|f\|_{p(\cdot),\Omega} = 0$, then $f = 0$ a.e. in $\Omega$ and $A_{p(\cdot)}(f) = 0$ by the property (2) of the modular. □

**Corollary 1.3** *The previous assertion can be represented in the equivalent form: if conditions* (1.3) *and* (1.5) *are fulfilled, then for every $f \in L^{p(\cdot)}(\Omega)$*

$$\min\left\{A^{\frac{1}{p^-}}_{p(\cdot)}(f), A^{\frac{1}{p^+}}_{p(\cdot)}(f)\right\} \le \|f\|_{p(\cdot),\Omega} \le \max\left\{A^{\frac{1}{p^-}}_{p(\cdot)}(f), A^{\frac{1}{p^+}}_{p(\cdot)}(f)\right\}. \tag{1.7}$$

*If the exponent $p$ is constant, then $p^+ = p^-$, which converts* (1.6) *and* (1.7) *into the equality $A_p(f) = \|f\|^p_{p,\Omega}$.*

**Corollary 1.4** *It follows from* (1.6)–(1.7) *that if conditions* (1.3) *and* (1.5) *are fulfilled, then for every $f \in L^{p(\cdot)}(\Omega)$*

$$\begin{array}{llll} \text{(i)} & \|f\|_{p(\cdot),\Omega} = 1 & \Leftrightarrow & A_{p(\cdot)}(f) = 1, \\ \text{(ii)} & \|f\|_{p(\cdot),\Omega} < 1 & \Leftrightarrow & A_{p(\cdot)}(f) < 1, \\ \text{(iii)} & \|f\|_{p(\cdot),\Omega} > 1 & \Leftrightarrow & A_{p(\cdot)}(f) > 1. \end{array}$$

**Lemma 1.2** *Let $\{f_n\}$ be a sequence of functions $f_n \in L^{p(\cdot)}(\Omega)$ and $f \in L^{p(\cdot)}(\Omega)$ with $p(x)$ satisfying* (1.3) *and* (1.5). *Then*

$$\|f_n - f\|_{p(\cdot),\Omega} \to 0 \text{ if and only if } A_{p(\cdot)}(f_n - f) \to 0 \text{ as } n \to \infty.$$

*Proof* The assertion is a byproduct of Lemma 1.1. □

**Lemma 1.3** *Let the functions $p(x)$ and $q(x)$ satisfy conditions* (1.3) *and* (1.5). *If $p(x) \geq q(x)$ a.e. in $\Omega$, then $L^{p(\cdot)}(\Omega) \subset L^{q(\cdot)}(\Omega)$.*

*Proof* By virtue of Lemma 1.1, it suffices to check that the condition $A_{p(\cdot)}(u) < \infty$ yields $A_{q(\cdot)}(u) < \infty$. By Young's inequality

$$\begin{aligned} A_{q(\cdot)}(u) &\leq \int_\Omega \left( \frac{q(x)}{p(x)} |u|^{p(x)} + \frac{p(x) - q(x)}{p(x)} \right) dx \\ &\leq \int_\Omega \left(1 + |u|^{p(x)}\right) dx = |\Omega| + A_{p(\cdot)}(u). \end{aligned} \tag{1.8}$$

□

**Corollary 1.5** *Under the conditions of* Lemma 1.3 *the norm of the embedding operator $L^{p(x)}(\Omega) \mapsto L^{q(x)}(\Omega)$ does not exceed $1 + |\Omega|$. It suffices to notice that by virtue of* (1.8)

$$\sup_{A_{p(\cdot),\Omega}(u)=1} \frac{A_{q(\cdot),\Omega}(u)}{A_{p(\cdot),\Omega}(u)} = \sup_{A_{p(\cdot),\Omega}(u)=1} A_{q(\cdot),\Omega}(u) \leq 1 + |\Omega|.$$

**Lemma 1.4** (The Hölder inequality) *Let conditions* (1.3) *and* (1.5) *be fulfilled. For every $f \in L^{p(x)}(\Omega)$ and $g \in L^{p'(x)}(\Omega)$ with $p'(x) = \dfrac{p(x)}{p(x) - 1}$ the following inequality holds:*

$$\int_\Omega |f\, g|\, dx \leq \left( \frac{1}{p^-} + \frac{1}{(p')^-} \right) \|f\|_{p(\cdot),\Omega} \|g\|_{p'(\cdot),\Omega} \leq 2 \|f\|_{p(\cdot),\Omega} \|g\|_{p'(\cdot),\Omega}. \tag{1.9}$$

*Proof* Let us denote $\|f\|_{p(\cdot),\Omega} = \lambda$, $\|g\|_{p'(\cdot),\Omega} = \mu$ and assume that $\lambda \neq 0$, $\mu \neq 0$. By Young's inequality, for a.e. $x \in \Omega$

$$|f(x)g(x)| = \lambda\, \mu \left| \frac{f(x)}{\lambda} \right| \left| \frac{g(x)}{\mu} \right|$$

$$\le \lambda\mu \left( \frac{1}{p(x)} \left| \frac{f(x)}{\lambda} \right|^{p(x)} + \frac{1}{p'(x)} \left| \frac{g(x)}{\mu} \right|^{p'(x)} \right)$$

$$\le \lambda\mu \left( \frac{1}{p^-} \left| \frac{f(x)}{\lambda} \right|^{p(x)} + \frac{1}{(p')^-} \left| \frac{g(x)}{\mu} \right|^{p'(x)} \right). \tag{1.10}$$

By Proposition 1.3

$$A_{p(\cdot)}(f/\lambda) = 1, \qquad A_{p(\cdot)}(g/\mu) = 1, \tag{1.11}$$

Integrating (1.10) over $\Omega$ and applying (1.11) we obtain (1.9):

$$\int_\Omega |f(x)g(x)|\, dx \le \lambda\, \mu \left( \frac{1}{p^-} A_{p(\cdot)}(f/\lambda) + \frac{1}{(p')^-} A_{p(\cdot)}(g/\mu) \right)$$
$$= \left( \frac{1}{p^-} + \frac{1}{(p')^-} \right) \|f\|_{p(\cdot),\Omega} \|g\|_{p'(\cdot),\Omega}.$$

Let $\lambda\mu = 0$. Assume, for the sake of definiteness, that $\lambda = 0$. Then $f = 0$ a.e. in $\Omega$ and the conclusion follows:

$$\int_\Omega |f(x)g(x)|\, dx = 0.$$

□

**Lemma 1.5** *Let $\Omega$ satisfy condition* (1.2), *$p(x)$ satisfy conditions* (1.3) *and* (1.5) *and $q = const \ge 1$. If $q \le p(x)$ a.e. in $\Omega$, then*

$$\|f\|_{q,\Omega} \le C\, \|f\|_{p(\cdot),\Omega} \quad \textit{with the constant} \quad C = (1 + |\Omega|)^{\frac{1}{q}}.$$

*Proof* Let $f \in L^{p(\cdot)}(\Omega)$ with $\|f\|_{p(\cdot),\Omega} = \lambda < \infty$. We may assume that $\lambda > 0$, otherwise the required inequality is obvious. Set $g(x) = f(x)/\lambda$. By Lemma 1.3 $g \in L^{p(\cdot)}(\Omega) \subset L^q(\Omega)$ with $\|g\|_{p(\cdot),\Omega} = A_{p(\cdot)}(g) = 1$. By Proposition 1.3

$$\frac{1}{\lambda} \|f\|_{q,\Omega} = \|g\|_{q,\Omega} \le A_q^{\frac{1}{q}}(g). \tag{1.12}$$

Applying Young's inequality and Corollary 1.4 (i) we find that

$$A_q(g) = \int_\Omega \left| \frac{f(x)}{\lambda} \right|^q dx$$
$$\le \int_\Omega \left( \frac{p(x) - q}{p(x)} + \frac{q}{p(x)} \left| \frac{f(x)}{\lambda} \right|^{p(x)} \right) dx$$
$$\le |\Omega| + A_{p(\cdot)}(g) = 1 + |\Omega|.$$

Plugging this inequality into (1.12) we arrive at the inequality

$$\|f\|_{q,\Omega} \leq (1+|\Omega|)^{\frac{1}{q}}\lambda = (1+|\Omega|)^{\frac{1}{q}}\|f\|_{p(\cdot),\Omega}.$$

□

**Lemma 1.6** *Let the exponents $p(x)$, $q(x)$ satisfy conditions* (1.3) *and* (1.5) *and $p(x) \geq q(x)$ a.e. in $\Omega$. Then the embedding $L^{p(x)}(\Omega) \subset L^{q(x)}(\Omega)$ is continuous. The norm of the embedding operator depends only on $|\Omega|$, $p^{\pm}$ and $q^{\pm}$:*

$$\|f\|_{q(\cdot),\Omega} \leq C\|f\|_{p(\cdot),\Omega}, \qquad C = C(|\Omega|, p^{\pm}, q^{\pm}).$$

*Proof* The embedding $L^{p(x)}(\Omega) \subset L^{q(x)}(\Omega)$ follows from Lemma 1.3. Let $f \in L^{p(x)}(\Omega)$ with $\|f\|_{p(\cdot),\Omega} = \lambda > 0$. Denote

$$\Omega_2 = \{x \in \Omega : p(x) = q(x)\}, \qquad \Omega_1 = \Omega \setminus \Omega_2$$

and consider the function $h = f/\lambda$. By Proposition 1.3 $A_{p(\cdot)}(h) = 1$ and

$$\begin{aligned}
A_{q(\cdot)}(h) &= \int_{\Omega_1} |h|^{q(x)}\,dx + \int_{\Omega_2} |h|^{q(x)}\,dx \\
&= \int_{\Omega_1} \left(|h|^{p(x)}\right)^{\frac{q(x)}{p(x)}}\,dx + \int_{\Omega_2} |h|^{p(x)}\,dx \\
&\leq 2\|1\|_{\frac{p(\cdot)}{p(\cdot)-q(\cdot)},\Omega_1} \left\||h|^{q(x)}\right\|_{\frac{p(\cdot)}{q(\cdot)},\Omega_1} + \int_{\Omega_2} |h|^{p(x)}\,dx \\
&\leq 2\max\left\{|\Omega|^{1-\frac{q^+}{p^-}}, |\Omega|^{1-\frac{q^-}{p^+}}\right\} \max\left\{A_{\frac{p(\cdot)}{q(\cdot)}}^{\frac{q^-}{p^+}}(|h|^{q(x)}),\ A_{\frac{p(\cdot)}{q(\cdot)}}^{\frac{q^+}{p^-}}(|h|^{q(x)})\right\} + A_{p(\cdot)}(h) \\
&= 2\max\left\{|\Omega|^{1-\frac{q^+}{p^-}}, |\Omega|^{1-\frac{q^-}{p^+}}\right\} \max\left\{A_{p(\cdot)}^{\frac{q^-}{p^+}}(h),\ A_{p(\cdot)}^{\frac{q^+}{p^-}}(h)\right\} + A_{p(\cdot)}(h) \\
&= 2\max\left\{|\Omega|^{1-\frac{q^+}{p^-}}, |\Omega|^{1-\frac{q^-}{p^+}}\right\} + 1.
\end{aligned}$$

On the other hand, by (1.7)

$$\frac{1}{\lambda}\|f\|_{q(\cdot),\Omega} = \|h\|_{q(\cdot),\Omega} \leq \max\left\{A_{q(\cdot)}^{\frac{1}{q^-}}(h),\ A_{q(\cdot)}^{\frac{1}{q^+}}(h)\right\}.$$

□

### 1.2.2 Equivalent Norm and Completeness of $L^{p(\cdot)}(\Omega)$

Given $f \in L^{p(\cdot)}(\Omega)$, we define the functional

$$|f|_{p(\cdot),\Omega} = \sup_{A_{p'(\cdot)}(g)\leq 1} \int_\Omega f(x)g(x)\,dx, \qquad p'(x) = \frac{p(x)}{p(x)-1}. \tag{1.13}$$

**Proposition 1.4** (The Orlicz norm of $L^{p(\cdot)}(\Omega)$) *The functional*

$$|\cdot|_{p(\cdot),\Omega} : L^{p(\cdot)}(\Omega) \mapsto \mathbb{R}$$

*defines a norm of* $L^{p(x)}(\Omega)$.

*Proof* We have to check that

1. $|f|_{p(\cdot),\Omega} \geq 0$ for every $f \in L^{p(x)}(\Omega)$;
2. $|f|_{p(\cdot),\Omega} = 0 \quad \Leftrightarrow \quad f = 0$ a.e. in $\Omega$;
3. $\forall\, \mu \in \mathbb{R},\ f \in L^{p(x)}(\Omega) \quad |\mu f|_{p(\cdot),\Omega} = |\mu||f|_{p(\cdot),\Omega}$;
4. $\forall\, f,\ g \in L^{p(x)}(\Omega) \quad |f+g|_{p(\cdot),\Omega} \leq |f|_{p(\cdot),\Omega} + |g|_{p(\cdot),\Omega}$.

(1) We argue by contradiction. Let us assume that $|f|_p < 0$. Let $g \in L^{p'(x)}(\Omega)$ be an arbitrary function such that $A_{p'(\cdot)}(g) \leq 1$. By (1.13)

$$0 > |f|_{p(\cdot),\Omega} \geq \frac{1}{2}\int_\Omega f(x)g(x)\,dx.$$

Since $A_{p'(\cdot)}(g) = A_{p'(\cdot)}(-g)$, we also have

$$0 > |f|_{p(\cdot),\Omega} \geq \frac{1}{2}\int_\Omega f(x)\,(-g(x))\,dx = -\frac{1}{2}\int_\Omega f(x)g(x)\,dx > 0,$$

which is impossible.

(2) For $f = 0$ a.e. in $\Omega$ the equality $|f|_p = 0$ follows from the definition of $|f|_{p(\cdot),\Omega}$. Let us assume that $|f|_{p(\cdot),\Omega} = 0$ but $f \neq 0$ a.e. in $\Omega$, i.e., $A_{p(\cdot)}(f) \neq 0$. Let us introduce the function

$$g(x) = |f(x)|^{p(x)-1}\,\mathrm{sign}\ f(x) \in L^{p'(x)}(\Omega).$$

We have

$$0 = |f|_{p(\cdot),\Omega} \geq \int_\Omega f(x)g(x)\,dx = \int_\Omega |f(x)|^{p(x)}\,dx = A_{p(\cdot)}(f) \neq 0,$$

which is impossible.

(3)–(4) This property is an immediate byproduct of the properties of the supremum. □

**Proposition 1.5** *If* $|f|_{p(\cdot),\Omega} < \infty$ *and* $A_{p'(\cdot)}(g) < \infty$, *then*

$$\left|\int_\Omega f(x)g(x)\,dx\right| \le \begin{cases} |f|_{p(\cdot),\Omega} & \text{if } A_{p'(\cdot)}(g) \le 1,\\ |f|_{p(\cdot),\Omega} A_{p'(\cdot)}(g) & \text{otherwise}. \end{cases}$$

*Proof* The first case is an immediate by product of (1.13). Let $A_{p'(\cdot)}(g) > 1$. By virtue of convexity of the modular $A_{p'(\cdot)}(\cdot)$

$$A_{p'(\cdot)}\left((A_{p'(\cdot)}(g))^{-1}g\right) \le (A_{p'(\cdot)}(g))^{-1}A_{p'(\cdot)}(g) = 1,$$

whence

$$\left|\int_\Omega f(x)g(x)\,dx\right| = A_{p'(\cdot)}(g)\left|\int_\Omega f(x)g(x)/A_{p'}(g)\,dx\right| \le A_{p'(\cdot)}(g)|f|_{p(\cdot),\Omega}.$$

□

**Proposition 1.6** *Let* $p(x)$ *satisfy* (1.3) *and* (1.5). *If* $A_{p(\cdot)}(f) < \infty$ *and* $|f|_{p(\cdot),\Omega} \le 1$, *then* $A_{p(\cdot)}(f) \le 1$.

*Proof* We argue by contradiction: let us assume that $A_{p(\cdot)}(f) > 1$. By the property (6) of the modular (continuity of $A_{p(\cdot)}(f/\lambda)$ with respect to $\lambda$) there is $\lambda > 1$ such that $A_{p(\cdot)}(f/\lambda) = 1$. Set

$$g(x) = |f(x)/\lambda|^{p(x)-1}\operatorname{sign} f(x), \qquad x \in \Omega.$$

Then

$$A_{p'(\cdot)}(g) = \int_\Omega \left(|f(x)/\lambda|^{p(x)-1}\right)^{\frac{p(x)}{p(x)-1}} dx = \int_\Omega |f/\lambda|^p\,dx = 1,$$

whence

$$|f|_{p(\cdot),\Omega} \ge \int_\Omega f(x)g(x)\,dx = \lambda\int_\Omega |f/\lambda|^{p(x)}dx = \lambda > 1,$$

a contradiction. □

**Proposition 1.7** *Let* $p(x)$ *satisfy* (1.3) *and* (1.5). *If* $|f|_{p(\cdot),\Omega} \le 1$, *then* $A_{p(\cdot)}(f) \le |f|_{p(\cdot),\Omega}$.

*Proof* Let us assume first that $A_p(f) < \infty$. Introduce the function

$$g(x) = |f(x)|^{p(x)-1}\operatorname{sign} f(x), \qquad x \in \Omega.$$

Then

$$A_{p(\cdot)}(f) = \int_\Omega |f(x)|^{p(x)}\,dx = \int_\Omega f(x)g(x)\,dx \le |f|_{p(\cdot),\Omega}.$$

To avoid the assumption $A_{p(\cdot)}(f) < \infty$ we consider the sequence of truncations

$$f_k(x) = \min\{k, |f(x)|\}\,\chi_{G_k}, \qquad k \in \mathbb{N},$$

where $\{G_k\}$ is a sequence of sets $G_k \subset G_{k+1} \subset \Omega$ such that $\Omega = \bigcup_{k=1}^\infty G_k$ and $\chi_{G_k}$ is the characteristic function of $G_k$. For every $k \in \mathbb{N}$ $A_p(f_k) < \infty$ and

$$|f_k|_{p(\cdot),\Omega} \le |f|_{p(\cdot),\Omega} \le 1.$$

The assertion follows now from the Lebesgue monotone convergence theorem. □

**Theorem 1.1** (Equivalence between the norms of Luxemburg and Orlicz) *Let conditions* (1.3) *and* (1.5) *be fulfilled. Then*

$$L^{p(\cdot)}(\Omega) = \left\{f : |f|_{p(\cdot),\Omega} < \infty\right\}$$

*and there exist constants* $C_*$, $C^*$ *such that*

$$C_*\|f\|_{p(\cdot),\Omega} \le |f|_{p(\cdot),\Omega} \le C^*\|f\|_{p(\cdot),\Omega} \qquad \forall f \in L^{p(\cdot)}(\Omega).$$

*Proof* Let $f \in L^{p(\cdot)}(\Omega)$. By Corollary 1.4, the inequality $A_{p'(\cdot)}(g) \le 1$ implies $\|g\|_{p'(\cdot),\Omega} \le 1$, and by Hölder's inequality

$$\int_\Omega f(x)g(x)\,dx \le \left(\frac{1}{p^-} + \frac{1}{(p')'}\right)\|f\|_{p(\cdot),\Omega}\|g\|_{p'(\cdot),\Omega} \le 2\|f\|_{p(\cdot),\Omega}.$$

Let us assume now that $0 < |f|_{p(\cdot),\Omega} < \infty$. Since

$$\left|\frac{f}{|f|_{p(\cdot),\Omega}}\right|_{p(\cdot),\Omega} = \sup_{A_{p'(\cdot)}(g)\le 1} \int_\Omega \frac{f(x)g(x)}{|f|_{p(\cdot),\Omega}}dx = \frac{|f|_{p(\cdot),\Omega}}{|f|_{p(\cdot),\Omega}} = 1,$$

it follows from Proposition 1.7 that $A_p(f/|f|_{p(\cdot),\Omega}) \le 1$. By Corollary 1.4

$$\|f/|f|_{p(\cdot),\Omega}\|_{p(\cdot),\Omega} \le 1,$$

whence $\|f\|_{p(\cdot),\Omega} \le |f|_{p(\cdot),\Omega}$ by the definition of the Luxemburg norm. □

**Lemma 1.7** *If* $p(x)$ *satisfies* (1.3) *and* (1.5), *the space* $L^{p(\cdot)}(\Omega)$ *is complete.*

*Proof* Let $\{f_k\}$ be a Cauchy sequence of functions $f_k \in L^{p(\cdot)}(\Omega)$: for every $\varepsilon > 0$ there exists $k_0 \in \mathbb{N}$ such that

$$\int_{\Omega} |f_m(x) - f_n(x)||g(x)|\,dx < \varepsilon \qquad \forall\, m, n \geq k_0 \tag{1.14}$$

and every $g$ with $A_{p'(\cdot)}(g) \leq 1$. Let us decompose $\Omega$ into pairwise disjoint subsets $G_k$ with $|G_k| < \infty$ and define the functions

$$g_k = \frac{\chi_{G_k}}{1 + |G_k|}, \qquad k \in \mathbb{N},$$

for which

$$A_{p'(\cdot)}(g_k) = \int_{\Omega} \frac{\chi_{G_k}\,dx}{(1 + |G_k|)^{p(x)}} \leq \frac{|G_k|}{1 + |G_k|} \leq 1.$$

Substituting $g = g_k$ into (1.14) we obtain the inequality

$$\int_{G_k} \frac{|f_m - f_n|}{1 + |G_k|}\,dx \leq \int_{\Omega} |f_m(x) - f_n(x)||g_k(x)|\,dx < \varepsilon,$$

whence

$$\int_{G_k} |f_m(x) - f_n(x)|\,dx < \varepsilon(1 + |G_k|).$$

It follows that $\{f_k\}$ is a Cauchy sequence in $L^1(G_k)$ for every $k$. Let us extract from $\{f_k\}$ subsequences $\{f_k^{(j)}\}$ and find functions $f^{(j)} \in L^1(G_j)$ such that

$$\begin{aligned}
\{f_k^{(1)}\} \subset \{f_k\} : f_k^{(1)} &\to f^{(1)} \text{ a.e. in } G_1, \quad f^{(1)} \in L^1(G_1),\\
\{f_k^{(2)}\} \subset \{f_k^{(1)}\} : f_k^{(2)} &\to f^{(2)} \text{ a.e. in } G_2, \quad f^{(2)} \in L^1(G_2),\\
&\cdots\\
\{f_k^{(m)}\} \subset \{f_k^{(m-1)}\} : f_k^{(m)} &\to f^{(m)} \text{ a.e. in } G_m, \ f^{(m)} \in L^1(G_m),\\
&\cdots
\end{aligned}$$

Consider the diagonal sequence $\{f_m^{(m)}\}$. Since it is a subsequence of every $\{f_m^{(j)}\}$, then

$$f_m^{(m)}(x) \to \sum_{k=1}^{\infty} f^{(k)}(x)\chi_{G_k} := f(x) \quad \text{for a.e. } x \in \Omega.$$

By virtue of (1.14)

$$\int_{\Omega} |f_m^{(m)} - f_n||g(x)|\,dx \leq \varepsilon$$

for every $m, n \geq n_0$ and $g$ such that $A_{p'(\cdot)}(g) \leq 1$. Applying the Fatou Lemma we pass to the limit and conclude that

$$\int_{\Omega} |f_n(x) - f(x)||g(x)|\,dx \leq \sup_{m \in \mathbb{N},\, m \geq n_0} \int_{\Omega} |f_m^{(m)}(x) - f_n(x)||g(x)|\,dx \leq \varepsilon$$

for every $n \geq n_0$ and every $g$ such that $A_{p'}(g) \leq 1$. Hence, $|f - f_n|_p \leq \varepsilon$. □

**Corollary 1.6** *If conditions* (1.3) *and* (1.5) *are fulfilled,* $L^{p(\cdot)}(\Omega)$ *is a Banach space.*

**Lemma 1.8** *If* $p(x)$ *satisfy* (1.3) *and* (1.5), *then* $L^{p(\cdot)}(\Omega)$ *is reflexive and separable.*

*Proof* By Lemma 1.7 the space $L^{p(\cdot)}(\Omega)$ is a complete and closed subspace of $L^{p^-}(\Omega)$. For $p^- > 1$ the space $L^{p^-}(\Omega)$ is a reflexive and separable Banach space. It follows that $L^{p(\cdot)}(\Omega)$ is also reflexive and separable (see, e.g., [5, Chap. 1]). □

Let $p(x)$ satisfy (1.3) and (1.5). By Corollary 1.4 $A_p(f) \leq 1$ if and only if $\|f\|_p \leq 1$. It follows from Theorem 1.1 that for every $g \in L^{p'(x)}(\Omega)$ the mapping $L^{p(x)}(\Omega) \mapsto \mathbb{R}$ defined by

$$G(f) = \int_\Omega f(x)g(x)\,dx, \qquad f \in L^{p(x)}(\Omega), \tag{1.15}$$

is a linear continuous functional on $L^{p(x)}(\Omega)$ with the norm satisfying the inequalities $C_*\|g\|_{p'} \leq \|G\| \leq C^*\|g\|_{p'}$.

**Lemma 1.9** *The following conditions are equivalent:*

(i) $p \in L^\infty(\Omega)$,
(ii) *for every linear continuous functional* $G$ *on* $L^{p(x)}(\Omega)$ *there exists a unique function* $g \in L^{p'(x)}(\Omega)$ *such that* (1.15) *holds.*

The proof can be found in [184] or [115, Part I, Sect. 2.7].

**Corollary 1.7** *The dual space to* $L^{p(x)}(\Omega)$ *is* $L^{p'(x)}(\Omega)$ *if and only if* $p \in L^\infty(\Omega)$.

### *1.2.3 Dense Sets in $L^{p(\cdot)}(\Omega)$*

**Lemma 1.10** *If* $p(x)$ *satisfies* (1.3) *and* (1.5), *the set of measurable and bounded in* $\Omega$ *functions is dense in* $L^{p(\cdot)}(\Omega)$.

*Proof* Given $f \in L^{p(\cdot)}(\Omega)$, let us consider the sequence of functions $f_k \in L^{p(\cdot)}(\Omega) \cap L^\infty(\Omega)$ defined by the formulas

$$f_k(x) = \begin{cases} f(x) & \text{if } |f(x)| \leq k, \\ k \text{ sign } f(x) & \text{if } |f(x)| > k. \end{cases}$$

Since $|f_k| \leq |f|$ and $f_k \to f$ a.e. in $\Omega$, by the Lebesgue dominated convergence theorem $A_{p(\cdot)}(f_k - f) \to 0$ as $k \to \infty$, and the assertion follows from Lemma 1.2. □

**Theorem 1.2** *Let $p(x)$ satisfy* (1.3) *and* (1.5). *Then the set $C(\Omega)\cap L^{\infty}(\Omega)$ is dense in $L^{p(\cdot)}(\Omega)$.*

*Proof* Let $f \in L^{p(\cdot)}(\Omega)$. By Lemma 1.10 for every fixed $\varepsilon > 0$ there exists a function $g \in L^{\infty}(\Omega) \cap L^{p(\cdot)}(\Omega)$ such that

$$\|f - g\|_{p(\cdot),\Omega} < \varepsilon. \tag{1.16}$$

By Luzin's theorem there exists a function $h(x) \in C(\Omega)$ and an open set $U$ such that

$$|U| \leq \min\left\{1, \left(\frac{\varepsilon}{2\,|g\|_{\infty}}\right)^{p^+}\right\},$$

$g(x) = h(x)$ everywhere in $\Omega\setminus U$ and $\sup|h(x)| = \sup\limits_{\Omega\setminus U}|g(x)| \leq \|g\|_{\infty}$. It follows that

$$\begin{aligned}
A_{p(\cdot)}((g-h)/\varepsilon) &= \int_{\Omega}\frac{|g-h|^{p(x)}}{\varepsilon^{p(x)}}\,dx \leq \int_{U}\frac{|g-h|^{p(x)}}{\varepsilon^{p(x)}}dx \\
&\leq \int_{U}\left(\frac{|g|+|h|}{\varepsilon}\right)^{p(x)} dx \\
&\leq \int_{U}\left(\frac{2\|g\|_{\infty}}{\varepsilon}\right)^{p^+} dx \\
&\leq |U|\max\left\{1, \left(\frac{2\|g\|_{\infty}}{\varepsilon}\right)^{p^+}\right\} \leq 1.
\end{aligned}$$

For every $\varepsilon < 1$

$$\frac{1}{\varepsilon^{p^-}}\int_{\Omega}|g(x)-h(x)|^{p(x)}\,dx \leq A_{p(\cdot)}((g-h)/\varepsilon) \leq 1 \quad\Rightarrow\quad A_{p(\cdot)}(g-h) \leq \varepsilon^{p^-},$$

and by Corollary 1.4

$$\|g-h\|_{p(\cdot),\Omega} \leq \max\left\{A_{p(\cdot)}^{\frac{1}{p^-}}(g-h),\ A_{p(\cdot)}^{\frac{1}{p^+}}(g-h)\right\} \leq \max\left\{\varepsilon, \varepsilon^{\frac{p^-}{p^+}}\right\} = \varepsilon^{\frac{p^-}{p^+}}.$$

Plugging this inequality into (1.16) we find that

$$\|f-h\|_{p(\cdot),\Omega} \leq \varepsilon + \varepsilon^{\frac{p^-}{p^+}}.$$

□

**Corollary 1.8** *Under the conditions of* Lemma 1.10 *the space $C_0^{\infty}(\Omega)$ is dense in $L^{p(\cdot)}(\Omega)$: $C_0^{\infty}(\Omega)$ is dense in the set of simple functions, simple functions are dense in $C(\Omega)\cap L^{\infty}(\Omega)$, and $C(\Omega)\cap L^{\infty}(\Omega)$ is dense in $L^{p(\cdot)}(\Omega)$.*

**Corollary 1.9** *Under the conditions of* Lemma 1.10 *the space $L^{p(\cdot)}(\Omega)$ is separable: the set of polynomials with rational coefficients is countable and dense in $C_0^\infty(\Omega)$.*

## 1.3 Sobolev Spaces with Variable Exponents

Let the domain $\Omega$ satisfy (1.2) and the exponent $p(x)$ satisfy (1.3) and (1.5). The Banach space $W_0^{1,\,p(\cdot)}(\Omega)$ is defined by

$$\begin{cases} W_0^{1,\,p(\cdot)}(\Omega) = \left\{u \in L^{p(\cdot)}(\Omega) : \, |\nabla u|^{p(x)} \in L^1(\Omega), \; u = 0 \text{ on } \partial\Omega\right\}, \\ \|u\|_{W_0^{1,p(\cdot)}(\Omega)} = \|u\|_{p(\cdot),\Omega} + \|\nabla u\|_{p(\cdot),\Omega}. \end{cases} \tag{1.17}$$

**Lemma 1.11** (The Poincaré inequality) *Let $\Omega$ and $p(x)$ satisfy conditions* (1.2), (1.3) *and* (1.5). *If $p(x) \in C^0(\overline{\Omega})$, then there exists a finite constant $C > 0$ such that for every $u \in W_0^{1,p(\cdot)}(\Omega)$*

$$\|u\|_{p(\cdot),\Omega} \le C\, \|\nabla u\|_{p(\cdot),\Omega}. \tag{1.18}$$

*Proof* It is sufficient to prove (1.18) for a set $B \cap \Omega$, where $B$ is a ball of radius so small that

$$\max_{B\cap\Omega} p(x) \le \frac{n+1}{n} \min_{B\cap\Omega} p(x). \tag{1.19}$$

The choice of $B$ depends on the modulus of continuity of $p(x)$ in $\Omega$. Denote

$$p^+ = \max_{B\cap\Omega} p(x), \qquad p^- = \min_{B\cap\Omega} p(x).$$

By the Sobolev embedding theorem, the following chain of embeddings holds:

$$W_0^{1,p(\cdot)}(\Omega) \subset W^{1,p(\cdot)}(B\cap\Omega) \subset W^{1,p^-}(B\cap\Omega) \subset L^{p^+}(B\cap\Omega) \subset L^{p(\cdot)}(B\cap\Omega).$$

Since $p(x)$ is uniformly continuous on $\Omega$, the domain $\Omega$ can be covered by a finite number of the balls satisfying (1.19), whence (1.18). □

An immediate consequence of inequality (1.18) is the possibility to define the equivalent norm of the space $W_0^{1,p(\cdot)}(\Omega)$ by the relation

$$\|u\|_{W_0^{1,p(\cdot)}(\Omega)} = \|\nabla u\|_{p(\cdot),\Omega}. \tag{1.20}$$

Let us denote by $C_{\log}(\Omega)$ the set of functions $p(x)$ that satisfy conditions (1.3) and (1.5) and are continuous in $\Omega$ with the logarithmic modulus of continuity:

$$C_{\log}(\Omega) := \left\{ p \in C^0(\overline{\Omega}) \;\middle|\; \begin{array}{l} \forall\, x,\ y \in \Omega,\, |x-y| < 1/2, \\ |p(x) - p(y)| \le \omega(|x-y|) \end{array} \right\}, \tag{1.21}$$

where $\omega$ is a continuous function such that

$$\overline{\lim_{\tau \to 0^+}}\ \omega(\tau) \ln \frac{1}{\tau} = C < +\infty, \quad C = const.$$

Let $\rho(x)$ be the Friedrichs mollifying kernel

$$\rho(x) = \kappa \begin{cases} \exp\left(\frac{-1}{1-|x|^2}\right) & \text{if } |x| < 1, \\ 0 & \text{if } |x| > 1, \end{cases} \qquad \kappa = const \text{ such that } \int_{\mathbb{R}^n} \rho(x)\, dx = 1.$$

Denote

$$\rho_\varepsilon(x) = \varepsilon^{-n} \rho\left(\frac{x}{\varepsilon}\right), \qquad \varepsilon > 0.$$

Given $f \in W_0^{1,p(\cdot)}(\Omega)$, we continue it by zero to the whole of $\mathbb{R}^n$ and use the same notation for the continued function. Let us consider the sequence of functions

$$f_\varepsilon(x) = f * \rho_\varepsilon \equiv \int_{\mathbb{R}^n} f(y)\, \rho_\varepsilon(y - x)\, dy, \quad \varepsilon > 0. \tag{1.22}$$

It is clear that since $\Omega$ is bounded, there a ball $B_R(0) = \{x \in \mathbb{R}^n : |x| < R\}$ such that $\Omega \subset B_R(0)$, and that $\operatorname{supp} f_\varepsilon \subset B_{R+1}(0) \equiv B$ for all $\varepsilon \in (0, 1)$.

**Lemma 1.12** *Let $\Omega' \subset\subset \Omega$ (i.e., $\operatorname{dist}(\partial\Omega, \partial\Omega') > 0$). If $p(x)$ satisfies conditions* (1.21), *then for every $f \in L^{p(\cdot)}(\Omega)$*

$$\|f_\varepsilon\|_{p(\cdot),\Omega'} \le C \left(\|f\|_{p(\cdot),\Omega} + \|f\|_{1,\Omega}\right), \tag{1.23}$$

$$\|f_\varepsilon - f\|_{p(\cdot),\Omega'} \to 0 \quad as\ \varepsilon \to 0 \tag{1.24}$$

*with an independent of $f$ constant $C$.*

*Proof* It is sufficient to study the case $\|f\|_{1,\Omega} + \|f\|_{p(\cdot),\Omega} = 1$, otherwise we may consider the scaled function $\widetilde{f} = f / \left(\|f\|_{1,\Omega} + \|f\|_{p(\cdot),\Omega}\right)$. By the definition

$$|f_\varepsilon(x)| \le \varepsilon^{-n} \sup_{x \in \mathbb{R}^n} \rho(x) \|f\|_{1,\Omega} \le \varepsilon^{-n}. \tag{1.25}$$

Introduce the functions

$$p_\varepsilon(x) = \min\{p(y) : |x - y| < \varepsilon\} \le p(y).$$

According to (1.21)

$$0 \le p(x) - p_\varepsilon(x) \le \omega(\varepsilon) \le \frac{2C}{\ln \frac{1}{\varepsilon}}.$$

Further, for all sufficiently small $\varepsilon > 0$ the inequality $\varepsilon^{-n\,\omega(\varepsilon)} \le \mathrm{e}^{2nC}$ holds with the constant $C$ from (1.21). For every $0 < \mu < \varepsilon^{-n}$

$$\mu^{p(x)} = \mu^{p(x)-p_\varepsilon(x)}\mu^{p_\varepsilon(x)} \le \mu^{\omega(x)}\mu^{p_\varepsilon(x)} \le \mathrm{e}^{2nC}\mu^{p_\varepsilon(x)}.$$

Gathering this inequality with (1.25) and using Hölder's inequality we find that

$$|f_\varepsilon(x)|^{p(x)} \le \mathrm{e}^{2nC}|f(x)|^{p_\varepsilon(x)} \le C_1\mathrm{e}^{2nC}\left(1 + |f(x)|^{p(x)}\right).$$

It follows that

$$A_{p(\cdot),\Omega'}(f_\varepsilon) \le C_1\left(|\Omega| + A_{p(\cdot),\Omega}(f)\right) \le C_1\left(1 + |\Omega|\right),$$

whence (1.23).

To prove (1.24) we consider the sequence of smoothing operators

$$S_\varepsilon : L^{p(\cdot)}(\Omega) \mapsto L^{p(\cdot)}(\Omega')$$

defined by formulas (1.22). By (1.23) the operators $S_\varepsilon$ are uniformly bounded in the operator norm and $S_\varepsilon f = f_\varepsilon \to f$ uniformly in $\Omega'$ for $f \in C_0^\infty(\Omega)$. Since the set $C_0^\infty(\Omega)$ is dense in $L^{p(\cdot)}(\Omega)$ (see Corollary 1.8), it follows from the Banach-Steinhaus theorem that $S_\varepsilon f = f_\varepsilon \to f$ in the norm of $L^{p(\cdot)}(\Omega')$ as $\varepsilon \to 0$ for every $f \in L^{p(\cdot)}(\Omega)$. □

**Lemma 1.13** *Let $\Omega$ satisfy* (1.2). *If $p(x) \in C_{\log}(\Omega)$, then the set $C_0^\infty(\Omega)$ is dense in $W_0^{1,p(\cdot)}(\Omega)$.*

*Proof* The assertion follows from Lemma 1.12 because for the continued function

$$\begin{aligned}
D_{x_i}(f * \rho_\varepsilon) &= \int_{\mathbb{R}^n} f(y) D_{x_i}\rho_\varepsilon(x-y)\,dy \\
&= -\int_{\mathbb{R}^n} f(y) D_{y_i}\rho_\varepsilon(x-y)\,dy \\
&= \int_{\mathbb{R}^n} D_{y_i} f(y)\rho_\varepsilon(x-y)\,dy \\
&= (D_{x_i} f) * \rho_\varepsilon \to D_{x_i} f \quad \text{as } \varepsilon \to 0 \text{ in } L^{p(x)}(\Omega).
\end{aligned}$$

□

Lemma 1.13 allows one to define the space $W_0^{1,p(\cdot)}(\Omega)$ as the closure of $C_0^\infty(\Omega)$ in the norm (1.20), provided that $p(x) \in C_{\log}(\Omega)$. Let us denote by

$$p^*(x) = \begin{cases} \dfrac{n\,p(x)}{n - p(x)} & \text{if } n > p(x), \\ \text{any number from } [1, \infty) & \text{if } n \le p(x) \end{cases}$$

the Sobolev conjugate exponent.

**Lemma 1.14** *Let $\Omega$ satisfy* (1.2). *Assume that $p(x)$ satisfies conditions* (1.3) *and* (1.5). *If $p(x), q(x) \in C(\overline{\Omega})$ and $q(x) < p^*(x)$ in $\overline{\Omega}$, then for every $u \in W_0^{1,p(\cdot)}(\Omega)$*

$$\|u\|_{q(\cdot),\Omega} \le C\,\|\nabla u\|_{p(\cdot),\Omega}$$

*with a constant $C$ depending on $p^{\pm}$, $n$, the properties of $\partial\Omega$ and the modulus of continuity of $p(x)$. The embedding $W_0^{1,p(\cdot)}(\Omega) \subset L^{q(\cdot)}(\Omega)$ is continuous and compact.*

*Proof* Let us consider first the case $p^+ < n$. For an arbitrary fixed $x \in \overline{\Omega}$ there is a neighborhood $U_x$ such that

$$\max_{\overline{U}_x} q(x) < \frac{\min_{\overline{U}_x} p(x)}{n - \min_{\overline{U}_x} p(x)}$$

Let $\{U_x\}_{x\in\Omega}$ be an open covering of the compact $\overline{\Omega}$. Choose a finite subcover $\{U_i : i = 1, 2, \ldots, k\}$ and denote

$$p_i^- = \min_{\overline{U}_i} p(x), \qquad q_i^+ = \max_{\overline{U}_i} q(x).$$

By Lemma 1.5 $u \in W^{1,p_i^-}(U_i)$, while by the Sobolev embedding theorem for the constant exponent spaces there is continuous and compact embedding

$$W^{1,p_i^-}(U_i) \hookrightarrow L^{q_i^+}(U_i).$$

By Lemma 1.6 $u \in L^{q_i^+}(U_i) \subset L^{q(x)}(U_i)$, whence $u \in L^{q(x)}(\Omega)$.

Let us now remove the assumption $p^+ < n$. Notice that $W_0^{1,p(\cdot)}(\Omega) \subset W_0^{1,r(\cdot)}(\Omega)$ with $r(x) = \min\{p(x), r\}$, where $r$ is a constant satisfying the inequalities

$$\frac{1}{n} < \frac{1}{r} < \frac{1}{n} + \frac{1}{q^+}.$$

Since the embedding $L^{q(\cdot)}(\Omega) \subset W_0^{1,r(\cdot)}(\Omega)$ is continuous and compact, the result follows. □

*Remark 1.1* It is known that the embedding $L^{q(\cdot)}(\Omega) \subset W_0^{1,p(\cdot)}(\Omega)$ may hold true even in the critical case $q(x) \le p^*(x)$ [127]. We refer to [115, 129, 268] for a further discussion of this issue.

## 1.4 Anisotropic Spaces of Functions Depending on $x$ and $t$

Let $\Omega \subset \mathbb{R}^n$ be a domain satisfying conditions (1.2), $Q = \Omega \times (0, T]$ be the generic cylinder of an arbitrary finite height $T$, $z = (x, t) \in Q$. Recall that the notation $Q_\tau = \Omega \times (0, \tau]$ is used whenever the height of the cylinder is of importance. The lateral boundary of $Q$ is denoted by $\Gamma = \partial\Omega \times (0, T]$ (or $\Gamma_\tau = \partial\Omega \times (0, \tau]$).

Let $\mathbf{p}(z) = (p_1(z), \dots, p_n(z))$ be a vector-valued function defined on $Q$. We assume that the components of $\mathbf{p}(z)$ satisfy the conditions

$$\begin{cases} p_i(z) \text{ are measurable functions defined on Q,} \\ p_i(z) : \ Q \mapsto (1, \infty), \\ \text{there exist constants } p_i^{\pm},\ p^{\pm} \text{ such that} \\ p_i(z) \in [p_i^-, p_i^+] \subseteq [p^-, p^+] \subset (1, \infty). \end{cases} \tag{1.26}$$

For a.e. $t \in (0, T)$ we introduce the anisotropic Banach space

$$\begin{aligned} &\mathbf{V}_t(\Omega) = \left\{u(x) : u(x) \in L^2(\Omega) \cap W_0^{1,1}(\Omega),\ |D_i u(x)|^{p_i(x,t)} \in L^1(\Omega)\right\}, \\ &\|u\|_{\mathbf{V}_t(\Omega)} = \|u\|_{2,\Omega} + \sum_{i=1}^n \|D_i u\|_{p_i(\cdot,t),\Omega}. \end{aligned} \tag{1.27}$$

The elements of the space $\mathbf{V}_t(\Omega)$ depend on $t \in (0, T)$ as a parameter and the norms $\|u\|_{\mathbf{V}_t(\Omega)}$ are functions of $t$. By $\mathbf{V}_t'(\Omega)$ we denote the dual space to $\mathbf{V}_t(\Omega)$ with respect to the scalar product in $L^2(\Omega)$. For a.e. $t \in (0, T)$ the inclusion

$$\mathbf{V}_t(\Omega) \subset \mathbf{X} = W_0^{1,p^-}(\Omega) \cap L^2(\Omega)$$

holds, which is why $\mathbf{V}_t(\Omega)$ is reflexive and separable as a closed subspace of $\mathbf{X}$. The elememts of $\mathbf{X}$ have zero traces on $\partial\Omega$. By $\mathbf{W}(Q)$ we denote the Banach space

$$\mathbf{W}(Q) = \left\{u : (0, T) \mapsto \mathbf{V}_t(\Omega)|\ u \in L^2(Q),\ |D_i u|^{p_i(x,t)} \in L^1(Q),\ u = 0 \text{ on } \Gamma\right\},$$

$$\|u\|_{\mathbf{W}(Q)} = \|u\|_{2,Q} + \sum_{i=1}^n \|D_i u\|_{p_i(\cdot),Q}.$$

$\mathbf{W}'(Q)$ is the dual of $\mathbf{W}(Q)$ (the space of linear functionals over $\mathbf{W}(Q)$):

$$w \in \mathbf{W}'(Q) \ \Leftrightarrow \ \begin{cases} \exists\,(w_0, w_1, \dots, w_n), \quad w_0 \in L^2(Q), \quad w_i \in L^{p_i'(\cdot)}(Q), \\ \forall \phi \in \mathbf{W}(Q) \quad \langle\langle w, \phi \rangle\rangle = \displaystyle\int_Q \left(w_0 \phi + \sum_{i=1}^n w_i D_i \phi\right) dz. \end{cases}$$

The norm in $\mathbf{W}'(Q)$ is defined by

$$\|v\|_{\mathbf{W}'(Q)} = \sup\left\{\langle\langle v, \phi\rangle\rangle \ : \ \phi \in \mathbf{W}(Q),\ \|\phi\|_{\mathbf{W}(Q)} \le 1\right\}.$$

Let $\mathbf{v} = (v_1, \dots, v_n)$ be a vector-valued function defined on $Q$. Assume that $p_i(z)$ satisfy conditions (1.3) and (1.26). Introduce the modular

$$A_{\mathbf{p}(\cdot)}(\mathbf{v}) = \sum_{i=1}^{n} \int_Q |v_i|^{p_i(z)}\, dz. \tag{1.28}$$

For the elements of $\mathbf{W}(Q)$ the following counterpart of inequalities (1.6) is true:

$$\begin{aligned} &\min\left\{\sum_{i=1}^{n} \|D_i u\|^{p^+}_{p_i(\cdot,\cdot),Q}, \sum_{i=1}^{n} \|D_i u\|^{p^-}_{p_i(\cdot,\cdot),Q}\right\} \\ &\qquad \le A_{\mathbf{p}(\cdot,\cdot),Q}(\nabla u) \le \max\left\{\sum_{i=1}^{n} \|D_i u\|^{p^-}_{p_i(\cdot,\cdot),Q}, \sum_{i=1}^{n} \|D_i u\|^{p^+}_{p_i(\cdot,\cdot),Q}\right\}. \end{aligned} \tag{1.29}$$

### *1.4.1 Approximation by Smooth Functions*

Set

$$\mathbf{V}_+(\Omega) = \left\{u(x)|\ u \in L^2(\Omega) \cap W_0^{1,1}(\Omega),\ |\nabla u| \in L^{p^+}(\Omega)\right\}.$$

Since $\mathbf{V}_+(\Omega)$ is separable, it is a span of a countable set of linearly independent functions $\{\psi_k\} \subset \mathbf{V}_+(\Omega)$. It is convenient to choose for $\{\psi_k\}$ the system of eigenfunctions of the operator

$$(\psi_j, w)_{H_0^s(\Omega)} = \lambda_j(\psi_j, w)_{L^2(\Omega)} \quad \forall w \in H_0^s(\Omega) \tag{1.30}$$

with $s$ chosen according to the inequality

$$\frac{s-1}{n} \ge \frac{1}{2} - \frac{1}{p^+}.$$

For such $s$ the embedding $H_0^s(\Omega) \subset W_0^{1,p^+}(\Omega)$ is continuous. Let $\rho$ be the Friedrichs mollifying kernel

$$\rho(s) = \begin{cases} \kappa \exp\left(-\frac{1}{1-|s|^2}\right) & \text{if } |s| < 1, \\ 0 & \text{if } |s| > 1, \end{cases} \qquad \kappa = const: \int_{\mathbb{R}^{n+1}} \rho(z)\, dz = 1.$$

Given a function $v \in L^1(Q)$, we extend it to the whole $\mathbb{R}^{n+1}$ by a function with compact support (keeping the same notation for the continued function) and then define

$$v_\varepsilon(z) = \int_{\mathbb{R}^{n+1}} v(s)\rho_\varepsilon(z-s)\,ds \qquad \text{with} \quad \rho_\varepsilon(s) = \frac{1}{\varepsilon^{n+1}}\rho\left(\frac{s}{\varepsilon}\right), \quad \varepsilon > 0.$$

For the exponents $p_i(x,t)$ depending on $(x,t) \in Q$ we will use the notation $p_i \in C_{\log}(Q_T)$ if $p_i$ satisfies condition (1.26) in the cylinder $Q$:

$$C_{\log}(Q) := \left\{ p \in C^0(\overline{Q}) \;\middle|\; \begin{array}{l} \forall\, z=(x,t),\ \zeta=(y,\tau)\in Q \\ \text{such that } |x-y|+|t-\tau|<\frac{1}{2}, \\ |p(z)-p(\zeta)| \le \omega(|z-\zeta|) \end{array} \right\} \tag{1.31}$$

with a continuous function $\omega$ satisfying the condition

$$\overline{\lim_{\tau\to 0^+}}\, \omega(\tau)\ln\frac{1}{\tau} = C < +\infty.$$

**Lemma 1.15** *If $u \in \mathbf{W}(Q)$ with the exponents $p_i(z) \in C_{\log}(Q)$, then*

$$\|u_\varepsilon\|_{\mathbf{W}(Q)} \le C\left(1+\|u\|_{\mathbf{W}(Q)}\right) \quad \textit{and} \quad \|u_\varepsilon - u\|_{\mathbf{W}(Q)} \to 0 \textit{ as } \varepsilon \to 0.$$

The assertion follows from Lemmas 1.12 and 1.13.

**Lemma 1.16** *If $p_i(z) \in C_{\log}(Q)$, then the set $\{\psi_k\}$ is dense in $\mathbf{V}_t(\Omega)$ for every $t \in [0,T]$.*

*Proof* For every $u \in \mathbf{V}_t(\Omega)$ there is a sequence $u_\delta(\cdot,t) \in C^\infty(\Omega)$ such that

$$\operatorname{supp} u_\delta(\cdot,t) \subset\subset \Omega \quad \text{and} \quad \|u-u_\delta\|_{\mathbf{V}_t(\Omega)} \to 0 \quad \text{as } \delta \to 0. \tag{1.32}$$

Such a sequence is obtained via convolution of $u$ with the Friedrichs mollifiers. Since $u_\delta \in C_0^\infty(\Omega) \subset \mathbf{V}_+(\Omega)$ and $\{\psi_m\}$ is dense in $\mathbf{V}_+(\Omega)$, one may choose constants $c_m$ such that

$$u_\delta^{(k)} \equiv \sum_{m=1}^{k} c_m\psi_m(x) \to u_\delta \quad \text{strongly in } \mathbf{V}_+(\Omega) \text{ as } k \to \infty.$$

Given an arbitrary $\varepsilon > 0$, $\|u_\delta - u_\delta^{(k)}\|_{\mathbf{V}_+(\Omega)} < \varepsilon$ for all $k \in \mathbb{N}$ from some $k(\varepsilon)$ on. By Lemma 1.5

$$\|u_\delta - u_\delta^{(k)}\|_{\mathbf{V}_t(\Omega)} \le C\,\|u_\delta - u_\delta^{(k)}\|_{\mathbf{V}_+(\Omega)} \le C\,\varepsilon \tag{1.33}$$

with an independent of $\varepsilon$ constant $C = C(n, |\Omega|, p^{\pm})$. Gathering (1.32) with (1.33) we conclude that for all sufficiently large $k$ and small $\delta$

$$\|u - u_\delta^{(k)}\|_{\mathbf{V}_t(\Omega)} \le \|u - u_\delta\|_{\mathbf{V}_t(\Omega)} + \|u_\delta - u_\delta^{(k)}\|_{\mathbf{V}_t(\Omega)} < 2\varepsilon \quad \forall t \in [0, T].$$

□

**Lemma 1.17** *Let $p_i(z) \in C_{\log}(Q)$. For every $u \in \mathbf{W}(Q)$ there is a sequence $\{d_k(t)\}$, $d_k(t) \in C^1[0, T]$, such that*

$$\left\| u - \sum_{k=1}^{m} d_k(t)\psi_k(x) \right\|_{\mathbf{W}(Q)} \to 0 \quad \textit{as } m \to \infty.$$

*Proof* (a) Let us first prove the assertion for $u \in C^\infty(0, T; C_0^\infty(\Omega))$. Let us represent

$$u = \sum_{i=1}^{\infty} u_i(t)\psi_i(x), \quad u_i(t) = (u(x,t), \psi_i(x))_{2,\Omega}. \tag{1.34}$$

For every $t \in [0, T]$

$$\|u\|^2_{H_0^s(\Omega)}(t) = \sum_{i=1}^{\infty} \lambda_i u_i^2(t) < \infty. \tag{1.35}$$

Consider the sequence $u^{(m)} = \sum_{i=1}^{m} u_i(t)\psi_i(x)$. For every $t \in [0, T]$

$$\phi_m(t) \equiv \|u - u^{(m)}\|^2_{H_0^s(\Omega)}(t) = \sum_{i=m+1}^{\infty} \lambda_i u_i^2(t) \to 0 \quad \text{as } m \to \infty$$

because of (1.35). The sequences $\{\phi_m(t)\}$ and $\left\{\phi_m^{p^+/2}(t)\right\}$ are monotone decreasing, nonnegative, and tend to zero as $m \to \infty$. Since $\phi_m(t), \phi_m^{\frac{p^+}{2}}(t) \in L^1(0, T)$, it follows from the monotone convergence theorem

$$\left\|u - u^{(m)}\right\|^{p^+}_{L^{p^+}(0,T;H_0^s(\Omega))} = \int_0^T \phi_m^{\frac{p^+}{2}}(\tau)\,d\tau \to 0, \qquad \int_0^T \phi_m(t)\,dt \to 0 \quad \text{as } m \to \infty.$$

Let $v \in C^\infty([0, T]; C_0^\infty(\Omega))$. Applying Lemma 1.6 we estimate

$$\|D_i v\|^{p^+}_{p_i(\cdot),Q} \le C\|D_i v\|^{p^+}_{p^+,Q} \le C\|\nabla v\|^{p^+}_{p^+,Q} = C\int_0^T \|\nabla v\|^{p^+}_{p^+,\Omega}(t)\,dt, \tag{1.36}$$

whence

$$\|v\|_{\mathbf{W}(Q)} \leq \|v\|_{L^2(0,T;H_0^s(\Omega))} + C_1 \left( \int_0^T \|v\|_{H_0^s(\Omega)}^{p^+} dt \right)^{\frac{1}{p^+}}. \tag{1.37}$$

Letting $v = u - u^{(m)}$ we finally have:

$$\left\| u - u^{(m)} \right\|_{\mathbf{W}(Q)} \leq \left( \int_0^T \phi_m(t)\, dt \right)^{\frac{1}{2}} + \widetilde{C}_1 \left( \int_0^T \phi_m^{\frac{p^+}{2}}(t)\, dt \right)^{\frac{1}{p^+}} \to 0 \quad \text{as } m \to \infty.$$

(b) Let $u \in \mathbf{W}(Q)$. Denote by $\{u_\delta\}$ the sequence of mollifications of $u$, $u_\delta \in C^\infty([0,T]; C_0^\infty(\Omega))$. Given an arbitrary $\varepsilon > 0$, we take $\delta$ such that $\|u - u_\delta\|_{\mathbf{W}(Q)} < \varepsilon$ and approximate $u_\delta$ using item (a):

$$\left\| u - u_\delta^{(m)} \right\|_{\mathbf{W}(Q)} \leq \|u - u_\delta\|_{\mathbf{W}(Q)} + \left\| u_\delta - u_\delta^{(m)} \right\|_{\mathbf{W}(Q)} < 2\varepsilon \tag{1.38}$$

with

$$u_\delta^{(m)} = \sum_{i=1}^m d_i(t)\psi_i(x), \qquad d_i(t) = (u_\delta, \psi_i)_{H_0^s(\Omega)} \in C^\infty[0,T].$$

□

**Lemma 1.18** *Let $u \in \mathbf{W}(Q)$ with $p_i(z) \in C_{\log}(Q)$. If $u_t \in \mathbf{W}'(Q)$, then $(u_h)_t \in \mathbf{W}'(Q)$ and for every $\psi \in \mathbf{W}(Q)$*

$$\langle\langle (u_h)_t, \psi \rangle\rangle \to \langle\langle u_t, \psi \rangle\rangle \quad \text{as } h \to 0.$$

*Proof* By the definition of $\mathbf{W}'(Q)$ there exist $\phi_0 \in L^2(Q)$, $\phi_i \in L^{p_i'(\cdot)}(Q)$ such that

$$\langle\langle u_t, \psi \rangle\rangle = (\phi_0, \psi)_{2,Q} + \sum_i (\phi_i, D_i\psi)_{2,Q} \qquad \forall\, \psi \in \mathbf{W}(Q).$$

Without loss of generality we assume that the functions $u$ and $\phi_i$, $i = 0, 1, \ldots, n$, are continued to the whole of $\mathbb{R}^{n+1}$ by functions with compact supports. By the definition

$$\begin{aligned} |(u_t)_h| &= |\langle\langle u_t, \rho_h \rangle\rangle| \leq \|u_t\|_{\mathbf{W}'} \|\rho_h\|_{\mathbf{W}} \leq C(h), \\ |(u_h)_t| &= |(u, (\rho_h)_t)_2| \leq \|u\|_2 \|(\rho_h)_t\|_2 \leq C(h), \end{aligned}$$

whence $\partial_\tau\, (u(y,\tau)\, \rho_h(x-y, t-\tau)) \in L^1(\mathbb{R}^{n+1})$ for every $(x,t) \in \mathbb{R}^{n+1}$ and

$$\int_{\mathbb{R}^{n+1}} \partial_\tau\, (u(y,\tau)\, \rho_h(x-y, t-\tau))\, dy d\tau = 0.$$

It follows that for every $(x, t) \in \mathbb{R}^{n+1}$

$$
\begin{aligned}
(u_h)_t &= \partial_t (u * \rho_h)(x, t) = u * (\rho_h)_t (x, t) \\
&= -\int u(y, \tau)(\rho_h)_\tau (x - y, t - \tau)\, dy d\tau \\
&= -\int \partial_\tau (u\, \rho_h)\, dy d\tau + \int u_\tau(y, \tau)\, \rho_h(x - y, t - \tau)\, dy d\tau \\
&= (u_t * \rho_h)(x, t) = (u_t)_h.
\end{aligned}
$$

For every $\psi \in \mathbf{W}(Q)$ (continued to $\mathbb{R}^{n+1}$)

$$
\begin{aligned}
\langle\langle (u_h)_t, \psi \rangle\rangle &= \int (u_t)_h \psi\, dz = \int u_t \psi_h\, dz \\
&= \int \left( \phi_0 \psi_h + \sum_i \phi_i\, D_i \psi_h \right) dz \\
&= \int \left( (\phi_0)_h \psi + \sum_i (\phi_i)_h\, D_i \psi \right) dz \to \langle\langle u_t, \psi \rangle\rangle \quad \text{as } h \to 0
\end{aligned}
$$

by virtue of Lemma 1.15. □

### 1.4.2 A Subspace of $\mathbf{W}(Q)$

Let $\mathbf{V}_t(\Omega)$ be the isotropic space defined by (1.27) with the exponents $p_i \equiv p(x, t) \in C_{\log}(Q)$. Let us consider the subspace of $\mathbf{W}(Q)$

$$
\mathbf{U}(Q) = \left\{ u : [0, T] \mapsto \mathbf{V}_t(\Omega) |\ u,\ u_t,\ |\nabla u_t| \in L^2(Q),\ |\nabla u|^{p(x,t)} \in L^1(Q) \right\}.
$$

The set $\mathbf{U}(Q)$ endowed with the norm

$$
\begin{aligned}
\|u\|_{\mathbf{U}(Q)} &= \|u\|_{\mathbf{W}(Q)} + \|u_t\|_{L^2(0,T;H_0^1(\Omega))} \\
&\equiv \|\nabla u\|_{p(\cdot,\cdot),Q} + \|u\|_{2,Q} + \|u_t\|_{2,Q} + \|\nabla u_t\|_{2,Q}
\end{aligned}
$$

becomes a Banach space. The dual space $\mathbf{U}'(Q)$ is the space of linear functionals over $\mathbf{U}(Q)$: if $u \in \mathbf{U}'(Q)$, there are functions $\phi_0, \phi_1 \in L^2(Q)$, $\psi_i \in L^2(Q)$ and $\chi_i \in L^{p'(\cdot,\cdot)}(Q)$, $i = 1, 2, \ldots, n$, such that for every $\Phi \in \mathbf{U}(Q)$

$$
\langle u, \Phi \rangle = (\Phi, \phi_0)_{2,Q} + (\Phi_t, \phi_1)_{2,Q} + \sum_{i=1}^{n} (D_i \Phi_t, \psi_i)_{2,Q} + \sum_{i=1}^{n} (D_i \Phi, \chi_i)_{2,Q}.
$$

Let $\{\psi_k\}$ be the set of eigenfunctions of problem (1.30) with

$$\frac{s-1}{n} \geq \frac{1}{2} - \frac{1}{\max\{2, p^+\}}.$$

**Lemma 1.19** *Let $p(z) \in C_{\log}(Q)$. For every $u \in \mathbf{U}(Q)$ there is a sequence $\{d_k(t)\}$, $d_k(t) \in C^2[0, T]$, such that*

$$\|u - \sum_{k=1}^{m} d_k(t)\psi_k(x)\|_{\mathbf{U}(Q)} \to 0 \quad \text{as } m \to \infty.$$

We omit the detailed proof because it follows immediately from the proofs of Lemmas 1.16 and 1.17. Given a function $u \in C^\infty(0, T; C_0^\infty(\Omega))$, we represent it by the series

$$u = \sum_{i=1}^{\infty} u_i(t)\psi_i(x), \quad u_i(t) = (u(x,t), \psi_i(x))_{2,\Omega}.$$

Consider the sequence $u^{(m)} = \sum_{i=1}^{m} u_i(t)\psi_i(x)$. Using (1.37) to estimate $\|u\|_{\mathbf{W}(Q)}$, we find that

$$\begin{aligned}
\|u\|_{\mathbf{U}(Q)} &\leq \|u\|_{\mathbf{W}(Q)} + C\left(\int_0^T \|u_t\|^2_{H_0^s(\Omega)}\,dt\right)^{\frac{1}{2}} \\
&= \|u\|_{\mathbf{W}(Q)} + C\left(\int_0^T \sum_{i=1}^{\infty} \lambda_i (u_i'(t))^2\,dt\right)^{\frac{1}{2}} \\
&\leq \|u\|_{L^2(0,T; H_0^s(\Omega)} + C\left(\int_0^T \|u\|^{p^+}_{H_0^s(\Omega)}(t)\,dt\right)^{\frac{1}{2}} + C_1\left(\int_0^T \sum_{i=1}^{\infty} \lambda_i (u_i'(t))^2\,dt\right)^{\frac{1}{2}}.
\end{aligned}$$

For every $t \in [0, T]$ the sequence

$$\phi_m(t) \equiv \|u - u^{(m)}\|^2_{H_0^s(\Omega)} + \|u_t - u_t^{(m)}\|^2_{H_0^s(\Omega)}$$

is monotone decreasing, nonnegative, and tends to zero as $m \to \infty$. By the Beppo Levi theorem

$$\|u - u^{(m)}\|_{\mathbf{U}(Q)} \leq C\left(\int_0^T \phi_m(t)\,dt\right)^{\frac{1}{2}} + C_1\left(\int_0^T \phi_m^{\frac{p^+}{2}}(t)\,dt\right)^{\frac{1}{p^+}} \to 0 \quad \text{as } m \to \infty.$$

In the general case when $u \in \mathbf{U}(Q)$ these arguments are applied to the sequence of mollifications that approximates $u$—see (1.38).

## 1.5 Formulas of Integration by Parts

The main assertion of this section is the rule of integration by parts in $t$ for the functions from Sobolev's spaces with variable exponents.

**Lemma 1.20** *Let $p_i(z) \in C_{\log}(Q)$. If $v, w \in \mathbf{W}(Q)$ and $v_t, w_t \in \mathbf{W}'(Q)$, then*

$$\forall \text{ a.e. } t_1, t_2 \in [0, T] \quad \int_{t_1}^{t_2} \int_{\Omega} v\, w_t\, dz + \int_{t_1}^{t_2} \int_{\Omega} v_t\, w\, dz = \int_{\Omega} v\, w\, dx \Big|_{t=t_1}^{t=t_2}.$$

*Proof* Let $t_1 < t_2$. Consider the function

$$\chi_k(t) = \begin{cases} 0 & \text{for } t \le t_1, \\ k(t - t_1) & \text{for } t_1 \le t \le t_1 + \frac{1}{k}, \\ 1 & \text{for } t_1 + \frac{1}{k} \le t \le t_2 - \frac{1}{k}, \\ k(t_2 - t) & \text{for } t_2 - \frac{1}{k} \le t \le t_2, \\ 0 & \text{for } t \ge t_2. \end{cases} \tag{1.39}$$

By $u_h, v_h$ we denote the mollifications of $u$ and $v$. For every $k \in \mathbb{N}$ and $h > 0$

$$0 = \int_Q (v_h w_h \chi_k)_t\, dz \equiv \int_Q (v_h w_h)_t \chi_k\, dz - k \int_{\theta - \frac{1}{k}}^{\theta} \int_{\Omega} v_h w_h\, dz \Big|_{\theta=t_1}^{\theta=t_2}.$$

The last two integrals on the right-hand side exist because $v_h, w_h \in L^2(Q)$. Letting $h \to 0$, we obtain the equality

$$\lim_{h\to 0} \int_Q (v_h\, (w_h)_t + (v_h)_t w_h)\, \chi_k(t)\, dz = k \int_{t_2 - \frac{1}{k}}^{t_2} \int_{\Omega} v\, w\, dz - k \int_{t_1}^{t_1 + \frac{1}{k}} \int_{\Omega} v\, w\, dz.$$

According to Lemmas 1.15 and 1.18 $v_h \to v$ in $\mathbf{W}(Q)$, $(w_h)_t = (w_t)_h \rightharpoonup w_t$ in $\mathbf{W}'(Q)$ as $h \to 0$, and $\|v\|_{\mathbf{W}}$, $\|(w_h)_t\|_{\mathbf{W}'}$ are uniformly bounded. It follows that

$$\begin{aligned} \lim_{h\to 0} \int_Q v_h\, (w_h)_t\, \chi_k(t)\, dz &= \lim_{h\to 0} \int_Q (v_h - v)(w_h)_t \chi_k(t)\, dz \\ &\quad + \lim_{h\to 0} \int_Q v\, ((w_h)_t - w_t) \chi_k(t)\, dz + \int_Q v\, w_t\, \chi_k(t)\, dz \\ &= \int_Q v\, w_t \chi_k(t)\, dz. \end{aligned}$$

In the same way we check that

$$\lim_{h\to 0} \int_Q (v_h)_t\, w_h\, \chi_k(t)\, dz = \int_Q v\, w_t \chi_k(t)\, dz.$$

By the Lebesgue differentiation theorem

$$\forall \text{ a.e. } \theta > 0 \qquad \lim_{k\to 0} k \int_{\theta-\frac{1}{k}}^{\theta} \left( \int_{\Omega} v\, w\, dx \right) dt = \int_{\Omega} v\, w\, dx,$$

whence for almost every $t_1,\ t_2 \in [0, T]$

$$\begin{aligned}\int_{t_1}^{t_2} \int_{\Omega} (v\, w_t + v_t w)\, dz &= \lim_{k\to\infty} \int_{Q} (v\, w_t + v_t w) \chi_k(t)\, dz \\ &= \lim_{k\to\infty} k \int_{\theta-\frac{1}{k}}^{\theta} \int_{\Omega} v\, w\, dx \Big|_{t=t_1}^{t=t_2} \\ &= \int_{\Omega} v\, w\, dx \Big|_{t=t_1}^{t=t_2}.\end{aligned}$$

□

**Corollary 1.10** *Let $u \in \mathbf{W}(Q)$ and $u_t \in \mathbf{W}'(Q)$ with the exponents $p_i(z) \in C_{\log}(Q)$. Then*

$$\forall \text{ a.e. } t_1, t_2 \in (0, T] \quad \int_{t_1}^{t_2} \int_{\Omega} u\, u_t\, dz = \frac{1}{2} \|u\|_{2,\Omega}^2 \Big|_{t=t_1}^{t=t_2}.$$

We will need the following elementary inequalities.

**Proposition 1.8** ([109]) *For every $p \geq 2$, $|a| \geq |b| \geq 0$*

$$||a|^{p-2}a - |b|^{p-2}b| \leq C(p)|a-b|(|a|+|b|)^{p-2}.$$

This proposition is an immediate byproduct of the easily verified relation

$$1 - t^{p-1} \leq C(p)(1-t)(1+t)^{p-2} \quad \forall\, p \geq 2, \quad t \in [0,1].$$

**Proposition 1.9** ([109]) *For $2 - p < \beta < 1$ and $|a| \geq |b| \geq 0$*

$$||a|^{p-2}a - |b|^{p-2}b| \leq C(p)|a-b|^{1-\beta}(|a|+|b|)^{p-2+\beta}.$$

The assertion follows from the inequality

$$1 - t^{p-1} \leq C(p)(1-t)^{1-\beta}(1+t)^{p-2+\beta}, \quad t \in [0,1]$$

with the same $p$ and $\beta$.

**Lemma 1.21** *Let $\mathbf{W}(Q)$ be the isotropic space with $p_i(z) \equiv p(z)$. Let $u \in \mathbf{W}(Q) \cap L^{\infty}(Q)$ with $u_t \in \mathbf{W}'(Q)$. Denote*

$$V_\varepsilon(u, z) := \int_0^u (\varepsilon + |s|)^{\gamma(z)} \, ds, \qquad \varepsilon > 0, \quad z = (x, t),$$

*with the exponent* $\gamma(z) \geq \gamma^- > -1$. *If*

$$p(z) \in C_{\log}(Q), \quad \gamma_t \in L^1(Q), \quad |\nabla \gamma(z)|^{p(z)} \in L^1(Q),$$

*then for a.e.* $t_1, \ t_2 \in [0, T]$

$$\begin{aligned}
\int_{t_1}^{t_2} \int_\Omega u_t \, V_\varepsilon(u, z) \, dz &= \int_\Omega \int_0^u V_\varepsilon(s, z) \, ds dx \Big|_{t=t_1}^{t=t_2} \\
&\quad - \int_{t_1}^{t_2} \int_\Omega \gamma_t A(\gamma) \left( (\varepsilon + |u|)^{\gamma+2} - \varepsilon^{\gamma+2} \right) dz \\
&\quad + \int_{t_1}^{t_2} \int_\Omega \gamma_t B(\gamma) \left( (\varepsilon + |u|)^{\gamma+2} \ln(\varepsilon + |u|) - \varepsilon^{\gamma+2} \ln \varepsilon \right) dz \\
&:= \mu_\varepsilon(u_\varepsilon, V_\varepsilon)
\end{aligned} \tag{1.40}$$

*with the coefficients*

$$B(\gamma) = \frac{1}{(\gamma + 1)(\gamma + 2)}, \qquad A(\gamma) = B(\gamma) \left[ \frac{1}{\gamma + 1} + \frac{1}{\gamma + 2} \right].$$

*Proof* Let $u_h \in C^\infty(Q)$ be the mollification of $u \in \mathbf{W}(Q)$ and

$$V_\varepsilon(u_h, z) = \int_0^{u_h} (\varepsilon + |s|)^{\gamma(z)} \, ds \equiv \frac{\operatorname{sign} u_h}{\gamma + 1} \left( (\varepsilon + |u_h|)^{\gamma+1} - \varepsilon^{\gamma+1} \right).$$

Since $u$ and $u_h$ are bounded by a constant $1 + K_0$, and $\gamma(z) \geq \gamma^- > -1$, it follows from Propositions 1.8 and 1.9 that

$$|V_\varepsilon(u_h, z) - V_\varepsilon(u, z)| \leq C \max \left\{ |u_h - u|, \ |u_h - u|^{1+\min\{0, \gamma^-\}} \right\}$$

with a constant $C \equiv C(\varepsilon, \ p^\pm, \ \gamma^\pm, \ K_0)$. The inclusion $u \in L^\infty(Q)$ entails the convergence $\| V_\varepsilon(u_h, z) - V_\varepsilon(u, z) \|_{r,Q} \to 0$ as $h \to 0$ for every $r > 1$.

Explicitly calculating the primitive, in the same way we check that for every $r > 1$

$$\left\| \int_u^{u_h} (\varepsilon + |s|)^{\gamma(z)} \ln(\varepsilon + |s|) \, ds \right\|_{r,Q} \to 0 \quad \text{as } h \to 0.$$

Let $\chi_k(t)$ be the function introduced in (1.39). On one hand

$$0 = \int_Q \partial_t \left( \psi_k(t) \int_0^{u_h} V_\varepsilon(s, z) \, ds \right) dz$$

$$
\begin{aligned}
&= \int_Q \psi_k(t)(u_h)_t V_\varepsilon(u_h, z)\, dz + \int_Q \psi_k'(t) \int_0^{u_h} V_\varepsilon(s, z)\, ds\, dz \\
&\quad - \int_Q \psi_k(t)\gamma_t(z) \left( \int_0^{u_h} \int_0^{w} (\varepsilon + |s|)^\gamma \ln(\varepsilon + |s|)\, ds dw \right) dz,
\end{aligned} \tag{1.41}
$$

one the other hand

$$
\begin{aligned}
\int_0^{u_h} \int_0^{w} (\varepsilon + |s|)^\gamma \ln(\varepsilon + |s|)\, ds dw &= -A(\gamma) \left( (\varepsilon + |u_h|)^{\gamma+2} - \varepsilon^{\gamma+2} \right) \\
&\quad + B(\gamma) \left( (\varepsilon + |u_h|)^{\gamma+2} \ln(\varepsilon + |u_h|) - \varepsilon^{\gamma+2} \ln \varepsilon \right).
\end{aligned} \tag{1.42}
$$

Since $u \in \mathbf{W}(Q) \cap L^\infty(Q)$ and $\gamma^- > -1$, $V_\varepsilon(u, z) \in \mathbf{W}(Q)$ for every $\varepsilon > 0$. Indeed: the estimate $\|u\|_{L^\infty(Q)} \leq M$ yields the inequalities

$$
\|V_\varepsilon(u, z)\|_{\infty, Q} \leq M_1(\gamma^\pm, M), \qquad \left\| \int_0^{|u|} (\varepsilon + s)^\gamma \, |\ln(\varepsilon + s)|\, ds \right\|_{\infty, Q} \leq M_2(\gamma^\pm, M),
$$

which entail the estimate

$$
|\nabla V_\varepsilon(u, z)| \leq (\varepsilon + |u|)^{\gamma(z)} |\nabla u| + |\nabla \gamma| \int_0^{|u|} (\varepsilon + s)^{\gamma(z)} |\ln(\varepsilon + s)|\, ds \quad \text{a.e. in } Q
$$

and the inclusion $|\nabla V_\varepsilon(u, z)|^{p(z)} \in L^1(Q)$.

Following the proof of Lemma 1.20 we may now pass to the limit as $h \to 0$ in every term of the right-hand sides in (1.41) and (1.42), and then pass to the limit as $k \to \infty$, which leads to (1.40). □

**Corollary 1.11** *Let $u \in \mathbf{W}(Q) \cap L^\infty(Q)$ and $u_t \in \mathbf{W}'(Q)$. Denote $v = \dfrac{u\,|u|^\gamma}{\gamma + 1}$ and assume that $v \in \mathbf{W}(Q)$. Under the foregoing conditions on the exponents $p(z)$ and $\gamma(z)$ the following formula of integration by parts holds:* $\forall$ a.e. $t_1, t_2 \in [0, T]$

$$
\begin{aligned}
\int_{t_1}^{t_2} \int_\Omega u_t\, v\, dz &= \int_\Omega \frac{u\, v}{\gamma + 2}\, dx \Big|_{t=t_1}^{t=t_2} \\
&\quad + \int_{t_1}^{t_2} \int_\Omega \frac{u\, v}{\gamma + 2}\, \gamma_t \left[ \frac{1}{\gamma + 1} \frac{1}{\gamma + 2} - \ln|u| \right] dz \equiv \mu(u, v).
\end{aligned}
$$

In the study of uniqueness and comparison of solutions of equations with double nonlinearity we will rely on the following version of the integration-by-parts formula. Let us define the functions

$$
\begin{aligned}
T_\delta(s) &= \frac{s}{\sqrt{\delta^2 + s^2}}, \quad \delta > 0, \\
\phi_{k,\delta,\theta}(v) &= \chi_{k,\theta}(t)\, T_\delta(v(z))
\end{aligned} \tag{1.43}
$$

with

$$\chi_{k,\theta}(t) = \begin{cases} 0 & \text{for } t \le 0, \\ k\,t & \text{for } 0 \le t \le \frac{1}{k}, \\ 1 & \text{for } \frac{1}{k} \le t \le \theta - \frac{1}{k}, \\ k\,(\theta - t) & \text{for } \theta - \frac{1}{k} \le t \le \theta, \\ 0 & \text{for } t \ge \theta, \end{cases} \qquad \frac{1}{k} < \theta \le T.$$

It is easy to see that

$$T_\delta(s) \to \ \operatorname{sign} s \text{ as } \delta \to 0,$$
$$T'_\delta(s) = \frac{\delta^2}{(\delta^2 + s^2)^{\frac{3}{2}}} > 0, \quad -1 \le s T'_\delta(s) \le 1 \text{ for } s \in \mathbb{R}.$$

**Lemma 1.22** *Let $\Phi(z, s)$ be a nonnegative function defined on $Q \times \mathbb{R}$, monotone increasing with respect to $s$. Assume that the functions $v_i$ and $w_i \equiv \Phi(z, v_i)$, $(i = 1, 2)$, possess the properties*

$$v_i \in \mathbf{W}(Q) \cap L^\infty(Q), \quad \partial_t w_i \in L^1(Q) \cap \mathbf{W}'(Q)$$

*and denote*

$$v = v_1 - v_2, \quad w = w_1 - w_2 \equiv \Phi(z, v_1) - \Phi(z, v_2).$$

*For a.e. $\theta \in (0, T)$ there exists the limit*

$$\lim_{\delta \to 0} \lim_{k \to \infty} \int_Q \phi_{k,\delta,\theta}(v)\, \partial_t w\, dz = \int_\Omega |w|\, dx \Big|_{t=0}^{t=\theta}.$$

*Proof* Recall the notation

$$Q_\tau = Q \cap \{t < \tau\}, \qquad \tau \in (0, T).$$

Since $w \in L^\infty(Q)$ and $\phi_{k,\delta,\theta}(v)$ are uniformly bounded, it follows from the dominated convergence theorem that

$$\int_Q \chi_{k,\theta}(t)\, T_\delta(v)\, \partial_t w\, dz \to \int_{Q_\theta} T_\delta(v)\, \partial_t w\, dz \quad \text{as } k \to \infty$$

and, because $\operatorname{sign} v = \operatorname{sign} w$,

$$\lim_{k \to \infty} \int_Q \phi_{k,\delta,\theta}(v)\, \partial_t w\, dz = \int_Q T_\delta(v)\, \partial_t w\, dz$$
$$\to \int_{Q_\theta} \operatorname{sign} v\, \partial_t w\, dz \equiv \int_{Q_\theta} \operatorname{sign} w\, \partial_t w\, dz = J \quad \text{as } \delta \to 0.$$

On the other hand, repeating the same arguments with the test-function $\phi_{k,\delta,\theta} \equiv \chi_{k,\theta}(t)\, T_\delta(w)$, we find that

$$J = \lim_{\delta\to 0}\lim_{k\to\infty} \int_Q T_\delta(w)\,\chi_{k,\theta}(t)\,\partial_t w\,dz.$$

The straightforward computation shows that

$$\begin{aligned}\int_Q T_\delta(w)\,\chi_{k,\theta}(t)\,\partial_t w\,dz &= \int_Q \chi_{k,\theta}(t)\,\partial_t\left(\int_0^w T_\delta(s)\,ds\right) dz \\ &= k\int_{\theta-1/k}^{\theta} dt \int_\Omega \left(\int_0^w T_\delta(s)\,ds\right) dx \\ &\quad - k\int_0^{1/k} dt \int_\Omega \left(\int_0^w T_\delta(s)\,ds\right) dx,\end{aligned}$$

where

$$\int_0^w T_\delta(s)\,ds = \sqrt{\delta^2+w^2}-\delta \to \sqrt{w^2} = |w| \quad \text{as } \delta\to 0.$$

Letting $k\to\infty$, $\delta\to 0$ and applying the Lebesgue differentiation theorem, we find that for a.e. $\theta\in(0,T)$

$$\begin{aligned}\int_Q T_\delta(w)\,\chi_{k,\theta}(t)\,\partial_t w\,dz &= \int_\Omega \left(\int_0^w T_\delta(s)\,ds\right) dx\Big|_{t=0}^{t=\theta} \\ &= \int_\Omega \sqrt{\delta^2+w^2}\,dx\Big|_{t=0}^{t=\theta} \to J = \int_\Omega |w|\,dx\Big|_{t=0}^{t=\theta}.\end{aligned}$$

□

## 1.6 Embedding Theorems in Anisotropic Spaces

To deal with anisotropic PDEs with variable nonlinearity we need specific embedding theorems. We adapt the known anisotropic inequalities of the Gagliargo-Nirenberg-Sobolev type. Let

$$\mathbf{V}(\Omega) = \{v(x) : v\in L^2(\Omega)\cap W_0^{1,1}(\Omega),\ |D_i v|^{p_i}\in L^1(\Omega)\}$$

be the anisotropic space with constant exponents of nonlinearity $p_i$. The set $\mathbf{V}(\Omega)$ endowed with the norm

$$\|v\|_{\mathbf{V}(\Omega)} = \|v\|_{2,\Omega} + \sum_{i=1}^n \|D_i v\|_{p_i,\Omega}$$

becomes a Banach space. Let $\mathbf{p} = (p_1, \ldots, p_n) \in \mathbb{R}^n$ be a constant vector with the components $1 < p_i < \infty$. By $W^{1,\mathbf{p}}(\mathbb{R}^n)$ we denote the Banach space obtained as the completion of the set $C_0^\infty(\mathbb{R}^n)$ (smooth functions with finite support) with respect to the norm

$$\|u\|_{W^{1,\mathbf{p}}(\mathbb{R}^n)} = \sum_{i=1}^{n} \|D_i u\|_{p_i,\mathbb{R}^n}.$$

**Lemma 1.23** *Let*

$$1 < p_i < \infty \quad \text{and} \quad \sum_{i=1}^{n} \frac{1}{p_i} > 1. \tag{1.44}$$

*There exists a constant $C$ such that for every $v \in W^{1,\mathbf{p}}(\mathbb{R}^n)$*

$$\|v\|_{r,\mathbb{R}^n} \leq C \left( \prod_{i=1}^{n} \| D_i v\|_{p_i,\mathbb{R}^n} \right)^{\frac{1}{n}}$$

*with*

$$r = \begin{cases} \dfrac{n\,p}{n-p} & \text{if } p < n, \\ \text{any number from } [1,\infty) & \text{if } p \geq n, \end{cases} \qquad \frac{1}{p} = \frac{1}{n} \sum_{i=1}^{n} \frac{1}{p_i}. \tag{1.45}$$

Lemma 1.23 was proved in paper [251], bounded domains were considered later in the papers [126, 224, 225]. It turns out that if the domain $\Omega$ is bounded, then the corresponding embedding takes place under specific conditions on the geometry of $\Omega$—see [131, 126, 158] for a discussion of this issue. We are interested solely in the case of zero boundary traces, for this reason we may overcome the difficulty by considering the functions continued by zero to the whole of $\mathbb{R}^n$ or to the minimal parallelepiped in $\mathbb{R}^n$ that contains $\Omega$. We will use Lemma 1.23 in the following form.

**Lemma 1.24** *Let $\Omega$ satisfy condition* (1.2) *and $\mathbf{p} = (p_1, \ldots, p_n)$ satisfy* (1.44). *Then there exists a constant $C$ such that for every $v \in \mathbf{V}(\Omega)$*

$$\|v\|_{r,\Omega} \leq C \left( \prod_{i=1}^{n} |D_i v\|_{p_i,\Omega} \right)^{\frac{1}{n}}$$

*with the exponent $r$ from* (1.45). *The constant $C$ depends on $n$, diam $\Omega$,* $\max_i p_i$ *and* $\min_i p_i$.

Given a function $u \in \mathbf{W}(Q)$, we introduce the functions

$$\begin{aligned} \Theta(t) &= \int_\Omega u^2(z)\,dx, \\ \Lambda(t) &= \int_\Omega \sum_{i=1}^n |D_i u(z)|^{p_i(z)}\,dx, \\ \Lambda_\sigma(t) &= \Lambda(t) + \int_\Omega |u(z)|^{\sigma(z)}\,dx \quad \text{if } u \in \mathbf{W}(Q) \cap L^{\sigma(\cdot,\cdot)}(Q). \end{aligned} \tag{1.46}$$

**Lemma 1.25** *Let the exponents $p_i(z)$, $\sigma(z)$ satisfy the conditions*

$$p_i(z) \in [p^-, p^+] \subset (1, \infty), \quad \sigma(z) \in (1, \infty), \quad p_i(z) \in C_{\log}(Q), \quad \sigma(z) \in C(\overline{Q}).$$

*For $u \in \mathbf{W}(Q)$ the functions $\Theta(t)$, $\Lambda(t)$ exist for a.e. $t \in (0, T)$. If $u \in \mathbf{W}(Q) \cap L^{\sigma(\cdot)}(Q)$, then for a.e. $t \in (0, T)$ there exist $\Theta(t)$ and $\Lambda_\sigma(t)$.*

*Proof* The functions $\Theta(t)$, $\Lambda(t)$ are nonnegative and, by the definition of $\mathbf{W}(Q)$, $\Theta(t), \Lambda(t) \in L^1(0, T)$. It follows that the functions

$$\int_0^t \Theta(\tau)\,d\tau, \quad \int_0^t \Lambda(\tau)\,d\tau$$

are differentiable for a.e. $t \in (0, T)$. If $u \in L^{\sigma(\cdot)}(Q)$, the same is true for the function $\Lambda_\sigma(t)$. □

**Lemma 1.26** *Let $u \in \mathbf{W}(Q)$, and let $p_i$, $\sigma$ be constant. If $p_i > 1$ and*

$$\frac{1}{p} \equiv \frac{1}{n}\sum_{i=1}^n \frac{1}{p_i} \le \frac{1}{2} + \frac{1}{n},$$

*then for a.e. $t \in (0, T)$*

$$\Theta(t) \le C\,\Lambda^{\frac{1}{\nu}}(t) \quad \text{with} \quad \nu = \frac{p}{2}. \tag{1.47}$$

*Let $u \in \mathbf{W}(Q) \cap L^\sigma(Q)$. If $p_i > 1$ and*

$$\begin{cases} \frac{1}{\sigma} + \frac{1}{p} \ge 1 + \frac{1}{n} & \text{if } n > p, \\ \sigma > 1 & \text{if } n \le p, \end{cases}$$

*then for a.e. $t \in (0, T)$*

$$\Theta(t) \le C\,\Lambda_\sigma(t)^{\frac{1}{\nu}} \quad \text{with} \quad \frac{1}{\nu} = \frac{1}{\sigma} + \frac{1}{p}. \tag{1.48}$$

*Proof* We start by proving (1.47). By Hölder's inequality and due to Lemma 1.24 with

$$r = \frac{np}{n-p} \geq 2 \quad \Leftrightarrow \quad p \geq \frac{2n}{n+2}$$

we have

$$\Theta(t) \equiv \int_{\Omega} u^2 \, dx \leq C' \, \|u\|^2_{r,\Omega} \leq C \prod_{i=1}^{n} \|D_i u\|^{\frac{2}{n}}_{p_i,\Omega} \leq C \, \Lambda(t)^{\frac{1}{\nu}}. \tag{1.49}$$

To prove (1.48) we use Hölder's inequality and Lemma 1.24 with

$$\sigma' = \frac{\sigma}{\sigma - 1} \leq r = \frac{np}{n-p} \quad \text{if } n > p \quad \Leftrightarrow \quad \frac{1}{\sigma} + \frac{1}{p} \geq 1 + \frac{1}{n}.$$

Then

$$\begin{aligned} \Theta(t) = \int_{\Omega} u^2 \, dx &\leq \|u\|_{\sigma,\Omega} \|u\|_{\sigma',\Omega} \leq C \, \|u\|_{\sigma,\Omega} \left( \sum_{i=1}^{n} \int_{\Omega} |D_i u|^{p_i} \, dx \right)^{\frac{1}{p}} \\ &\leq C \, \Lambda_{\sigma}^{\frac{1}{\sigma} + \frac{1}{p}}(t) = C \Lambda_{\sigma}^{\frac{1}{\nu}}(t). \end{aligned}$$

□

Let

$$\begin{aligned} &1 < p^- \leq p_i^-(t) \leq p_i(z) \leq p_i^+(t) \leq p^+ < \infty, \\ &1 < \sigma^- \leq \sigma^-(t) \leq \sigma(z) \leq \sigma^+(t) \leq \sigma^+ < \infty, \end{aligned} \tag{1.50}$$

where $p^{\pm}, \sigma^{\pm}$ are given constants, $p_i^{\pm}(t)$ and $\sigma^{\pm}(t)$ are defined in (1.1). Accept the notation

$$\frac{1}{p^{\pm}(t)} = \frac{1}{n} \sum_{i=1}^{n} \frac{1}{p_i^{\pm}(t)}.$$

**Lemma 1.27** *Let $p_i(z), \sigma(z) \in C_{\log}(Q)$ satisfy conditions* (1.50) *and*

$$\frac{1}{\nu^-(t)} = \frac{1}{\sigma^-(t)} + \frac{1}{p^-(t)} \leq \frac{n+1}{n} \quad \textit{for } \frac{1}{n} < \frac{1}{p^-(t)}. \tag{1.51}$$

*Then for every $u \in \mathbf{W}(Q) \cap L^{\sigma(\cdot)}(Q)$ and a.e. $t \in (0, T)$*

$$\min \left\{ \Theta^{\nu^+(t)}(t), \, \Theta^{\nu^-(t)} \right\} \leq C \, \Lambda_{\sigma}(t) \quad \textit{with} \quad \frac{1}{\nu^+(t)} = \frac{1}{\sigma^+(t)} + \frac{1}{p^+(t)}. \tag{1.52}$$

*Proof* Following the proof of Lemma 1.26 we find that for every fixed $t \in (0, T)$

$$\int_\Omega u^2\,dx \le C\,\|u\|_{\sigma^-(t),\Omega}\|u\|_{(\sigma^-(t))',\Omega}$$
$$\le C\,\|u\|_{\sigma^-(t),\Omega}\left(\prod_{i=1}^n \|D_i u\|_{p_i^-(t),\Omega}\right)^{\frac{1}{n}}, \tag{1.53}$$

provided that (1.51) is fulfilled. Let us denote $\rho(t) = \|u(\cdot,t)\|_{\sigma(\cdot,t),\Omega}$. Applying Young's inequality and then using the definition of the norm of $L^{\sigma(\cdot,t)}(\Omega)$ we estimate

$$\begin{aligned}\int_\Omega \left(\frac{|u(z)|}{\rho(t)}\right)^{\sigma^-(t)} dx &\le \int_\Omega \left(\frac{\sigma^-(t)}{\sigma(z)}\left(\frac{|u(z)|}{\rho(t)}\right)^{\sigma(z)} + 1 - \frac{\sigma^-(t)}{\sigma(z)}\right) dx\\ &\le \int_\Omega \left(\frac{|u(z)|}{\rho(t)}\right)^{\sigma(z)} dx + \left(1 - \frac{\sigma^-(t)}{\sigma^+(t)}\right)|\Omega|\\ &\le 1 + \left(1 - \frac{\sigma^-(t)}{\sigma^+(t)}\right)|\Omega|.\end{aligned}$$

It follows that

$$\begin{aligned}\|u(\cdot,t)\|_{\sigma^-(t),\Omega} &\equiv \left(\int_\Omega |u(z)|^{\sigma^-(t)}\,dx\right)^{\frac{1}{\sigma^-(t)}}\\ &\le \left(1 + \left(1 - \frac{\sigma^-(t)}{\sigma^+(t)}\right)|\Omega|\right)^{\frac{1}{\sigma^-(t)}} \rho(t)\\ &\le \widetilde{C}\,\|u(\cdot,t)\|_{\sigma(\cdot,t),\Omega}\end{aligned}$$

with the coefficient

$$\widetilde{C} = \left(1 + \left(1 - \frac{\sigma^-}{\sigma^+}\right)|\Omega|\right)^{\frac{1}{\sigma^-}}.$$

Combining this inequality with (1.6), we have

$$\begin{aligned}\|u\|_{\sigma^-(t),\Omega} &\le \widetilde{C}\,\|u(\cdot,t)\|_{\sigma(\cdot,t),\Omega}\\ &\le C\,\max\left\{\left(\int_\Omega |u|^{\sigma(z)}\,dx\right)^{\frac{1}{\sigma^+(t)}}, \left(\int_\Omega |u|^{\sigma(z)}\,dx\right)^{\frac{1}{\sigma^-(t)}}\right\}\end{aligned}$$

or, equivalently,

$$C'\int_\Omega |u|^{\sigma(z)}\,dx \ge \begin{cases} \|u\|^{\sigma^+(t)}_{\sigma^-(t),\Omega} & \text{if } \|u\|_{\sigma^-(t),\Omega} \le 1,\\ \|u\|^{\sigma^-(t)}_{\sigma^-(t),\Omega} & \text{if } \|u\|_{\sigma^-(t),\Omega} > 1.\end{cases}$$

For every $i = 1, 2, \dots, n$

$$C'' \int_{\Omega} |D_i u|^{p_i(z)}\, dx \geq \begin{cases} \|D_i u\|^{p_i^+(t)}_{p_i^-(t),\Omega} & \text{if } \|D_i u\|_{p_i^-(t),\Omega} \leq 1, \\ \|D_i u\|^{p_i^-(t)}_{p_i^-(t),\Omega} & \text{if } \|D_i u\|_{p_i^-(t),\Omega} > 1. \end{cases}$$

It follows that

$$\|u\|_{\sigma^-(t),\Omega} \left( \prod_{i=1}^{n} \|D_i u\|_{p_i^-(t),\Omega} \right)^{\frac{1}{n}} \leq C\, \Lambda_\sigma^{\frac{1}{\mu(t)}}(t)$$

with the exponents

$$\frac{1}{\mu(t)} = \frac{1}{\widetilde{\sigma}(t)} + \frac{1}{n} \sum_{i=1}^{n} \frac{1}{\widetilde{p}_i(t)} \in \left[ \frac{1}{\nu^+(t)},\ \frac{1}{\nu^-(t)} \right]$$

and

$$\widetilde{p}_i(t) = \begin{cases} p_i^-(t) & \text{if } \|D_i u\|_{p_i^-(t),\Omega} > 1, \\ p_i^+(t) & \text{if } \|D_i u\|_{p_i^-(t),\Omega} \leq 1, \end{cases} \qquad \widetilde{\sigma}(t) = \begin{cases} \sigma^-(t) & \text{if } \|u\|_{\sigma^-(t),\Omega} > 1, \\ \sigma^+(t) & \text{if } \|u\|_{\sigma^-(t),\Omega} \leq 1. \end{cases}$$

Plugging these formulas to (1.53) we find that

$$\Theta^{\mu(t)}(t) \leq C\, \Lambda_\sigma(t).$$

The assertion now follows because

$$\Theta^{\mu(t)}(t) \geq \min\left\{ \Theta^{\nu^+(t)}(t),\ \Theta^{\nu^-(t)}(t) \right\} = \begin{cases} \Theta^{\nu^+(t)} & \text{if } \Theta(t) \leq 1, \\ \Theta^{\nu^-(t)} & \text{if } \Theta(t) \geq 1. \end{cases}$$

□

**Lemma 1.28** *Let $u \in \mathbf{W}(Q)$. If*

$$p^-(t) \geq \frac{2n}{n+2}, \tag{1.54}$$

*then for a.e. $t \in (0, T)$*

$$\min\left\{ \Theta^{\nu^+(t)}(t),\ \Theta^{\nu^-(t)}(t) \right\} \leq C\, \Lambda(t), \qquad \frac{1}{\nu^\pm(t)} = \frac{2}{p^\pm(t)}, \tag{1.55}$$

*with a constant $C = C\left(\Omega,\ p^+,\ p^-,\ n\right)$.*

*Proof* By Lemma 1.24

$$\Theta(t) \leq C \left( \prod_{i=1}^{n} \|D_i u\|_{p_i^-(t),\Omega} \right)^{\frac{2}{n}}$$

and the conclusion follows as in the proof of Lemma 1.27. □

**Corollary 1.12** *Let $\Theta(t) \leq 1$. Since $\nu^+(t) \geq \nu^-(t)$, in this case the left-hand sides of inequalities* (1.55) *and* (1.52) *can be changed to $\Theta^{\nu^+(t)}(t)$.*

*Remark 1.2* In Chap. 10 similar embedding properties are proved for the elements of the space

$$\{v(x) : v \in L^{m(\cdot)+1}(\Omega) \cap W_0^{1,1}(\Omega), \ |D_i v|^{p_i} \in L^1(\Omega)\}$$

with $m(\cdot)$ and $p_i(\cdot)$ subject to suitable regularity conditions.

## 1.7 Interpolation Inequalities

The elements of conventional Sobolev spaces $W^{1,p}(\Omega)$ with constant exponents $p \in (1, \infty)$ satisfy the interpolation inequalities of Gagliardo-Nirenberg-Sobolev type [5, 133, 243].

**Lemma 1.29** *Let $\Omega \subset \mathbb{R}^n$ be an open bounded set with the boundary $\Gamma = \partial\Omega \in C^1$. For every $u \in W^{1,p}(\Omega)$*

$$\|u\|_{q,\Omega} \leq C \left( \|\nabla u\|_{p,\Omega} + \|u\|_{\gamma,\Omega} \right)^{\theta} \|u\|_{r,\Omega}^{1-\theta} \tag{1.56}$$

*with $1 \leq \gamma < \infty$, a constant $C = C(p, q, r, \gamma, n, |\Omega|)$ and the exponents*

$$\begin{gathered} \theta = \frac{\frac{1}{r} - \frac{1}{q}}{\frac{1}{r} - \frac{1}{p^*}}, \qquad p^* = \frac{np}{n-p}, \\ \text{if } p < n : \ C = \left( \frac{n-1}{n} p^* \right)^{\theta}, \quad q \in \begin{cases} [r, p^*] & \text{if } r \leq p^* \\ [p^*, r] & \text{if } r > p^*, \end{cases} \\ \text{if } p \geq n : \ C = \left( \max\left\{ \frac{n-1}{n} q, 1 + \frac{p-1}{p} r \right\} \right)^{\theta}, \quad q \in [r, \infty). \end{gathered} \tag{1.57}$$

*In the case $p > n$ inequality* (1.56) *is fulfilled with $q = \infty$, $\theta = \frac{np}{np+(p-n)}$, and an independent of $\Omega$ constant $C$.*

**Corollary 1.13** *The term $\|\nabla u\|_{p,\Omega} + \|u\|_{\gamma,\Omega}$ in* (1.56) *can be substituted by any other norm of $W^{1,p}(\Omega)$. Assume that in the conditions of* Lemma 1.29 *$\partial\Omega \in Lip$ and*

*$u \in W_0^{1,p}(\Omega)$. Let us continue $u$ by zero to a smooth domain $\Omega'$ containing $\Omega$ and denote the continued functions by $\tilde{u}$. Since $\|u\|_{W_0^{1,p}(\Omega)} = \|\tilde{u}\|_{W_0^{1,p}(\Omega')} \le C\|\nabla\tilde{u}\|_{p,\Omega'}$ and $\|\nabla\tilde{u}\|_{p,\Omega'} = \|\nabla u\|_{p,\Omega}$, applying to the continued function (1.56) we transform it to the form*

$$\|u\|_{q,\Omega} \le C\|\nabla u\|_{p,\Omega}^{\theta}\|u\|_{r,\Omega}^{1-\theta} \tag{1.58}$$

*with the same exponents $q$, $r$, $\theta$.*

**Lemma 1.30** *Under the conditions of* Lemma 1.29, *for every $u \in W^{1,p}(\Omega)$*

$$\|u\|_{q,\Gamma} \le C\left(\|\nabla u\|_{p,\Omega} + \|u\|_{\gamma,\Omega}\right)^{\theta}\|u\|_{r,\Omega}^{1-\theta} \tag{1.59}$$

*with a constant $C = C(p, q, r, \gamma, n, |\Omega|)$ and the exponents*

$$\gamma \in [1, \infty), \qquad \theta = \frac{qn - r(n-1)}{p(n+r) - nr}\frac{p}{q} \in (0, 1), \tag{1.60}$$

$$\begin{array}{lll} 1 \le r < \dfrac{np}{n-p}, & 1 \le q < \dfrac{p(n-1)}{n-p} & \text{if } n > p, \\ 1 \le r < \infty, & 1 \le q < \infty & \text{if } n = p, \\ 1 \le r \le \infty, & 1 \le q \le \infty & \text{if } n < p. \end{array} \tag{1.61}$$

We will need these inequalities in the special situation when $\Omega = B_\rho(x_0)$. Let us assume first $\Omega = B_1(x_0)$. In this case inequalities (1.56) and (1.59) hold with independent of $\Omega$ constants $C$. Let $\Omega = B_\rho(x_0)$ and $\Gamma = \partial B_\rho(x_0)$. Set $x = \rho y$ and consider the functions $u(x)$ and $v(y) = u(x)$. If $u(x) \in W^{1,p}(B_\rho(x_0))$, then $v(x) \in W^{1,p}(B_1(x_0))$, but

$$\begin{aligned} &\|\nabla v\|_{p,B_1(x_0)} = \rho^{1-\frac{n}{p}}\|\nabla u\|_{p,B_\rho(x_0)}, &&\|v\|_{q,\partial B_1(x_0)} = \rho^{-\frac{n-1}{q}}\|u\|_{q,\partial B_\rho(x_0)} \\ &\|v\|_{\gamma,B_1(x_0)} = \rho^{-\frac{n}{\gamma}}\|u\|_{\gamma,B_\rho(x_0)}, &&\|v\|_{r,B_1(x_0)} = \rho^{-\frac{n}{r}}\|u\|_{r,B_\rho(x_0)} \end{aligned}$$

(under the foregoing restrictions on $p, r, n, q$). Plugging these equalities into (1.56) and (1.59) for $v$ in $B_1(x_0)$ and simplifying, we find that

$$\begin{aligned} \|u\|_{q,B_\rho(x_0)} &\le C\rho^{\frac{n}{q}-(1-\theta)\frac{n}{r}}\left(\rho^{1-\frac{n}{p}}\|\nabla u\|_{p,\Omega} + \rho^{-\frac{n}{\gamma}}\|u\|_{\gamma,B_\rho(x_0)}\right)^{\theta}\|u\|_{r,B_\rho(x_0)}^{1-\theta} \\ &= C\rho^{\alpha}\left(\|\nabla u\|_{p,\Omega} + \rho^{-1+\frac{n}{p}-\frac{n}{\gamma}}\|u\|_{\gamma,B_\rho(x_0)}\right)^{\theta}\|u\|_{r,B_\rho(x_0)}^{1-\theta}, \end{aligned}$$

$$\begin{aligned} \|u\|_{q,\partial B_\rho(x_0)} &\le C\rho^{\frac{n-1}{q}-(1-\theta)\frac{n}{r}}\left(\rho^{1-\frac{n}{p}}\|\nabla u\|_{p,\Omega} + \rho^{-\frac{n}{\gamma}}\|u\|_{\gamma,B_\rho(x_0)}\right)^{\theta}\|u\|_{r,B_\rho(x_0)}^{1-\theta} \\ &= C\rho^{\beta}\left(\|\nabla u\|_{p,\Omega} + \rho^{-1+\frac{n}{p}-\frac{n}{\gamma}}\|u\|_{\gamma,B_\rho(x_0)}\right)^{\theta}\|u\|_{r,B_\rho(x_0)}^{1-\theta} \end{aligned}$$

with $\alpha = \beta = 0$ by virtue of (1.57) and (1.60).

**Lemma 1.31** *For every $u \in W^{1,p}(B_\rho(x_0))$*

$$\|u\|_{q,\partial B_{\rho(x_0)}} \leq C\left(\|\nabla u\|_{p,B_\rho(x_0)} + \rho^{-\delta}\|u\|_{\gamma,B_\rho(x_0)}\right)^\theta \|u\|^{1-\theta}_{r,B_\rho(x_0)\Omega} \tag{1.62}$$

*with $\delta = 1 - \frac{n}{p} + \frac{1}{\gamma}$, $\theta$ given in* (1.60)*, an independent of $\rho$ constant $C$, and the exponents $r$, $p$, $q$, $\gamma$ satisfying inequalities* (1.61)*.*

The proofs of the trace-interpolation inequality in the form (1.62) can be found in [24, 36, 113].

**Lemma 1.32** *Let $u(z) \in \mathbf{W}(Q) \cap L^\infty(0,T;L^2(\Omega))$. Assume that there is a finite constant $M$ such that for every cylinder $Q_R(x_0) = B_R(x_0) \times (0,T) \subseteq Q$*

$$ess\sup_{(0,T)} \|u(\cdot,\tau)\|^2_{L^2(B_R(x_0))} + \int_{Q_R(x_0)} |\nabla u|^{p(z)}\,dz \leq M. \tag{1.63}$$

*If $p(z) \in C(\overline{Q})$, then for every $Q_R(x_0) \subseteq Q$*

$$\|u\|_{p(\cdot),Q_R(x_0)} \leq C(M, p^\pm, n, \omega), \tag{1.64}$$

*where $\omega(z)$ is the modulus of continuity of $p$ in $Q$.*

*Proof* Let us take a finite cover of $Q_R(x_0)$ by the cylinders

$$D_i = B_\rho(x_i) \times (\tau_i, \tau_i + \tau) \subset Q, \quad i = 1, \ldots, k,$$

and denote

$$p_i^+ = \sup_{D_i} p(z), \qquad p_i^- = \inf_{D_i} p(z).$$

Since $p \in C(\overline{Q})$, the parameters $\rho$ and $\tau$ can be chosen so small that

$$\frac{p_i^+}{p_i^-} < 1 + \frac{2}{n} \quad \text{for every } D_i. \tag{1.65}$$

Let us consider an arbitrary $D_i$. There are two possibilities. If $p_i^+ \leq 2$, we apply Young's inequality and (1.63) to conclude that

$$\|u\|^2_{p(\cdot),D_i} \leq C\left(1 + \|u\|^2_{2,D_i}\right) \leq C\,(1 + M\tau)\,.$$

Let us assume now that $p_i^+ > 2$. According to (1.65), in this case

$$0 < \theta_i = \frac{\frac{1}{2} - \frac{1}{p_i^+}}{\frac{1}{2} - \frac{n-p_i^-}{np_i^-}} < \frac{p_i^-}{p_i^+}. \tag{1.66}$$

Let us apply the interpolation inequality (1.56) with $\gamma = r = 2$, $q = p_i^+$, $p = p_i^-$ and the exponent $\theta_i$ from (1.66): for a.e. $t \in (\tau_i, \tau_i + \tau)$

$$\|u\|_{p_i^+, B_\rho(x_i)}^{p_i^+} \le C \left( \|\nabla u\|_{p_i^-, B_\rho(x_i)}^{p_i^+ \theta_i} + \|u\|_{2, B_\rho(x_i)}^{p_i^+ \theta_i} \right) \|u\|_{2, B_\rho(x_i)}^{p_i^+(1-\theta_i)}. \tag{1.67}$$

Since $\theta_i p_i^+ < p_i^-$, integrating (1.67) in $t$ over the interval $(0, T)$ and applying Hölder's inequality we obtain the estimate

$$\begin{aligned} \|u\|_{p_i^+, D_i}^{p_i^+} &\le C(\tau) \left( \|\nabla u\|_{p_i^-, D_i}^{p_i^+ \theta_i} + \|u\|_{L^\infty(\tau_i, \tau_i+\tau); L^2(B_\rho(x_i)))}^{p_i^+ \theta_i} \right) \\ &\quad \times \|u\|_{L^\infty(\tau_i, \tau_i+\tau); L^2(B_\rho(x_i)))}^{p_i^+(1-\theta_i)} \\ &\le C(\tau) \left( \|\nabla u\|_{p_i^-, D_i}^{p_i^+ \theta_i} + M^{\frac{p_i^+ \theta_i}{2}} \right) M^{\frac{p_i^+}{2}(1-\theta_i)}. \end{aligned} \tag{1.68}$$

By Lemma 1.5, (1.6) and (5.8)

$$\|\nabla u\|_{p_i^-, D_i} \le C \|\nabla u\|_{p(\cdot), D_i} \le C' \|\nabla u\|_{p(\cdot), Q_R(x_0)} \le C'' \max\left\{ M^{\frac{1}{p^+}}, M^{\frac{1}{p^-}} \right\}.$$

Plugging this inequality into (1.68) we conclude that (1.64) is fulfilled in every cylinder $D_i$ with a constant $C$ depending only on $\omega(z)$, $p_i^\pm$, $n$ and $M$. To obtain (1.64) it is sufficient to sum up these estimates:

$$\|u\|_{p(\cdot), Q_R(x_0)} \le \sum_{i=1}^{k} \|u\|_{p(\cdot), D_i} \le C \sum_{i=1}^{k} \|u\|_{p_i^+, D_i}.$$

□

# Chapter 2
# A Porous Medium Equation with Variable Nonlinearity

## 2.1 Introduction

We devote this chapter to study the homogeneous Dirichlet problem for the semilinear parabolic equation

$$u_t = \operatorname{div}\left(a\,|u|^{\gamma(x,t)}\nabla u\right) + f(x,t,u,\nabla u) \tag{2.1}$$

with a given coefficient $a > 0$, an exponent $\gamma(x,t) > -1$ and the lower-order term $f$. Equation (2.1) is formally parabolic, but it degenerates or becomes singular wherever $u = 0$ and $\gamma \neq 0$. Equations of this type appear in the most natural way in various physical contexts such as mechanics of fluids and gases, the theory of heat propagation or diffusion processes. There exists an abundant literature devoted to study the questions of existence, uniqueness and qualitative properties of solutions to nonlinear parabolic equations of the type (2.1) with constant exponent of nonlinearity. The most studied prototypes are the famous porous medium equation, $u_t = \Delta u^m$, or the signed porous medium equation

$$u_t = \operatorname{div}(|u|^{m-1}\nabla u) \tag{2.2}$$

with constant exponent $m > 0$. Notwithstanding the fact that these equations have played the role of a "touchstone" for numerous methods in the theory of degenerate parabolic equations—see, e.g. [36, 64, 171, 197, 254]—not much is known about how their solutions respond to the variation of the exponent of nonlinearity. Under the assumption that the exponent $m$ may vary from one point to another, equation (2.2) ceases to be invariant with respect to scaling, a fact which makes inapplicable many of well-developed methods in the theory of nonlinear PDEs.

We start by the analysis of the model equation

$$u_t = \operatorname{div}\left(a(x,t)\,|u|^{\gamma(x,t)}\nabla u\right) + f(x,t) \tag{2.3}$$



S. Antontsev and S. Shmarev, *Evolution PDEs with Nonstandard Growth Conditions*, Atlantis Studies in Differential Equations 4,
DOI 10.2991/978-94-6239-112-3_2

with a given variable exponent $\gamma > -1$. The existence and uniqueness theorems for the model equation (2.3) are extended then to the complete equation (2.1) with lower-order terms. Another generalization consists in extending the existence result to the class of semilinear anisotropic equations

$$u_t - \sum_{i=1}^{n} D_i \left( |u|^{\gamma_i(x,t)} D_i u \right) = f, \qquad D_i = \frac{\partial}{\partial x_i} \quad (i = 1, \ldots, n),$$

with given exponents $\gamma_i(x, t) > -1$. We study also the question of existence of bounded stationary weak solutions of quasi-linear anisotropic equations

$$-\sum_{i=1}^{n} D_i \left( a_i |u|^{\alpha_i(x)} D_i u \right) + f(x, u) = 0.$$

## 2.2 Model Equation: Assumptions and Results

Let us consider the following problem:

$$u_t - \operatorname{div} (|u|^{\gamma(x,t)} \nabla u) = f \quad \text{in } Q_T, \tag{2.4}$$

$$u = 0 \text{ on } \Gamma_T = \partial\Omega \times [0, T], \qquad u(x, 0) = u_0(x) \text{ in } \Omega. \tag{2.5}$$

It is always assumed that $\gamma$ is measurable and bounded in $Q_T$, and

$$-1 < \gamma^- \le \gamma(x, t) \le \gamma^+ < \infty \quad \text{a.e. in } Q_T \tag{2.6}$$

with given constants $\gamma^-$ and $\gamma^+$.

**Definition 2.1** A locally integrable bounded function $u(x, t)$ is called *weak solution* of problem (2.4)–(2.5) if:

(i) $u \in L^\infty(Q_T)$, $|u|^{\gamma(x,t)/2} \nabla u \in L^2(Q_T)$, $u_t \in L^2(0, T; H^{-1}(\Omega))$,
(ii) $u = 0$ on $\partial\Omega \times (0, T)$ in the sense of traces,
(iii) for every test-function $\zeta(x, t) \in C^\infty(0, T; C_0^\infty(\Omega))$, $\zeta(x, T) = 0$,

$$\int_0^T \int_\Omega (-u\zeta_t + |u|^{\gamma(x,t)} \nabla u \nabla \zeta - f\zeta) dxdt = \int_\Omega u_0 \zeta(x, 0)\, dx. \tag{2.7}$$

The main existence result is given in the following theorem.

**Theorem 2.1** *Let $\gamma(x, t) : Q_T \mapsto \mathbb{R}$ be a measurable function satisfying condition (2.6). Assume that $\nabla \gamma \in L^2(Q_T)$. If $f \in L^2(Q_T) \cap L^1(0, T; L^\infty(\Omega))$ and*

$$\|u_0\|_{\infty,\Omega} + \int_0^T \|f(\cdot,t)\|_{\infty,\Omega}\,dt = K(T) < \infty, \tag{2.8}$$

*then problem* (2.4) *and* (2.5) *has at least one weak solution in the sense of* Definition 2.1. *The solution is bounded and satisfies the estimate* $\|u\|_{\infty,Q_T} \le K(T)$ *with the constant* $K(T)$ *from* (2.8).

The uniqueness theorem is proved under stronger assumptions on the data.

**Theorem 2.2** *Let the conditions of* Theorem 2.1 *be fulfilled. If the exponent* $\gamma$ *satisfy the additional assumptions*

$$\gamma(x,t) \ge \gamma^- > 0 \text{ in } Q_T, \quad \sup_{x\in\overline{\Omega}} |\nabla\gamma| \in L^2(0,T), \tag{2.9}$$

*then the solution of problem* (2.4) *and* (2.5) *is unique.*

## 2.3 Regularization

Let us consider the auxiliary nonlinear parabolic problem

$$\begin{cases} Lu \equiv u_t - \operatorname{div}\left(a(\varepsilon,u,M,x,t))\nabla u\right) = f & \text{in } Q_T, \\ u(x,0) = u_0 \text{ in } \Omega, \qquad u = 0 \text{ on } \Gamma_T, \end{cases} \tag{2.10}$$

which depends on positive parameters $\varepsilon > 0$ and $M$. The coefficient $a$ has the form

$$a = \left(\varepsilon^2 + \min\{u^2, M^2\}\right)^{\gamma(x,t)/2} \tag{2.11}$$

and (2.10) is a nondegenerate quasilinear parabolic equation because

$$0 < C'(\varepsilon,M,\gamma^{\pm}) \le a = \left(\varepsilon^2 + \min\{u^2, M^2\}\right)^{\gamma(x,t)/2} \le C(\varepsilon,M,\gamma^{\pm}).$$

A weak solution of problem (2.10) is constructed by means of the Schauder Fixed Point Theorem [187, Chap. 4, Sect. 8]. Let us consider the linear problem

$$\begin{cases} u_t - \operatorname{div}\left(\left(\varepsilon^2 + \min\{v^2, M^2\}\right)^{\gamma(x,t)/2}\nabla u\right) = \tau f & \text{in } Q_T, \\ u(x,0) = \tau\, u_0 \text{ in } \Omega, \qquad u = 0 \text{ on } \Gamma_T, \end{cases} \tag{2.12}$$

where $\tau \in [0,1]$ is a real parameter, and $v \in L^2(Q_T)$ is a given function. Let us denote $\mathscr{B}_R = \{v : \|v\|_{2,Q_T} < R\}$ and consider the solution $u$ of problem (2.12) as a solution of the functional equation

$$u = \tau\,\Phi(v) \qquad \text{with} \quad \tau\,\Phi(u) : \mathscr{B}_R \times [0,1] \mapsto L^2(Q_T).$$

Solvability of problem (2.10) in a ball $\mathscr{B}_R = \{v : \|v\|_{2,Q_T} \leq R\}$ will follow if we prove that

(1) the mapping $\Phi(v) : \mathscr{B}_R \mapsto \mathscr{B}_R$ is continuous and compact,
(2) for every $\tau \in [0, 1]$ the fixed points of the mapping $u = \tau \Phi(v)$ satisfy the estimate $\|v\|_{2,Q_T} \leq R'$.

Let us define the Banach space $\mathbf{V}$ as the completion of $C^\infty(0, T; C_0^\infty(\Omega))$ in the norm

$$\|v\|_{\mathbf{V}} = \max_{[0,T]} \|v\|_{2,\Omega} + \|\nabla v\|_{2,Q_T}, \tag{2.13}$$

By $\mathbf{V}_0$ we denote the subset of the elements of $\mathbf{V}$ for which

$$\frac{1}{h}\int_0^{T-h} \|v(x, t+h) - v(x, t)\|^2_{2,\Omega} dt \to 0 \quad \text{as } |h| \to 0. \tag{2.14}$$

Compactness of the mapping $\tau\Phi(v)$ follows from the classical results on solvability of linear parabolic equations with measurable coefficients [185, Chap. 3]: for every $u_0 \in L^2(\Omega)$ and $f, v \in L^2(Q_T)$, a measurable exponent $\gamma(x, t)$ satisfying condition (2.6), and any $\tau \in [0, 1]$ problem (2.12) has a unique solution $u \in \mathbf{V}_0$. Moreover, by [185, Chap. 3, Theorem 4.5] the mapping $\tau\Phi(v)$ is continuous in $\mathscr{B}_R$. For every test-function $\zeta$ satisfying the conditions of Definition 2.1

$$\int_{t_1}^{t_2}\int_\Omega (-u\zeta_t + a\,\nabla u \nabla\zeta - \tau\, f\zeta)dxdt = \int_\Omega u_0\zeta(x, 0)\,dx, \tag{2.15}$$

and $u_t \in L^2(0, T; H^{-1}(\Omega))$, $\|u\|_{L^\infty(Q_T)} \leq C(R)$.

To check the fulfillment of the second condition with some $R' > 0$, it amounts to derive the a priori estimates for all possible solutions of the nonlinear problem (2.12) with $v = u$ and $\tau \in [0, 1]$. The next section is entirely devoted to deriving these estimates.

## 2.4 A Priori Estimates

We consider the nonlinear problems

$$\begin{cases} u_t - \text{div}\ (a(\varepsilon, u, M, x, t)\nabla u) = \tau\, f & \text{in } Q_T, \\ u(x, 0) = \tau\, u_0 \text{ in } \Omega, \qquad u = 0 \text{ on } \Gamma_T, \end{cases} \tag{2.16}$$

with $\tau \in [0, 1]$ and the coefficient $a(\varepsilon, u, M, x, t)$ defined in (2.11).

**Lemma 2.1** *The solution of problem* (2.16) *satisfies the estimate*

$$\|u\|_{\infty,Q_T} \le \tau \left( \int_0^T \|f(\cdot,t)\|_{\infty,\Omega}\, dt + \|u_0\|_{\infty,\Omega} \right) \le \tau K(T). \tag{2.17}$$

*Proof* Multiplying Eq. (2.16) by $u^{2k-1}$ with an arbitrary $k \in \mathbb{N}$ and integrating over $\Omega$, we arrive at the relation

$$\begin{aligned}\frac{1}{2k}\frac{d}{dt}\left(\|u(\cdot,t)\|_{2k,\Omega}^{2k}\right) + (2k-1)\int_\Omega a|\nabla u|^2 u^{2(k-1)} dx \\ = \tau \int_\Omega u^{2k-1} f\, dx.\end{aligned} \tag{2.18}$$

By Hölder's inequality

$$\tau \left| \int_\Omega u^{2k-1} f\, dx \right| \le \tau\, \|u(\cdot,t)\|_{2k,\Omega}^{2k-1} \|f(\cdot,t)\|_{2k,\Omega},$$

whence

$$\begin{aligned}\|u(\cdot,t)\|_{2k,\Omega}^{2k-1} \frac{d}{dt}\left(\|u(\cdot,t)\|_{2k,\Omega}\right) + (2k-1)\int_\Omega a|\nabla u|^2 u^{2(k-1)} dx \\ \le \tau\, \|u(\cdot,t)\|_{2k,\Omega}^{2k-1} \|f(\cdot,t)\|_{2k,\Omega}.\end{aligned}$$

Simplifying and then integrating the last inequality in $t$, we obtain the following estimates for the solutions of problem (2.16):

$$\|u(\cdot,t)\|_{2k,\Omega} \le \tau \left( \int_0^t \|f(\cdot,t)\|_{2k,\Omega}\, dt + \|u_0\|_{2k,\Omega} \right), \qquad k = 1, 2, \dots.$$

Estimate (2.17) follows from this inequality as $k \to \infty$. □

**Corollary 2.1** *If we choose $M > K(T)$, then*

$$\min\{u^2, M^2\} = u^2 \quad \text{and} \quad a = (\varepsilon^2 + u^2)^{\gamma(x,t)/2}. \tag{2.19}$$

**Corollary 2.2** *Let $u_0 \ge 0$ and $f \ge 0$. In this special case the solution $u(x,t)$ is nonnegative in $\overline{Q}_T$.*

*Proof* Set $u^- = \min\{u, 0\} \le 0$. Multiplying (2.16) by $u^-$, integrating over $\Omega$ and taking into account the equalities $u^-(x,0) = 0$, $u^-\big|_{\Gamma_T} = 0$ we find that

$$\frac{1}{2}\frac{d}{dt}\left\|u^-(\cdot,t)\right\|_{2,\Omega} + \int_\Omega a|\nabla u^-|^2 dx \le 0.$$

It follows that for every $t > 0$

$$0 \le \|u^-(\cdot, t)\|_{2,\Omega} \le \|u^-(x, 0)\|_{2,\Omega} = 0,$$

whence the assertion. □

**Lemma 2.2** *The solutions of problem* (2.16) *satisfy the estimates*

$$\left\| \sqrt{a}\, \nabla u \right\|_{2,Q_T} \le C \tag{2.20}$$

*with an independent of* $\varepsilon$ *constant* $C$.

*Proof* It follows from (2.18) with $k = 1$ that for every $t \in (0, T)$ and $\tau \in [0, 1]$

$$\begin{aligned} \frac{1}{2} \|u(\cdot, t)\|_{2,\Omega}^2 + \int_{Q_t} a\, |\nabla u|^2 dxdt &\le \frac{\tau}{2} \|u_0\|_{2,\Omega}^2 + \tau \int_{Q_T} |u|\, |f|\, dxdt \\ &\le \frac{1}{2} \|u_0\|_{2,\Omega}^2 + K(T) \int_0^T \|f(\cdot, t)\|_{\infty,\Omega}\, dt \\ &\le (|\Omega| + K(T))K(T). \end{aligned}$$

□

**Corollary 2.3** *The solutions of problem* (2.16) *satisfy the estimates*

$$\left\| (\varepsilon^2 + u^2)^{\gamma^+/4} \nabla u \right\|_{2,Q_T} \le C \tag{2.21}$$

*with a constant* $C$ *not depending on* $\varepsilon$.

*Proof* The inequality $|u| \le K$ a.e. in $Q_T$ yields

$$\left( \frac{\varepsilon^2 + u^2}{\varepsilon^2 + K^2} \right)^{\gamma(x,t)} \ge \left( \frac{\varepsilon^2 + u^2}{\varepsilon^2 + K^2} \right)^{\gamma^+}. \tag{2.22}$$

The assertion immediately follows now from Lemma 2.2. □

**Lemma 2.3** *The solutions of problem* (2.16) *satisfy the estimates*

$$\|a\, \nabla u\|_{2,Q_T} \le C$$

*with a constant* $C$ *not depending on* $\varepsilon$. *Moreover,* $\|\nabla u\|_{2,Q_T} \le C$ *if* $\gamma^+ \le 0$.

*Proof* Let us consider the function

$$\psi(u) \equiv \int_0^u (\varepsilon^2 + s^2)^{\gamma^-/2} ds, \qquad \psi = 0 \text{ on } \Gamma_T.$$

Notice that

$$\begin{aligned} I &= \int_{Q_T} u\, \psi_t \, dxdt \\ &= \int_{Q_T} u\, u_t\, (\varepsilon^2 + u^2)^{\gamma^-/2} dxdt \\ &= \int_{Q_T} \frac{\partial}{\partial t} \left( \int_0^u s\, \left(\varepsilon^2 + s^2\right)^{\gamma^-/2} ds \right) dxdt \\ &= \int_{\Omega} \left( \int_{\tau\, u_0(x)}^{u(x,T)} s\, \left(\varepsilon^2 + s^2\right)^{\gamma^-/2} ds \right) dx. \end{aligned}$$

By virtue of Lemma 2.1 $|I| \leq M \equiv M(\gamma^-, K)$. Choosing in (2.15) $\zeta = \psi(u)$ for the test-function we deduce that

$$\int_{Q_T} a\, \psi'(u)\, |\nabla u|^2 \, dxdt \leq M + \tau \int_{Q_T} |f|\, \psi(u)\, dxdt \leq M + \tau M_1 \int_{Q_T} |f|\, dxdt \tag{2.23}$$

with the constant

$$M_1 = \max \psi = 2 \int_0^K s\, \left(\varepsilon^2 + s^2\right)^{\gamma^-/2} ds.$$

It is easy to see [cf. with (2.22)] that

$$\begin{aligned} a &= (\varepsilon^2 + u^2)^{\frac{\gamma}{2}} \leq (\varepsilon^2 + K^2)^{\frac{\gamma}{2} - \frac{\gamma^-}{2}} (\varepsilon^2 + K^2)^{\frac{\gamma^-}{2}} \\ &= M_2(K, \gamma^-) \psi'(u), \qquad M_2 = const. \end{aligned}$$

Gathering this inequality with (2.23), we obtain

$$\int_{Q_T} a^2\, |\nabla u|^2 \, dxdt \leq M_2 \int_{Q_T} a \psi'(u) |\nabla u|^2 \, dxdt \leq C(K, \gamma^-).$$

□

**Corollary 2.4** *The solutions of problem* (2.16) *satisfy the estimates*

$$\left\| (\varepsilon^2 + u^2)^{\gamma^+/2} \nabla u \right\|_{2, Q_T} \leq C, \qquad \left\| |u|^{\gamma^+} \nabla u \right\|_{2, Q_T} \leq C \tag{2.24}$$

*with an independent of* $\varepsilon$ *constant* $C$.

**Lemma 2.4** *For the solutions of problem* (2.16) $u_t \in L^2(0, T; H^{-1}(\Omega))$ *and*

$$\|u_t\|_{L^2(0,T;H^{-1}(\Omega))} \leq M$$

*with an independent of* $\varepsilon$ *constant* $M$.

*Proof* It is sufficient to show that for every $\zeta \in C^{\infty}(0,T;C_0^{\infty}(\Omega))$, $\zeta(x,0)=\zeta(x,T)=0$,

$$I=\left|\int_0^T\int_{\Omega} u\,\zeta_t\,dxdt\right|\leq M\,\|\zeta\|_{L^2(0,T;H_0^1(\Omega))}$$

with a constant $M$ independent of $\zeta$, $\varepsilon$ and $u$. By virtue of identity (2.15) and the uniform estimates of Lemma 2.3

$$I\leq\left(\|a\,\nabla\,u\|_{2,Q_T}+\tau\,C(n,\Omega)\|f\|_{2,Q_T}\right)\,\|\nabla\,\zeta\|_{2,Q_T},$$

where $C(n,\Omega)$ is the best constant in the Poincaré inequality:

$$\forall\,\eta\in W_0^{1,2}(\Omega)\qquad\|\eta\|_{2,\Omega}\leq C(n,\Omega)\|\nabla\,\eta\|_{2,\Omega}.$$

□

Let us represent Eq. (2.16) in the form

$$u_t=\operatorname{div}\mathbf{G},\tag{2.25}$$

with

$$\mathbf{G}=a\nabla u+\tau\,\mathbf{f}\in L^2(Q_T),\quad\operatorname{div}\mathbf{f}=f,\quad\mathbf{f}|_{\Gamma_T}=0.$$

For the vector-valued function $\mathbf{f}$ we take the potential vector $\mathbf{f}=\nabla\theta$, where $\theta(x,t)$ is the solution of the Dirichlet problem for the Laplace equation

$$\Delta\theta=f\,\text{in}\,\Omega,\qquad\theta=0\,\text{on}\,\partial\Omega,\quad t\in(0,T).\tag{2.26}$$

The solution of problem (2.26) is considered as a function of $x\in\Omega$ depending on $t$ as a parameter. Since $f\in L^2(Q_T)$ by assumption, $f(x,t)\in L^2(\Omega)$ for a.e. $t\in(0,T)$. For a.e. $t\in(0,T)$ problem (2.26) has a unique solution $\theta\in H_0^1(\Omega)$ and

$$\|\theta\|_{2,\Omega}^2+\|\nabla\theta\|_{2,\Omega}^2\leq C\,\|f\|_{2,\Omega}^2.$$

Integrating this estimate in $t$, we find that

$$\|\theta\|_{2,Q_T}^2+\|\nabla\theta\|_{2,Q_T}^2\leq C\,\|f\|_{2,Q_T}^2.$$

Equation (2.25) is fulfilled in the sense of distributions. If $\partial_{x_i}u\in L^2(Q_T)$, which corresponds to the case $\gamma^+\leq 0$ (see Lemma 2.3), then according to [185, Chap. III, Lemma 4.1] the function $u(x,t)$ satisfies the estimates

$$\begin{cases}\|u(x+\mathbf{e}_i h,t)-u(x,t)\|_{2,Q_T^h}^2\leq C|h|^2,\quad i=1,\dots,n,\\ \|u(x,t+h)-u(x,t)\|_{2,Q_T^h}^2\leq C|h|,\end{cases}\tag{2.27}$$

where

$$Q_T^h = \{(x,t) \in Q_T : (x + \mathbf{e}_i h, t), (x, t+h) \in Q_T, \ i = 1, \ldots, n\},$$

$h$ is a scalar parameter, $\mathbf{e}_i$ are the unit vectors of the axes $x_i$.

**Lemma 2.5** *Let $u(x,t)$ be a solution of problem* (2.16) *with $\gamma^+ \geq 0$. The function*

$$z(x,t) = z[u(x,t)] \equiv \int_0^{u(x,t)} |s|^{\gamma^+} ds$$

*satisfies the estimates*

$$\begin{cases} \|z(x + \mathbf{e}_i h, t) - z(x,t)\|^2_{2,Q_T^h} \leq C|h|^2, \quad i = 1, \ldots, n, \\ \|z(x, t+h) - z(x,t)\|^2_{2,Q_T^h} \leq C|h|. \end{cases} \tag{2.28}$$

*Proof* Since $\nabla z = |u|^{\gamma^+} \nabla u$, estimates (2.24) yield the inclusions $\partial_{x_i} z \in L^2(Q_T)$, and the first of estimates (2.28) stems from the inequality

$$\int_{Q_T^h} \frac{|z(x + \mathbf{e}_i h, t) - z(x,t)|^2}{|h|^2} dx \leq C, \quad i = 1, \ldots, n.$$

To derive the second estimate we introduce the functions

$$\delta u = u(x, t+h) - u(x,t), \qquad \delta z = z(x, t+h) - z(x,t).$$

Notice that sign $\delta u =$ sign $z$ by the definition of $z$,

$$|\delta z| = \left| \int_{u(x,t)}^{u(x,t+h)} |s|^{\gamma^+} ds \right| \leq M^{\gamma^+} |\delta u| \quad \text{with } M > K(T), \tag{2.29}$$

$$|\delta z|^2 \leq (1 + M)^{\gamma^+} |\delta z||\delta u| = (1 + M)^{\gamma^+} \delta z \cdot \delta u,$$
$$\|\nabla(\delta z)\|_{L^2(Q_T^h)} \leq C.$$

Then

$$\begin{aligned} (1 + M)^{-\gamma^+} \|\delta z\|^2_{2,\Omega} &\leq (\delta z, \ \delta u)_{2,\Omega} \\ &= \left( \delta z, \int_t^{t+h} \nabla \mathbf{G}(\cdot, \tau) d\tau \right)_{2,\Omega} \\ &= - \left( \nabla(\delta z), \int_t^{t+h} \mathbf{G}(\cdot, \tau) d\tau \right)_{2,\Omega}. \end{aligned} \tag{2.30}$$

Integrating (2.30) in $t \in (0, T-h)$, applying Hölder's inequality and then (2.29), we obtain the estimate

$$
\begin{aligned}
(1+M)^{-\gamma^+}\|\delta z\|^2_{2,Q_T^h} &\le \int_0^{T-h}\int_\Omega |\nabla(\delta z)|^2\,dxdt \int_0^{T-h}\int_\Omega\left(\int_t^{t+h}\mathbf{G}(x,\tau)d\tau\right)^2 dxdt\\
&\le C\int_0^{T-h}\int_\Omega\left(\int_t^{t+h}|\mathbf{G}(x,\tau)|\,d\tau\right)^2 dxdt\\
&\le C\,|h|\left(\int_0^{T-h}\int_\Omega\int_t^{t+h}|\mathbf{G}(x,\tau)|^2\,d\tau dxdt\right)\le C\,|h|.
\end{aligned}
$$

□

## 2.5 Passage to the Limit

In the proof of convergence of the sequence $\{u_\varepsilon\}$ we rely on the following result.

**Lemma 2.6** ([235, Theorem 5]) *Let $X$, $B$ and $Y$ be Bahach spaces, $X \subset B \subset Y$ with compact embedding $X \subset B$. If a family of functions $\mathscr{F}$ possesses the properties*

1. *$\mathscr{F}$ is bounded in $L^p(0,T;X)$ with $1 \le p \le \infty$,*
2. *$\|f(x,t+h) - f(x,t)\|_{L^p(0,T-h;Y)} \to 0$ as $h \to 0$ uniformly for $f \in \mathscr{F}$,*

*then $\mathscr{F}$ is relatively compact in $L^p(0,T;B)$(in $C(0,T;B)$ if $p=\infty$).*

Let $\{u_\varepsilon\}$ be the sequence of solutions of problem (2.10). If $\gamma^+ \le 0$, the sequence $\{u_\varepsilon\}$ satisfy the uniform estimates (2.27). Since the embedding $H^1_0(\Omega) \subset L^2(\Omega)$ is compact, it follows from Lemma 2.6 that $\{u_\varepsilon\}$ is precompact in $L^2(Q_T)$. Let $\gamma^+ \ge 0$. By (2.28)

$$
\|z[u_\varepsilon(x,t+h)] - z[u_\varepsilon(x,t)]\|^2_{2,Q_T^h} \to 0 \quad \text{as } h \to 0, \qquad \|\nabla z[u_\varepsilon]\|_{2,Q_T} \le C,
$$

which yields precompactness of the sequence $z[u_\varepsilon(x,t)]$ in $L^2(Q_T)$. Not loosing generality we may assume that the converging subsequences coincide with the entire sequences. Convergence of $\{z_\varepsilon[u_\varepsilon]\}$ in $L^2(Q_T)$ yields convergence of $\{z[u_\varepsilon]\}$ and $\{u_\varepsilon\}$ a.e. in $Q_T$. Since $\|u_\varepsilon\|_{\infty,Q_T} \le K(T)$, by the dominated convergence theorem $\{u_\varepsilon\}$ converges strongly in $L^p(Q_T)$ with any $p > 1$. Finally, by Lemma 2.4 $u_t$ are uniformly bounded in $L^2(0,T;H^{-1}(\Omega))$, while $a|\nabla u_\varepsilon|^2$ are uniformly bounded by Lemma 2.3. Thus, there exist functions $u$ and $\chi \in (L^2(Q_T))^n$ such that

$$
\begin{aligned}
&u_\varepsilon \to u \quad \text{a.e. in } Q_T,\\
&u_\varepsilon \to u \quad \text{in } L^p(Q_T) \text{ with any } 1 < p < \infty,\\
&u_{\varepsilon t} \rightharpoonup u_t \quad \text{in } L^2(0,T;H^{-1}(\Omega)),\\
&a\nabla u_\varepsilon \rightharpoonup \chi \quad \text{in } L^2(Q_T).
\end{aligned}
\tag{2.31}
$$

Let us consider the auxiliary function $g[u] = u(\varepsilon^2+u^2)^{\gamma/2}$. It is easy to calculate that

$$\nabla g = \left[\gamma \frac{u^2}{\varepsilon^2+u^2} + 1\right] (\varepsilon^2+u^2)^{\gamma/2} \nabla u + \frac{1}{2} \nabla \gamma \left[u\,(\varepsilon^2+u^2)^{\gamma/2}\right] \ln(\varepsilon^2+u^2).$$

If $\gamma(x,t)+1>0$, then $\nabla g \in L^2(Q_T)$ because

$$\begin{aligned}|\nabla g|^2 &\le \left((\gamma+1)|a \nabla u| + \frac{1}{2}(\varepsilon^2+u^2)^{(\gamma+1)/2} |\ln(\varepsilon^2+u^2)||\nabla\gamma|\right)^2 \\ &\le C_1 |a\nabla u|^2 + C_2|\nabla\gamma|^2\end{aligned}$$

with constants $C_i$ depending on $\gamma^{\pm}$ and $\|u\|_{\infty,Q_T}$. It follows that there exists a function $g^*$ such that

$$\nabla g_\varepsilon \rightharpoonup \nabla g^* \text{in} L^2(Q_T),$$

while by virtue of (2.31)

$$g_\varepsilon \equiv g[u_\varepsilon] = u_\varepsilon(\varepsilon^2+u_\varepsilon^2)^{\gamma/2} \to u|u|^\gamma \text{ a.e. in } Q_T \text{ and in } L^q(Q_T) \text{ with } 1<q<\infty.$$

Let us identify the limit $\nabla g^*$. Notice that

$$\begin{aligned}&\frac{(1+\gamma)u_\varepsilon^2+\varepsilon^2}{\varepsilon^2+u_\varepsilon^2}(\varepsilon^2+u_\varepsilon^2)^{\gamma/2}\nabla u_\varepsilon \\ &\qquad = \nabla g_\varepsilon - \frac{1}{2}\nabla\gamma\left[u_\varepsilon\,(\varepsilon^2+u_\varepsilon^2)^{\gamma/2}\right]\ln(\varepsilon^2+u_\varepsilon^2). \qquad (2.32)\end{aligned}$$

For every smooth test-function $\psi$

$$\begin{aligned}\int_{Q_T} \nabla g^* \cdot \psi\, dxdt &= \lim_{\varepsilon\to 0}\int_{Q_T} \nabla g_\varepsilon \cdot \psi\, dxdt \\ &= -\lim_{\varepsilon\to 0}\int_{Q_T} g_\varepsilon \cdot \nabla\psi\, dxdt = -\int_{Q_T} u\,|u|^\gamma\,\nabla\psi\, dxdt,\end{aligned}$$

whence

$$\nabla g^* = \nabla(u|u|^\gamma) = (1+\gamma)|u|^\gamma\nabla u + \frac{1}{2}u|u|^\gamma(\ln u^2)\nabla\gamma \quad \text{a.e. in } Q_T. \qquad (2.33)$$

For every test-function $\zeta$ from the conditions of Definition 2.1 we can pass to the limit as $\varepsilon\to 0$ in each term of the integral relation

$$\sum_{i=1}^4 I_i^{(\varepsilon)} \equiv \int_{Q_T}\left(-u_\varepsilon\zeta_t + (\varepsilon^2+u_\varepsilon^2)^{\gamma(x,t)/2}\nabla u_\varepsilon\nabla\zeta - f\,\zeta\right)dxdt - \int_\Omega u_0\,\zeta(x,0)\,dx = 0.$$

By virtue of (2.31)

$$I_1^{(\varepsilon)} \to -\int_{Q_T} u\,\zeta_t\,dxdt \qquad \text{as } \varepsilon \to 0.$$

According to (2.32) and (2.33)

$$\begin{aligned} I_2^{(\varepsilon)} &= \int_{Q_T} \frac{\varepsilon^2 + u_\varepsilon^2}{(1+\gamma)u_\varepsilon^2 + \varepsilon^2} \\ &\qquad \times \left[\nabla g_\varepsilon - \frac{1}{2}\left[u_\varepsilon\,(\varepsilon^2 + u_\varepsilon^2)^{\gamma/2}\right] \ln(\varepsilon^2 + u_\varepsilon^2)\,\nabla\gamma\right] \nabla\zeta\,dxdt \\ &\to \int_{Q_T} \frac{1}{\gamma+1}\left[\nabla(u|u|^\gamma) - \frac{1}{2}u|u|^\gamma\left(\ln u^2\right)\nabla\gamma\right]\nabla\zeta\,dxdt \\ &= \int_{Q_T} |u|^\gamma \nabla u\,\nabla\zeta\,dxdt \qquad \text{as } \varepsilon \to 0. \end{aligned}$$

Gathering these formulas we conclude that the limit function $u(x,t)$ is a weak solution of problem (2.4)–(2.5). The proof of Theorem 2.1 is completed.

## 2.6 Uniqueness of Weak Solutions

For the proof of uniqueness of weak solutions of the model problem (2.4)–(2.5) we assume that the data satisfy the conditions of Theorem 2.1 and the additional assumptions (2.9).

Let us notice that since $C^\infty(0,T; C_0^\infty(\Omega))$ is dense in the set

$$\mathscr{M} = \left\{\zeta \in L^2(0,T; H_0^1(\Omega)) : \zeta_t \in L^2(0,T; H^{-1}(\Omega))\right\},$$

the weak solutions of problem (2.4) and (2.5) satisfy identity (2.7) with the test-functions $\zeta \in \mathscr{M}$, $\zeta(x,T) = 0$ a.e. in $\Omega$.[1]

Let us assume that there are two different solutions of problem (2.4) and (2.5) $u_1$ and $u_2$ and consider the function $u = u_1 - u_2$. This function satisfies the identity

$$\int_{Q_T} (-u\zeta_t + (|u_1|^{\gamma(x,t)}\nabla u_1 - |u_2|^{\gamma(x,t)}\nabla u_2)\nabla\zeta)\,dxdt = 0 \tag{2.34}$$

[1] Such a choice of the test-function is meaningful because every function $\zeta \in L^2(0,T; H_0^1(\Omega))$ with $\zeta_t \in L^2(0,T; H^{-1}(\Omega))$ belongs to $C([0,T]; L^2(\Omega))$ after possible redefining on a set of zero measure in $(0,T)$—see [197, p. 156] or Lemma 4.5 below.

for every $\zeta \in \mathscr{M}$, $\zeta(x,T) = 0$ a.e. in $\Omega$. Let us introduce the functions

$$W(x,t) \equiv W[u(x,t)] = \frac{1}{1+\gamma(x,t)} u\,|u|^{\gamma(x,t)},$$
$$F(u,x,t) = u\,|u|^{\gamma} \frac{\nabla \gamma}{1+\gamma} \left( \frac{\ln u^2}{2} - \frac{1}{1+\gamma} \right).$$

It is straightforward to check that

$$\begin{aligned} \nabla W(x,t) &= |u|^{\gamma} \nabla u + u\,|u|^{\gamma} \frac{\nabla \gamma}{1+\gamma} \left( \frac{\ln u^2}{2} - \frac{1}{1+\gamma} \right) \\ &\equiv |u|^{\gamma} \nabla u + F(u,x,t). \end{aligned}$$

Since

$$|u_1|^{\gamma} \nabla u_1 - |u_2|^{\gamma} \nabla u_2 = \nabla(W[u_1] - W[u_2]) - (F(u_1,x,t) - F(u_2,x,t)),$$

for every smooth test-function $\zeta(x,t)$, vanishing on $\partial\Omega$,

$$\begin{aligned} &\int_{Q_T} \left( |u_1|^{\gamma} \nabla u_1 - |u_2|^{\gamma} \nabla u_2 \right) \nabla \zeta \, dxdt \\ &\quad = \int_{Q_T} \left( \nabla \left( W[u_1] - W[u_2] \right) \nabla \zeta + \left( F(u_2,x,t) - F(u_1,x,t) \right) \nabla \zeta \right) dxdt. \end{aligned}$$

Integrating by parts in the first term on the right-hand side, we arrive at the following representation:

$$\begin{aligned} &\int_{Q_T} \left( |u_1|^{\gamma} \nabla u_1 - |u_2|^{\gamma} \nabla u_2 \right) \nabla \zeta \, dxdt \\ &\quad = \int_{Q_T} (u_1 - u_2) \left[ -A\,\Delta \zeta + B \nabla \zeta \right] dxdt \end{aligned}$$

with the vector-valued function $B$ of the form

$$B = A \nabla \ln(1+\gamma) + D \nabla \gamma.$$

Here

$$A = \frac{W[u_1] - W[u_2]}{u_1 - u_2} = \frac{1}{1+\gamma} \frac{u_1|u_1|^{\gamma} - u_2|u_2|^{\gamma}}{u_1 - u_2} \geq 0,$$
$$D = \frac{F(u_1,x,t) - F(u_2,x,t)}{u_1 - u_2} = -\frac{\nabla \gamma}{2(1+\gamma)} \frac{u_1\,|u_1|^{\gamma(x,t)} \ln u_1^2 - u_2\,|u_2|^{\gamma(x,t)} \ln u_2^2}{u_1 - u_2}.$$

It is easy to verify that for $\gamma(x,t) \geq \gamma^- > 0$

$$
\begin{aligned}
&0 \leq A \leq C, \qquad |D| \leq C\,|\nabla \gamma|, \\
&\frac{D^2}{A} = \frac{|\nabla\gamma|^2}{4(1+\gamma)} \frac{\left(u_1\,|u_1|^{\gamma(x,t)} \ln u_1^2 - u_2\,|u_2|^{\gamma(x,t)} \ln u_2^2\right)^2}{(u_1-u_2)\,(u_1|u_1|^{\gamma} - u_2|u_2|^{\gamma})} \leq C|\nabla\gamma|^2, \\
&\frac{D^2}{A} \leq C, \qquad \frac{B^2}{A} \leq C|\nabla\gamma|^2
\end{aligned}
\tag{2.35}
$$

with a constant $C$ depending on $\gamma^-$ and $\max|u_i|$. Using this notation, we rewrite identity (2.34) in the form

$$
\int_{Q_T} [-\zeta_t - A\,\Delta\,\zeta + B\,\nabla\,\zeta]\,u\,dxdt = 0. \tag{2.36}
$$

Let us set $\zeta(x,t) = \eta(x, T-t)$, where $\eta(x,t)$ is the solution of the parabolic problem

$$
\begin{cases}
\eta_t - (A+\varepsilon)\,\Delta\,\eta + B\,\nabla\,\eta = h & \text{in} Q_T, \\
\eta(x,0) = 0, \qquad \eta = 0 \,\text{on}\, \Gamma_T
\end{cases}
\tag{2.37}
$$

with an arbitrary small parameter $\varepsilon > 0$ and an arbitrary $h \in L^2(Q_T)$. For every $h$ and $\varepsilon$ problem (2.37) has a unique continuous weak solution $\eta$ such that $\eta_t,\ D^2_{ij}\eta \in L^2(Q_T)$. Instead of identity (2.36) we have now:

$$
\int_{Q_T} u\,[h + \varepsilon\Delta\,\eta]\,dxdt = 0. \tag{2.38}
$$

We proceed to derive a priori estimates on the derivatives of $\eta$. Let us assume that $h \in C(0,T; C^1_0(\Omega))$. Multiplying Eq. (2.37) by $\Delta\eta$ and integrating over $\Omega$ we obtain the equality

$$
\frac{1}{2}\frac{d}{dt}\int_\Omega |\nabla\eta|^2 dx + \int_\Omega (A+\varepsilon)|\Delta\,\eta|^2 dx = I_1 + I_2 \tag{2.39}
$$

with

$$
I_1 = \int_\Omega B\,\nabla\eta\Delta\eta\,dx, \qquad I_2 = -\int_\Omega h\Delta\eta\,dx = \int_\Omega \nabla h\,\nabla\eta\,dx. \tag{2.40}
$$

By Hölder's inequality

$$
|I_1| \leq \frac{1}{2}\int_\Omega (A+\varepsilon)|\Delta\eta|^2\,dx + \frac{1}{2}\int_\Omega \frac{|B|^2}{(A+\varepsilon)}|\nabla\eta|^2\,dx, \tag{2.41}
$$

$$
|I_2| \leq \frac{1}{2}\int_\Omega |\nabla\eta|^2\,dx + \frac{1}{2}\int_\Omega |\nabla h|^2\,dx. \tag{2.42}
$$

Gathering (2.39)–(2.42) we have:

$$
\begin{aligned}
\frac{d}{dt}\int_\Omega |\nabla\eta|^2\,dx &+ \int_\Omega (A+\varepsilon)|\Delta\eta|^2\,dx \\
&\le \int_\Omega \left[\frac{|B|^2}{(A+\varepsilon)}+1\right]|\nabla\eta|^2\,dx + \int_\Omega |\nabla h|^2\,dx \\
&\equiv J. \qquad (2.43)
\end{aligned}
$$

By virtue of (2.35)

$$
\frac{|B|^2}{(A+\varepsilon)} \le \frac{|B|^2}{A} \le C\ \sup_\Omega\,(|\nabla\gamma|)^2 \qquad (2.44)
$$

with a constant $C$ not depending on $\varepsilon$. The right-hand side of (2.43) is estimated as follows:

$$
J \le \lambda(t)\int_\Omega |\nabla\eta|^2dx + \int_\Omega |\nabla h|^2dx
$$

with $\lambda(t) = C\,(1+\sup_\Omega|\nabla\gamma|^2)$. Plugging this estimate into (2.43) and applying Gronwall's Lemma, we have:

$$
\int_\Omega |\nabla\eta|^2dx + \int_0^t\int_\Omega (A+\varepsilon)|\Delta\eta|^2dxdt \le C\int_0^t\int_\Omega |\nabla h|^2dxdt. \qquad (2.45)
$$

Gathering (2.38) with (2.45) and applying Hölder's inequality we conclude that

$$
\begin{aligned}
\left|\int_0^T\int_\Omega uh\,dxdt\right| &= \left|\int_0^T\int_\Omega \varepsilon\,\Delta\eta\,dxdt\right| \\
&\le T\sqrt{\varepsilon}\sqrt{|\Omega|}\left(\int_0^T\int_\Omega \varepsilon|\Delta\eta|^2dxdt\right)^{\frac{1}{2}} \\
&\le T\sqrt{\varepsilon}\sqrt{|\Omega|}\left(\int_0^T\int_\Omega (\varepsilon+A)|\Delta\eta|^2dxdt\right)^{\frac{1}{2}} \to 0 \quad \text{when}\ \varepsilon\to 0,
\end{aligned}
$$

whence

$$
\int_0^T\int_\Omega uh\,dxdt = 0.
$$

Since the last relation is true for any smooth function $h$, it follows that $u\equiv 0$.

*Remark 2.1* We leave a gap between the conditions sufficient for the existence and uniqueness of weak solutions for problem (2.4) and (2.5). The existence is proved under the assumption $-1 < \gamma(x,t) < \infty$, while the uniqueness theorem is true if

$0 < \gamma^- \leq \gamma(x,t) < \infty$. It is known that the weak solution of the Cauchy problem for Eq. (2.4) with the constant exponent $\gamma(x,t) \equiv \gamma \in (-1,0)$ need not be unique [230, 253]. Conditions of non-uniqueness in the limit case $\gamma = -1$ are given in [162].

Similar arguments show that the solutions of problem (2.4) and (2.5) obey the comparison principle.

**Lemma 2.7** *Let $u_1, u_2$ be the weak solutions of problem* (2.4)–(2.5) *corresponding to the initial functions $u_{10}, u_{20}$. If the problem data satisfy the conditions of* Theorem 2.2 *and $u_{10} \geq u_{20}$ a.e. in $\Omega$, then $u_1 \geq u_2$ a.e. in $Q_T$.*

*Proof* Following the proof of Theorem 2.2 we find that the difference $u = u_1 - u_2$ satisfies, instead of (2.36), the equality

$$\int_{Q_T} [-\zeta_t - A\,\Delta\,\zeta + B\,\nabla\,\zeta]\, u\, dxdt = \int_\Omega u_0(x)\zeta(x,0)\,dx$$

with the right-hand side nonnegative by assumption. Fix an arbitrary $h \geq 0$ and take for $\eta$ the solution of problem (2.37). Since $\eta \geq 0$ in $Q_T$ by the maximum principle, this leads to the inequality

$$\int_{Q_T} u\;[h + \varepsilon\Delta\,\eta]\; dxdt = \int_\Omega u_0(x)\eta(x,T)\,dx \geq 0,$$

which substitutes (2.38). Then

$$\int_0^T \int_\Omega u\,h\,dxdt \geq -C\sqrt{\varepsilon},$$

and the assertion follows because $h \geq 0$ and $\varepsilon > 0$ are arbitrary. □

## 2.7 Equations with Lower-Order Terms

Let us consider the problem

$$\begin{cases} u_t = \operatorname{div}\left(a|u|^{\gamma(x,t)}\nabla u\right) + \mathbf{b}|u|^{\gamma(x,t)/2}\nabla u - c|u|^{\sigma(x,t)-2}u + d \quad \text{in } Q_T, \\ u = 0 \text{ on } \Gamma \times [0,T], \\ u(x,0) = u_0(x) \text{ in } \Omega \end{cases} \tag{2.46}$$

where the domain $\Omega$ and the boundary $\Gamma = \partial\Omega$ satisfy the foregoing assumptions and $T < \infty$. We assume that the scalar functions $a(x,t,r)$, $c(x,t,r)$ and the vector $\mathbf{b}(x,t,r)$ are given functions of the arguments $(x,t) \in Q_T = \Omega \times (0,T]$ and $r \in \mathbb{R}$.

The exponents of nonlinearity $\gamma$ and $\sigma$ are given measurable functions subject to the conditions

$$\forall \text{ a.e. } (x,t) \in Q_T \quad \begin{cases} \gamma(x,t) \in [\gamma^-, \gamma^+] \subset (-1, \infty), \\ \sigma(x,t) \in [\sigma^-, \sigma^+] \subseteq [1, \infty) \end{cases} \tag{2.47}$$

with some constants $\gamma^{\pm}$ and $\sigma^{\pm}$. We assume that $a$, $\mathbf{b}$, $c$ and $d(x,t,r)$ are Caratheòdory functions defined on $\overline{Q}_T \times \mathbb{R}$ and that there exist constants $a_0 > 0$, $c_0 \geq 0$, $D \geq 0$, $a_1, b_1, c_1$ such that $\forall\,(x,t,r) \in \overline{Q}_T \times \mathbb{R}$

$$\begin{cases} a_0 \leq a(x,t,r) \leq a_1, \qquad c_0 \leq c(x,t,r) \leq c_1 \\ |\mathbf{b}(x,t,r)| \leq b_1, \\ |d(x,t,r)| \leq D|r| + f(x,t) \quad \text{with some } f(x,t) \geq 0. \end{cases} \tag{2.48}$$

**Definition 2.2** A locally integrable function $u(x,t)$ is called weak solution of problem (2.46) if:

(i) $u \in L^{\infty}(Q_T)$, $|u|^{\gamma(x,t)/2}\nabla u \in L^2(Q_T)$, $u_t \in L^2(0,T; H^{-1}(\Omega))$,
(ii) $u = 0$ on $\Gamma_T$ in the sense of traces,
(iii) for every test-function $\zeta(x,t) \in C^{\infty}(0,T; C_0^{\infty}(\Omega))$, $\zeta(x,T) = 0$ in $\Omega$,

$$\int_{Q_T} \Big(-u\zeta_t + a\,|u|^{\gamma(x,t)}\nabla u \nabla\zeta - \zeta\,\mathbf{b}|u|^{\gamma(x,t)/2}\nabla u + c\,|u|^{\sigma(x,t)-2}u\,\zeta - d\zeta\Big)\,dxdt = \int_{\Omega} u_0\zeta(x,0)dx. \tag{2.49}$$

**Theorem 2.3** *Let $\gamma(x,t)$ and $\sigma(x,t)$ be measurable in $Q_T$ functions satisfying conditions* (2.47). *Assume that $\nabla\gamma \in L^2(Q_T)$. Let conditions* (2.48) *be fulfilled. If*

$$\|u_0\|_{\infty,\Omega} + \frac{e^{TD}-1}{D}\,\|f(x,t)\|_{\infty,Q_T}\,dt = K(T) < \infty, \tag{2.50}$$

*then problem* (2.46) *has a weak solution in the sense of* Definition 2.2. *This solution is bounded and satisfies the inequality $\|u\|_{\infty,Q_T} \leq K(T)$ with the constant $K(T)$ from condition* (2.50).

The proof of Theorem 2.3 is an adaptation of the proof of Theorem 2.1. A solution of problem (2.46) is obtained as the limit of the sequence of solutions of the regularized problems

$$\begin{cases} u_t - \operatorname{div}(a_{\varepsilon,M}(x,t,u))\nabla u) - \mathbf{b}_{\varepsilon,M}(x,t,u)\,\nabla u \\ \qquad\qquad + c_{\varepsilon,M}(x,t,u)\,u = d(x,t,u) \qquad \text{in } Q_T, \\ u(x,0) = u_0 \text{ in } \Omega, \qquad u = 0 \text{ on } \Gamma_T \end{cases} \tag{2.51}$$

with the coefficients

$$a_{\varepsilon,M}(x,t,u) = a(x,t,u)\left(\varepsilon^2 + \min\{u^2, M^2\}\right)^{\gamma(x,t)/2},$$
$$\mathbf{b}_{\varepsilon,M}(x,t,u) = \mathbf{b}(x,t,u)\left(\varepsilon^2 + \min\{u^2, M^2\}\right)^{\gamma(x,t)/4},$$
$$c_{\varepsilon,M}(x,t,u) = c(x,t,u)\left(\varepsilon^2 + \min\{u^2, M^2\}\right)^{(\sigma(x,t)-2)/2}$$

and positive parameters $\varepsilon$ and $M$ (the parameter $M$ will be chosen later). Let denote $\mathscr{B}_R = \{v : \|v\|_{2,Q_T} < R\}$. For every fixed $v \in \mathscr{B}_R$ and $\tau \in [0,1]$ there exists a unique solution $u$ of the linear problem

$$\begin{cases} u_t - \operatorname{div}\left(a_{\varepsilon,M}(x,t,v))\nabla u\right) - \mathbf{b}_{\varepsilon,M}(x,t,v)\,\nabla u \\ \qquad\qquad +c_{\varepsilon,M}(x,t,v)\,u = \tau\, d(x,t,v) \qquad \text{in } Q_T, \\ u(x,0) = \tau\, u_0 \text{ in } \Omega, \qquad u = 0 \text{ on } \Gamma_T. \end{cases} \tag{2.52}$$

The solution of problem (2.51) is sought as a fixed point of the mapping $u = \tau\,\Phi(v)$ with $\tau = 1$. The existence of a fixed point in a ball $\mathscr{B}_R = \{v : \|v\|_{2,Q_T} \le R\}$ follows from the Schauder fixed point principle and reduces to checking continuity and compactness of the mapping $\Phi(v) : \mathscr{B}_R \mapsto \mathscr{B}_R$ and to deriving suitable a priori estimates for the fixed points of the mapping $u = \tau\,\mathscr{B}(v)$ in a ball $\|v\|_{2,Q_T} \le R$. Almost all arguments used in the proof of Theorem 2.1 are applicable in the present case and can be omitted. We will present in detail only those that require modifications.

For every $v \in L^2(Q_T)$ and $\tau \in [0,1]$ problem (2.52) admits a weak solution $u \in \mathbf{V}_0$ (the space $\mathbf{V}_0$ is defined in (2.13) and (2.14). Compactness of the mapping $\tau\Phi(v)$ follows from results on the solvability of linear parabolic equations with measurable coefficients [185, Chap. 3], the mapping $\tau\Phi(v)$ is continuous in $\mathscr{B}_R$—see [185, Chap. 3, Theorem 4.5].

**Lemma 2.8** *The solution of problem* (2.51) *satisfies the inequality*

$$\|u\|_{\infty,Q_T} \le \tau\,\|u_0\|_{\infty,\Omega} + \frac{e^{\tau T D} - 1}{D}\,\|f\|_{\infty,Q_T}\,. \tag{2.53}$$

*Proof* Multiplying Eq. (2.51) by $u^{2k-1}$ and integrating over $\Omega$ we arrive at the equality

$$\begin{aligned} \frac{1}{2k}\frac{d}{dt}\left(\|u(\cdot,t)\|_{2k,\Omega}^{2k}\right) + (2k-1)\int_\Omega a_{\varepsilon,M}|\nabla u|^2 u^{2(k-1)}dx + \int_\Omega c_{\varepsilon,M}|u|^{2k}dx \\ = \int_\Omega \mathbf{b}_{\varepsilon,M}\nabla u\, u^{2k-1}\,dx + \tau\int_\Omega d\,u^{2k-1}dx \equiv I_1 + I_2, \quad k = 1,2,\ldots. \end{aligned} \tag{2.54}$$

By Hölder's inequality, the terms $I_1$ and $I_2$ are estimated as follows:

$$
\begin{aligned}
|I_1| &\le \int_\Omega \frac{|\mathbf{b}|}{\sqrt{a}}|u|^k \sqrt{a_{\varepsilon,M}}|\nabla u|\,|u|^{k-1}dx \\
&\le \frac{(2k-1)}{2}\int_\Omega a_{\varepsilon,M}|\nabla u|^2|u|^{2(k-1)}dx + \frac{1}{2(2k-1)}\int_\Omega \frac{|\mathbf{b}|}{a}|u|^{2k}dx \\
&\le \frac{(2k-1)}{2}\int_\Omega a_{\varepsilon,M}|\nabla u|^2|u|^{2(k-1)}dx + \lambda\,\|u(\cdot,t)\|^{2k}_{2k,\Omega}\,, \qquad (2.55)
\end{aligned}
$$

where

$$
\frac{1}{2(2k-1)}\frac{|\mathbf{b}|^2}{\widetilde{a}} \le \frac{1}{2(2k-1)}\frac{|b_1|^2}{a_0} \stackrel{def}{\equiv} \lambda \to 0 \quad \text{for}\, k \to \infty,
$$

$$
|I_2| \le \tau D\,\|u(\cdot,t)\|^{2k}_{2k,\Omega} + \tau\,\|u(\cdot,t)\|^{2k-1}_{2k,\Omega}\,\|f(\cdot,t)\|_{2k,\Omega}\,. \qquad (2.56)
$$

Substituting (2.55) and (2.56) into (2.54), we come to the inequality

$$
\begin{aligned}
&\frac{1}{2k}\frac{d}{dt}\left(\|u(\cdot,t)\|^{2k}_{2k,\Omega}\right) + \frac{(2k-1)}{2}\int_\Omega a_{\varepsilon,M}|\nabla u|^2\,u^{2(k-1)}dx \\
&\quad \le (\lambda+\tau D)\,\|u(\cdot,t)\|^{2k}_{2k,\Omega} + \tau\,\|u(\cdot,t)\|^{2k-1}_{2k,\Omega}\,\|f(\cdot,t)\|_{2k,\Omega}\,,
\end{aligned}
$$

whence

$$
\frac{d}{dt}\left(\|u(\cdot,t)\|_{2k,\Omega}\right) \le (\lambda+\tau D)\,\|u(\cdot,t)\|_{2k,\Omega} + \tau\,\|f(\cdot,t)\|_{2k,\Omega}\,. \qquad (2.57)
$$

Integrating (2.57) in $t$, for every $k = 1, 2, \ldots$ we get

$$
\|u(\cdot,t)\|_{2k,\Omega} \le \tau\left(\|u_0\|_{2k,\Omega} + e^{t(\lambda+\tau D)}\int_0^t e^{-s(\lambda+\tau D)}\,\|f(\cdot,s)\|_{2k,\Omega}\,ds\right).
$$

Passing in this inequality to the limit as $k \to \infty$, we obtain the needed estimate:

$$
\|u\|_{\infty,Q_T} \le \tau\,\|u_0\|_{\infty,\Omega} + \frac{e^{\tau TD}-1}{D}\,\|f\|_{\infty,Q_T}\,.
$$

□

**Corollary 2.5** *If $D = 0$, inequality* (2.53) *becomes*

$$
\|u\|_{\infty,Q_T} \le \tau\,\|u_0\|_{\infty,\Omega} + \tau T\,\|f\|_{\infty,Q_T}\,.
$$

**Corollary 2.6** *Choosing $M > K(T)$, we have*

$$\min\{u^2, M^2\} = u^2,$$

*which renders* (2.51) *a problem with the single regularization parameter $\varepsilon$.*

**Lemma 2.9** *The solution of problem* (2.51) *satisfies the inequalities*

$$\left\| \left(\varepsilon^2 + u^2\right)^{\gamma(x,t)/4} \nabla u \right\|_{2,Q_T} \le C \tag{2.58}$$

*with a constant $C$ not depending on $\varepsilon$.*

*Proof* For $k = 1$ equality (2.54) takes on the form

$$\begin{aligned} \frac{1}{2}\frac{d}{dt}\left(\|u(\cdot,t)\|_{2,\Omega}^2\right) + \int_\Omega a_{\varepsilon,M}|\nabla u|^2\,dx + \int_\Omega c_{\varepsilon,M}|u|^2 dx \\ = \int_\Omega \mathbf{b}_{\varepsilon,M}\nabla u\, u dx + \tau \int_\Omega d\, u\, dx \equiv I_1 + I_2. \end{aligned} \tag{2.59}$$

Applying Hölder's inequality and the already derived estimate on the maximum of the solution, we obtain the inequality

$$\begin{aligned} |I_1| &\le \int_\Omega \frac{|\mathbf{b}|}{\sqrt{a}}|u|\,\sqrt{a}|\nabla u|\,dx \le \frac{1}{2}\int_\Omega a_{\varepsilon,M}|\nabla u|^2 dx + \frac{1}{2}\int_\Omega \frac{|\mathbf{b}|^2}{a}|u|^2 dx \\ &\le \frac{1}{2}\int_\Omega a_{\varepsilon,M}|\nabla u|^2 dx + C(K). \end{aligned}$$

Substituting this inequality into (2.59) and integrating in $t$, we have

$$\begin{aligned} &\frac{1}{2}\ \|u(\cdot,t)\|_{2,\Omega}\Big|_0^T \\ &+ \frac{1}{2}\int_{Q_T} a_{\varepsilon,M}|\nabla u|^2 dxdt + \int_{Q_T} c_{\varepsilon,M}|u|^2 dxdt \le C(K, D, a_1, T), \end{aligned}$$

whence (2.58). □

*Remark 2.2* The solutions of problem (2.51) satisfy the estimates

$$\left\| (\varepsilon^2 + u^2)^{\gamma^+/4}\nabla u \right\|_{2,Q_T} \le C \tag{2.60}$$

with a constant $C$ independent of $\varepsilon$.

The proof is identical to that of (2.21).

**Lemma 2.10** *The solutions of problem* (2.51) *satisfy the inequalities*

$$\|a_{\varepsilon,M}\, \nabla u\|_{2,Q_T} \leq C$$

*with a constant* $C$ *not depending on* $\varepsilon$ *constant* $C$.

*Proof* Let us introduce the function

$$\psi(u) \equiv \int_0^u (\varepsilon^2 + s^2)^{\gamma^-/2} ds, \qquad \psi = 0 \,\text{on}\, \Gamma_T,$$

which can be taken for the test-function in the integral identity (2.49). Notice that

$$\begin{aligned} I &= \int_{Q_T} u\, \psi_t \, dxdt = \int_{Q_T} u\, u_t\, (\varepsilon^2 + u^2)^{\gamma^-/2} dxdt \\ &= \int_{Q_T} \frac{\partial}{\partial t} \left( \int_0^u s\, \left(\varepsilon^2 + s^2\right)^{\gamma^-/2} ds \right) dxdt \\ &= \int_{\Omega} \left( \int_{\tau\, u_0(x)}^{u(x,T)} s\, \left(\varepsilon^2 + s^2\right)^{\gamma^-/2} ds \right) dx. \end{aligned}$$

By virtue of Lemma 2.8 $|I| \leq M \equiv M(\gamma^-, K)$. Multiplying (2.51) by the function $\zeta = \psi(u)$ and integrating by parts we find that

$$\begin{aligned} \int_{Q_T} a_{\varepsilon,M}\, \psi'(u)\, |\nabla u|^2 \, dxdt &\leq M + \int_{Q_T} |\psi(u)|\, |\mathbf{b}_{\varepsilon,M}| |\nabla u| \, dxdt \\ &- \int_{Q_T} c_{\varepsilon,M}\, u\, \psi(u)\, dxdt + \tau \int_{Q_T} (D\,|u| + |f|)\, \psi(u)\, dxdt \\ &\leq M + \tau \left( T \text{ meas}\, \Omega\, D\, K(T) + M_1 \int_{Q_T} |f|\, dxdt \right) \end{aligned}$$

with the constant

$$M_1 = \max \psi = 2 \int_0^K s\, \left(\varepsilon^2 + s^2\right)^{\gamma^-/2} ds.$$

Proceeding as in the proof of Lemma 2.3 we find that

$$\begin{aligned} a_{\varepsilon,M}\, \psi'(u) &= a \left(\varepsilon^2 + u^2\right)^{\gamma(x,t)/2} \psi'(u) \\ &\geq a_0 \left(\varepsilon^2 + u^2\right)^{\gamma(x,t)/2} \left(\varepsilon^2 + u^2\right)^{\gamma^-/2} \geq a_0 \left(\varepsilon^2 + u^2\right)^{\gamma(x,t)}, \end{aligned}$$

and the assertion follows. □

**Corollary 2.7** *The solutions of problems* (2.51) *satisfy the estimates*

$$\left\|(\varepsilon^2+u^2)^{\gamma^+/2}\nabla u\right\|_{2,Q_T} \leq C, \qquad \left\||u|^{\gamma^+}\nabla u\right\|_{2,Q_T} \leq C \tag{2.61}$$

*with an independent of* $\varepsilon$ *constant* $C$.

**Corollary 2.8** *If* $\gamma^+ \leq 0$, *then* $\|\nabla u\|_{2,Q_T} \leq C$.

**Lemma 2.11** *The solutions of problem* (2.51) *have the weak derivatives* $u_t \in L^2(0,T;H^{-1}(\Omega))$ *and*

$$\|u_t\|_{L^2(0,T;H^{-1}(\Omega))} \leq M$$

*with an independent of* $\varepsilon$ *constant* $M$.

*Proof* It is sufficient to check that for every $\zeta \in C^\infty(0,T;C_0^\infty(\Omega))$, $\zeta(x,0) = \zeta(x,T) = 0$,

$$I = \left|\int_{Q_T} u\,\zeta_t\,dxdt\right| \leq M\,\|\zeta\|_{L^2(0,T;H_0^1(\Omega))}$$

with a constant $M$ not depending on $\zeta$ and $u$. Choosing $\zeta$ for the test-function in (2.49) and using the uniform estimates of Lemmas 2.8–2.10 we find:

$$\begin{aligned}
I &\leq \left\|(\varepsilon^2+u^2)^{\frac{\gamma}{2}}\nabla u\right\|_{2,Q_T}\|\nabla\zeta\|_{2,Q_T} + b_1\left\|(\varepsilon^2+u^2)^{\frac{\gamma}{4}}\nabla u\right\|_{2,Q_T}\|\zeta\|_{2,Q_T}\\
&\quad + c_1\left\|(\varepsilon^2+u^2)^{\frac{\sigma-1}{2}}\right\|_{2,Q_T}\|\zeta\|_{2,Q_T} + \left(D\|u\|_{2,Q_T}+\|f\|_{2,Q_T}\right)\|\zeta\|_{2,Q_T}\\
&\leq C\|\zeta\|_{L^2(0,T;H_0^1(\Omega))}
\end{aligned}$$

with a constant $C$ depending on the norms of $u_0$, $f$ and constants in the structural assumptions (2.47) and (2.48). □

Let us introduce the notation

$$Q_T^h = \{(x,t)\in Q_T : (x+\mathbf{e}_i h,t),(x,t+h)\in Q_T,\ i=1,\dots,n\},$$

where $h$ is a scalar parameter, $\mathbf{e}_i$ are the unit vectors of the axes $x_i$.

**Lemma 2.12** *Let* $u(x,t)$ *be a solution of problem* (2.51) *and* $\gamma^+ \geq 0$. *Then the function*

$$z(x,t) = z[u(x,t)] \equiv \int_0^{u(x,t)} |s|^{\gamma^+}\,ds$$

*satisfies the estimates*

$$\|z(x+\mathbf{e}_i\, h,t)-z(x,t)\|^2_{2,Q^h_T}\leq C|h|^2,\qquad \|z(x,t+h)-z(x,t)\|^2_{2,Q^h_T}\leq C|h|. \tag{2.62}$$

*Proof* The inclusion $\nabla z\in L^2(Q_T)$ follows from the equality $\nabla z=|u|^{\gamma^+}\nabla u$ and estimates (2.61). The first of estimates (2.62) is then an immediate byproduct of the inequality

$$\int_{Q^h_T}\frac{|z(x+\mathbf{e}_i\, h,t)-z(x,t)|^2}{|h|^2}dx\leq C,\qquad i=1,\dots,n.$$

For the functions $\delta u=u(x,t+h)-u(x,t)$, $\delta z=z(x,t+h)-z(x,t)$ we have

$$\begin{aligned}&\int_\Omega \delta z\delta u\,dx\\ &=\int_\Omega\delta z\int_t^{t+h}\left(\operatorname{div}(a_{\varepsilon,M}\nabla u)+\mathbf{b}_{\varepsilon,M}\nabla u-c_{\varepsilon,M}u+\tau\, d(x,\theta,u)\right)d\theta dx,\end{aligned} \tag{2.63}$$

$$\delta z=\int_{u_2}^{u_1}|s|^{\gamma^+}ds\leq\delta u,\qquad \|\nabla(\delta z)\|_{L^2(Q^h_T)}\leq C. \tag{2.64}$$

Let us integrate (2.63) over the interval $(0,T-h)$ and apply (2.64):

$$\begin{aligned}C\,\|\delta z\|^2_{2,Q^h_T}\leq&\int_0^{T-h}\int_\Omega|\nabla(\delta z)|\int_t^{t+h}|a_{\varepsilon,M}\nabla u|\,d\theta dxdt\\ &+\int_0^{T-h}\int_\Omega|\delta z|\int_t^{t+h}|\mathbf{b}_{\varepsilon,M}||\nabla u|\,d\theta dxdt\\ &-\int_0^{T-h}\int_\Omega|\delta z|\int_t^{t+h}|c_{\varepsilon,M}||u|\,d\theta dxdt\\ &+\tau\int_0^{T-h}\int_\Omega|\delta z|\int_t^{t+h}|d(x,\theta,u)|\,d\theta dxdt\\ \equiv&\, I_1+I_2+I_3+I_4.\end{aligned}$$

To estimate $I_i$ we use Hölder's inequality, the estimates

$$|\mathbf{b}_{\varepsilon,M}|\leq b_1\left(\varepsilon^2+u^2\right)^{\gamma(x,t)/4},\quad |c_{\varepsilon,M}||u|\leq c_1\left(\varepsilon^2+u^2\right)^{(\sigma(x,t)-1)/2},$$

and Lemmas 2.8 and 2.10. □

**Lemma 2.13** *Let* $\gamma^+\leq 0$. *Then*

$$\|u(x,t+h)-u(x,t)\|^2_{2,Q^h_T}\leq C\,|h|,\qquad \|z(x+\mathbf{e}_i h,t)-z(x,t)\|^2_{2,Q^h_T}\leq C|h|^2.$$

*Proof* It is sufficient to repeat the proof of Lemma 2.12 substituting $\delta z$ by $\delta u$ and taking into account the fact that $\|\nabla(\delta u)\|_{2,Q_T}$ is already estimated in Corollary 2.5. □

Let us introduce the functions $g[u] = u(\varepsilon^2 + u^2)^{\gamma/2}$ and $h[u] = u(\varepsilon^2 + u^2)^{\gamma/4}$. Arguing as in the proof of Theorem 2.1 and relying on Lemma 2.6 we extract from the sequence $\{u_\varepsilon\}$ a subsequence, for which we keep the same name, such that

$$\begin{aligned} &u_\varepsilon \to u \text{ a.e. in } Q_T, \\ &u_{\varepsilon t} \rightharpoonup u_t \text{ in } L^2(0,T; H^{-1}(\Omega)), \\ &u_\varepsilon \to u \text{ in } L^p(Q) \text{ with any } 1 < p < \infty, \\ &a\nabla u_\varepsilon \rightharpoonup \chi \text{ in } L^2(Q_T) \end{aligned} \tag{2.65}$$

and

$$\begin{aligned} &g_\varepsilon \equiv g[u_\varepsilon] = u_\varepsilon(\varepsilon^2 + u_\varepsilon^2)^{\gamma/2} \to u|u|^\gamma \quad \text{a.e. in } Q_T, \\ &h_\varepsilon \equiv h[u_\varepsilon] = u_\varepsilon(\varepsilon^2 + u_\varepsilon^2)^{\gamma/4} \to u|u|^{\gamma/2} \quad \text{a.e. in } Q_T \text{ and in } L^q(Q_T),\, 1 < q < \infty, \\ &\nabla g_\varepsilon \rightharpoonup \nabla g^*,\, \nabla h_\varepsilon \rightharpoonup \nabla h^* \text{ in } L^2(Q_T) \end{aligned}$$

with some functions $u$, $g^*$, $h^*$, $\chi$. It remains to identify the limit functions $\nabla g^*$ and $\nabla h^*$. Notice that

$$\begin{aligned} &\frac{(1+\gamma)u_\varepsilon^2 + \varepsilon^2}{\varepsilon^2 + u_\varepsilon^2}(\varepsilon^2 + u_\varepsilon^2)^{\gamma/2}\,\nabla u_\varepsilon \\ &\qquad = \nabla g_\varepsilon - \frac{1}{2}\nabla\gamma\left[u_\varepsilon\,(\varepsilon^2 + u_\varepsilon^2)^{\gamma/2}\right]\ln(\varepsilon^2 + u_\varepsilon^2) \end{aligned} \tag{2.66}$$

For every test-function $\psi$

$$\lim_{\varepsilon\to 0}\int_{Q_T}\nabla g_\varepsilon\,\psi = \int_{Q_T}\nabla g^*\,\psi = -\lim_{\varepsilon\to 0}\int_{Q_T} g_\varepsilon\,\nabla\psi = -\int_{Q_T} u\,|u|^\gamma\,\nabla\psi,$$

whence

$$\nabla g^* = \nabla(u|u|^\gamma) = (1+\gamma)|u|^\gamma\,\nabla u + \frac{1}{2}u|u|^\gamma(\ln u^2)\,\nabla\gamma \quad \text{a.e. in } Q_T. \tag{2.67}$$

In the same way we check that

$$\nabla h^* = \nabla(u|u|^{\gamma/2}) = \left(1 + \frac{\gamma}{2}\right)|u|^{\gamma/2}\,\nabla u + \frac{1}{4}u|u|^{\gamma/2}(\ln u^2)\,\nabla\gamma \quad \text{a.e. in } Q_T.$$

For every test-function from the conditions of Definition 2.2 one may pass to the limit when $\varepsilon \to 0$ in each term of the integral identity

$$\sum_{i=1}^{6} I_i^{(\varepsilon)} \equiv \int_{Q_T} \Big( - u_\varepsilon \zeta_t + a(x,t,u_\varepsilon)(\varepsilon^2 + u_\varepsilon^2)^{\gamma(x,t)/2} \nabla u_\varepsilon \nabla \zeta$$
$$- \zeta\, \mathbf{b}(x,t,u)(\varepsilon^2 + u_\varepsilon^2)^{\gamma(x,t)/4} \nabla u_\varepsilon$$
$$+ \zeta\, c(x,t,u)(\varepsilon^2 + u_\varepsilon^2)^{(\sigma(x,t)-2)/2} u_\varepsilon - f\,\zeta \Big) dxdt$$
$$- \int_\Omega u_0 \zeta(x,0)\, dx = 0.$$

By virtue of (2.65)

$$I_1^{(\varepsilon)} \to - \int_{Q_T} u\, \zeta_t\, dxdt, \qquad I_6^{(\varepsilon)} \to - \int_\Omega u_0\, \zeta(x,0)\, dx \qquad \text{as } \varepsilon \to 0,$$

while by virtue of (2.66) and (2.67)

$$I_3^{(\varepsilon)} \to \int_{Q_T} \mathbf{b}(x,t,u)|u|^{\gamma/2} \nabla u\, \zeta\, dxdt,$$
$$I_4^{(\varepsilon)} \to \int_{Q_T} c(x,t,u)|u|^{\sigma-2} u\, \zeta\, dxdt,$$

$$I_2^{(\varepsilon)} = \int_{Q_T} a(x,t,u) \frac{\varepsilon^2 + u_\varepsilon^2}{(1+\gamma)u_\varepsilon^2 + \varepsilon^2}$$
$$\times \left[ \nabla g_\varepsilon - \frac{1}{2} \left[ u_\varepsilon (\varepsilon^2 + u_\varepsilon^2)^{\gamma/2} \right] \ln (\varepsilon^2 + u_\varepsilon^2) \nabla \gamma \right] \nabla \zeta\, dxdt$$
$$\to \int_{Q_T} \frac{a(x,t,u)}{\gamma+1} \left[ \nabla (u|u|^\gamma) - \frac{1}{2} u|u|^\gamma \left( \ln u^2 \right) \nabla \gamma \right] \nabla \zeta\, dxdt$$
$$= \int_{Q_T} a(x,t,u)|u|^\gamma \nabla u\, \nabla \zeta\, dxdt \qquad \text{as } \varepsilon \to 0.$$

Gathering these formulas we conclude that the limit function is a weak solution of problem (2.46).

## 2.8 Equations with Anisotropic Nonlinearity

A revision of the proof of Theorem 2.1 shows that the same arguments provide existence of a weak solution of problem (2.4)–(2.5) for equations with anisotropic nonlinearity. Let us consider the following problem:

$$u_t - \sum_{i=1}^{n} D_i \left( |u|^{\gamma_i(x,t)} D_i u \right) = f(x,t) \quad \text{in } Q_T, \tag{2.68}$$

$$u = 0 \text{ on } \Gamma_T = \partial\Omega \times [0,T], \qquad u(x,0) = u_0(x) \text{ in } \Omega, \tag{2.69}$$

where $\gamma_i(x,t)$ are given bounded functions defined on $Q_T$. It is assumed that

$$\forall \text{a.e. } (x,t) \in Q_T \qquad \gamma_i(x,t) \in [\gamma^-, \gamma^+] \subset (-1, \infty), \quad i = 1, \ldots, n, \tag{2.70}$$

with given constants $\gamma^{\pm}$.

**Definition 2.3** A locally integrable bounded function $u(x,t)$ is called *weak solution* of problem (2.68)–(2.69) if:

(i) $u \in L^\infty(Q_T)$, $|u|^{\gamma_i(x,t)/2}\nabla u \in L^2(Q_T)$, $u_t \in L^2(0,T; H^{-1}(\Omega))$,
(ii) $u = 0$ on $\Gamma_T$ in the sense of traces,
(iii) for every test-function $\zeta(x,t) \in C^\infty(0,T; C_0^\infty(\Omega))$, $z(x,T) = 0$,

$$\int_{Q_T} (-u\zeta_t + \sum_{i=1}^{n} |u|^{\gamma_i(x,t)} D_i u \, D_i \zeta - f\,\zeta) \, dxdt = \int_\Omega u_0 \zeta(x,0) \, dx. \tag{2.71}$$

**Theorem 2.4** *Let $\gamma_i(x,t)$ be measurable in $Q_T$ functions. Assume that $\gamma_i$ satisfy condition* (2.70) *and $\nabla \gamma_i \in L^2(Q_T)$. If $f \in L^2(Q_T) \cap L^1(0,T; L^\infty(\Omega))$ and*

$$\|u_0\|_{\infty,\Omega} + \int_0^T \|f(\cdot,t)\|_{\infty,\Omega} = K(T) < \infty, \tag{2.72}$$

*then problem* (2.68)–(2.69) *has at least one weak solution in the sense of* Definition 2.3. *The solution is bounded and satisfies the estimate $\|u\|_{\infty,Q_T} \le K(T)$ with the constant $K(T)$ from condition* (2.72).

*Remark 2.3* The question of uniqueness of weak solution to the anisotropic problem (2.68)–(2.69) is left open.

## 2.9 Stationary Solutions

Let us consider the problem

$$\begin{cases} -\sum_i D_i(a_i(x,u)|u|^{\alpha_i(x)} D_i u) + c(x,u)|u|^{\sigma(x)-2} u = f(x) \quad \text{in } \Omega, \\ u = 0 \quad \text{on } \partial\Omega. \end{cases} \tag{2.73}$$

About the domain $\Omega$ and the coefficients in Eq. (2.73) we assume the following:

1. $a_i(x,r)$ and $c(x,r)$ are Carathéodory functions,

$$\begin{aligned} &\forall x \in \Omega,\ r \in \mathbb{R} \quad 0 < a_0 \le a_i(x,r) \le A_0 < \infty, \\ &0 \le c_0 \le c(x,r) \le C_0 < \infty \end{aligned} \tag{2.74}$$

   with some constants $a_0$, $c_0$, $A_0$, $C_0$.
2. $\alpha_i(x)$ and $\sigma(x)$ are continuous functions satisfying the conditions

$$\alpha_i(x) \in \left(\alpha_i^-, \alpha_i^+\right) \subseteq (\alpha^-, \alpha^+) \subset (-1, \infty), \quad \sigma(x) \in (\sigma^-, \sigma^+) \subset (1, \infty), \tag{2.75}$$

   $\alpha_i^{\pm}$, $\alpha^{\pm}$, $\sigma^{\pm}$ are known constants,

**Definition 2.4** A locally integrable in $\Omega$ function $u(x)$ is called *weak solution of problem* (2.73), if

1. $u \in L^\infty(\Omega)$, $|u|^{\alpha_i(x)/2}|D_i u| \in L^2(\Omega)$, $(i = 1, \dots, n)$,
2. $u = 0$ on $\Gamma$ in the sense of traces,
3. for every test-function $\eta \in W_0^{1,2}(\Omega) \cap L^{\sigma(x)}(\Omega)$ the integral identity holds

$$\sum_i \int_\Omega a_i(x,u)|u|^{\alpha_i(x)} D_i u D_i \eta \, dx + \int_\Omega c(x,u)|u|^{\sigma(x)-2} u \eta \, dx = \int_\Omega f \eta \, dx. \tag{2.76}$$

About the function $f$ we assume that

$$f \in \begin{cases} L^p(\Omega) \text{ with } p > n/2 & \text{if } \alpha_i(x) \ge 0 \text{ in } \Omega, \\ L^\infty(\Omega) & \text{if } \min_i \inf_\Omega \alpha_i(x) < 0. \end{cases} \tag{2.77}$$

**Theorem 2.5** *Let conditions* (2.74) *and* (2.75) *be fulfilled and, additionally to these conditions,*

$$\|D_i \alpha_i(x)\|_{2,\Omega} \le C, \quad i = 1, 2, \dots, n. \tag{2.78}$$

*Let us assume that either $\alpha_i(x) \ge 0$ in $\Omega$, or $\alpha_i(x) > -1$ in $\Omega$ and $c_0 > 0$. Then for every right-hand side $f$, satisfying condition* (2.77), *problem* (2.73) *has a weak solution. The solution satisfies estimate*

$$\sum_i \| |u|^{\alpha_i(x)/2} D_i u \|_{2,\Omega} + \sum_i \| |u|^{\alpha_i(x)} D_i u \|_{2,\Omega} + c_0 \|u\|_{\sigma(\cdot),\Omega} + \|u\|_{\infty,\Omega} \le \Lambda \tag{2.79}$$

*with a constant $\Lambda$ depending on $\|f\|$, $|\Omega|$, $n$ and the constants in conditions* (2.74) *and* (2.75).

*Remark 2.4* Condition (2.78) can be relaxed in the following way:

$$\|D_i\alpha_i(x)\|_{2,\Omega_i^-} \le C, \quad \Omega_i^- = \{x \in \Omega : \alpha_i(x) \le 0\}.$$

—see Lemma 2.20 below.

A weak solution of problem (2.73) is obtained as the limit of the sequence of solutions of the regularized problems

$$\begin{cases} -\sum_i D_i(A_i(\varepsilon, M, x, u)D_i u) + C(\varepsilon, M, x, u)u = f(x) & \text{in } \Omega, \\ u = 0 \quad \text{on } \Gamma \end{cases} \tag{2.80}$$

with positive parameters $\varepsilon$, $M$ and the coefficients

$$\begin{aligned} 0 < C_i'(\varepsilon, M, \alpha^\pm) &\le A_i \equiv a_i(x,u)(\varepsilon^2 + \min\{u^2, M^2\})^{\alpha_i(x)/2} \\ &\le C_i''(\varepsilon, M, \alpha^\pm) < \infty, \\ 0 \le C'(\varepsilon, M, \sigma^\pm) &\le C \equiv c(x,u)(\varepsilon^2 + \min\{u^2, M^2\})^{(\sigma(x)-2)/2} \\ &\le C''(\varepsilon, M, \sigma^\pm) < \infty. \end{aligned}$$

**Lemma 2.14** *For every $\varepsilon > 0$, $\tau \in [0,1]$, $f(x) \in L^p(\Omega)$ with $p > n/2$ problem* (2.80) *has a solution $u \in W_0^{1,2}(\Omega) \cap L^\infty(\Omega) \cap L^{\sigma(x)}(\Omega)$ satisfying the estimate*

$$c_0\|u\|_{\sigma(\cdot)} + \|u\|_{\infty,\Omega} \le C \tag{2.81}$$

*with a constant $C$ independent of $M$ and $\varepsilon$.*

The existence of a weak solution of problem (2.80) is proved by means of the Schauder fixed point principle [187, Chap. 4, § 8]. Let us consider the linear problem

$$\begin{cases} -\sum_i D_i(A_i(\varepsilon, M, x, v)D_i u) + C(\varepsilon, M, x, v)u = \tau f(x) & \text{in } \Omega, \\ u = 0 \text{ on } \Gamma, \quad \tau \in [0,1], \quad v \in L^2(\Omega). \end{cases} \tag{2.82}$$

For every given $v \in L^2(\Omega)$ and every $\varepsilon > 0$, $M \ge 1$ and $\tau \in [0,1]$ problem (2.82) has a unique solution $u \in W_0^{1,2}(\Omega)$, [187, Chap. 3], satisfying the integral identity

$$\forall \eta \in W_0^{1,2}(\Omega) \quad \sum_i \int_\Omega A_i D_i u D_i \eta \, dx + \int_\Omega C u \eta \, dx = \int_\Omega f \eta \, dx. \tag{2.83}$$

Let us denote $\mathscr{B}_R = \{v : \|v\|_{L^2(\Omega)} < R\}$. Problem (2.82) defines the mapping $(v, \tau) \mapsto u$ which can be represented in the form

$$u = \tau \Phi(v) : \mathscr{B}_R \times [0,1] \mapsto L^2(\Omega)$$

because of the linearity with respect to $\tau$. The solution of problem (2.80) is then a fixed point of the mapping $\Phi(\cdot)$. According to the Schauder principle the mapping $\Phi(v)$ has at least one fixed point in the ball $\mathscr{B}_R$, if

1. the mapping $\Phi(v) : \mathscr{B}_R \mapsto \mathscr{B}_R$ is continuous and compact,
2. for every $\tau \in [0, 1)$ all possible solutions of the equation $u = \tau\Phi(v)$ satisfy the estimate $\|u\|_{2,\Omega} < R$.

**Lemma 2.15** *The mapping $\Phi(v) : \mathscr{B}_R \mapsto \mathscr{B}_R$ is continuous and compact.*

*Proof* Since the embedding $W_0^{1,2}(\Omega) \subset L^2(\Omega)$ is compact, so is the mapping $\Phi(v)$. The continuity of $\Phi$ follows from the continuity of the coefficients $A_i$ and $C$ with respect to the variable $v$. □

**Lemma 2.16** *Let $\alpha_i(x) \geq 0$ and $f \in L^p(\Omega)$ with $p > n/2$. Then the solution of the regularized problem* (2.82) *with the parameter $M \geq 1$ satisfies the inequality*

$$\|u\|_{\infty,\Omega} \leq K \equiv \left(1 + \frac{C}{a_C}\|f\|_{p,\Omega}^{\frac{2}{n}-\frac{1}{p}}\right)^{\frac{1+\frac{2}{n}-\frac{1}{p}}{\frac{2}{n}-\frac{1}{p}}} \tag{2.84}$$

*with a constant $C$ not depending on $M$.*

*Proof* Take an arbitrary number $k \geq 1$ and consider the function $\zeta = \max\{0, u-k\}$. The function $\zeta$ is an admissible test-function in identity (2.83). Besides,

$$\nabla\zeta = \begin{cases} 0, & \text{if } u \leq k, \\ \nabla u, & \text{if } u > k. \end{cases}$$

Let us denote $\Omega_k = \Omega \cap \{u > k\}$. Substituting $\zeta$ into the integral identity (2.83) and taking into account the inequality $u \cdot \zeta \geq 0$, we get

$$I := \sum_i \int_{\Omega_k} a_i(x, u)(\varepsilon^2 + \min\{u^2, M^2\})^{\alpha_i(x)/2}|D_i u|^2\,dx \leq \int_{\Omega_k} f(u-k)\,dx.$$

On the set $\Omega_k$

$$\varepsilon^2 + \min\{u^2, M^2\} \geq \varepsilon^2 + \min\{k^2, M^2\} \geq 1,$$

which yields the inequality

$$(\varepsilon^2 + \min\{u^2, M^2\})^{\alpha_i(x)/2} \geq \frac{1}{k}.$$

Hence,

$$\int_{\Omega_k} |\nabla u|^2\,dx \le \frac{k}{a_0}\int_{\Omega_k} f(u-k)\,dx \equiv \frac{k}{a_0} I_3.$$

Estimating $I_3$ by Hölder's inequality and then applying the embedding theorem, we arrive at the estimate

$$|I_3| \le |\Omega_k|^{\frac{1}{2}-\frac{1}{p}} \|u-k\|_{2,\Omega_k} \|f\|_{p,\Omega} \le C|\Omega_k|^{\frac{1}{2}+\frac{2}{n}-\frac{1}{p}} \|f\|_{p,\Omega} \|\nabla u\|_{2,\Omega_k}.$$

Thus,

$$\int_{\Omega_k} |\nabla u|^2\,dx \le k^2 \left(\frac{C}{a_0}\right)^2 |\Omega_k|^{1+\frac{2}{n}-\frac{2}{p}} \|f\|^2_{p,\Omega}$$

with a constant $C$ not depending on $M$. Using [187, Chap. 2, Lemma 5.3], we conclude that for $p > n/2$ the inequality $\|u\|_{\infty,\Omega} \le K$ holds with a constant $K$ which depends on $n$, $p$, $1/a_0$ and $\|f\|_{p,\Omega}$ in the following way [187, Chap. 2, Lemma 5.1]: [187]

$$1 \le K \le \left(1 + \frac{C}{a_0}\|f\|^{\frac{2}{n}-\frac{1}{p}}_{p,\Omega}\right)^{\frac{1+\frac{2}{n}-\frac{1}{p}}{\frac{2}{n}-\frac{1}{p}}}.$$

□

**Lemma 2.17** *Let $c_0 > 0$ and $f \in L^\infty(\Omega)$. Then the solution of the regularized problem* (2.82) *with the parameter $M \ge 1$ satisfies the inequality*

$$\|u\|_{\infty,\Omega} \le K \equiv \max\left\{1, \left(\frac{1}{c_0}\|f\|_{\infty,\Omega}\right)^{\frac{1}{\sigma^- - 1}}\right\}. \tag{2.85}$$

*Proof* Arguing as in the proof of Lemma 2.16 we take for the test-function $\zeta = \max\{u-k, 0\}$ with the parameter $k \ge 1$. Substituting $\zeta$ into the integral identity (2.83), we obtain the inequality

$$\sum_i \int_{\Omega_k} A_i |D_i u|^2\,dx + c_0 \int_{\Omega_k} \min\{k, M\}^{\sigma(x)-1}(u-k)\,dx \le \tau \int_{\Omega_k} f(u-k)\,dx,$$

whence the estimate

$$0 \ge (c_0 k^{\sigma^- -1} - \tau\|f\|_{\infty,\Omega}) \int_{\Omega_k} (u-k)\,dx.$$

Increasing $k$, we conclude that necessarily $|\Omega_k| = 0$ for

$$k \geq k_0 \equiv (\frac{1}{c_0}\|f\|_{\infty,\Omega})^{1/(\sigma^- - 1)}.$$

The inequality $-u \leq k_0$ a. e. in $\Omega$ is proved likewise. □

The obtained estimates on the maximum of the solution of problem (2.80) do not depend on $M$, which allows us to choose $M = K$. Then the coefficients in equation (2.80) take on the form

$$A_i(\varepsilon, M, x, w) = (\varepsilon^2 + w^2)^{\alpha_i(x)/2}, \quad C(\varepsilon, M, x, w) = (\varepsilon^2 + w^2)^{(\sigma(x)-2)/2}$$

and (2.80) transforms into a problem with the single regularization parameter $\varepsilon$. Without special mentioning, in what follows we tacitly assume that $M = K$ with the constant $K$ from (2.84) or (2.85).

**Lemma 2.18** *For $c_0 > 0$ the fixed points of the mapping $\Phi$ satisfy the inequality*

$$c_0 \int_\Omega |u|^{\sigma(x)}\,dx \leq K\|f\|_{\infty,\Omega}|\Omega|$$

*with the constant $K$ from* (2.85).

*Proof* Let us substitute $u$ into (2.83) as the test-function and then drop the nonnegative term on the left-hand side of the appearing equality:

$$c_0 \int_\Omega (\varepsilon^2 + u^2)^{(\sigma(x)-2)/2} u^2\,dx \leq \int_\Omega |f||u|\,dx \leq K|\Omega|\|f\|_{\infty,\Omega}.$$

□

The proof of Lemma 2.14 is completed.

**Lemma 2.19** *If condition* (2.77) *is fulfilled, then the solution of problem* (2.80) *satisfies the inequalities*

$$\sum_i \|\sqrt{A_i}D_i u\|^2_{2,\Omega} \leq C, \quad \sum_i \|A_i D_i u\|^2_{2,\Omega} \leq C \tag{2.86}$$

*with a constant $C$ independent of $\varepsilon$.*

*Proof* The former of inequalities (2.86) follows from (2.83) with $\eta = u$. To prove the latter we make use of the following inequality, which is a byproduct of (2.84) and (2.85):

$$\left(\frac{\varepsilon^2 + u^2}{\varepsilon^2 + K^2}\right)^{\alpha_i(x)/2} \leq \left(\frac{\varepsilon^2 + u^2}{\varepsilon^2 + K^2}\right)^{\alpha^-/2} \quad \text{a. e. in } \Omega.$$

Let us introduce the function

$$\phi(u) \equiv A_0 \int_0^u (\varepsilon^2 + s^2)^{\alpha^-/2}\, ds \in L^\infty(\Omega) \cap W_0^{1,2}(\Omega).$$

For $\alpha^- > -1$ the function $|\phi(u)|$ is uniformly with respect to $\varepsilon$ bounded by the constant $\phi(K)$. Substituting $\phi(u)$ into (2.83) as the test-function, we obtain the inequality

$$\sum_i \int_\Omega A_i (\varepsilon^2 + u^2)^{\alpha^-/2} |D_i u|^2\, dx + c_0 \int_\Omega (\varepsilon^2 + u^2)^{(\sigma(x)-2)/2} u\phi(u)\, dx \le \int_\Omega |f||\phi(u)|\, dx.$$

Noting that $u\phi(u) \ge 0$, we have

$$\begin{aligned}\int_\Omega A_i^2 |D_i u|^2\, dx &\le \sum_i \int_\Omega a_i^2(x,u)(\varepsilon^2 + u^2)^{\alpha_i(x)} |D_i u|^2\, dx \\ &\le \phi(K) \begin{cases} |\Omega| \|f\|_{\infty,\Omega} & \text{for } f \in L^\infty(\Omega), \\ |\Omega|^{1/p'} \|f\|_{p,\Omega} & \text{for } f \in L^p(\Omega). \end{cases}\end{aligned}$$

□

Denote by $\{u_\varepsilon\}$ the sequence of solutions of the regularized problems (2.80). Estimates (2.81) and (2.86) allow us to extract from $\{u_\varepsilon\}$ a subsequence (which will be assumed to coincide with the whole sequence), that possesses the properties:

$$\begin{aligned}& A_i D_i u_\varepsilon \rightharpoonup B_i(x) \quad \text{in } L^2(\Omega), \\ & u_\varepsilon \to u \quad \text{a.e. in } \Omega, \\ & \left(\varepsilon^2 + u_\varepsilon^2\right)^{(\sigma(x)-2)/2} u_\varepsilon \rightharpoonup |u|^{\sigma(x)-2} u \quad \text{in } L^2(\Omega)\end{aligned} \tag{2.87}$$

with some functions $u$ and $B_i \in L^2(\Omega)$. Taking into account (2.87) and passing to the limit as $\varepsilon \to 0$ in identity (2.83), we have that for every $\eta \in W_0^{1,2}(\Omega)$

$$\sum_i \int_\Omega B_i(x) D_i \eta\, dx + \int_\Omega c(x,u)|u|^{\sigma(x)-2} u\eta\, dx = \int_\Omega f\eta\, dx.$$

It remains to show that $B_i(x) = |u|^{\alpha_i(x)} D_i u$. Consider the function

$$G_i(u) = (\varepsilon^2 + u^2)^{\alpha_i(x)} u.$$

It is easy to calculate that

$$D_i G_i(u) = \left[1 + \alpha_i \frac{u^2}{\varepsilon^2 + u^2}\right] (\varepsilon^2 + u^2)^{\alpha_i/2} D_i u$$
$$+ \frac{1}{2} u(\varepsilon^2 + u^2)^{\alpha_i/2} D_i \alpha_i \ln(\varepsilon^2 + u^2),$$
$$|D_i G_i(u)|^2 \leq C[|A_i|^2 |D_i u|^2 + |D_i \alpha_i(x)|^2].$$

By virtue of (2.86), $\|D_i G_i(u_\varepsilon)\|_{2,\Omega} \leq C$ uniformly with respect to $\varepsilon$. This allows us to extract a subsequence $\{u_\varepsilon\}$ such that $D_i G_i(u_\varepsilon) \rightharpoonup D_i H_i$ in $L^2(\Omega)$. Hence, for every test–function $\zeta$

$$\lim_{\varepsilon \to 0} \int_\Omega D_i G_i(u_\varepsilon) \zeta \, dx = \int_\Omega D_i H_i \zeta \, dx = -\lim_{\varepsilon \to 0} G_i(u_\varepsilon) D_i \zeta \, dx$$
$$- \int_\Omega |u|^{\alpha_i(x)} u D_i \zeta \, dx,$$

whence the equality

$$D_i H_i = D_i(|u|^{\alpha_i(x)} u) = (1 + \alpha_i(x)) |u|^{\alpha_i(x)} D_i u + \frac{1}{2} |u|^{\alpha_i(x)} u (\ln u^2) D_i \alpha_i(x).$$

Further,

$$A_i(\varepsilon, M, x, u_\varepsilon) D_i u_\varepsilon = a_i(x, u_\varepsilon) \frac{\varepsilon^2 + u_\varepsilon^2}{(1 + \alpha_i(x)) u_\varepsilon^2 + \varepsilon^2}$$
$$\times \left[ D_i G_i(u_\varepsilon) - \frac{1}{2} u \left(\varepsilon^2 + u_\varepsilon^2\right)^{\alpha_i(x)/2} D_i \alpha_i(x) \ln \left(\varepsilon^2 + u_\varepsilon^2\right) \right],$$

which yields that for every $\eta \in W_0^{1,2}(\Omega)$

$$\sum_i \int_\Omega A_i(\varepsilon, M, x, u_\varepsilon) D_i u_\varepsilon D_i \eta \, dx \to$$
$$\sum_i \int_\Omega a_i(x, u) |u|^{\alpha_i(x)} D_i u D_i \eta \, dx \quad \text{as } \varepsilon \to 0.$$

**Lemma 2.20** *The assertion of* Theorem 2.5 *remains valid if condition* (2.78) *is removed and substituted by the following one:*

$$\|D_i \alpha_i(x)\|_{2,\Omega_i^-} \leq C, \quad \Omega_i^- = \{x \in \Omega : \alpha_i(x) \leq 0\}.$$

*Proof* Condition (2.78) was solely used to justify the limit as $\varepsilon \to 0$ in problem (2.80). Let us show that on the set $\{x \in \Omega : \alpha_i(x) > 0\}$ one may pass to the limit without this condition. Not loosing generality we assume that $\alpha^- > 0$. Let us represent the first term on the left–hand side of (2.83) as follows:

$$
\begin{aligned}
I &:= \sum_i \int_\Omega a_i(x, u_\varepsilon) \left(\varepsilon^2 + u_\varepsilon^2\right)^{\alpha_i(x)/2} D_i u_\varepsilon D_i \eta \, dx = I^{(1)}_{\delta,\varepsilon} + I^{(2)}_{\delta,\varepsilon} \\
&\equiv \sum_i \int_{\Omega_\delta} \cdots + \sum_i \int_{\Omega \setminus \Omega_\delta} \cdots,
\end{aligned}
$$

where $\Omega_\delta = \{x \in \Omega : |u_\varepsilon| > \delta\}$ and $\delta > 0$ is an arbitrary small number. It is easy to see that if (2.74), (2.75) and (2.86) are fulfilled, then

$$
\mu(\delta) \sum_i \int_{\Omega_\delta} |D_i u_\varepsilon|^2 \, dx \le \sum_i \int_{\Omega_\delta} a_i(x, u_\varepsilon) \left(\varepsilon^2 + u_\varepsilon^2\right)^{\alpha_i(x)/2} |D_i u_\varepsilon|^2 \, dx \le C
$$

with a constant $\mu(\delta) > 0$. Hence, the sequences $\{D_i u_\varepsilon\}$ converge weakly in $L^2(\Omega_\delta)$, while the sequences

$$
\left\{ a_i(x, u_\varepsilon) \left(\varepsilon^2 + u_\varepsilon^2\right)^{\alpha_i(x)/2} \right\}, \quad i = 1, \ldots, n,
$$

converge almost everywhere. Therefore, for every fixed $\delta > 0$ there exists

$$
\lim_{\varepsilon \to 0} I^{(1)}_{\delta,\varepsilon} = \sum_i \int_{\Omega_\delta} a(x, u) |u|^{\alpha_i(x)} D_i u D_i \eta \, dx.
$$

Taking into account (2.74) and (2.86), we estimate $I^{(2)}_{\delta,\varepsilon}$ as follows:

$$
\left| I^{(2)}_{\delta,\varepsilon} \right| \le \sqrt{A_0} (\delta^2 + \varepsilon^2)^{\alpha^-/2} \| \sqrt{A_i} D_i u_\varepsilon \|_{2,\Omega} \| D_i \eta \|_{2,\Omega} \le C (\delta^2 + \varepsilon^2)^{\alpha^-/2}.
$$

Thus, $\lim_{\varepsilon \to 0} \left| I^{(2)}_{\varepsilon,\delta} \right| \le C \delta^{\alpha^-}$ and for every $\delta > 0$

$$
\lim_{\varepsilon \to 0} I = \sum_i \int_{\Omega_\delta} a_i(x, u) |u|^{\alpha_i(x)} D_i u D_i \eta \, dx + o(\delta^{\alpha^-}).
$$

Plugging estimates (2.86) and applying the Lebesgue dominated convergence theorem, we finally obtain:

$$
\lim_{\delta \to 0} \lim_{\varepsilon \to 0} |I| = \sum_i \int_\Omega a_i(x, u) |u|^{\alpha_i(x)} D_i u D_i \eta \, dx.
$$

□

## 2.10 Remarks

The presentation of this chapter is based on results of [42, 43, 45, 46], see also [195] for a study of isotropic equation (2.46) with $\mathbf{b} \equiv 0$. The proof of uniqueness follows ideas of [33, 34, 37, 62, 63]. We were interested here in *energy solutions*, whose properties are studied in detail in the next chapter.

A different approach to the study of anisotropic porous medium equation with constant exponents is developed in [245, 246]. It is shown that for every nonnegative $u_0 \in L^\infty(\mathbb{R}^n) \cap C(\mathbb{R}^n)$ the equation

$$u_t = \sum_{i=1}^{n} \left( (u^{m_i})_{x_i} + b_i u^{n_i} \right)_{x_i} - cu^r, \qquad m_i, n_i, r > 0, \quad c \geq 0,$$

has a nonnegative solution $u \in C^0(\overline{Q}_T)$ in a domain $Q_T = \Omega \times (0, T]$. The solution is understood in the following sense: for every subdomain $\Omega' \subset \Omega$, $T' \in (0, T)$ and every nonnegative test-function $\phi \in C^{2,1}(\Omega' \times (0, T'])$, vanishing on the lateral boundary of the cylinder $Q'_{T'}$,

$$\begin{aligned} &\int_{Q'_{T'}} \left( u\phi_t + \sum_{i=1}^{n} \left( u^{m_i} \phi_{x_i x_i} - b_i u^{n_i} \phi_{x_i} \right) - cu^r \phi \right) dxdt \\ &\qquad - \int_{\Omega'} u\phi \, dx \Big|_{t=0}^{t=T'} - \int_0^{T'} \int_{\partial\Omega'} \sum_{i=1}^{n} u^{m_i} \phi_{x_i} \cos(x_i, \boldsymbol{\nu}) \, d\Gamma dt = 0, \end{aligned}$$

where $\boldsymbol{\nu}$ denotes the unit normal vector to $\partial\Omega'$. The proof requires restrictions on the geometry of $\partial\Omega$: it is assumed that either $\Omega$ is strictly convex, or that $\partial\Omega$ satisfies the exterior ball condition. Besides, the oscillations $\max m_i - \min m_i$ should be suitably small. The existence result is extended to the Dirichlet problem in a bounded cylinder, the uniqueness theorem is proved under stronger restrictions on the data.

The Cauchy problem for the anisotropic equation

$$u_t = \sum_{i=1}^{n} \left( u^{m_i} \right)_{x_i x_i}$$

with the initial data in $L^1(\mathbb{R}^n)$ is studied in [168]. The equation is fulfilled in the sense of distributions and the solution belongs to $C([0, \infty); L^1(\mathbb{R}^n)) \cap C(\mathbb{R}^n \times (0, \infty)) \cap L^\infty(\mathbb{R}^n \times [\varepsilon, \infty))$, $\varepsilon > 0$. Another approach to the study of the anisotropic PDEs of the type

$$\partial_t(|u|^\rho u) - \sum_{i=1}^{n} D_i \left[ \left( a_i |u|^{\gamma_i} + b_i |D_i u|^{p_i - 2} \right) D_i u + B_i(z, u) \right] + f(z, u) = 0$$

was proposed in [244], see also [119].

Local continuity of bounded energy solutions of the model equation (2.3) with $a \equiv 1$ and $f \equiv 0$ was established in [159, 161]. For the anisotropic equation (2.69) with constant exponents this fact was proved in [160].

A numerical study of problem (2.4) and (2.5) is performed in [121, 122]. The authors propose and justify a moving mesh algorithm and present results of numerical simulation.

# Chapter 3
# Localization of Solutions of the Generalized Porous Medium Equation

Solutions of nonlinear parabolic equations may possess the properties not displayed by the solutions of any linear equation: the finite speed of propagation of disturbances from the problem data in space and vanishing in a finite time. Both effects are well-studied for equations which admit explicit solutions (or sub/supersolutions), or possess the property of invariance with respect to scaling. If the methods based on comparison or scaling fail, the localization properties can be derived from the behavior of local energy functions associated with the solution. This is the case of equations with variable and anisotropic nonlinearity. We devote this chapter to study the following localization properties of solutions of the generalized porous medium Eq. (2.4).

- *Finite speed of propagation.* Let $u_0 = 0$ in $B_r(x_0) = \{x \in \Omega : |x - x_0| < r\}$. We say that the solution possesses the property of finite speed of propagation of disturbances from the data if there exist a function $\rho(t) > 0$ and $t_* > 0$ such that

$$u(x,t) \equiv 0 \quad \text{in } B_{\rho(t)}(x_0) \text{ for all } t \in [0, t_*].$$

  We show that the weak solution of Eq. (2.4) possesses this property if $\gamma(x,t)$ is continuous at the point $(x_0, 0)$ and $\gamma(x_0, 0) > 0$.
- *The waiting time effect.* Let $u_0 = 0$ in a ball $B_r(x_0)$ and assume that $B_r(x_0)$ touches the set $\overline{\operatorname{supp} u_0}$ at a point $\xi$. We say that the solution $u(x,t)$ has the waiting time property at the point $\xi$ if $u(x,t) = 0$ in $B_r(x_0)$ for all $t \in [0, t_*]$. We establish conditions on the data $u_0$ and $f$ sufficient for the existence of this effect.
- *Extinction in a finite time.* Let us say that the solution has the property of extinction in a finite time if there exists $t_*$ such that $u = 0$ in $\Omega$ for all $t \geq t_*$. We show that this is true if $\gamma(x,t) \leq \gamma^+ < 0$ in $Q_T$. We also derive estimates on the rate of vanishing of the norm of the weak solution when $t \to \infty$.
- *Extinction in a borderline case.* It turns out that the condition $-1 < \gamma(x,t) < 0$ is sufficient but not necessary for the finite time extinction. We present an example of the situation when that the solution vanishes in a finite time even in the borderline case $\gamma^+ = 0$.



S. Antontsev and S. Shmarev, *Evolution PDEs with Nonstandard Growth Conditions*, Atlantis Studies in Differential Equations 4,
DOI 10.2991/978-94-6239-112-3_3

## 3.1 Notation and Assumptions

Let us introduce the following notation. Given a point $(x_0, t_0) \in Q_T$, we denote

$$B_\rho \equiv B_\rho(x_0) = \{x \in \mathbb{R}^n : |x - x_0| < \rho\} \subset \Omega,$$
$$S_\rho \equiv S_\rho(x_0) = \partial B_\rho(x_0).$$

We introduce the energy functions

$$b(\rho, t) = ||u(\cdot, t)||^2_{2,B_\rho}, \qquad \overline{b}(\rho) = \sup_{0 \le t \le t_0} b(\rho, t),$$

$$\begin{cases} E(\rho, t) = \displaystyle\int_0^t \int_{B_\rho} u^{\gamma(x,\tau)} |\nabla u|^2 dx d\tau, \\ \overline{E}(\rho) = \displaystyle\sup_{0 \le t \le t_0} E(\rho, t) = \int_0^{t_0} \int_{B_\rho} u^{\gamma(x,\tau)} |\nabla u|^2 dx d\tau, \end{cases}$$

$$\begin{cases} E^+(\rho, t) = \displaystyle\int_0^t \int_{B_\rho} u^{\gamma^+} |\nabla u|^2 dx d\tau, \\ \overline{E}^+(\rho) = \displaystyle\sup_{0 \le t \le t_0} E^+(\rho, t) = \int_0^{t_0} \int_{B_\rho} u^{\gamma^+} |\nabla u|^2 dx dt \equiv E^+(\rho, t_0). \end{cases}$$

Since $E(\rho, t)$ is monotone non-decreasing in $\rho$ and $t$, there exist the weak derivatives

$$\begin{cases} E_\rho(\rho, t) = \displaystyle\int_0^t \int_{S_\rho} u^{\gamma(x,\tau)} |\nabla u|^2 dS d\tau, \\ E_t(\rho, t) = \displaystyle\int_{B_\rho} u^{\gamma(x,t)} |\nabla u|^2 dx, \qquad E_{\rho t}(\rho, t) = \int_{S_\rho} u^{\gamma(x,t)} |\nabla u|^2 dS, \end{cases}$$

$$\begin{cases} \overline{E}_\rho(\rho) = E_\rho(\rho, t_0), \\ E^+_\rho(\rho, t) = \displaystyle\int_0^t \int_{S_\rho} u^{\gamma^+} |\nabla u|^2 dS dt, \qquad \overline{E}^+_\rho(\rho) = E^+_\rho(\rho, t_0), \\ E^+_t(\rho, t) = \displaystyle\int_{B_\rho} u^{\gamma^+} |\nabla u|^2 dx, \qquad E^+_{\rho t}(\rho, t) = \int_{S_\rho} |u|^{\gamma^+} |\nabla u|^2 dS. \end{cases}$$

It will be assumed that

$$\begin{cases} \text{there is } \rho_0 > 0 \text{ such that } B_{\rho_0}(x_0) \subset \Omega \text{ and} \\ b(\rho, 0) = \displaystyle\int_{B_{\rho_0}(x_0)} u^2(x, 0) dx = 0, \qquad f(x, t) \equiv 0 \text{ in } B_{\rho_0}(x_0) \times [0, T]. \end{cases} \tag{3.1}$$

## 3.2 Finite Speed of Propagation

The considerations of this section are confined to the cylinder

$$G_\delta = B_\delta(x_0) \times \{t \in [0, \delta]\} \quad \text{with } \delta = const \geq \rho_0$$

and some $\rho_0 > 0$. Let $u(x, t)$ be a function satisfying the conditions

$$u_t = \operatorname{div}\left(|u|^{\gamma(x,t)} \nabla u\right) + f \quad \text{in } G_\delta, \tag{3.2}$$

$$u(x, 0) = u_0(x) \text{ in } B_\delta(x_0). \tag{3.3}$$

**Definition 3.1** A locally integrable in $G_\delta$ bounded function $u(x, t)$ is called *local weak solution of* Eq. (3.2) *in* $G_\delta$ if:

(i) $u \in L^\infty(G_\delta)$, $|u|^{\frac{\gamma(x,t)}{2}} \nabla u \in L^2(G_\delta)$,
(ii) for any test-function $\zeta \in L^2(0, \delta; H_0^1(B_\delta(x_0))) \cap L^\infty(G_\delta)$ satisfying the conditions

$$\zeta_t \in L^2(0, \delta; H^{-1}(B_\delta(x_0)), \quad \zeta(x, \delta) = 0 \text{ a.e. in } B_\delta(x_0),$$

$$\int_{G_\delta} \left(-u\zeta_t + |u|^{\gamma(x,t)} \nabla u \nabla \zeta - f\zeta\right) dxdt = \int_{B_\delta(x_0)} u_0 \zeta(x, 0)\, dx. \tag{3.4}$$

Since the set of test functions in Definition 2.1 with $T = \delta$ is dense in the set of test-function admitted in Definition 3.1, every weak solution of problem (2.4) and (2.5) is a local weak solution of Eq. (3.2) in the sense of Definition 3.1. For the sake of simplicity we will assume that

$$|u| \leq 1. \tag{3.5}$$

Since $|u| \leq M$ a.e. in $G_\delta$ by assumption, the function $v = u/M$ satisfies the conditions $|v| \leq 1$ a.e. in $G_\delta$ and

$$v_t = \operatorname{div}\left(a(x, t)|v|^{\gamma(x,t)} \nabla v\right) + \frac{f}{M}, \qquad a(x, t) = M^{\gamma(x,t)}.$$

The equation for $v$ is a partial case of Eq. (2.46) studied in the previous chapter. It is easy to check that all further results remain true if the diffusion coefficient $a = 1$ in Eq. (3.2) is substituted by a function $a(x, t)$ such that $0 < a_0 \leq a(x, t) \leq a_1 < \infty$.

**Theorem 3.1** (Finite speed of propagation) *Let condition* (3.1) *be fulfilled,*

$$\gamma(x_0, 0) > 0 \quad \text{and} \quad \gamma(x, t) \in C^0(\overline{Q_T}).$$

*Then there exists* $\delta_0 > 0$ *such that every local weak solution of* Eq. (3.2) *in* $G_{\delta_0}$, *satisfying* (3.5) *and the condition*

$$E_0 = \overline{b}(\rho_0) + \overline{E}(\rho_0) < \infty,$$

*possesses the property of finite speed of propagation:*

$$u(x,t) = 0 \quad for\ (x,t) \in B_{\rho(t)}(x_0) \times [0, \delta_0].$$

*The function* $\rho(t)$ *is defined by the formula*

$$\rho^{\nu}(t) = \rho_0^{\nu} - Ct^{\lambda} E_0^{\sigma},$$

*with positive constants* $\nu$, $\lambda$, $\sigma$, *and* $C = C(\overline{b}(\rho_0), \delta_0, \gamma^{\pm}, \rho_0, N)$.

*Remark 3.1* Since $\rho(t)$ is a monotone decreasing continuous function with $\rho(0) = \rho_0 > 0$, the set $B_{\rho(t)}(x_0)$ is nonempty for small $t$.

The proof follows the ideas of monograph [36] devoted to study the localization properties of solutions to nonlinear equations with constant exponents of nonlinearity. The proof of Theorem 3.1 is split into three steps:

1. justification of the integration-by-parts formula for local weak solutions of Eq. (3.2),
2. derivation of a nonlinear ordinary differential inequality for a suitable energy function,
3. analysis of the ordinary differential inequality.

## 3.3 Formula of Integration by Parts: The Energy Relations

**Lemma 3.1** *Let* $G_\delta = B_\delta(x_0) \times \{t \in (0, \delta]\} \subset Q_T$. *Under the conditions of* Theorem 2.1 *every local weak solution of* Eq. (3.2) *in* $G_\delta$ *satisfies the energy relation*

$$\frac{1}{2} b(\rho, \theta)\Big|_{\theta=0}^{\theta=\tau} + E(\rho, \tau) + \int_0^{\tau} \int_{B_\rho} f\, u\, dx d\theta = - \int_0^{\tau} \int_{S_\rho} u\, |u|^{\gamma(x,t)} |\nabla u|^2\, dx dt \tag{3.6}$$

*for a.e.* $\rho \in (0, \delta)$ *and* $\tau \in (0, \delta]$.

*Proof* Let $u$ be a local weak solution of Eq. (3.2) in $G_\delta \subset Q_T$ with some $\delta > 0$. Let us denote

$$u^+ = \max\{u, 0\} \geq 0, \qquad u^- = \max\{-u, 0\} \geq 0, \qquad u = u^+ - u^-.$$

Fix an arbitrary $\varepsilon > 0$, introduce the functions

$$u_\varepsilon^\pm = \begin{cases} u^\pm & \text{if } u^\pm > \varepsilon, \\ \varepsilon & \text{if } u^\pm \leq \varepsilon, \end{cases}$$

and then set $u_\varepsilon = u_\varepsilon^+ - u_\varepsilon^-$. For every $\varepsilon > 0$ the functions $u_\varepsilon$ possess the properties [185, Chap. 2, §4]

$$\nabla u_\varepsilon = \begin{cases} \nabla u & \text{if } |u| > \varepsilon \\ 0 & \text{if } |u| < \varepsilon, \end{cases} \qquad u_{\varepsilon\, t} = \begin{cases} u_t & \text{if } |u| > \varepsilon \\ 0 & \text{if } |u| < \varepsilon. \end{cases}$$

Moreover, by virtue of (3.5) $\nabla u_\varepsilon \in L^2(G_\delta)$ because

$$\begin{aligned} \int_{G_\delta} |\nabla u_\varepsilon|^2\, dxdt &= \int_{G_\delta^{(\varepsilon)}} |\nabla u|^2\, dxdt \\ &\leq \varepsilon^{-\gamma^+} \int_{G_\delta} u^{\gamma(x,t)} |\nabla u|^2\, dxdt \leq C\, \varepsilon^{-\gamma^+} \end{aligned}$$

under the notation $G_\delta^{(\varepsilon)} = G_\delta \cap \{|u| > \varepsilon\}$. Given $\delta > 0$ and $\rho \in (0, \delta)$, we define the function $\zeta_k(r)$, $r = |x - x_0|$:

$$\zeta_k(r) = \begin{cases} 1 & \text{if } r \in \left[0, \rho - \frac{1}{k}\right], \\ k(\rho - r) & \text{if } r \in \left[\rho - \frac{1}{k}, \rho\right], \\ 0 & \text{if } r \in [\rho, \delta], \end{cases} \qquad k \in \mathbb{N},$$

and

$$\theta_{m,\tau}(t) = \begin{cases} 1 & \text{if } 0 \leq t < \tau - \frac{1}{m}, \\ m(\tau - t) & \text{if } \tau - \frac{1}{m} \leq t \leq \tau, \\ 0 & \text{if } t > \tau, \end{cases} \qquad m \in \mathbb{N}, \quad \tau \in (0, \delta]. \tag{3.7}$$

For every $\varepsilon > 0$ and all sufficiently big natural $k$, $m$ the function $\eta = u_\varepsilon\, \zeta_k(r)\, \theta_m(t)$ can be taken for the test-function in (3.4), which leads to the equality

$$\begin{aligned} \frac{m}{2} \int_{\tau - \frac{1}{m}}^{\tau} \int_{B_\rho} \zeta_k(x) u_\varepsilon^2\, dxdt + \int_{G_\delta^{(\varepsilon)}} \theta_{m,\tau}(t) (\zeta_k |u|^\gamma |\nabla u|^2 + u\, |u|^\gamma \nabla u \nabla \zeta_k + f \zeta_k u)\, dxdt \\ = \int_{B_\rho} \zeta_k(x)\, u_0\, u_\varepsilon(x, 0)\, dx. \end{aligned}$$

Letting $\varepsilon \to 0$ we obtain

$$\frac{m}{2} \int_{\tau - \frac{1}{m}}^{\tau} \int_{B_\rho} \zeta_k(x) u^2\, dxdt - \frac{1}{2} \int_{B_\delta} \zeta_k(x) u_0^2\, dx$$

$$+\int_0^\tau \int_{B_\rho} \theta_{m,\tau}(t)\zeta_k \left(|u|^\gamma |\nabla u|^2 + fu\right) dxdt = -J_k \tag{3.8}$$

with

$$J_k := \int_0^\delta \int_{B_\rho} u\,|u|^\gamma \nabla u \nabla \zeta_k dxdt.$$

By the Lebesgue differentiation theorem, for a.e. $\tau \in (0, \delta)$ there exists the limit

$$\lim_{m\to\infty} \frac{m}{2} \int_{\tau-\frac{1}{m}}^\tau \int_{B_\rho} \zeta_k(x) u^2\, dxdt = \frac{1}{2}\int_{B_\rho} \zeta_k(x) u^2(x,\tau)\, dx$$

and equality (3.8) takes the form

$$\frac{1}{2}\int_{B_\rho} \zeta_k(x) u^2(x,\tau)\, dx - \frac{1}{2}\int_{B_\delta} \zeta_k(x) u_0^2\, dx$$
$$+\int_0^\tau \int_{B_\rho} \zeta_k \left(|u|^\gamma |\nabla u|^2 + fu\right) dxdt = -J_k. \tag{3.9}$$

In the spherical coordinate system with the origin at the point $x_0$, the function $J_k$ becomes

$$\begin{aligned} J_k &= \int_0^\tau \int_{B_\rho} |u|^{\gamma(x,t)} \nabla u\, \nabla \zeta_k dxdt \\ &= k\int_0^\tau \int_{\rho-\frac{1}{k}<|x-x_0|<\rho} u\,|u|^\gamma\, \nabla u \cdot \frac{x-x_0}{r}\, dxdt, \\ &= k\int_{\rho-\frac{1}{k}}^\rho \left(\int_0^\tau \int_{\omega_n} u\,|u|^\gamma\, \nabla u \cdot \frac{x-x_0}{r} r^{n-1}\, d\omega dt\right) dr, \end{aligned}$$

where $\omega_n$ is the surface of the unit sphere in $\mathbb{R}^n$. In our assumptions on $u$

$$\int_0^\tau \int_{\omega_n} u\,|u|^\gamma\, |\nabla u| r^{n-1}\, d\omega dt \in L^1(0,\rho),$$

and it follows from the Lebesgue differentiation theorem that for a.e. $\rho \in (0, \delta)$ there exists

$$\begin{aligned} &\lim_{k\to\infty} k\int_0^\tau \int_{\rho-\frac{1}{k}<|x-x_0|<\rho} u\,|u|^\gamma\, \nabla u \cdot \frac{x-x_0}{|x-x_0|}\, dxdt \\ &\qquad = \int_0^\tau \int_{S_\rho} u\,|u|^\gamma\, \nabla u \cdot \mathbf{n}\, dSdt := I(\rho,\tau), \end{aligned}$$

where $\mathbf{n}$ denotes the exterior unit normal to $S_\rho(x_0)$. We may now pass in (3.9) to the limit as $k \to \infty$: for a.e. $\rho \in (0, \delta)$ and a.e. $\tau \in (0, \delta)$

$$\frac{1}{2}\int_{B_\rho} u^2 dx\Bigg|_{t=0}^{t=\tau} + \int_0^\tau \int_{B_\rho} (|u|^{\gamma(x,t)}|\nabla u|^2 + fu)dxdt$$
$$= -\int_0^\tau \int_{S_\rho} u\,|u|^{\gamma(x,t)}\,\nabla u \cdot \mathbf{n}\, dSdt \equiv -I(\rho,\tau). \tag{3.10}$$

□

**Lemma 3.2** *Let under the conditions of* Theorem 2.1 $\gamma^+ \le 0$. *Then the weak solution of problem* (2.4)–(2.5) *satisfies the energy relation: for a.e.* $\tau \in (0, T)$

$$\frac{1}{2}\int_\Omega u^2 dx\Bigg|_{t=0}^{t=\tau} + \int_0^\tau \int_\Omega |u|^{\gamma(x,t)}|\nabla u|^2\, dxdt = \int_0^\tau \int_\Omega f\,u\; dxdt. \tag{3.11}$$

*Proof* Consider the sequences

$$u_\varepsilon^+ = \max\{u;\ \varepsilon\} - \varepsilon, \qquad u_\varepsilon^- = \max\{-u;\ \varepsilon\} - \varepsilon.$$

Notice that since $\gamma(x,t) \le \gamma^+ \le 0$ and $|u| \le 1$

$$\begin{aligned}\int_{Q_T} |\nabla\, u_\varepsilon^\pm|^2 dxdt &= \int_{Q_T} |u|^{-\gamma(x,t)} \left(|u|^{\gamma(x,t)}|\nabla\, u_\varepsilon^\pm|^2\right) dxdt \\ &= \int_{Q_T^\varepsilon} |u|^{-\gamma(x,t)} \left(|u|^{\gamma(x,t)}|\nabla\, u|^2\right) dxdt \\ &\le \||u|^{\gamma/2}\nabla\, u\|^2_{2,Q_T},\end{aligned}$$

where $Q_T^\varepsilon = \{(x,t) \in Q_T : |u| > \varepsilon\}$. Let us consider the functions $u_\varepsilon^\pm \theta_{k,\tau}(t)$ with $\theta_{k,\tau}(t)$ defined in (3.7). Since $u_\varepsilon^\pm \in L^2(0,T; H_0^1(\Omega))$, $u_{\varepsilon t}^\pm \in L^2(0,T; H^{-1}(\Omega))$ and $\theta_{k,\tau}(T)) = 0$, these functions can be taken for the test function in Definition 2.1. Following the proof of Lemma 3.1, we may pass to the limit as $k \to \infty$ in the corresponding integral identities:

$$\frac{1}{2}\int_\Omega u\, u_\varepsilon^\pm dx\Bigg|_{t=0}^{t=\tau} + \int_{Q_\tau} |u|^{\gamma(x,t)}\left|\nabla u_\varepsilon^\pm\right|^2 dxdt = \pm\int_{Q_\tau} f u_\varepsilon^\pm\, dxdt.$$

Letting $\varepsilon \to 0$ we obtain the energy relations for the functions $u^+ = \max\{u;\ 0\}$ and $u^- = \max\{-u; 0\}$:

$$\frac{1}{2}\int_\Omega \left(u^\pm\right)^2 dx\Bigg|_{t=0}^{t=\tau} + \int_{Q_\tau} |u^\pm|^{\gamma(x,t)}\left|\nabla u^\pm\right|^2 dxdt = \pm\int_{Q_\tau} f u^\pm\, dxdt.$$

Summing these relations and taking into account the formulas

$$u^+ + u^- = |u|, \qquad u^+ - u^- = u,$$

$$u^2 = \left(u^+\right)^2 + \left(u^-\right)^2, \quad |\nabla u|^2 = |\nabla u^+|^2 + |\nabla u^-|^2,$$

we obtain (3.11). □

## 3.3.1 Ordinary Differential Inequality for the Energy Function

The energy relation (3.10) holds true for a.e. $t \in (0, T)$ and a.e. $\rho > 0$ such that $B_\rho(x_0) \subset \Omega$. If assumptions (3.1) are fulfilled, equality (3.10) can be specified as follows: for a.e. $\rho \in (0, \rho_0)$, $t \in (0, T)$

$$\frac{1}{2}\int_{B_\rho(x_0)} u^2(x,t)dx + \int_0^t \int_{B_\rho(x_0)} |u|^{\gamma(x,t)}|\nabla u|^2 dxdt = -I(\rho,t) \tag{3.12}$$

with

$$|I| \le \int_0^t \int_{S_\rho(x_0)} |u|^{\gamma(x,\theta)+1}|\nabla u|\, dSd\theta \equiv \Lambda.$$

According to (3.5) $|u|^{\gamma(x,t)} \ge |u|^{\gamma^+}$, and since $t \in (0, T)$ is arbitrary, we may write

$$E(\rho,t) \ge \frac{1}{2}\left(E(\rho,t) + E^+(\rho,t)\right) \ge \frac{1}{2}\left(\overline{E}(\rho) + \overline{E}^+(\rho)\right).$$

Formula (3.12) can now be continued as follows:

$$\overline{b}(\rho,t) + \overline{E}(\rho,t) + \overline{E}^+(\rho,t) \le 6\Lambda. \tag{3.13}$$

We proceed to evaluate $\Lambda$. Applying Hölder's inequality and using (3.5) we have:

$$\begin{aligned} \Lambda^2 &\le \overline{E}_\rho \int_0^t \int_{S_\rho(x_0)} |u|^{\gamma(x,t)+2} dSdt \\ &\le \overline{E}_\rho \int_0^t \int_{S_\rho(x_0)} |u|^{2+\gamma^-} dSdt \equiv \overline{E}_\rho\, \Theta^2. \end{aligned} \tag{3.14}$$

Let us apply to the function $|u|^{1+\gamma^+/2}$ the trace-interpolation inequality (1.62) with $p = \gamma = 2$: for a.e. fixed $t \in (0, T)$

$$\begin{aligned} &\int_{S_\rho} |u|^{2+\gamma^+} dS \\ &\quad \le C\left[\int_{B_\rho} |u|^{\gamma^+}|\nabla u|^2 dx + \rho^{-\beta}\left(b(\rho,t)\right)^{\frac{2+\gamma^+}{2}}\right]^\theta \left(b(\rho,t)\right)^{\frac{2+\gamma^+}{2}(1-\theta)} \end{aligned}$$

with the exponents

$$\theta = \frac{N\gamma^+ + 2}{N\gamma^+ + 4} \in (0, 1), \qquad \beta = \left(2 + \frac{N\gamma^+}{2}\right) > 1.$$

In our notation this inequality can be written as

$$\begin{aligned}\int_{S_\rho} |u|^{2+\gamma^+}\, dS &\le C\left[E_t^+(\rho, t) + \rho^{-\beta} b(\rho, t)(\overline{b}(\rho_0))^{\frac{\gamma^+}{2}}\right]^\theta (b(\rho, t))^{\frac{2+\gamma^+}{2}(1-\theta)} \\ &\le C_1 \rho^{-\beta\theta}\left[E_t^+(\rho, t) + b(\rho, t)\right]^\theta (b(\rho, t))^{\frac{2+\gamma^+}{2}(1-\theta)}\end{aligned}$$

with the constant

$$C_1 = C\left(\rho_0^\beta + \overline{b}(\rho_0)^{\frac{\gamma^+}{2}}\right)^\theta.$$

By Hölder's inequality

$$\int_{S_\rho(x_0)} |u|^{2+\gamma^-}\, dS \le \rho^\kappa \left(\int_{S_\rho(x_0)} |u|^{2+\gamma^+}\, dS\right)^{\frac{2+\gamma^-}{2+\gamma^+}},$$

with the exponent

$$\kappa = (N-1)\frac{\gamma^+ - \gamma^-}{2+\gamma^+} \ge 0.$$

Gathering the above inequalities we obtain the estimate

$$\begin{aligned}&\int_0^{t_0} \rho^\kappa \left(\int_{S_\rho(x_0)} |u|^{2+\gamma^+}\, dS\right)^{\frac{2+\gamma^-}{2+\gamma^+}} dt \\ &\le \int_0^{t_0} \rho^\kappa \left(C_1 \rho^{-\beta\theta}\left[E_t^+ + b\right]^\theta b^{\frac{2+\gamma^+}{2}(1-\theta)}\right)^{\frac{2+\gamma^-}{2+\gamma^+}} dt \\ &\le C_2\, \rho^{1-\nu} \int_0^{t_0} \left(C_1\left[E_t^+ + b\right]^\theta\, b^{\frac{2+\gamma^+}{2}(1-\theta)}\right)^{\frac{2+\gamma^-}{2+\gamma^+}} dt \\ &\le C_2 \rho^{1-\nu}\overline{b}^{(2+\gamma^-)(1-\theta)/2} \int_0^{t_0} \left(E_t^+ + b\right)^{\frac{\theta(2+\gamma^-)}{2+\gamma^+}} dt \\ &\le C_3 \rho^{1-\nu} t_0^\lambda \overline{b}^{(2+\gamma^-)(1-\theta)/2}\left(\overline{E}^+ + \overline{b}\right)^{\theta\frac{2+\gamma^-}{2+\gamma^+}} \\ &\equiv C_3 \rho^{1-\nu} t_0^\lambda \left(\overline{E}^+ + \overline{b}\right)^{2\mu} \\ &\le C_3 \rho^{1-\nu} t_0^\lambda \left(\overline{b} + \overline{E} + \overline{E}^+\right)^{2\mu}\end{aligned} \tag{3.15}$$

with the constants

$$C_2 = C_1^{\frac{2+\gamma^-}{2+\gamma^+}}, \qquad C_3 = C_2(1+t_0)^{\theta\frac{(2+\gamma^-)}{2(2+\gamma^+)}},$$

$$\begin{aligned}\nu &= 1-\kappa+\beta\theta\frac{2+\gamma^-}{2+\gamma^+} \\ &= -(N-1)\frac{\gamma^+-\gamma^-}{2+\gamma^+}+\left(2+\frac{N\gamma^+}{2}\right)\frac{N\gamma^++2}{N\gamma^++4}\frac{2+\gamma^-}{2+\gamma^+},\end{aligned}$$

$$\mu = \frac{2+\gamma^-}{4}(1-\theta)+\frac{\theta}{2}\frac{2+\gamma^-}{2+\gamma^+} = \frac{2+\gamma^-}{2}\left(\frac{(1-\theta)}{2}+\frac{\theta}{2+\gamma^+}\right),$$

$$\lambda = \frac{1}{4}\left(1-\frac{2+\gamma^-}{2+\gamma^+}\theta\right) > 0.$$

Let us claim that

$$\nu > 0, \qquad \mu \in \left(\frac{1}{2}, 1\right) \text{ in } G_{\delta_0}. \tag{3.16}$$

Since $\gamma(x,t) \in C^0(\overline{Q_T})$, $\gamma(x_0,0) > 0$, and condition (3.16) is true for $\gamma^+ = \gamma^- = \gamma(x_0,0)$, we may always choose $\delta_0$ so small that (3.16) is fulfilled in $G_{\delta_0}$ by continuity.

Gathering (3.13)–(3.15) and simplifying, we arrive at the nonlinear ordinary differential inequality

$$\left(\overline{b}(\rho)+\overline{E}(\rho)+\overline{E}^+(\rho)\right)^{1-\mu} \le C_3\rho^{(1-\nu)/2}t_0^{\lambda/2}\overline{E}_\rho^{1/2}. \tag{3.17}$$

Keeping on the left-hand side only the term $\overline{E}$ and dropping the nonnegative terms $\overline{E}^+$ and $\overline{b}$, we obtain the simplified inequality

$$\overline{E}^{1-\sigma} \le C_4 t_0^{\lambda}\rho^{1-\nu}\overline{E}_\rho \tag{3.18}$$

with $C_4 = C_3^2$ and

$$0 < 1-\sigma = 2(1-\mu) < 1. \tag{3.19}$$

### 3.3.2 Analysis of the Differential Inequality

Integration of (3.18) over the interval $(\rho, \rho_0)$ leads to the inequality

$$0 \le \overline{E}^{\sigma}(\rho) \le \overline{E}^{\sigma}(\rho_0) - C_5 t_0^{-\lambda}(\rho_0^{\nu}-\rho^{\nu}), \qquad C_5 = \frac{\sigma}{\nu C_4}.$$

Since $\overline{E}$ is nonnegative by definition, this inequality entails

$$\overline{E} = \sup_{0 \le t \le t_0} E(\rho, t) = \int_0^{t_0} \int_{B_\rho} u^{\gamma(x,t)} |\nabla u|^2 dx dt = 0,$$

provided that

$$\rho^\nu \le \rho^\nu(t) \le \rho_0^\nu - \frac{t^\lambda}{C_5} \overline{E}^C (\rho_0), \quad t \in [0, t_0].$$

By virtue of (3.17) we also have

$$\overline{b}(\rho) = \sup_{0 \le t \le t_0} b(\rho, t) = 0,$$

and the conclusion follows:

$$u(x,t) = 0 \quad \text{for } x \in B_{\rho(t)}(x_0), t \in [0, t_0],$$

The proof of Theorem 3.1 is completed.

## 3.4 The Waiting Time Effect

Let us assume that in addition to condition (3.1)

$$\Sigma \equiv b(\rho, 0) + \int_0^t \int_{B_\rho(x_0)} f^2 dx dt \le \varepsilon(\rho - \rho_0)_+^{1/\sigma}, \quad \varepsilon = const > 0, \qquad (3.20)$$

for $\rho \in [\rho_0, \rho_1]$ with the constant $\sigma \in (0, 1)$ defined in (3.19). Inequality (3.20) controls the rate of vanishing of the right-hand side $f(x,t)$ and the initial datum $u_0(x)$ near the lateral boundary of the cylinder $B_{\rho_0}(x_0) \times (0, t_0)$ and the ball $B_{\rho_0}(x_0)$.

**Theorem 3.2** (The waiting time property) *Let the conditions of* Theorem 3.1 *be fulfilled. Assume that condition* (3.20) *holds. If a local weak solution $u(x,t)$ of* Eq. (3.2) *has a finite energy,*

$$E_1 = \sup_{0 \le t \le \delta_0} b(\rho_1, \delta_0) + E(\rho_1, \delta_0) < \infty,$$

*then there exist positive constants $\varepsilon_* > 0$ and $t^* \le t_0 \le T$ such that for every $\varepsilon \in (0, \varepsilon^*]$*

$$u(x,t) = 0 \quad in \quad B_{\rho_0}(x_0) \times [0, t^*].$$

*Proof* The energy relation (3.6) with $\delta = \rho > \rho_0$ has the form

$$\frac{1}{2}\int_{B_\rho(x_0)} u^2(x,t)dx + \int_0^t \int_{B_\rho(x_0)} |u|^{\gamma(x,t)}|\nabla u|^2\,dxdt = -I(\rho,t) + J(\rho,t), \tag{3.21}$$

where

$$J(\rho,t) = \frac{1}{2}\int_{B_\rho(x_0)} u^2(x,0)dx - \int_0^t \int_{B_\rho(x_0)} f u\,dxdt$$

and $I(\rho,t)$ is defined in (3.10). Repeating the estimates derived in the proof of Theorem 3.1, we arrive at the inequality

$$\overline{b}(\rho) + \overline{E}(\rho) + \overline{E}^+(\rho) \le C_3 \rho^{-\frac{\nu-1}{2}} t_0^{\frac{\lambda}{2}} \left(\overline{E}_\rho\right)^{\frac{1}{2}} \left(\overline{E}^+ + \overline{b}\right)^\mu + |J|. \tag{3.22}$$

The integral $J$ is estimated in the following way:

$$2|J| \le b(\rho,0) + \int_0^t \int_{B_\rho(x_0)} f^2 dxdt + t_0\,\overline{b}(\rho), \qquad t \le t_0.$$

For small $t_0$ (3.22) yields the inequality

$$\overline{b}(\rho) + \overline{E}(\rho) + \overline{E}^+(\rho) \le C_4 \left(\rho^{-\frac{\nu-1}{2}} t_0^{\frac{\lambda}{2}} \left(\overline{E}_\rho\right)^{\frac{1}{2}} \left(\overline{E}^+ + \overline{b}\right)^\mu + \Sigma\right).$$

By Young's inequality

$$\overline{b}(\rho) + \overline{E}(\rho) + \overline{E}^+(\rho) \le C_5 \left(\left(\rho^{-\frac{\nu-1}{2}} t_0^{\frac{\lambda}{2}} \left(\overline{E}_\rho\right)^{\frac{1}{2}}\right)^{\frac{1}{1-\mu}} + \Sigma\right),$$

whence

$$\left(\overline{b}(\rho) + \overline{E}(\rho) + \overline{E}^+(\rho)\right)^{2(1-\mu)} \le C_6 \left(\rho^{-(\nu-1)} t_0^{\lambda} \overline{E}_\rho + \Sigma^{2(1-\mu)}\right).$$

Set $2(1-\mu) = 1-\sigma$. Plugging (3.20) into the last inequality we arrive at the nonhomogeneous ordinary differential inequality for $\overline{E}$:

$$\overline{E}^{1-\sigma} \le C \left(t_0^{\nu-1} \rho_0^{-\lambda} \overline{E}_\rho + \varepsilon^{2(1-\mu)} (\rho-\rho_0)_+^{\frac{1-\sigma}{\sigma}}\right). \tag{3.23}$$

Let $t^*$ be chosen from the condition

$$A^{1-\sigma} = C\,\rho_0^{-\lambda}(t^*)^{\nu-1}\frac{A}{\sigma} + C\varepsilon^{1-\sigma}$$

with a constant

$$A > A_0 = \max\left\{C\varepsilon^{1-\sigma}, E_1\,(\rho_1-\rho_0)^{1/\sigma}\right\}.$$

It is easy to check that the function

$$\mathscr{E}(\rho) \equiv A\,(\rho - \rho_0)^{1/\sigma}$$

is a solution of the nonlinear ODE

$$\mathscr{E}^{1-\sigma} = C\left(t_0^{\nu-1}\rho_0^{-\lambda}\mathscr{E}' + \varepsilon^{2(1-\mu)}(\rho-\rho_0)_+^{\frac{1-\sigma}{\sigma}}\right), \qquad \mathscr{E}(\rho_1) \geq E_1.$$

The function $\mathscr{E}(\rho)$ is a majorant for $\overline{E}(\rho)$, and the result follows:

$$\overline{E}(\rho) \leq A(\rho-\rho_0)_+^{\frac{1}{\sigma}}, \qquad \rho \in [\rho_0, \rho_1].$$

□

## 3.5 Asymptotic Behavior for Large *t*

**Theorem 3.3** *Let $u(x,t)$ be a weak solution of problem* (2.4)–(2.5). *Let*

$$\gamma^+ > 0. \tag{3.24}$$

(i) *If in* Eq. (3.2) $f(x,t) \equiv 0$, *then the solution $u(x,t)$ satisfies the estimate*

$$y^{\mu-1}(t) \leq \frac{y_0^{\mu-1}}{1 + t\,C_0(\mu-1)y_0^{\mu-1}}$$

*where*

$$y(t) = \|u(\cdot,t)\|_{2,\Omega}^2, \quad y_0 = y(0), \quad \mu = 1 + \frac{\gamma^+}{2} > 1,$$

*and $C_0$ is an absolute constant depending on $\gamma^+$, $\Omega$ and $n$.*

(ii) *If $f \not\equiv 0$, there exist positive constants $K$ and $B$, depending on $\Omega$, $\gamma^+$ and $n$, such that if*

$$||f(\cdot,t)||_{2,\Omega} \leq \frac{K}{(Bt+1)^{(2\mu-1)/2(\mu-1)}},$$

*then*

$$y(t) \leq \frac{y_0}{(Bt+1)^{1/(\mu-1)}}.$$

*Remark 3.2* If in the conditions of Theorem 2.1

$$-1 < \gamma^- \leq \gamma(x,t) \leq \gamma^+ \leq 0,$$

in condition (3.24) we may take for $\gamma^+$ any positive number.

*Proof* Let $u(x,t)$ be a weak solution of problem (3.2) and (3.3). For every $t \in (0,T)$ the energy relation (3.11) holds:

$$\frac{1}{2}\int_\Omega u^2 dx\Big|_{t=t_1}^{t=t_2} + \int_{t_1}^{t_2}\int_\Omega |u|^{\gamma(x,t)}|\nabla u|^2\, dxdt = \int_{t_1}^{t_2}\int_\Omega f u\, dxdt. \tag{3.25}$$

By the method of construction, every weak solution satisfies

$$\int_\Omega (|u|^{\gamma(x,t)}|\nabla u|^2 - fu)dx \in L^1(0,T),$$

which allows us to apply to (3.25) the Lebesgue differentiation theorem: for a.e. $t \in (0,T)$

$$\frac{1}{2}\frac{d}{dt}\int_\Omega u^2(x,t)dx + \int_\Omega |u(x,t)|^{\gamma(x,t)}|\nabla u(x,t)|^2 dx = -\int_\Omega fudx. \tag{3.26}$$

Recall that the weak solutions are assumed to satisfy condition (3.5). Applying the embedding theorem [147, Theorem 7.10], we have: for every

$$1 \le p \le \begin{cases} \frac{2n}{n-2} & \text{if } n \ge 2, \\ \infty & \text{if } n = 1 \end{cases}$$

there is a constant $C \equiv C(p, n, \gamma^+, \Omega)$ such that

$$\begin{aligned}\left(\int_\Omega |u|^{p(\gamma^+ +2)/2}dx\right)^{2/p} &\le C\int_\Omega |\nabla(u^{(\gamma^+ +2)/2})|^2 dx \\ &= C\int_\Omega |u|^{\gamma^+}|\nabla u|^2 dx \\ &\le C\int_\Omega |u|^{\gamma(x,t)}|\nabla u|^2 dx.\end{aligned}$$

By Hölder's inequality

$$\begin{aligned} y(t) \equiv \int_\Omega u^2 dx &\le C\left(\int_\Omega u^{p(\gamma^+ +2)/2}dx\right)^{4/p(2+\gamma^+)} \\ &\le C_0\left(\int_\Omega |u|^{\gamma(x,t)}|\nabla u|^2 dx\right)^{2/(2+\gamma^+)}\end{aligned}$$

with a new constant $C_0 = C_0(p, \gamma^+, \Omega)$. Substituting this inequality into (3.26), we obtain the differential inequality

$$y' + C_0 y^{\mu}(t) \le \Lambda, \tag{3.27}$$

with

$$\Lambda = \left| \int_\Omega f u dx \right|, \qquad \mu = \frac{\gamma^+ + 2}{2} > 1.$$

(i) Let $f \equiv 0$. Integrating inequality (3.27), we come to the following estimate on the rate of vanishing of the $L^2(\Omega)$-norm of the weak solution:

$$y^{\mu-1}(t) \le \frac{y_0^{\mu-1}}{1 + tC_0(\mu - 1)y_0^{\mu-1}} \to 0 \quad \text{when}\, t \to \infty,$$

where $y_0 = \|u(x,0)\|^2_{2,\Omega}$.

(ii) Let $f \not\equiv 0$. By Hölder's inequality

$$\Lambda \le \sqrt{y(t)}\, \|f(\cdot,t)\|_{2,\Omega},$$

whence the nonhomogeneous ordinary differential inequality:

$$y' + C_0 y^\mu(t) \le \sqrt{y}\|f(\cdot,t)\|_{2,\Omega}.$$

Let us introduce the new sought function $u(t) = \sqrt{y(t)}$. This function satisfies the ordinary differential inequality

$$u' + \frac{1}{2} C_0\, u^{2\mu-1} \le \frac{1}{2}\|f(\cdot,t)\|_{2,\Omega}.$$

Let

$$\|f(\cdot,t)\|_{2,\Omega} \le \frac{K}{(Bt+1)^{(2\mu-1)/2(\mu-1)}}$$

with the constants

$$K < C_0\, y_0^{(2\mu-1)/2}, \quad B = (C_0\, y_0^{(2\mu-1)/2} - K)(\mu - 1)y_0^{-1/2} > 0.$$

The function

$$w(t) = \sqrt{y_0}\,(Bt+1)^{-\frac{1}{2(\mu-1)}}$$

solves the nonlinear ODE

$$w' + \frac{1}{2} C w^{2\mu-1}(t) = 0, \quad w(0) = \sqrt{y_0},$$

and majorates the function $u$: $u(t) \le w(t)$. Reverting to the function $y(t)$ we get the estimate

$$y(t) = \|u(\cdot,t)\|^2_{2,\Omega} \le y_0\,(Bt+1)^{-\frac{1}{(\mu-1)}}.$$

□

## 3.6 Vanishing in a Finite Time

**Theorem 3.4** *Let the conditions of* Theorem 2.1 *be fulfilled and $u(x,t)$ be a weak solution of problem* (2.4)–(2.5). *Assume that*

$$-1 < \gamma^- \le \gamma(x,t) \le \gamma^+ < 0.$$

(i) *If $f(x,t) \equiv 0$, then $u(x,t)$ vanishes in the finite time:*

$$u(x,t) = 0 \quad in\, \Omega \ for\, t \ge t^*,$$

*where $t^*$ depends only on $\|u_0\|_{2,\Omega}$, $\gamma^+$, and $\Omega$.*

(ii) *Let $f \not\equiv 0$. There exist $\varepsilon' > 0$ and $t' > t^*$ such that if*

$$\|f(\cdot,t)\|_{2,\Omega} \le \varepsilon \left[1 - \frac{t}{t_f}\right]_+^{-(1+\gamma^+)/\gamma^+} \tag{3.28}$$

*with $t_f \ge t'$ and $0 < \varepsilon < \varepsilon'$. Then*

$$u(x,t) \equiv 0 \qquad for\, t \ge t_f.$$

*Proof* Arguing as in the previous section, for the function $y(t) = \|u(\cdot,t)\|^2_{2,\Omega}$ we derive the ordinary differential inequality

$$y'(t) + Cy^{\mu}(t) \le \Lambda \tag{3.29}$$

with the right-hand side and the exponent

$$\Lambda = \left|\int_\Omega f\,u\,dx\right|, \qquad \mu = 1 + \frac{\gamma^+}{2} < 1,$$

and with a finite constant $C$ depending on the constant in the embedding theorem.

(i) Let $f \equiv 0$. A straightforward integration of (3.29) leads to the inequality

$$0 \le y^{1-\mu}(t) \le y^{1-\mu}(0) - C(1-\mu)t,$$

whence

$$y(t) \equiv 0 \quad \text{for}\, t \ge t^* = \frac{y^{1-\mu}(0)}{C(1-\mu)} < \infty. \tag{3.30}$$

It is then necessary that $u(x,t) = 0$ a.e. in $\Omega$ for $t \ge t^*$.

(ii) For $f \not\equiv 0$ the ordinary differential inequality for $y(t)$ becomes

$$y' + Cy^{\mu} \le \sqrt{y}\|f(\cdot,t)\|_{2,\Omega}.$$

Let us introduce the function $u(t) = \sqrt{y(t)}$ and write the previous inequality in the form

$$u' + \frac{1}{2} C u^{\nu}(t) \le \frac{1}{2} \|f(\cdot, t)\|_{2,\Omega}, \qquad \nu = 2\mu - 1 = \gamma^+ + 1 < 1, \tag{3.31}$$

where $f$ satisfies condition (3.28). Let us consider the ordinary differential equation

$$v' + \frac{1}{2} C\, v^{\nu}(t) = \frac{\varepsilon}{2} \left[1 - \frac{t}{t_f}\right]_+^{-(1+\gamma^+)/\gamma^+}. \tag{3.32}$$

The solutions of Eq. (3.32) are majorants for $u(t)$. It is easy to check that the function

$$u(t) = u(0) \left[1 - \frac{t}{t_f}\right]_+^{-1/\gamma^+}$$

satisfies (3.32), provided that the parameters $u(0)$, $\varepsilon$, and $t_f$ are chosen according to the condition

$$-\frac{1}{1-\nu} u(0) \frac{1}{t_f} + \frac{1}{2} C u^{-(1+\gamma^+)/\gamma^+}(0) - \frac{\varepsilon}{2} \ge 0. \tag{3.33}$$

Given $u(0)$, we fulfill this condition increasing $t_f$ and decreasing $\varepsilon$. □

*Remark 3.3* The effect of vanishing in a finite time is guaranteed by condition (3.33) which connects three parameters: $\varepsilon$—the "intensity" of the right-hand side (the source) in Eq. (3.2), $t_f$—the moment when this source vanishes, and $\|u_0\|_{2,\Omega}$. The assertion of Theorem 3.4 remains true if one of the three parameters is given whereas the other two are varied in order to satisfy condition (3.33).

## 3.7 A Limit Case in the Finite Time Vanishing: $\gamma^+ = 0$

The condition $\gamma^+ < 0$ plays a crucial role in the proof of Theorem 3.4. However, this condition is sufficient but not necessary for the finite time extinction. A solution of problem (2.4)–(2.5) may extinct in a finite time even if $\gamma^+ = \sup \gamma(x, t) = 0$. Let us consider the following problem:

$$\begin{cases} u_t - \left(u^{\gamma(x)} u_x\right)_x = 0 & \text{in } Q_T = (0, 1) \times (0, T), \\ u(0, t) = u(1, t) = 0, & u(x, 0) = u_0(x) \ge 0, \end{cases} \tag{3.34}$$

with the exponent $\gamma \equiv \gamma(x)$ satisfying the conditions

$$\begin{cases} -1 < \gamma^- \leq \gamma(x) \leq \gamma^+ = 0 \quad \text{for } x \in [0,1], \\ \gamma(x_0) = 0 \text{ where either } x_0 = 0, \text{ or } x_0 = 1, \\ (\gamma'(x))^2 \geq \gamma_0^2 > 0, \quad \gamma''(x) \leq 0. \end{cases} \tag{3.35}$$

By Theorem 2.1, under these conditions on $\gamma(x)$ problem (3.34) has a nonnegative weak solution $u(x,t)$ for every nonnegative $u_0 \in L^2(0,1)$.

**Theorem 3.5** *Let conditions* (3.35) *be fulfilled. Assume that* $u_0 \in L^2(0,1)$ *and* $0 \leq u_0 \leq 1$. *Then every nonnegative weak solution* $u(x,t)$ *of problem* (3.34) *vanishes in the finite time:*

$$u(x,t) \equiv 0 \; for \; x \in [0,1] \, and \; t \geq t_* = -\frac{8}{\gamma_0^2} \frac{1}{\ln y(0)}.$$

*Proof* The nonnegative weak solution of problem (3.34) satisfies the energy relation (3.26)

$$\frac{1}{2}\frac{d}{dt}\int_0^1 u^2\,dx + \int_0^1 u^{\gamma(x)} u_x^2\,dx = 0, \tag{3.36}$$

whence

$$y(t) \equiv \int_0^1 u^2(x,t)\,dx \leq y(0) < 1.$$

Moreover, $0 \leq u(x,t) \leq 1$ in $Q_T$ by vertue of Theorem 2.1. Let us introduce the function

$$\Phi(x,t) \equiv \Phi[u(x,t)] = \int_0^{u^{1+\gamma(x)/2}} \tau \, \ln \tau \, d\tau \leq 0, \quad \Phi(0,t) = \Phi(1,t) = 0. \tag{3.37}$$

A straightforward calculations lead to the formulas

$$\left(1+\frac{\gamma}{2}\right) u^{\gamma}\, u_x^2 = \left(u^{1+\gamma/2}\right)_x^2 - \gamma'\, u^{1+\gamma/2} \left(u^{1+\gamma/2}\right)_x \ln u + \frac{(\gamma')^2}{4} u^{2+\gamma} \ln^2 u,$$

$$\Phi'_x = \left(1+\frac{\gamma}{2}\right) \left(u^{1+\gamma/2}\right)_x u^{1+\gamma/2} \ln u.$$

Gathering them, we may represent

$$u^{\gamma}\, u_x^2 = \frac{1}{(1+\gamma/2)^2} \left[ \left(u^{1+\gamma/2}\right)_x^2 + \frac{(\gamma')^2}{4} u^{2+\gamma} \ln^2 u - \frac{\gamma'}{1+\gamma/2} \Phi'_x \right].$$

Since $\gamma(x)$ is convex and $\Phi \leq 0$, we also have that

$$\int_0^1 \Phi'_x \frac{\gamma'}{(1+\gamma/2)^3}\,dx = -\int_0^1 \Phi \left(\frac{\gamma'}{(1+\gamma/2)^3}\right)' dx \leq 0.$$

Writing the energy relation in the form

$$\frac{1}{2}\frac{d}{dt}\int_0^1 u^2\,dx + \int_0^1 \frac{1}{(1+\gamma/2)^2}\left[\left(u^{1+\gamma/2}\right)_x^2 + \frac{(\gamma')^2}{4}\,u^{2+\gamma}\,\ln^2 u\right]dx$$
$$+\int_0^1 \left(\frac{\gamma'}{(1+\gamma/2)^3}\right)'\,\Phi\,dx = 0$$

and then dropping the nonnegative terms, we arrive at the inequality

$$\frac{1}{2}\frac{d}{dt}\int_0^1 u^2(x,t)dx + \int_0^1 \frac{1}{(1+\gamma/2)^2}\,\frac{(\gamma')^2}{4}\,u^{2+\gamma}\,\ln^2 u\,dx \le 0.$$

The assumptions on $\gamma(x)$ and $u(x,t)$ allow one to continue this inequality as follows:

$$\frac{1}{2}\frac{d}{dt}\int_0^1 u^2(x,t)dx + \frac{\gamma_0^2}{16}\int_0^1 u^2\,\ln^2 u^2\,dx \le 0. \tag{3.38}$$

Let us consider now the function $\Psi(s) \equiv s\,\ln^2 s$. Since

$$\Psi'(s) = \ln^2 s + 2\ln s, \qquad \Psi''(s) = \frac{2}{s}\,(\ln s + 1)\,,$$

this function is convex on the interval $s \in (0, 1/e)$. By Jensen's inequality

$$\Psi\left(\int_0^1 u^2\,dx\right) \equiv \left(\int_0^1 u^2\,dx\right)\,\ln^2\left(\int_0^1 u^2\,dx\right)$$
$$\le \int_0^1 \Psi(u^2)\,dx \equiv \int_0^1 u^2\,\ln^2\,u^2\,dx,$$

and (3.38) becomes

$$\frac{1}{2}\frac{d}{dt}\int_0^1 u^2\,dx + \frac{\gamma_0^2}{16}\left(\int_0^1 u^2\,dx\right)\,\ln^2\left(\int_0^1 u^2\,dx\right) \le 0.$$

Thus, the function $y(t) = \|u(\cdot,t)\|^2_{2,(0,1)}$ satisfies the ordinary differential inequality

$$y' + \frac{\gamma_0^2}{8}\,y\,\ln^2 y \le 0, \qquad y(0) < 1. \tag{3.39}$$

Any nonnegative solution of the ODE

$$z' + \frac{\gamma_0^2}{8}\,z\,\ln^2 z = 0, \qquad z(0) = y(0) < 1,$$

is a majorant for the function $y(t)$. The function $z(t)$ can be constructed explicitly:

$$z(t) = \exp\left(\frac{1}{\dfrac{1}{\ln\, y(0)} + \dfrac{\gamma_0^2}{8}\,t}\right).$$

This is a monotone decreasing function that vanishes at the instant

$$t_* = -\frac{8}{\gamma_0^2}\frac{1}{\ln\, y(0)} \in (0, \infty).$$

Since $z(t) \geq y(t) \geq 0$, the assertion follows. □

*Remark 3.4* If $\gamma(x) \equiv 0$ on an interval $I = [a, b] \subset [0, 1]$, the assertion of Theorem 3.5 is false unless we impose some additional assumptions on the data. Indeed, let us assume the contrary: $\gamma(x) \in (-1, 0]$, $\gamma(x) \equiv 0$ for $x \in [a, b] \subset [0, 1]$, and $u(x, t) \equiv 0$ for all $x \in [0, 1], t \in [t_*, T]$. In the rectangle $(a, b) \times (0, t_*)$ the function $u(x, t)$ can be considered as a nonnegative solution of the Dirichlet problem for the heat equation, so that the assumption $u(x, t_*) \equiv 0$ on $[a, b]$ contradicts the strong maximum principle.

## 3.8 Nonlinearity of Other Type: An Example

Similar arguments apply to derive localization properties of solutions to equations with another nonlinear structure. As an example, let us consider a motion of an ideal barotropic gas in a porous medium. Denote by $\rho$ the gas density, by $V$ the velocity, and let $p$ stands for the pressure. The motion is governed by the mass conservation law,

$$\rho_t + \operatorname{div}(\rho V) = 0, \tag{3.40}$$

the Darcy law, which for the anisotropic nonhomogeneous medium has the form

$$V = -k(x)\nabla\, p \quad \text{with a given matrix}\, k(x), \tag{3.41}$$

and the equation of state $p = P(\rho)$. The last dependence is usually given by the expression $P(s) = \mu\, s^{\gamma}$ with some known constants $\mu$ and $\gamma$. Plugging the Darcy law (3.41) and the state equation into the mass conservation law (3.40), we arrive at the semilinear parabolic equation for the density $\rho$:

$$\rho_t = \frac{\mu\,\gamma}{1+\gamma}\operatorname{div}\left(k(x)\nabla\,\rho^{1+\gamma}\right). \tag{3.42}$$

Let us assume now that the exponent $\gamma$ is not a constant but may vary from point to point and is a given function of the independent variables $(x, t)$. This hypothesis drastically changes the mathematical model of the process: the continuity equation (3.40) for the density $\rho$ transforms into the equation

$$\rho_t = \mu \operatorname{div} \left(k(x)\, \rho \nabla\, \rho^{\gamma(x,t)}\right)$$

and can be formally written in the form

$$\rho_t = \mu \operatorname{div} \left(k(x) \gamma\, \rho^{\gamma} \nabla\, \rho + (\rho\, \ln\, \rho)\, k(x) \cdot \nabla \gamma\right). \tag{3.43}$$

Since Eq. (3.43) is not of the form (3.2), the prior localization results are not applicable. In the special situation when $\gamma$ is independent of $x$, $\gamma(x, t) = \gamma(t)$, Eq. (3.43) simplifies and reads

$$\rho_t = \mu \gamma(t) \operatorname{div} \left(k(x) \rho^{\gamma(t)} \nabla \rho\right). \tag{3.44}$$

Equation (3.44) is a partial case of Eq. (2.4). In the other special case, $\gamma(x, t) = \gamma(x)$, from relations (3.40) and (3.41) and the inverse of the state equation, $\rho = (\mu^{-1} p)^{\frac{1}{\gamma(x)}}$, we derive the following equation for the pressure $p$:

$$\partial_t \left(p^{\frac{1}{\gamma(x)}}\right) = \mu^{\frac{1}{\gamma(x)}} \operatorname{div} \left(k(x) \left(\frac{p}{\mu}\right)^{\frac{1}{\gamma(x)}} \nabla\, p\right). \tag{3.45}$$

This equation is not of the form (2.4) or (2.46), and cannot be transformed into an equation of one of these types by means of a change of the unknown function. Although a straightforward application of the already established results is impossible, the method of local energy estimates proves to be versatile enough to deal with this new situation. Let us consider the problem

$$\partial_t \left(p^{\frac{1}{\gamma(x)}}\right) = \mu \operatorname{div} \left(k(x) p^{\frac{1}{\gamma(x)}} \nabla\, p\right) \tag{3.46}$$

$$p = 0 \text{ on } \Gamma_T, \qquad p(x, 0) = p_0(x) \geq 0 \text{ in } \Omega. \tag{3.47}$$

We assume that the variable exponent $\gamma$ satisfies the inequalities

$$1 \leq \gamma_- \leq \gamma(x) \leq \gamma_+ < \infty. \tag{3.48}$$

For the sake of simplicity of presentation, let us also assume that in Eq. (3.46) $k = \mu = 1$. Let $p(x, t)$ be a bounded solution of problem (3.46)–(3.47):

$$0 \leq p(x, t) \leq \sup_{\Omega} \rho_0^{\gamma(x,0)} = p^+ \tag{3.49}$$

(see Lemma 2.1). The further proceeding is formal, we tacitly assume that the weak solutions of problem (3.46)–(3.47) are as smooth as is needed to justify all the required operations.

### *3.8.1 Asymptotic Behavior for Large t*

Let us multiply Eq. (3.46) by $p$, integrate over the sliced cylinder $(t, t+h) \times \Omega$ and integrate by parts in $x$. Dividing the resulting relation by $h$ and letting $h \to 0$ we arrive at the energy equality: for a.e. $t \in (0, T)$

$$\frac{d}{dt}\left(\int_\Omega \frac{p^{\frac{1+\gamma}{\gamma}}}{1+\gamma}\,dx\right) + \int_\Omega p^{\frac{1}{\gamma}}\,|\nabla p|^2\,dx = 0. \tag{3.50}$$

Due to (3.49), we may estimate

$$p^{\frac{1}{\gamma}} = \left(\frac{p}{p^+}\right)^{\frac{1}{\gamma}} (p^+)^{\frac{1}{\gamma}} \le p^{\frac{1}{\gamma^+}}\,(p^+)^{\frac{1}{\gamma}-\frac{1}{\gamma^+}}.$$

It follows that there exists a finite constant $C \equiv C(p^+, \gamma^\pm) \ge 1$ such that $\frac{1}{C} p^{\frac{1}{\gamma_-}} \le p^{\frac{1}{\gamma(x)}} \le C p^{\frac{1}{\gamma_+}}$, whence

$$\int_\Omega \frac{p^{\frac{1+\gamma}{\gamma}}}{1+\gamma}\,dx \le \frac{1}{1+\gamma^-}\int_\Omega p^{1+\frac{1}{\gamma}}\,dx \le C\int_\Omega p^{1+\frac{1}{\gamma^+}}\,dx \le C\left(\int_\Omega p^{2\frac{1+2\gamma^-}{2\gamma^-}}\,dx\right)^{\frac{1+\frac{1}{\gamma^+}}{2+\frac{1}{\gamma^-}}}$$

with a constant $C = C(p^+, \gamma^\pm, |\Omega|)$. By the Poincaré inequality

$$\int_\Omega \left(p^{\frac{1+2\gamma^-}{2\gamma^-}}\right)^2 dx \le C\int_\Omega p^{\frac{1}{\gamma^-}}\,|\nabla u|^2\,dx \le C\int_\Omega p^{\frac{1}{\gamma(x)}}\,|\nabla u|^2\,dx.$$

Let us introduce the function

$$Y(t) := \int_\Omega \frac{p^{\frac{1+\gamma}{\gamma}}}{1+\gamma}\,dx.$$

Combining the above inequalities, we estimate

$$Y^\mu(t) \le \int_\Omega p^{\frac{1}{\gamma_-}}\,|\nabla p|^2\,dx \le C\int_\Omega p^{\frac{1}{\gamma}}\,|\nabla p|^2\,dx, \qquad \mu = \frac{2+\frac{1}{\gamma^-}}{1+\frac{1}{\gamma^+}} > 1.$$

Equality (3.50) leads to the differential inequality for the function $Y(t)$:

$$Y'(t) + CY^{\mu}(t) \le 0 \quad \text{for } t > 0, \qquad Y(0) = \int_{\Omega} \frac{p_0^{\frac{1+\gamma}{\gamma}}}{1+\gamma} dx. \tag{3.51}$$

A straightforward integration gives the estimate

$$Y(t) \le \frac{Y(0)}{(1 + Ct(\mu - 1)\, Y^{\mu-1}(0))^{\frac{1}{\mu-1}}}. \tag{3.52}$$

**Proposition 3.1** *If $p$ is a bounded weak solution of problem* (3.46)–(3.48) *satisfying the energy equality* (3.50), *then the energy $Y(t)$ obeys inequality* (3.52).

### 3.8.2 Finite Speed of Propagation

Let $B_\rho \subset \Omega$ be an arbitrary ball. Assume that

$$p(x, 0) = 0 \text{ for } \rho \le \rho_0. \tag{3.53}$$

Arguing by analogy with Lemma 3.1 we derive the local energy relation

$$\int_{B_\rho} \frac{p^{\frac{1+\gamma}{\gamma}}}{1+\gamma} dx + \int_0^t \int_{B_\rho} p^{\frac{1}{\gamma}} |\nabla p|^2 dx d\tau = \int_0^t \int_{S_\rho} p^{\frac{1}{\gamma}+1} \nabla p \cdot \nu dS d\tau. \tag{3.54}$$

Let us introduce the parameter

$$\beta = 2\frac{\frac{\gamma_-}{\gamma} + \gamma_- - 1}{1 + \gamma_-}$$

and the new variable $u = p^{\frac{1+\gamma_-}{2\gamma_-}}$. It is straightforward to check that

$$p = u^{\frac{2\gamma_-}{1+\gamma_-}}, \qquad p^{\frac{1+\gamma_-}{\gamma_-}} = u^2,$$
$$p^{\frac{1}{\gamma}} |\nabla p|^2 = \left(\frac{2\gamma_-}{1+\gamma_-}\right)^2 u^{\beta} |\nabla u|^2, \qquad p^{\frac{1}{\gamma}+1} \nabla p = \frac{2\gamma_-}{1+\gamma_-} u^{\beta+1} \nabla u.$$

In the new notations (3.54) takes the form

$$\int_{B_\rho} \frac{u^{\frac{2\gamma_-}{1+\gamma_-}\frac{1+\gamma}{\gamma}}}{1+\gamma} dx + \int_0^t \int_{B_\rho} \left(\frac{2\gamma_-}{1+\gamma_-}\right)^2 u^{\beta} |\nabla u|^2 dx d\tau$$

$$= \int_0^t \int_{S_\rho} \frac{2\gamma_-}{1+\gamma_-} u^{\beta+1} \nabla u \cdot \nu dS d\tau.$$

We assume that the solution is bounded by a constant $M$, which can be taken equal to one. Then

$$0 \le u \le 1, \quad u^2 \le u^\lambda \quad \text{with}\, \lambda = \frac{2\gamma_-}{1+\gamma_-}\frac{1+\gamma}{\gamma} \le 2$$

and the following inequality holds (cf. with (3.10), (3.12)):

$$\int_{B_\rho} u^2 dx + \int_0^t \int_{B_\rho} u^\beta |\nabla u|^2 dx d\tau \le C \int_0^t \int_{S_\rho} u^{\beta+1} |\nabla u| \, dS d\tau \tag{3.55}$$

with an independent of $u$ constant $C$. The properties of the functions satisfying this inequality are studied in Theorem 3.1.

**Proposition 3.2** (Finite speed of propagation) *Let conditions* (3.48) and (3.49) *be fulfilled. Then there exists* $\delta_0 > 0$ *such that every local weak solution of Eq.* (3.46) *in*

$$G_\delta = B_\delta(x_0) \times [0, T] \quad with\, \delta = const \ge \rho_0.$$

*satisfying* (3.47) *and the condition*

$$\int_{B_{\rho_0}} \frac{p(\cdot,t)^{\frac{1+\gamma}{\gamma}}}{1+\gamma(\cdot)} dx + \int_0^t \int_{B_{\rho_0}} p^{\frac{1}{\gamma}} |\nabla p|^2 dx d\tau \le E_0 < \infty, \tag{3.56}$$

*possesses the property of finite speed of propagation:*

$$p(x,t) = 0 \quad for\, (x,t) \in B_{\rho(t)}(x_0) \times [0, \delta_0],$$

*where* $\rho(t)$ *given by the formula*

$$\rho^\nu(t) = \rho_0^\nu - C t^\lambda E_0^\sigma, \tag{3.57}$$

*with positive constants* $\nu$, $\lambda$, $\sigma$, *and* $C = C(\delta_0, \gamma_\pm, \rho_0, n)$.

## 3.9 Bibliographic Remarks

It is long known and well-documented in the literature that nonlinear evolutions equations admit solutions localized in space or in time—see the review paper [171] and papers [69, 70, 232, 254, 270]. These properties were first established by means of comparison with explicit solutions or suitable sub/super solutions—see [148, 149] for a thorough analysis of this property based on the study of the "travelling wave" solutions and a review of the relevant results. The large time behavior of solutions is well-understood for equations which possess the property of invariance with respect to scaling, [138, 175, 176, 252]. The method of local energy estimates [36], see also [24, 35, 102, 113], offers an alternative to the methods based on comparison or scaling. It is applicable in the situations when the above methods fail, as it happens in the case of an equation with variable nonlinearity.

The results of this chapter were partly published in [42]. Localization properties of stationary solutions of isotropic equation (3.2) and its anisotropic counterpart (2.68) were studied in [43–45].

The limit behavior of solutions of the porous medium equation with the constant exponent $m \to \infty$ was studied in [78, 124]. The singular limit of the porous medium equation as $m \to 0$ was studied in [163, 164, 231]. The behavior of solutions of "almost linear" parabolic equations or equations with the nonlinearity exponents singularly perturbed near certain critical values was analyzed in [172, 173, 174].

# Chapter 4
# Anisotropic Equations with Variable Growth and Coercivity Conditions

## 4.1 Introduction

This chapter is devoted to study the Dirichlet problem for the parabolic equation

$$\begin{cases} u_t - \sum_i \dfrac{d}{dx_i}\left[a_i(z,u)|D_i u|^{p_i(z)-2}D_i u + b_i(z,u)\right] + d(z,u) = 0 & \text{in } Q_T, \\ u = 0 \text{ on } \Gamma_T, \qquad u(x,0) = u_0(x) \text{ in } \Omega, \end{cases} \tag{4.1}$$

where $D_i$ denote the partial derivatives with respect to $x_i$ and

$$\frac{df(z,v)}{dx_i} = D_i f(z,v) + \frac{\partial f(z,v)}{\partial v} D_i v.$$

The coefficients $a_i(z,u)$, $b_i(z,u)$ and $d(z,u)$ may depend on $z = (x,t)$, $u(z)$ and obey the following conditions:

$$a_i(z,r),\, b_i(z,r),\, d(z,r) \text{ are Carathéodory functions,} \tag{4.2}$$

$$\forall\,(z,r) \in Q_T \times \mathbb{R} \qquad 0 < a_0 \leq a_i(z,r) \leq a_1 < \infty, \quad a_0,\, a_1 = const, \tag{4.3}$$

$$\begin{aligned} &\forall\,(z,r) \in Q_T \times \mathbb{R} \\ &\sum_i |b_i(z,r)|^{p_i'(z)} \leq b_0|r|^{\lambda} + h_b(z), \quad p_i' = \frac{p_i(z)}{p_i(z)-1}, \\ &|d(z,r)| \leq d_0|r|^{\lambda-1} + h_d(z) \end{aligned} \tag{4.4}$$

with positive constants $b, d_0, d_1, d_2, \lambda > 1$, and

$$h_b(z) \in L^1(Q_T), \quad h_d(z) \in L^{\lambda'}(Q_T), \quad \lambda' = \frac{\lambda}{\lambda - 1}. \tag{4.5}$$


S. Antontsev and S. Shmarev, *Evolution PDEs with Nonstandard Growth Conditions*, Atlantis Studies in Differential Equations 4,
DOI 10.2991/978-94-6239-112-3_4

The exponents $p_i(z)$ are given continuous in $Q_T$ functions such that

$$p_i(z) \subset (p_i^-, p_i^+) \subseteq [p^-, p^+] \subset (1, \infty), \tag{4.6}$$

with finite constants $p^{\pm}$, $p_i^{\pm} > 1$. Moreover, unless specially indicated we assume that $p_i$ satisfy the log-continuity condition (1.31).

The main feature of Eq. (4.1) is the variable character of nonlinearity which causes a gap between the growth and coercivity conditions. Because of this gap, equations of the type (4.1) are often called equations with nonstandard growth conditions. Equation (4.1) can be regarded as a generalization of the evolution $p$-Laplace equation

$$u_t = \Delta_p u, \qquad \Delta_p u \equiv \operatorname{div}(|\nabla u|^{p-2}\nabla u), \tag{4.7}$$

with the constant exponent $p \in (1, \infty)$. In course of the last decades Eq. (4.7) was intensively studied and has played an important role in development of methods of analysis of degenerate and singular PDEs. There is an extensive literature devoted to Eq. (4.7). We limit ourselves by referring here to monographs [114, 197], papers [16, 81, 169, 170] and the review paper [171] for an insight into the theory of evolution $p$-Laplace equation.

The main existence results of this chapter are stated in Theorems 4.1 and 4.2 of Sect. 4.2. It is proved in Theorem 4.1 that problem (4.1) has at least one global in time weak solution, provided that the growth conditions (4.4)–(4.6) are fulfilled with $2 \leq \lambda = \max\{2, p^- - \delta\}$ for some $\delta > 0$. The assertion remains true if $\lambda = \max\{2, p^-\}$, but under the additional condition of smallness of the data $u_0$, $h_d$ and $h_b$ in the corresponding norms. The case $\lambda > \max\{2, p^-\}$ is studied in Theorem 4.2. It is shown that in this range of exponents, and with the functions $h_b, h_d$ satisfying (4.5), problem (4.1) has a local in time solution if the parameters $\lambda$, $p^-$ and $n$ are subject to the conditions

$$\max\{2, p^-\} < \lambda < p^-\left(1 + \frac{2}{n}\right), \quad \max\left\{1, \frac{2n}{2+n}\right\} < p^-. \tag{4.8}$$

The proofs do not require monotonicity of the lower-order term $d(z, u)$.

Section 4.3 is devoted to derive the $L^\infty$ estimates for the solutions of problem (4.1). To this end we subject the functions $h_d$, $h_b$ to the stronger restrictions

$$\begin{aligned} &|d(z,r)| \leq d_0|r| + h_d, \\ &|b_i(z,r)| \leq b_0|r| + h_b, \qquad h_b,\ h_d \in L^1(0, T; L^\infty(\Omega)). \end{aligned} \tag{4.9}$$

Under these assumptions, it is stated in Theorem 4.3 that the weak solutions of problem (4.1) are globally bounded. The growth restriction can be relaxed for the terms $d(z, u)$ of a special form. Namely, if we assume that in the foregoing assumptions

$$d(z,u) = d_1(z,u)|u|^{\sigma(z)-2}u + d_2(z,u)|u|^{\lambda-2}u + h_d$$

with

$$1 < \lambda \le \inf_{Q_T} \sigma(z) < M, \quad d_1 \ge d_{01} = const > 0, \quad |d_2| \le d_{02} = const < \infty,$$

and that the inequality

$$d_{01}R^{\sigma(z)-1} - d_{02}R^{\lambda-1} - b_0 R - \sup_{Q_T} h_d(z) - \sup_{Q_T} |h_b(z)| \ge 0$$

holds in $Q_T$ for some $R > 0$, then the solutions of (4.1) are globally bounded. Moreover, once such a bound is established, the existence of a global weak solution is a byproduct of Theorem 4.1. Finally, we drop conditions (4.8) and show the under assumptions (4.9) problem (4.1) admits a local in time bounded solution for every $\lambda \ge 1$.

Uniqueness of weak solutions is studied in Sect. 4.4. It is shown that the weak solution of problem (4.1) is unique if the function $u \mapsto d(z,u)$ is monotone increasing and

$$|a_i(z,u) - a_i(z,v)| \le \omega(|u-v|)$$

with the modulus of continuity $\omega$ satisfying the condition

$$\int_\varepsilon \frac{ds}{\omega^\alpha(s)} \to \infty \quad \text{as } \varepsilon \to 0 \quad \text{for some } 1 < \alpha < \frac{p^+}{p^+-1}.$$

If the omit the condition of monotonicity of $d(z,u)$, the uniqueness of weak solutions still can be proved but under stronger continuity and growth assumptions: one has to claim that $d(z,u)$ is Lipschitz-continuous with respect to $u$ and $\omega^\alpha(s) = Cs^2$. In Sect. 4.5 we study the dependence of the regularity of solutions to problem (4.1) on the regularity properties of $p_i$, $a_i$ and $\sigma$ in the special case

$$a_i \equiv a_i(z), \quad d(z,u) = c(z)|u|^{\sigma(z)-2}u - f(z), \quad c(z) \le 0.$$

It turns out that if $u_0 \in L^{\sigma(x,0)}(\Omega)$, $D_i u_0 \in L^{p_i(x,0)}(\Omega)$, and if the exponents $p_i$ and $\sigma$ are nonincreasing functions of $t$, then the solutions of problem (4.1) possess better regularity properties:

$$u_t \in L^2(Q_T), \quad |u|^{\sigma(z)}, \ |D_i u|^{p_i(z)} \in L^\infty(0,T; L^1(\Omega)),$$
$$|D_i u|^{p_i} |\ln |D_i u|| p_{it}|, \ |u|^\sigma |\ln |u|| \sigma_t| \in L^1(Q_T).$$

In Sect. 4.6 we consider the homogeneous Dirichlet problem for the singularly perturbed equation

$$u_t^{(\varepsilon)} = \varepsilon \Delta_{p^+} u^{(\varepsilon)} + \Delta_{p(x,t)} u^{(\varepsilon)} + f, \qquad \varepsilon > 0,$$

where the $p^+$-Laplace operator is defined in (4.7) and

$$\Delta_{p(x,t)} v \equiv \operatorname{div}\left(|\nabla v|^{p(x,t)-2} \nabla v\right).$$

It is shown that the solution of the evolution $p(x,t)$-Laplace equation can be obtained as the limit of the sequence $\{u^{(\varepsilon)}\}$ as $\varepsilon \to 0$.

The Cauchy problem for the equation

$$u_t = \operatorname{div}\left(a(x,t)|\nabla u|^{p(x)-2} \nabla u\right) + f.$$

is studied in Sect. 4.7. It is proved that for $u_0 \in L^2(\mathbb{R}^n)$ and $f \in L^2(\mathbb{R}^n \times (0,T))$ the Cauchy problem has at least one weak solution. Uniqueness and boundedness of weak solutions is established under stronger assumptions on the data $u_0$, $f$ and the exponent $p(x)$. Unlike the rest of results of this chapter, the Cauchy problem is solved under the assumption that the exponent $p(x)$ is measurable and bounded in $\mathbb{R}^n$, but not necessarily continuous.

Existence of bounded stationary solutions of Eq. (4.1) is discussed in Sect. 4.8. In the concluding Sect. 4.9 we give an account of other approaches to solving the evolution equations of $p(x,t)$-Laplace type.

## 4.2 Existence Theorems

In this section we prove the existence of weak solutions to problem (4.1) under the general growth conditions (4.4).

**Definition 4.1** (*Energy solution*) A function $u : Q_T \mapsto \mathbb{R}$ is called weak (energy) solution of problem (4.1) if

1. $u \in \mathbf{W}(Q_T) \cap C([0,T]; L^2(\Omega))$, $u_t \in \mathbf{W}'(Q_T)$,
2. for every test-function

$$\zeta \in \mathbf{Z} \equiv \{\eta(z) : \eta \in \mathbf{W}(Q_T) \cap C([0,T]; L^2(\Omega)),\ \eta_t \in \mathbf{W}'(Q_T)\}$$

the integral identity holds

$$\int_{Q_T} \Big( u_t \zeta + \sum_i (a_i |D_i u|^{p_i - 2} D_i u + b_i) \cdot D_i \zeta + d(z, u)\zeta \Big) \, dz = 0, \quad (4.10)$$

3. for every $\eta(x) \in C_0^\infty(\Omega)$

$$\int_\Omega (u(x,t) - u_0(x)) \, \eta(x) \, dx \to 0 \quad \text{as}\, t \to 0.$$

The following are the main results of this section.

**Theorem 4.1** (a) *Let us assume that*

(1) *the coefficients $a_i(z,r)$, $b_i(z,r)$, $d(z,r)$ satisfy conditions* (4.2)–(4.4),
(2) *the exponents $p_i(z)$ satisfy* (4.6) *and $p_i \in C_{\log}(Q_T)$*
(3) *the constant $\lambda$ satisfies the condition*

$$\lambda = \max\{2, p^- - \delta\} \quad \textit{with some}\, \delta > 0. \quad (4.11)$$

*Then for every $u_0 \in L^2(\Omega)$ problem* (4.1) *has at least one weak solution $u \in \mathbf{W}(Q_T)$. This solution satisfies the estimate*

$$\|u\|^2_{L^\infty(0,T;L^2(\Omega))} + a_0 \sum_{i=1}^n \int_{Q_T} |D_i u|^{p_i} dz \le M(\|u_0\|^2_{2,\Omega} + K + 1) \qquad (4.12)$$

*with a constant $M$ independent of $u$ and*

$$K = \|h_b\|_{1,Q_T} + \|h_d\|_{\lambda', Q_T}.$$

*Moreover, $u_t \in \mathbf{W}'(Q_T)$.*

(b) *The assertion remains true if* (4.11) *is substituted by the condition $\lambda = \max\{2, p^-\}$ and the constant $b_0 + d_0$ in* (4.4) *is appropriately small in comparison with $a_0$.*

**Theorem 4.2** *Let us assume that in the conditions of* Theorem 4.1 *condition* (4.11) *is substituted by the following one:*

$$\max\{2, p^-\} < \lambda < p^- \left(1 + \frac{2}{n}\right), \qquad \max\left\{1, \frac{2n}{2+n}\right\} < p^-. \qquad (4.13)$$

*Then there exists $T_0 > 0$, defined through $\|u_0\|^2_{2,\Omega} + K$, such that problem* (4.1) *has at least one weak solution $u \in \mathbf{W}(Q_{T_0})$ satisfying estimate* (4.12) *in $Q_{T_0}$. The weak solution exists globally in time if $\|u_0\|^2_{2,\Omega} + K$ is sufficiently small.*

### 4.2.1 Galerkin's Approximations

A solution of problem (4.1) is constructed as the limit of the sequence of Galerkin's approximations. Let us define the operator $L:\ \mathbf{V}_t(\Omega)\mapsto \mathbf{V}_t'(\Omega)$:

$$\langle Lv,\,\phi\rangle_\Omega=\int_\Omega\Big(\sum_{i=1}^n a_i|D_iv|^{p_i-2}D_iv\cdot D_i\phi+d(z,v)\phi\Big)\,dx,\quad \phi\in\mathbf{V}_t(\Omega).$$

Let $\{\psi_k\}$ be the orthonormal basis of $H_0^s(\Omega)$ composed of the eigenfunctions of problem (1.30). The approximate solutions to problem (4.1) are sought in the form

$$u^{(m)}(z)=\sum_{k=1}^m c_k^{(m)}(t)\psi_k(x),$$

where the coefficients $c_k^{(m)}(t)$ are defined from the equalities

$$(u_t^{(m)},\,\psi_k)_{2,\Omega}=-\langle Lu^{(m)},\,\psi_k\rangle_\Omega=0,\quad k=1,\dots,m. \tag{4.14}$$

Equalities (4.14) generate the system of $m$ ordinary differential equations for the coefficients $c_k^{(m)}(t)$:

$$\begin{cases}\left(c_k^{(m)}\right)'=F_k\left(t,c_1^{(m)}(t),\dots,c_m^{(m)}(t)\right),\\ c_k^{(m)}(0)=\displaystyle\int_\Omega u_0(x)\psi_k\,dx\qquad k=1,\dots,m.\end{cases} \tag{4.15}$$

If the coefficients $a_i$, $b_i$, $d$ and the exponents $p_i$ satisfy the conditions of Theorem 4.1(a), the functions $F_k$ are continuous in all their arguments.

### 4.2.2 A Priory Estimates

**Lemma 4.1** *Let the conditions of* Theorem 4.1(a) *be fulfilled. Then for every* $T<\infty$ *and* $m\in\mathbb{N}$ *system* (4.15) *has a solution* $\{c_k^{(m)}(t)\}_{k=1}^m$ *on the interval* $(0,T)$ *and the corresponding function* $u^{(m)}$ *satisfies the estimate*

$$\|u^{(m)}(\cdot,t)\|^2_{L^\infty(0,T;L^2(\Omega))}+\int_{Q_T}a_0\sum_i|D_iu^{(m)}|^{p_i}\,dz\le M\left[\|u_0\|^2_{L^2(\Omega)}+K+1\right] \tag{4.16}$$

*with the constants* $M,\ K$ *defined in the conditions of* Theorem 4.1.

*Proof* By Peano's Theorem, for every finite $m$ system (4.15) has a solution $c_i^{(m)}(t)$, $i = 1, \dots, m$, on an interval $(0, T_m)$. Multiplying each of equalities (4.14) by $c_k^{(m)}(t)$ and summing over $k = 1, \dots, m$, we arrive at the relation

$$\frac{1}{2}\ \|u^{(m)}\|^2_{2,\Omega}\Big|_{t=0}^{t=\tau} + \int_{Q_\tau} \Big(\sum_i \Big[a_i(z, u^{(m)})|D_i u^{(m)}|^{p_i} dz + b_i(z, u^{(m)}) D_i u^{(m)}\Big] + d(z, u^{(m)}) u^{(m)}\Big)\, dz = 0, \quad \tau \in [0, T_m]. \tag{4.17}$$

Using (4.3), (4.4) and applying Young's inequality, we estimate: $\forall\, \varepsilon > 0$

$$\begin{aligned} |b_i(z, u^{(m)}) D_i u^{(m)}| &\leq \varepsilon\, a_0 |D_i u^{(m)}|^{p_i} + C\, |b_i(z, u^{(m)})|^{p_i'} \\ &\leq \varepsilon\, a_0 |D_i u^{(m)}|^{p_i} + C\, (b_0 |u^{(m)}|^{\lambda} + |h_b|), \end{aligned} \tag{4.18}$$

$$|d(z, u^{(m)}) u^{(m)}| \leq (d_0 + \varepsilon_d)|u^{(m)}|^{\lambda} + C |h_d|^{\lambda'}, \quad \varepsilon_d \in (0, 1), \tag{4.19}$$

with a constant $C$ depending on $\varepsilon, \varepsilon_d, a_0, p^-, p^+$. Plugging (4.18)–(4.19) into (4.17), choosing $\varepsilon$ sufficiently small and simplifying, we get the estimate

$$\begin{aligned} &\frac{1}{2}\ \|u^{(m)}\|^2_{2,\Omega}\Big|_{t=0}^{t=\tau} + \int_{Q_\tau} a_0 \sum_i |D_i u^{(m)}|^{p_i}\, dz \\ &\leq C \int_{Q_\tau} \Big((d_0 + b_0 + \varepsilon_d)|u^{(m)}|^{\lambda} + |h_b| + |h_d|^{\lambda'}\Big)\, dz \\ &\leq C(d_0 + b_0 + \varepsilon_d) \int_{Q_\tau} \|u^{(m)}\|^{\lambda} dz + CK. \end{aligned} \tag{4.20}$$

Let $\lambda = 2$. Using Gronwall's inequality to estimate the function $\|u^{(m)}(\cdot, t)\|^2_{2,\Omega}$ and then reverting to (4.17), we obtain the required estimate (4.16).

Let $2 < \lambda = p^- - \delta$. This assumption yields the inequality $\lambda < \frac{n(p^- - \delta)}{n - p^- + \delta}$, which allows one to make use of the embedding theorem in Sobolev spaces:

$$\|u^{(m)}(\cdot, t)\|^{\lambda}_{\lambda,\Omega} \leq C(\lambda, p^-, n) \|\nabla u^{(m)}(\cdot, t)\|^{\lambda}_{p^-,\Omega}. \tag{4.21}$$

Applying Young's inequality, we obtain the estimate

$$\begin{aligned} \int_\Omega |u^{(m)}|^{\lambda} dx &\leq C \left(\int_\Omega |\nabla u^{(m)}|^{p^-} dx\right)^{\frac{\lambda}{p^-}} \\ &\leq C \left[\left(\sum_i \int_\Omega |D_i u^{(m)}|^{p_i} dx\right)^{\frac{\lambda}{p^-}} + 1\right] \\ &\leq \varepsilon\, a_0 \sum_i \int_\Omega |D_i u^{(m)}|^{p_i} dx + C(\varepsilon, \delta, \Omega, a_0,\ p^{\pm}). \end{aligned} \tag{4.22}$$

Gathering these estimate with (4.17) and choosing $\varepsilon$ appropriately small, we find that

$$\frac{1}{2}\|u^{(m)}\|_{2,\Omega}^2\Big|_{t=0}^{t=\tau} + a_0\sum_i\int_{Q_\tau}|D_iu^{(m)}|^{p_i}\,dz \le C\,(K+1)\,.$$

The right-hand side of the obtained estimate does not depend on $m$, which is why the solution of system (4.15) can be continued to the maximal interval $[0,T]$. □

**Lemma 4.2** *The assertion of* Lemma 4.1 *remains true for* $\lambda = \max\{2, p^-\}$, *provided that the constant* $b_0 + d_0$ *is sufficiently small in comparison with* $a_0$.

*Proof* We only have to study the case $\lambda = p^-$. By the Poincaré inequality

$$\int_\Omega |u^{(m)}|^\lambda dx \le C\int_\Omega |\nabla u^{(m)}|^{p^-}dx, \quad C = C(n,\lambda).$$

Combining the last inequality with (4.20), we find that

$$\begin{aligned}\frac{1}{2}\|u^{(m)}\|_{2,\Omega}^2\Big|_{t=0}^{t=\tau} &+ a_0\sum_i\int_{Q_\tau}|D_iu^{(m)}|^{p_i}\,dz \\ &\le C(d_0+b_0+\varepsilon_d)\sum_i\int_{Q_\tau}|D_iu^{(m)}|^{p_i}\,dz + CK \quad \text{with } \varepsilon_d\in(0,1).\end{aligned} \tag{4.23}$$

The conclusion follows if we claim that $C(b_0+d_0) < a_0$ and choose $\varepsilon_d$ sufficiently small. □

**Lemma 4.3** *Let condition* (4.13) *be fulfilled. Then there exists* $T_0$, *depending on* $\|u_0\|_{2,\Omega}^2 + K$, *such that the assertion of* Lemma 4.1 *is true on every interval* $[0,T]$ *with* $T < T_0$.

*Proof* Instead of (4.21) we now make use of the interpolation inequality (1.56)

$$\|u^{(m)}\|_{\lambda,\Omega}^\lambda \le C(\lambda,p^-,n)\|\nabla u^{(m)}\|_{p^-,\Omega}^{\theta\lambda}\|u^{(m)}\|_{2,\Omega}^{(1-\theta)\lambda}, \tag{4.24}$$

with the exponent

$$\theta = \frac{\lambda-2}{\lambda}\,\frac{np^-}{np^- - 2(n-p^-)} \in (0,1). \tag{4.25}$$

The inclusion $\theta\in(0,1)$ follows from condition (4.13):

$$\frac{\theta\lambda}{p^-} = \frac{n(\lambda-2)}{np^- - 2(n-p^-)} < 1 \iff \lambda < p^-\left(1+\frac{2}{n}\right). \tag{4.26}$$

Applying (1.29) and Young's inequality we transform (4.24) into the inequality

$$
\begin{aligned}
\|u^{(m)}(\cdot,t)\|_{\lambda,\Omega}^{\lambda} &\le C\left(\max\left\{A_{\mathbf{p}}^{\frac{1}{p^+}}(\nabla u^{(m)}),\, A_{\mathbf{p}}^{\frac{1}{p^-}}(\nabla u^{(m)})\right\}\right)^{\theta\lambda}\|u^{(m)}\|_{2,\Omega}^{(1-\theta)\lambda} \\
&\le \varepsilon A_{\mathbf{p}}(\nabla u^{(m)}) + C\max\left\{\|u^{(m)}\|_{2,\Omega}^{2\gamma^+},\, \|u^{(m)}\|_{2,\Omega}^{2\gamma^-}\right\} \qquad (4.27)\\
&\le \varepsilon A_{\mathbf{p}}(\nabla u^{(m)}) + C\left\{\|u^{(m)}\|_{2,\Omega}^{2\gamma^+} + \|u^{(m)}\|_{2,\Omega}^{2\gamma^-}\right\},
\end{aligned}
$$

where the modular $A_{\mathbf{p}}(v)$ is defined in (1.28) and the constant exponents $\gamma^{\pm}$ are given by the equalities

$$
\gamma^- = \frac{(1-\theta)\lambda}{2}\frac{p^-}{p^- - \lambda\theta} \quad\ge\quad \gamma^+ = \frac{(1-\theta)\lambda}{2}\frac{p^+}{p^+ - \lambda\theta}.
$$

Gathering (4.27) with (4.17)–(4.19) and choosing $\varepsilon$ appropriately small, we have

$$
\begin{aligned}
&\frac{1}{2}\|u^{(m)}\|_{2,\Omega}^2\Big|_{t=0}^{t=\tau} + \int_{Q_\tau} a_0\sum_i |D_i u^{(m)}|^{p_i}\,dz \\
&\le C\left(\int_0^\tau \left\{\|u^{(m)}\|_{2,\Omega}^{\gamma^+} + \|u^{(m)}\|_{2,\Omega}^{\gamma^-}\right\}dt + K\right). \\
&\le C\left(1+\frac{\gamma^+}{\gamma^-}\right)\int_0^\tau \|u^{(m)}\|_{2,\Omega}^{\gamma^-}dt + \frac{\gamma^- - \gamma^+}{\gamma^+}t + CK.
\end{aligned}
$$

Introduce the function

$$
Y(t) = \|u^{(m)}(\cdot,t)\|_{2,\Omega}^2
$$

and continue the previous inequality as follows:

$$
\begin{aligned}
&Y(t) \le A + B\int_0^t Y^{\gamma^-}(\tau)\,d\tau, \\
&A = Y(0) + DT + 2CK, \quad B = 2C\left(1+\frac{\gamma^+}{\gamma^-}\right), \quad D = \frac{\gamma^- - \gamma^+}{\gamma^+}. \qquad (4.28)
\end{aligned}
$$

Let $Z(t)$ be the function satisfying the conditions

$$
Z(t) = Z(0) + B\int_0^t Z^{\gamma^-}(\tau)\,d\tau, \quad Z(0) > A.
$$

The function $Z(t)$ is given by the explicit formula:

$$
Z(t) = \frac{Z(0)}{\left(1 - BZ^{\gamma^- -1}(0)(\gamma^- - 1)t\right)^{\frac{1}{\gamma^- -1}}} \quad \text{for } t < t_0 = \frac{1}{BZ^{\gamma^- -1}(0)(\gamma^- - 1)}.
$$

It is easy to see that $Z(t)$ majorates $Y(t)$ on the interval $[0, t_0]$. Indeed: if we assume that

$$t^* = \sup\{t : Y(t) - Z(t) < 0\} < t_0,$$

it is necessary that at the instant $t = t^*$

$$0 = Y(t^*) - Z(t^*) \leq B \int_0^t (Y^{\gamma^-}(\tau) - Z^{\gamma^-}(\tau))\, d\tau < 0,$$

which is impossible. Thus, $Y(t)$ is bounded for $t < t_0$ and estimate (4.16) takes the form

$$\|u^{(m)}(\cdot, t)\|_{2,\Omega}^2 + \int_{Q_t} a_0 \sum_i |D_i u^{(m)}|^{p_i}\, dz \leq C'(1 + K), \quad t \in (0, t_0).$$

□

## 4.2.3 Compactness and Passage to the Limit

**Lemma 4.4** *Under the foregoing conditions* $u_t^{(m)} \in L^{(p^+)'}(0, T; H^{-s}(\Omega))$ *and*

$$\|u_t^{(m)}\|_{L^{(p^+)'}(0,T;H^{-s}(\Omega))} \leq C' \left[1 + \|h_d\|_{\lambda',Q}^{\lambda'} + \|u_0\|_{2,\Omega}^2\right].$$

*Proof* Let us introduce the subspaces of $\mathbf{W}(Q)$

$$\mathbf{Z}_m = \left\{ \eta(x, t) \,\middle|\, \eta = \sum_{k=1}^m d_k(t)\psi_k(x)\ ,\ d_k(t) \in C^1(0, T) \right\}. \tag{4.29}$$

By the choice of $s$ in (1.30)

$$\mathbf{Z}_m \subset L^{p^+}(0, T; H_0^s(\Omega)) \subset \mathbf{W}(Q).$$

Given $\phi \in L^{p^+}(0, T; H_0^s(\Omega))$, we denote

$$\phi^{(m)} = \sum_{i=1}^m \phi_k(t)\psi_k(x) \in \mathbf{Z}_m, \qquad \phi_k(t) = (\phi, \psi_k)_{2,\Omega}.$$

Since $\{\psi_k\}$ are orthogonal in $L^2(\Omega)$, the definition of $u^{(m)}$ yields

$$
\begin{aligned}
-\int_Q u_t^{(m)}\phi\,dz &= -\int_Q u_t^{(m)}\phi^{(m)}\,dz \\
&= \sum_{i=1}^n \int_Q |D_i u^{(m)}|^{p_i-2} D_i u^{(m)} \cdot D_i\phi^{(m)}\,dz \\
&\quad + \int_Q \sum_{i=1}^n b_i(z,u^{(m)}) D_i\phi^{(m)}\,dz + \int_Q d(z,u^{(m)})\phi^{(m)}\,dz.
\end{aligned}
$$

It follows from the estimates of the previous subsection that

$$
\begin{aligned}
&\| |D_i u^{(m)}|^{p_i-1} \|_{p_i',Q} \\
&\quad \le \max\left\{ \left(\int_Q |D_i u^{(m)}|^{p_i}\,dz\right)^{\frac{1}{(p_i')^-}}, \left(\int_Q |D_i u^{(m)}|^{p_i}\,dz\right)^{\frac{1}{(p_i')^+}} \right\} \le C
\end{aligned}
$$

with a constant $C$ depending only on the data. Due to the choice of basis $\{\psi_k\}$

$$
\begin{aligned}
\|D_i\phi^{(m)}\|_{p_i,Q} &\le C\|D_i\phi^{(m)}\|_{p^+,Q} \\
&\le C\left(\int_0^T \|\phi^{(m)}\|^{p^+}_{H_0^s(\Omega)}(t)\,dt\right)^{\frac{1}{p^+}} \le C\|\phi\|_{L^{p^+}(0,T;H_0^s(\Omega))}.
\end{aligned}
$$

Combining these inequalities we obtain the estimate

$$
\begin{aligned}
\left|\int_Q \sum_{i=1}^n |D_i u^{(m)}|^{p_i-2} D_i u^{(m)} D_i\phi^{(m)}\,dz\right| &\le C\sum_{i=1}^n \| |D_i u^{(m)}|^{p_i-1}\|_{p_i',Q} \|D_i\phi^{(m)}\|_{p_i,Q} \\
&\le C\,\|\phi\|_{L^{p^+}(0,T;H_0^s(\Omega))}.
\end{aligned}
$$

By virtue of the estimates on $\|u^{(m)}\|_{\lambda,Q}$

$$
\begin{aligned}
\left|\int_Q \sum_{i=1}^n b_i(z,u^{(m)}) D_i\phi^{(m)}\,dz\right| &\le C\sum_{i=1}^n \|b_i\|_{p_i',Q} \|D_i\phi^{(m)}\|_{p_i(\cdot),Q} \\
&\le \widetilde{C}\sum_{i=1}^n \|D_i\phi^{(m)}\|_{p_i(\cdot),Q} \\
&\le \widetilde{C}\|\phi\|_{L^{p^+}(0,T;H_0^s(\Omega))},
\end{aligned}
$$

$$
\begin{aligned}
\left|\int_Q d\left(z, u^{(m)}\right) \phi^{(m)}\, dz\right| &\leq \|d\|_{\lambda',Q} \|\phi^{(m)}\|_{\lambda,Q} \\
&\leq \left[C(d_0, \lambda) \| |u^{(m)}|^{\lambda-1} \|_{\lambda',Q} + \|h_d\|_{\lambda',Q}\right] \|\phi^{(m)}\|_{\lambda,Q} \\
&\leq C \|\phi^{(m)}\|_{\lambda,Q} \\
&\leq C \|\phi^{(m)}\|_{L^{p^+}(0,T;H_0^s(\Omega))} \\
&\leq C \|\phi\|_{L^{p^+}(0,T;H_0^s(\Omega))}
\end{aligned}
$$

with independent of $m$ constants $C$. It follows now that for every $\phi \in L^{p^+}(0, T; H_0^s(\Omega))$ and an arbitrary $m$

$$
\left|\int_Q u_t^{(m)} \phi\, dz\right| \leq C\, \|\phi\|_{L^{p^+}(0,T;H_0^s(\Omega))},
$$

and the assertion follows. □

The following inclusions hold:

$$
\begin{cases}
u^{(m)} \in \mathbf{W}(Q) \subseteq L^{p^-}(0, T;\ W_0^{1,p^-}(\Omega)), \\
u_t^{(m)} \in L^{(p^+)'}\left(0, T;\ H^{-s}(\Omega)\right), \\
W_0^{1,p^-}(\Omega) \subset L^r(\Omega) \subset H^{-s}(\Omega), \quad 1 < r < \dfrac{np^-}{n - p^-}
\end{cases}
$$

with compact embedding $W_0^{1,p^-}(\Omega) \subset L^r(\Omega)$. By Lemma 2.6 the sequence $\{u^{(m)}\}$ contains a subsequence which converges strongly in $L^q(Q)$ with some $q > 1$ and almost everywhere in $Q$ to a function $u$. Using the uniform in $m$ estimates of the previous section we extract from the sequence $\{u^{(m)}\}$ a subsequence (for the sake of simplicity we assume that it coincides with the whole sequence) such that

$$
\begin{aligned}
& u^{(m)} \to u \text{ weakly in } \mathbf{W}(Q), \text{ strongly in } L^q(Q), q > 1, \text{ and a.e. in } Q, \\
& u^{(m)}(x, T) \rightharpoonup U(x) \text{ in } L^2(\Omega), \\
& u^{(m)}(x, 0) \to u_0 \text{ in } L^2(\Omega), \\
& u_t^{(m)} \rightharpoonup u_t \quad \text{in } L^{(p^+)'}\left(0, T;\ H^{-s}(\Omega)\right), \\
& b_i\left(z, u^{(m)}\right) \to b_i\,(z, u) \text{ in } L^{p_i'(z)}(Q), \\
& d\left(z, u^{(m)}\right) \to d\,(z, u) \text{ in } L^{\lambda'}(Q), \\
& a_i(z, u^{(m)}) \left|D_i u^{(m)}\right|^{p_i(z)-2} D_i u^{(m)} \rightharpoonup A_i(z) \text{ in } L^{p_i'(\cdot)}(Q),
\end{aligned}
\tag{4.30}
$$

with some functions

$$U(x) \in L^2(\Omega), \qquad A_i(z) \in L^{p_i'(\cdot)}(Q), \qquad u \in \mathbf{W}(Q).$$

Fix an arbitrary $l \in \mathbb{N}$. Each of $u^{(m)}$ with $m > l$ satisfies the identity

$$\int_Q \Big[ u_t^{(m)} \eta_l + \sum_i \left( a_i |D_i u^{(m)}|^{p_i-2} D_i u^{(m)} + b_i \right) \cdot D_i \eta_l \\ + d(z, u^{(m)})\, \eta_l \Big]\, dxdt = 0, \qquad \eta_l \in \mathbf{Z}_l \subset \mathbf{Z}_m. \tag{4.31}$$

Letting $m \to \infty$ in (4.31) we conclude that for every fixed $l$ and $\eta \in \mathbf{Z}_l$

$$-\int_Q u_t \eta_l \, dz = \int_Q \left[ \sum_i (A_i(z) + b_i(z, u)) \cdot D_i \eta_l + d(z, u)\eta_l \right] dz. \tag{4.32}$$

Let $\eta \in \mathbf{W}(Q)$. Since $L^{p^+}(0, T; H_0^s(\Omega))$ is dense in $\mathbf{W}(Q)$, there is a sequence $\{\eta_l\} \in L^{p^+}(0, T; H_0^s(\Omega))$ such that $\|\eta_l - \eta\|_{\mathbf{W}(Q)} \to 0$ as $l \to \infty$. Then $D_i \eta_l \to D_i \eta$ weakly in $L^{p_i(\cdot)}(Q)$, which allows one to pass to the limit as $l \to \infty$ in the first term on the right-hand side of (4.32). Under the conditions of Theorem 4.1 $\|\eta_l\|_{\lambda, Q} \leq C \|\eta_l\|_{\mathbf{W}(Q)}$, whence $\eta_l \to \eta$ weakly in $L^{\lambda}(Q)$ (up to a subsequence). Since $d(z, u) \in L^{\lambda'}(Q)$, we may now pass to the limit as $l \to \infty$ in the second and third terms on the right-hand side of (4.32). This means that the left-hand side of (4.32) also has a limit as $l \to \infty$,

$$-\int_Q u_t \eta \, dz = \int_Q \left[ \sum_i (A_i(z) + b_i(z, u)) \cdot D_i \eta + d(z, u)\eta \right] dz \qquad \forall \eta \in \mathbf{W}(Q),$$

and $u_t \in \mathbf{W}'(Q)$. Moreover, by virtue of Lemmas 4.1, 4.2 and (4.30)

$$\left| \int_Q u_t \eta \, dz \right| \leq \left( \sum_i \|A_i\|_{p_i'(\cdot), Q} + \|d\|_{\lambda', Q} \right) \|\eta\|_{\mathbf{W}(Q)} \leq \widetilde{C}\, \|\eta\|_{\mathbf{W}(Q)}$$

with a constant $C$ depending on the norms of $u_0$, $h_b$ and $h_d$, whence $\|u_t\|_{\mathbf{W}'(Q)} \leq \widetilde{C}$.

**Lemma 4.5** *Let $u \in \mathbf{W}(Q)$, $u_t \in \mathbf{W}'(Q)$. There exists $\widehat{u} \in C([0, T]; L^2(\Omega)) \cap \mathbf{W}(Q)$ such that $\|u - \widehat{u}\|_{2,\Omega}(t) = 0$ for a.e. $t \in (0, T)$.*

*Proof* Let $u_h \in C^{\infty}(Q)$ be the sequence of mollifications of $u$. By Lemmas 1.17 and 1.18 $u_h \to u$ in $\mathbf{W}(Q)$, $(u_h)_t \to u_t$ in $\mathbf{W}'(Q)$ as $h \to 0$. For every $h_1, h_2 > 0$ and every $t \in [0, T]$

$$\|u_{h_1} - u_{h_2}\|_{2,\Omega}^2(t) = \|u_{h_1} - u_{h_2}\|_{2,\Omega}^2(0) + 2 \int_0^t \int_\Omega (u_{h_1} - u_{h_2})(u_{h_1} - u_{h_2})_t \, dz,$$

whence

$$\begin{aligned}\sup_{t\in(0,T)} \|u_{h_1}-u_{h_2}\|^2_{2,\Omega}(t) &\le \|u_{h_1}-u_{h_2}\|^2_{2,\Omega}(0)+2\int_Q (u_{h_1}-u_{h_2})(u_{h_1}-u_{h_2})_t\,dz\\ &\le \|u_{h_1}-u_{h_2}\|^2_{2,\Omega}(0)+2\,\|u_{h_1}-u_{h_2}\|_{\mathbf{W}(Q)}\|(u_{h_1}-u_{h_2})_t\|_{\mathbf{W}'(Q)}\to 0\end{aligned}$$

as $h_1, h_2 \to 0$. It follows that $\{u_h\} : [0, T] \mapsto L^2(\Omega)$ is a Cauchy sequence in $C([0, T]; L^2(\Omega))$ and contains a subsequence which converges to a function $\widehat{u} \in C([0, T]; L^2(\Omega))$. On the other hand, $u_h \to u \in \mathbf{W}(Q)$, which means that $\|u - \widehat{u}\|_{2,\Omega}(t) = 0$ for a.e. $t \in (0, T)$. □

From now on we identify the limit function $u \in \mathbf{W}(Q)$ in (4.30) with its continuous representative $\widehat{u} \in C([0, T]; L^2(\Omega)) \cap \mathbf{W}(Q)$, in particular, $U(x) = \widehat{u}(x, T)$ and $\widehat{u}(x, 0) = u_0(x) \in L^2(\Omega)$. Notice that by virtue of Lemma 4.5 the inclusion $\eta \in \mathbf{Z}$ yields $\eta \in C([0, T]; L^2(\Omega))$. By Lemmas 4.5 and 1.20, for every $t_1, t_2 \in [0, T]$ and every $\eta \in \mathbf{Z}$

$$\int_\Omega u\,\eta\,dx\Big|_{t=t_1}^{t=t_2} + \int_{t_1}^{t_2}\int_\Omega \left(-u\,\eta_t + \sum_{i=1}^n (A_i + b_i)\cdot D_i\eta + d(z,u)\,\eta\right) dz = 0. \tag{4.33}$$

Since $u \in \mathbf{Z}$, it can be taken for the test-function in (4.33), which leads to the energy equality: $\forall\ t_1, t_2 \in [0, T]$

$$\frac{1}{2}\int_\Omega u^2\,dx\Big|_{t=t_1}^{t=t_2} + \int_{t_1}^{t_2}\int_\Omega \left(\sum_{i=1}^n (A_i + b_i)\cdot D_i u + d(z,u)\,u\right) dz = 0. \tag{4.34}$$

The initial condition is fulfilled by virtue of Lemma 4.5. To complete the proofs of Theorems 4.1 and 4.2, it remains to identify the limit functions $A_i$.

**Lemma 4.6** *For almost all $z \in Q$*

$$A_i(z) = a_i(z,u)|D_i u|^{p_i(z)-2} D_i u, \qquad i = 1, \dots, n.$$

*Proof* The proof relies on the monotonicity of the operator $\mathscr{M}(s) = |s|^{p-2}s$: $\forall\ \xi$, $\eta \in \mathbb{R}^n$

$$\begin{aligned}&(\mathscr{M}(\xi)-\mathscr{M}(\eta))\,(\xi-\eta)\\ &\quad\ge \begin{cases} 2^{-p}\,|\xi-\eta|^p & \text{if } 2 \le p < \infty,\\ (p-1)\,|\xi-\eta|^2\left(|\xi|^p+|\eta|^p\right)^{\frac{p-2}{p}} & \text{if } 1 < p < 2.\end{cases}\end{aligned} \tag{4.35}$$

According to (4.35), for every $\zeta \in \mathbf{Z}_m$

$$X_m^{(i)} := \int_Q a_i \left( |D_i u^{(m)}|^{p_i - 2} D_i u^{(m)} - |D_i \zeta|^{p_i - 2} D_i \zeta \right) D_i (u^{(m)} - \zeta)\, dz \geq 0.$$

By virtue of (4.31) with $\eta_l = u^{(m)}$

$$\frac{1}{2} \|u^{(m)}(\cdot, t)\|_{2,\Omega}^2 \Big|_{t=0}^{t=T} + \int_Q \left( \sum_i a_i |D_i u^{(m)}|^{p_i} + b_i\, D_i u^{(m)} + u^{(m)}\, d(z, u^{(m)}) \right) dz = 0.$$

Notice that for every $\zeta \in \mathbf{Z}_s$ with $s \leq m$

$$\begin{aligned} |D_i u^{(m)}|^{p_i} &= |D_i u^{(m)}|^{p_i - 2} D_i u^{(m)} \cdot D_i (u^{(m)} - \zeta) - |D_i u^{(m)}|^{p_i - 2} D_i u^{(m)} \cdot D_i \zeta \\ &= \left( |D_i u^{(m)}|^{p_i - 2} D_i u^{(m)} - |D_i \zeta|^{p_i - 2} D_i \zeta \right) \cdot D_i (u^{(m)} - \zeta) \\ &\quad + |D_i \zeta|^{p_i - 2} D_i \zeta \cdot D_i (u^{(m)} - \zeta) + |D_i u^{(m)}|^{p_i - 2} D_i u^{(m)} \cdot D_i \zeta \end{aligned}$$

and rewrite this equality in the following form:

$$\begin{aligned} 0 \leq \sum_i X_m^{(i)} &= -\frac{1}{2} \|u^{(m)}(\cdot, T)\|_{2,\Omega}^2 + \frac{1}{2} \|u^{(m)}(\cdot, 0)\|_{2,\Omega}^2 - \int_Q u^{(m)} d(z, u^{(m)})\, dz \\ &- \sum_i \int_Q a_i \left( |D_i u^{(m)}|^{p_i - 2} D_i u^{(m)} - |D_i \zeta|^{p_i - 2} D_i \zeta \right) \cdot D_i \zeta\, dz \\ &- \sum_i \int_Q a_i |D_i \zeta|^{p_i - 2} D_i \zeta \cdot D_i u^{(m)}\, dz - \sum_i \int_Q b_i(z, u^{(m)}) D_i u^{(m)}\, dz. \end{aligned}$$

By the property of lower semicontinuity of the norm

$$-\liminf_{m \to \infty} \|u^{(m)}(\cdot, T)\|_{2,\Omega}^2 \leq -\|u(\cdot, T)\|_{2,\Omega}^2.$$

Letting $m \to \infty$ we now find that

$$\begin{aligned} 0 \leq \frac{1}{2} \|u_0\|_{2,\Omega}^2 - \frac{1}{2} \|u(\cdot, T)\|_{2,\Omega}^2 - \int_Q \left( \sum_i (A_i(z) + b_i) \cdot D_i u + u\, d(z, u) \right) dz \\ + \sum_i \int_Q A_i(z) \cdot D_i (u - \zeta)\, dz - \sum_i \int_Q a_i |D_i \zeta|^{p_i - 2} D_i \zeta \cdot D_i (u - \zeta)\, dz. \end{aligned}$$

Using (4.34) with $t_1 = 0$, $t_2 = T$, we finally have:

$$\sum_i \int_Q \left( A_i(z) - a_i |D_i \zeta|^{p_i - 2} D_i \zeta \right) \cdot D_i (u - \zeta)\, dz \geq 0 \quad \forall \zeta \in \mathbf{Z}_s,\ s \in \mathbb{N}.$$

We may now take $\zeta = u \pm \varepsilon\, \eta$ with arbitrary $\varepsilon > 0$ and $\eta \in \mathbf{W}(Q)$. Simplifying and letting $\varepsilon \to 0$ we arrive at the inequalities

$$\pm \sum_i \int_Q \left( A_i(z) - |D_i u|^{p_i(z)-2} D_i u \right) \cdot D_i \eta \, dz \geq 0 \qquad \forall\, \eta \in \mathbf{W}(Q),$$

which are impossible unless $A_i(z) = |D_i u|^{p_i(z)-2} D_i u$ a.e. in $Q$. □

This completes the proof of Theorems 4.1 and 4.2.

## 4.3 $L^\infty$ estimates

### *4.3.1 Global Estimates*

**Theorem 4.3** *Let the conditions of* Theorem 4.1 *be fulfilled and, additionally,*

$$\forall\, k \in \mathbb{N} \quad \sup \left\{ |s|^{p_i'(z)-1} \left| \frac{\partial b_i(z,s)}{\partial s} \right|^{p_i'(z)} : z \in Q_T,\ s \in [-k,k] \right\} = B_k < \infty, \tag{4.36}$$

$$\forall\, s \in \mathbb{R},\ z \in Q_T \qquad \begin{aligned} &(a) \quad |d(z,s)| \leq d_0 |s| + h_d(z), \\ &(b) \quad \left| \frac{\partial b_i(z,s)}{\partial x_i} \right| \leq b_0 |s| + h_b(z) \end{aligned} \tag{4.37}$$

*with nonnegative constants $d_0$, $b_0$. Then the weak solution of problem* (4.1) *is bounded and satisfies the estimate*

$$\begin{aligned} \|u(\cdot,t)\|_{\infty,\Omega} &\leq e^{C_0 t} \|u_0\|_{\infty,\Omega} \\ &\quad + e^{C_0 t} \int_0^t e^{-C_0 t} \left( \|h_b(\cdot,\tau)\|_{\infty,\Omega} + \|h_d(\cdot,\tau)\|_{\infty,\Omega} \right) d\tau \end{aligned} \tag{4.38}$$

*with $C_0 = b_0 + d_0$.*

*Proof* Let us fix $k \in \mathbb{N}$ and consider the auxiliary problem

$$\begin{cases} u_t - \displaystyle\sum_i \frac{d}{dx_i} \left[ a_i |D_i u|^{p_i - 2} D_i u + b_i \right] + d_K(z,u) = 0 \quad \text{in } Q_T, \\ u = 0 \text{ on } \Gamma_T, \qquad u(x,0) = u_0(x) \text{ in } \Omega \end{cases} \tag{4.39}$$

with

$$d_K(z,u) \equiv d(z, \min\{|u|;\ K\} \operatorname{sign} u).$$

Since for every finite $K$

$$|d\,(z, \min\{|u|;\ K\}\,\mathrm{sign}\,u) \leq d_0\,K^{\lambda-1} + h_d,$$

it follows from Theorem 4.1 that problem (4.39) has a weak solution $u(z)$ in the sense of Definition 4.1. Let us introduce the function

$$u_K = \min\{|u|,\ K\}\,\mathrm{sign}\,u \equiv \begin{cases} K & \text{if}\, u > K, \\ u & \text{if}\, |u| \leq K, \\ -K & \text{if}\, u < -K. \end{cases}$$

Notice that $d_K(z,u) = d(z,u_K)$. For every $m \in \mathbb{N}$ the function $u_K^{2m-1}$ can be taken for the test-function in the integral identity. It is straightforward to check that

$$\begin{aligned} \int_0^t \int_\Omega u_t u_K^{2m-1}\,dxd\tau &= \int_\Omega u u_K^{2m-1}\,dx\Big|_{\tau=0}^{\tau=t} - \int_{Q_t} u\partial_t \left(u_K^{2m-1}\right)\,dz \\ &= \int_\Omega u u_K^{2m-1}\,dx\Big|_{\tau=0}^{\tau=t} - \int_{Q_t} u_K\partial_t \left(u_K^{2m-1}\right)\,dz \\ &= \int_\Omega u u_K^{2m-1}\,dx\Big|_{\tau=0}^{\tau=t} - \frac{2m-1}{2m}\int_\Omega u_K^{2m}\,dx\Big|_{\tau=0}^{\tau=t}. \end{aligned}$$

Let us write this equality in the form

$$\begin{aligned} &\int_\Omega u u_K^{2m-1}\,dx\Big|_{\tau=0}^{\tau=t} - \frac{2m-1}{2m}\int_\Omega u_K^{2m}\,dx\Big|_{\tau=0}^{\tau=t} \\ &\quad + (2m-1)\sum_{i=1}^n \int_0^t \int_\Omega a_i u_K^{2(m-1)}|D_i u_K|^{p_i(x,t)}\,dz \\ &= -\int_0^t \int_\Omega d_K(z,u) u_K^{2m-1}\,dxd\tau + \sum_{i=1}^n \int_0^t \int_\Omega b_i(z,u) D_i u_K^{2m-1}\,dz \\ &\equiv -I + \sum_{i=1}^n J_i. \end{aligned} \tag{4.40}$$

Since $b_i(z,u)D_i u_K^{2m-1} = b_i(z,u_K)D_i u_K^{2m-1}$, integration by parts transforms $J_i$ into

$$\begin{aligned} J_i &= -\int_0^t \int_\Omega \frac{\partial b_i(z,u_K)}{\partial u} u_K^{2m-1} D_i u_K dz - \int_0^t \int_\Omega \frac{\partial b_i(z,u_K)}{\partial x_i} u_K^{2m-1} dz \\ &:= J_i^{(1)} + J_i^{(2)}. \end{aligned} \tag{4.41}$$

Applying Young's and Hölder's inequalities and plugging (4.36)–(4.37), we estimate $J_i^{(1)}$, $J_i^{(2)}$ as follows: for every $\varepsilon > 0$

$$
\begin{aligned}
|J_i^{(1)}| &\le \varepsilon a_0(2m-1)\int_0^t\int_\Omega |u_K|^{2(m-1)}|D_iu_K|^{p_i}\,dz \\
&\quad + \int_0^t \frac{C(p_i^\pm,\varepsilon)}{(2m-1)^{\frac{1}{p_i-1}}}\int_\Omega \left(|u_K|^{p_i'-1}\left|\frac{\partial b_i(z,u_K)}{\partial u}\right|^{p_i'}\right)|u_K|^{2m-1}\,dz \\
&\le \varepsilon a_0(2m-1)\int_0^t\int_\Omega |u_k|^{2(m-1)}|D_iu_K|^{p_i}\,dz \\
&\quad + B_K\int_0^t \frac{C(p_i^\pm,\varepsilon)|\Omega|^{\frac{1}{2m}}}{(2m-1)^{\frac{1}{p_i-1}}}\left(\int_\Omega |u_K|^{2m}dx\right)^{1-\frac{1}{2m}}\,d\tau,
\end{aligned}
$$

$$
\begin{aligned}
|J_i^{(2)}| &\le \left|\int_0^t\int_\Omega \frac{\partial b_i(z,u_K)}{\partial x_i}u_K^{2m-1}dz\right| \le \int_0^t\int_\Omega (b_0|u_K|+h_b)\,|u_K|^{2m-1}dz \\
&\le b_0\int_0^t\int_\Omega u_K^{2m}\,dz + \int_0^t\left(\int_\Omega h_b^{2m}dx\right)^{\frac{1}{2m}}\left(\int_\Omega u_K^{2m}dx\right)^{\frac{2m-1}{2m}}\,dt \\
&\le b_0\int_0^t\int_\Omega u_K^{2m}\,dz + |\Omega|^{\frac{1}{2m}}\int_0^t \|h_b\|_{\infty,\Omega}\left(\int_\Omega u_K^{2m}dx\right)^{\frac{2m-1}{2m}}\,dt
\end{aligned}
$$

and

$$
\begin{aligned}
\int_0^t\int_\Omega |d_K(z,u)||u_K|^{2m-1}\,dz &\le d_0\int_0^t\int_\Omega |u_K|^{2m}\,dz + \int_0^t\int_\Omega h_d|u_K|^{2m-1}\,dz \\
&\le d_0\int_0^t\int_\Omega |u_K|^{2m}\,dz + |\Omega|^{\frac{1}{2m}}\int_0^t \|h_d\|_{\infty,\Omega}\left(\int_\Omega u_K^{2m}dx\right)^{\frac{2m-1}{2m}}\,dt.
\end{aligned}
$$

Introduce the function

$$
y_m(t) := \|u_K\|_{2m,\Omega}(t).
$$

Substituting the derived estimates into (4.40) and using the fact that $uu_K^{2m-1} \ge u_K^{2m}$ and $u_0u_K^{2m-1}(x,0) = u_0^{2m}$, we arrive at the integral inequality for $y_m(t)$: for every $\varepsilon \in (0,1)$

$$
\frac{1}{2m}y_m^{2m}(t) \le \frac{1}{2m}y^{2m}(0) + \alpha\int_0^t y_m^{2m}(\tau)\,d\tau + |\Omega|^{\frac{1}{2m}}\int_0^t \beta(\tau)y_m^{2m-1}(\tau)\,d\tau \tag{4.42}
$$

under the notations

$$\alpha = b_0 + d_0, \qquad \beta(t) = B_K \frac{C(p_i^{\pm}, \varepsilon)}{(2m-1)^{\frac{1}{p_i-1}}} + \|h_b\|_{\infty,\Omega}(t) + \|h_d\|_{\infty,\Omega}(t).$$

Let $z(t)$ be the solution of the equation

$$\frac{1}{2m} z^{2m}(t) = \alpha \int_0^t z^{2m}(\tau)\,d\tau + |\Omega|^{\frac{1}{2m}} \int_0^t \beta(\tau) z^{2m-1}(\tau)\,d\tau + \frac{1}{2m} z^{2m}(0) \quad (4.43)$$

with $z(0) = \|u_0\|_{2m,\Omega} + \delta$ and an arbitrary $\delta > 0$. The function $z(t)$ can be constructed as the solution of the Cauchy problem for the ODE obtained from (4.43) by means of differentiation:

$$\begin{cases} z'(t) = \alpha\, z(t) + |\Omega|^{\frac{1}{2m}} \beta(t), & t > 0, \\ z(0) = \|u_0\|_{2m,\Omega} + \delta. \end{cases}$$

This equation can be explicitly integrated and the solution of (4.43) has the form

$$z(t) = \left(\delta + \|u_0\|_{2m,\Omega}\right) e^{\alpha t} + |\Omega|^{\frac{1}{2m}} e^{\alpha t} \int_0^t e^{-\alpha\tau} \beta(\tau) d\tau.$$

By the choice of the initial data $y_m(0) - z(0) = -\delta < 0$, which yields the inequality $y_m(t) < z(t)$ for all $t > 0$. Now we argue by contradiction. If the assertion is false, then

$$t^* = \sup\{t \geq 0 : \; y_m(t) < z(t)\} < \infty.$$

Since $0 \leq y_m(t) \leq K|\Omega|^{\frac{1}{2m}}$, it follows from (4.42) that $t^*$ is strictly positive. Subtracting (4.43) from (4.42) we find that

$$0 = \frac{1}{2m}\left(y_m^{2m}(t^*) - z^{2m}(t^*)\right) \leq \alpha \int_0^{t^*} \underbrace{\left(y_m^{2m}(\tau) - z^{2m}(\tau)\right)}_{<0} d\tau$$

$$+ |\Omega|^{\frac{1}{2m}} \int_0^{t^*} \beta(\tau) \underbrace{\left(y_m^{2m-1}(\tau) - z^{2m-1}(\tau)\right)}_{<0} d\tau + \frac{1}{2m} \underbrace{\left(y_m^{2m}(0) - z^{2m}(0)\right)}_{<0} < 0,$$

which is impossible. Thus, $y_m(t) \leq z(t)$ for every $m$ and $\delta$. Letting $m \to \infty$ we conclude that for every $\delta > 0$

$$\begin{aligned}\|u_K(t)\|_{\infty,\Omega} &\le \left(\delta + \|u_0\|_{\infty,\Omega}\right) e^{\alpha t} \\ &\quad + e^{\alpha t} \int_0^t e^{-\alpha\tau} \left(\|h_b\|_{\infty,\Omega}(\tau) + \|h_d\|_{\infty,\Omega}(\tau)\right) d\tau \\ &:= \kappa(t) \le \kappa(T),\end{aligned}$$

which yields estimate (4.38) for the solution of the auxiliary problem (4.39). To complete the proof we notice that under the choice $K = 1 + \kappa(T)$ the solution $u_K$ of the auxiliary problem does not exceed $K$. For this choice of $K$ we have $d_K(z, u_K) = d(z, u_K)$, which means that the solution of the auxiliary problem (4.39) coincides with the solution of problem (4.1). □

*Remark 4.1* The question of local boundedness of solutions to Eq. (4.1) with anisotropic but constant growth conditions was studied in paper [269] by means of suitable embedding theorems in the anisotropic Sobolev spaces.

### 4.3.2 Global Existence via Boundedness

Let us consider Eq. (4.1) with the term $d(z, u)$ of the special form:

$$d(z, u) = d_1(z, u)|u|^{\sigma(z)-2}u + d_2(z, u)|u|^{\lambda-2}u + h_d(z), \tag{4.44}$$

with

$$0 < d_{01} \le d_1(z, u) < \infty, \quad |d_2(z, u)| \le d_{02} < \infty, \quad d_{01}, d_{02}, \lambda = const > 0. \tag{4.45}$$

If $\sigma(z) \in C(\overline{Q}_T)$ and $1 \le \sigma^- \le \sigma^+ \le \lambda$, the existence of a weak solution follows from Theorems 4.1 and 4.2. If we additionally assume that the conditions of Theorem 4.3 are fulfilled, then this solution is bounded. We now turn to the case

$$2 < \lambda < \sigma^- \le \sigma(z) \le \sigma^+ < \infty, \tag{4.46}$$

which does not fall into the scope of Theorems 4.1–4.3. Let us take a positive number $R_0 < \infty$ such that

$$\mathscr{P}(z, R_0) \equiv d_{01} R_0^{\sigma(z)-1} - d_{02} R_0^{\lambda-1} - b_0 R_0 - \sup_{Q_T} |h_d(z)| - \sup_{Q_T} |h_b(z)| \ge 0. \tag{4.47}$$

Because of condition $\sigma^- > \lambda > 2$, such a number always exists, provided that

$$\sup_{Q_T} |h_b| + \sup_{Q_T} |h_d| < \infty. \tag{4.48}$$

**Theorem 4.4** *Let the coefficients $a_i$, $b_i$ and the exponents $p_i$ satisfy the conditions of* Theorem 4.3. *Assume that $d(z,u)$ satisfies condition* (4.44). *Let $\sigma(z)$ be measurable in $Q_T$ and let conditions* (4.44)–(4.47) *be fulfilled. Then problem* (4.1) *has in $Q_T$ at least one bounded weak solution satisfying the estimate*

$$\|u\|_{\infty,Q_T} \le \max\left\{\sup_\Omega |u_0|;\ R_0\right\}.$$

*Remark 4.2* The conditions of Theorem 4.4 are surely fulfilled for the diffusion–absorption equation

$$u_t = \Delta_{p(z)}u - |u|^{\sigma(z)-2}u + h_d(z), \quad \sigma(z) > 1.$$

In this case

$$\|u\|_{\infty,\Omega}(t) \le \sup_\Omega |u_0| + \int_0^t \|h_d\|_{\infty,\Omega}(\tau)\,d\tau.$$

*Proof* Fix an arbitrary finite number $R > 0$ and consider the regularized problem

$$\begin{cases} u_t - \sum_i \dfrac{d}{dx_i}\left[a_i|D_iu|^{p_i(z)-2}D_iu + b_{iR}(z,u)\right] + d_R(z,u) = 0 \quad \text{in } Q_T, \\ u = 0 \text{ on } \Gamma_T, \qquad u(x,0) = u_0(x) \text{ in } \Omega \end{cases} \tag{4.49}$$

with

$$d_R(z,u) = d_1(z,u)|u_R|^{\sigma(z)-2}u_R + d_2(z,u_R) + h_d(z), \quad b_{iR}(z,u) = b_i(z,u_R),$$

and

$$u_R = \min\{|u|, R\}\operatorname{sign} u, \qquad D_iu_R \equiv \begin{cases} 0 & \text{if } |u| > R, \\ D_iu & \text{if } |u| \le R. \end{cases}$$

The regularized problem (4.49) has a global weak solution. Moreover, since $b_i$ satisfy the conditions of Theorem 4.3, this solution is globally in time bounded: $\|u\|_{\infty,Q_T} \le C(R)$. The theorem will be proved if we show that the constant $C(R)$ is in fact independent of $R$. Let us set

$$R = \max\left\{R_0,\ \sup_\Omega |u_0|\right\}$$

with $R_0$ satisfying the inequality $\mathscr{P}(z,R_0) \ge 0$. Let us take for the test-function in (4.10) the function

$$u_+ = \max\{u - R, 0\}, \qquad D_iu_+ \equiv \begin{cases} D_iu & \text{if } u > R, \\ 0 & \text{if } u \le R. \end{cases}$$

Arguing as in the proof of Theorem 4.3 we arrive at the equality

$$\frac{1}{2}\frac{d}{dt}\int_\Omega u_+^2(x,t)\,dx + \sum_i \int_\Omega \left(a_i\,|D_i u_+|^{p_i} + b_i(z,u_R)D_i u_+\right)\,dx \\ + \int_\Omega d_R(z,u)u_+\,dx = 0 \quad \forall\,\text{a.e. } t \in (0,T), \tag{4.50}$$

which can be written in the form

$$\forall\,\text{a.e. } t \in (0,T) \quad \frac{1}{2}\frac{d}{dt}\int_\Omega u_+^2\,dx + \sum_i \int_\Omega a_i\,|D_i u_+|^{p_i}\,dx + I \equiv \sum_{i=1}^n \left(J_i^{(1)} + J_i^{(2)}\right).$$

In the last relation

$$I = \int_\Omega \Big(d_1(z,u)\,(\min\{|u|,R\})^{\sigma(z)-1}\,\mathrm{sign}\,u \\ + d_2\left(z,(\min\{|u|,R\})^{\lambda-1}\,\mathrm{sign}\,u\right) + h_d(z)\Big)u_+\,dx,$$

$$J_i^{(1)} = -\int_\Omega \frac{\partial b_i(z,u_R)}{\partial u}u_+ D_i u_R dx = 0, \qquad J_i^{(2)} = \int_\Omega \frac{\partial b_i(z,u_R)}{\partial x_i}u_+ dx.$$

The terms $J_i^{(2)}$ are estimated exactly as in the proof of Theorem 4.3,

$$\left|J_i^{(2)}\right| \le \left|\int_\Omega \frac{\partial b_i(z,u_R)}{\partial x_i}u_+ dx\right| \le \int_\Omega (b_0 R + |h_b|)\,u_+ dx,$$

while

$$I \ge \int_\Omega \left(d_{01}R^{\sigma(z)-1} - d_{02}R^{\lambda-1} - \sup_{Q_T}|h_d|\right)u_+\,dx.$$

Gathering these estimates we find that

$$\frac{1}{2}\frac{d}{dt}\int_\Omega u_+^2\,dx + \sum_i \int_\Omega a_i\,|D_i u_+|^{p_i}\,dx + \int_\Omega \mathscr{P}(z,R_0)u_+ \le 0$$

with the function $\mathscr{P}$ introduced in (4.47), Since $\mathscr{P}(z,R_0) \ge 0$ by the choice of $R_0$, and $u_+(x,0) = 0$ by the choice of $R$, the last inequality yields

$$u_+(z) = 0 \qquad \forall\,\text{a.e. } z \in Q_T,$$

whence $u(z) \le R$ a.e. in $Q_T$. The same argument shows that

$$u_-(z) = \max\{-u(z) - R, 0\} = 0$$

and, finally,

$$|u(z)| \le R = \max\left\{\sup_{\Omega} |u_0(x)|,\ R_0\right\}.$$

This inequality means that

$$b_{iR}(z,u) \equiv b_i(z,u), \qquad d_R(z,u) \equiv d(z,u),$$

which completes the proof. □

### 4.3.3 Local Existence via Boundedness

Let us consider problem (4.1) with the term $d(z,u)$ satisfying the growth condition

$$|d(z,u)| \le d_0|u|^{\lambda-1} + h_d(z), \quad \lambda = const > 2. \tag{4.51}$$

For $0 \le \lambda \le 2$ the existence of a global bounded solution to problem (4.1) is proved in Theorem 4.1. The next theorem asserts the existence of a local bounded solution in the case $\lambda > 2$.

**Theorem 4.5** *Let us assume that in the conditions of* Theorem 4.1 *and* (4.3) *the growth condition on the function $d(z,u)$ is substituted by* (4.51). *Then for every $u_0 \in L^\infty(\Omega)$ there exists $\theta \in (0,T]$ depending on*

$$\lambda,\ b_0,\ d_0,\ \|u_0\|_{L^\infty(\Omega)},\ \|h_d\|_{L^1(0,\theta;L^\infty(\Omega))} \text{ and } \|h_b\|_{L^1(0,\theta;L^\infty(\Omega))}$$

*such that in the cylinder $Q_\theta$ problem* (4.1) *has at least one weak solution $u \in \mathbf{W}(Q_\theta)$ with $u_t \in \mathbf{W}'(Q_\theta)$ and $\|u\|_{\infty,Q_\theta} < \infty$. This solution can be continued to the interval $[0,T^*]$ where*

$$T^* = \sup\{\theta \in [0,T] :\ \|u\|_{\infty,Q_\theta} < \infty\}.$$

*Proof* Let us consider the auxiliary problem

$$\begin{cases} u_t - \sum_i D_i\left(a_i|D_iu|^{p_i(z)-2}D_iu + b_i\right) + d_r(z,u) = 0 \ \text{ in } Q_T \\ u = 0 \text{ on } \Gamma, \qquad u(x,0) = u_0(x) \text{ in } \Omega \end{cases} \tag{4.52}$$

with the right-hand side

$$d_r(z,u) = d(z, \min\{|u|,r\}\,\mathrm{sign}\,u), \quad r = const > 1. \tag{4.53}$$

As in the proof of Theorem 4.3, we will make use of the fact that

$$|d_r(z,u)| \le d_0 r^{\lambda-1} + h_d(z), \quad d_r(z,u) = d(z,u) \quad \text{if } r \ge |u|.$$

By Theorems 4.1 and 4.3, for every $r > 1$ the regularized problem (4.52) has a global bounded weak solution $u(z)$. Let us show that the function $w(t) = \|u(\cdot, t)\|_{\infty,\Omega}$ can be estimated by a constant which does not depend on $r$. Following the proof of Theorem 4.3 we find that the solution of (4.52) satisfies inequality (4.38) with $C_0 = b_0$ and $h_d$ substituted by $h_d + d_0 r^{\lambda-1}$:

$$\|u(\cdot, t)\|_{\infty,\Omega} \leq e^{b_0 t}\|u_0\|_{\infty,\Omega} + e^{b_0 t}\int_0^t e^{-b_0 t}\|h_b(\cdot, t)\|_{\infty,\Omega}\, dt$$
$$+ e^{b_0 t}\int_0^t e^{-b_0 t}\|h_d(\cdot, t)\|_{\infty,\Omega}\, dt + d_0 r^{\lambda-1} t e^{b_0 t} \equiv \mathscr{R}(r, t).$$

For every fixed $r > 1$

$$\mathscr{R}(r, t) \to \|u_0\|_{\infty,\Omega} \quad \text{as } t \to 0,$$

whence for every $r \geq \|u_0\|_{\infty,\Omega}$ there is $t \equiv t(r)$ such that

$$\forall t \in [0, t(r)] \qquad \|u(\cdot, t)\|_{\infty,\Omega} \leq r.$$

It follows that for $r$ and $t(r)$ chosen in this way $\|u(\cdot, t)\|_{\infty,\Omega} \leq r$ for all $t \leq t(r)$, i.e. the constructed solution of the regularized problem (4.52) is a weak solution of problem (4.1) in the cylinder $Q_{t(r)}$. The possibility of continuation of this solution to the maximal interval $[0, T^*]$ follows from the fact that the function $u(x, t(r))$ possesses the same properties as the initial function $u_0$. □

## 4.4 Uniqueness Theorems

In this section we study the question of uniqueness of weak solutions to the problem

$$\begin{cases} u_t - \displaystyle\sum_i \frac{d}{dx_i}\left[a_i(z, u)|D_i u|^{p_i(z)-2} D_i u\right] + d(z, u) = 0 & \text{in } Q, \\ u = 0 \text{ on } \Gamma, \qquad u(x, 0) = u_0(x) \text{ in } \Omega. \end{cases} \tag{4.54}$$

The weak solution is understood in the sense of Definition 4.1.

Let us assume that the functions $a_i$ are continuous with the modulus of continuity $\omega$,

$$|a_i(z, u_1) - a(z, u_2)| \leq \omega(|u_1 - u_2|), \tag{4.55}$$

and claim that the function $\omega$ satisfies the condition

$$\int_\varepsilon^1 \frac{ds}{\omega^\alpha(s)} \to \infty \quad \text{as } \varepsilon \to 0+ \qquad \text{for some } 1 < \alpha < \frac{p^+}{p^+ - 1}. \tag{4.56}$$

Without loss of generality we may assume that $p^+ \geq 2$.

**Theorem 4.6** *Let*

$$1 < p^- \le p_i(z) \le p^+ < \infty, \quad 0 < a_0 \le a_i(z,u) \le a_1 < \infty, \quad i = 1, \dots, n.$$

*Problem* (4.54) *does not admit more than one solution* $u \in \mathbf{W}(Q_T)$ *if conditions* (4.55) *and* (4.56) *are fulfilled and*

$$u \mapsto d(z,u) \text{ is a nondecreasing function.} \tag{4.57}$$

*Proof* We argue by contradiction. Let us assume that problem (4.54) admits two different solutions $u_1, u_2 \in \mathbf{W}(Q_T)$ and there is $\delta > 0$ such that for some $\tau \in (0,T]$ $w = u_2 - u_1 > \delta$ on a set $\Omega_\delta$:

$$\Omega_\delta = \Omega \cap \{x \in \Omega : w(x,\tau) > \delta\} \quad \text{and} \quad |\Omega_\delta| = \mu > 0.$$

We will show that this assumption leads to a contradiction unless $\mu = 0$. Not loosing generality we assume that $t = T$. Set

$$d_i \equiv d(z,u_i), \qquad a_{ij} \equiv a_j(z,u_i), \qquad i = 1,2, \ j = 1, \dots, n.$$

By Definition 4.1, for every test-function $\zeta \in \mathbf{Z}$

$$\begin{aligned} &\int_{Q_\tau} \left( w_t\zeta + \sum_{i=1}^n a_{2i}(|D_i u_2|^{p_i-2} D_i u_2 - |D_i u_1|^{p_i-2} D_i u_1) D_i\zeta + (d_2 - d_1)\zeta \right) dz \\ &\quad + \int_{Q_\tau} \sum_{i=1}^n (a_{2i} - a_{1i}) |D_i u_1|^{p_i-2} D_i u_1)\, D_i\zeta \, dz = 0. \end{aligned} \tag{4.58}$$

Let us denote

$$A(u_2,u_1) = d_2 - d_1, \quad J(u_2,u_1,\zeta) = -\int_{Q_\tau} \sum_{i=1}^n (a_{2i} - a_{1i}) |D_i u_1|^{p_i-2} D_i u_1 \, D_i\zeta \, dz$$

and write (4.58) in the form

$$\begin{aligned} &\int_{Q_\tau} \left( w_t\zeta + \sum_{i=1}^n a_{2i}(|D_i u_2|^{p_i-2} D_i u_2 - |D_i u_1|^{p_i-2} D_i u_1) D_i\zeta + A(u_2,u_1)\zeta \right) dz \\ &\quad = J(u_2,u_1,\zeta). \end{aligned} \tag{4.59}$$

Introduce the functions

$$F_\varepsilon(\xi)=\begin{cases}\displaystyle\int_\varepsilon^\xi \frac{ds}{\omega^\alpha(s)} & \text{if } \xi>\varepsilon,\\ 0 & \text{if } \xi\le\varepsilon,\end{cases}\qquad G_\varepsilon(\eta)=\begin{cases}\displaystyle\int_\varepsilon^\eta F_\varepsilon(s)\,ds & \text{if } \eta>\varepsilon,\\ 0 & \text{if } \eta\le\varepsilon.\end{cases}\tag{4.60}$$

$F_\varepsilon$, $G_\varepsilon$ depend on the parameters $\delta\ge\varepsilon>0$ and the function $\omega(\cdot)$, which satisfies condition (4.56). The definition of $F_\varepsilon$ and (4.57) yield:

$$\forall\, u,v\in\mathbb{R}\qquad A(u,v)F_\varepsilon(u-v)\ge 0.\tag{4.61}$$

Set $Q_{\varepsilon,\tau}\equiv\{z\in Q_\tau\,:\; w>\varepsilon\}$. By virtue of (4.60)

$$D_iF_\varepsilon(w)=\begin{cases}\dfrac{D_iw}{\omega^\alpha(w)} & \text{in } Q_\varepsilon,\\ 0 & \text{in } Q\backslash Q_\varepsilon.\end{cases}$$

Letting in (4.59) $\zeta=F_\varepsilon(w)$, we obtain:

$$\begin{aligned}&\int_{\Omega_\varepsilon} G_\varepsilon(w(x,\tau))\,dx+\int_{\varepsilon,\tau} A(u_2,u_1)F_\varepsilon(w)\,dz\\ &\quad+\int_{Q_{\varepsilon,\tau}}\left(\sum_{i=1}^n a_{2i}(|D_iu_2|^{p_i-2}D_iu_2-|D_iu_1|^{p_i-2}D_iu_1)\frac{D_iw}{\omega^\alpha(w)}\right)dz\\ &\equiv J(u_2,u_1,F_\varepsilon(w)).\end{aligned}\tag{4.62}$$

Notice that since $\delta\ge\varepsilon$, then $\Omega_\delta\subseteq\Omega_\varepsilon$, $|\Omega_\varepsilon|\ge|\Omega_\delta|>\mu$ and, by virtue of (4.56),

$$\int_{\Omega_\varepsilon} G_\varepsilon(w(x,\tau))\,dx\ge\mu F_\varepsilon(\delta)\to\infty\quad\text{as } \varepsilon\to 0+.\tag{4.63}$$

Let us consider first the case $p_i\ge 2$. By virtue of (4.3) and the first inequality of (4.35)

$$a_0\frac{|D_iw|^{p_i}}{\omega^\alpha(w)}\le a_{2i}(|D_iu_1|^{p_i-2}D_iu_1-|D_iu_2|^{p_i-2}D_iu_2)\frac{D_iw}{\omega^\alpha(w)}.\tag{4.64}$$

According to (4.56)

$$\frac{p_i}{p_i-1}\ge\frac{p^+}{p^+-1}\ge\alpha>1.$$

Applying Young's inequality, we may estimate the integrand of $J$ in the following way:

$$
\begin{aligned}
\left|(a_{21i} - a_{1i})|D_i u_1|^{p_i-2} D_i u_1 \frac{D_i w}{\omega^\alpha(w)}\right| &\le \omega(w)|D_i u_1|^{p_i-1} \frac{|D_i w|}{\omega^\alpha(w)} \\
&\le \frac{a_0}{2} \frac{|D_i w|^{p_i}}{\omega^\alpha(w)} + C(a_0, p^+)|D_i u_1|^{p_i} \omega^{p_i'-\alpha}(w) \\
&\le \frac{a_0}{2} \frac{|D_i w|^{p_i}}{\omega^\alpha(w)} + C(a_0, p^+)\, |D_i u_1|^{p_i}.
\end{aligned}
\tag{4.65}
$$

Let now $1 < p^- \le p_i < 2$. Applying (4.3) and the second inequality of (4.35) we have

$$
\begin{aligned}
&a_0(p^- - 1)(|D_i u_1| + |D_i u_2|)^{p_i-2} \frac{|D_i w|^2}{\omega^\alpha(w)} \\
&\quad \le a_{2i}(|D_i u_2|^{p_i-2} D_i u_2 - |D_i u_1|^{p_i-2} D_i u_1) \frac{D_i w}{\omega^\alpha(w)}
\end{aligned}
\tag{4.66}
$$

and

$$
\begin{aligned}
&\left|(a_{2i} - a_{1i})|D_i u_1|^{p_i-2} D_i u_1 \frac{D_i w}{\omega^\alpha(w)}\right| \le \left|\omega(w)(|D_i u_1| + |D_i u_2|)^{p_i-1} \frac{D_i w}{\omega^\alpha(w)}\right| \\
&\le \left|\omega(w)(|D_i u_1| + |D_i u_2|)^{p_i-1} \frac{D_i w}{\omega^\alpha(w)}\right| \\
&\le \frac{a_0(p^- - 1)}{2} (|D_i u_1| + |D_i u_2|)^{p_i-2} \frac{|D_i w|^2}{\omega^\alpha(w)} \\
&\quad + C\omega^{2-\alpha}(w)(|D_i u_2| + |D_i u_1|)^{p_i} \\
&\le \frac{a_0(p^- - 1)}{2} (|D_i u_1| + |D_i u_2|)^{p_i-2} \frac{|D_i w|^2}{\omega^\alpha(w)} + \widetilde{C}(|D_i u_2| + |D_i u_1|)^{p_i}
\end{aligned}
\tag{4.67}
$$

with

$$
1 < \alpha \le \frac{p^+}{p^+ - 1} \le 2.
$$

Plugging the pointwise estimates (4.64), (4.65) and (4.66), (4.67) into (4.62) and dropping the nonnegative terms, we arrive at the inequality

$$
\int_{\Omega_\varepsilon} G_\varepsilon(w(x,\tau))\, dx \le \widetilde{C} \int_{Q_{\varepsilon,\tau}} \sum_{i=1}^n (|D_i u_1| + |D_i u_2|)^{p_i} dz = \widehat{C} \tag{4.68}
$$

with a constant $\widetilde{C}$ independent of $\varepsilon$. Gathering this inequality with (4.63) we obtain the needed contradiction. This means that $\mu = 0$ and $w \le 0$ a.e. in $Q_T$. To complete the proof it suffices to replace $u_1$ and $u_2$. □

These arguments extend to non-monotone functions $d(z, u)$. Let

$$u_t - \sum_{i=1}^{n} \frac{d}{dx_i}\left[a_i(z,u)(|D_i u_1|^{p_i-2} D_i u\right] + d(z,u) = 0. \tag{4.69}$$

We assume that

$$\begin{cases} |a_i(z,u_1) - a_i(z,u_2)| \le C_1 \omega(|u_1 - u_2|), \\ \omega^{\alpha}(s) = s^2, \\ |d(z,u_1) - d_i(z,u_2)| \le C_2 |u_1 - u_2|, \end{cases} \tag{4.70}$$

It is easy to calculate that in this case

$$F_\varepsilon(s) = \begin{cases} \dfrac{1}{\varepsilon} - \dfrac{1}{s} & \text{for } s > \varepsilon, \\ 0 & \text{otherwise,} \end{cases} \qquad G_\varepsilon(s) = \begin{cases} \dfrac{s}{\varepsilon} - 1 - \ln\left(\dfrac{s}{\varepsilon}\right) & \text{for } s > \varepsilon, \\ 0 & \text{otherwise.} \end{cases}$$

**Proposition 4.1** *There exists a positive number $\mu > 2$ such that*

$$sF_\varepsilon(s) \le \begin{cases} 2G_\varepsilon(s) & \textit{for } s \ge \mu\varepsilon \\ const & \textit{for } \varepsilon \le s \le \mu\varepsilon. \end{cases} \tag{4.71}$$

*Proof* Set $z = s/\varepsilon$ and introduce the function

$$f(z) = 2G_\varepsilon(s) - sF_\varepsilon(s) \equiv z - 1 - 2\ln z.$$

Obviously,

$$f(1) = 0, \quad f(z) \to \infty \text{ as } z \to \infty, \quad f'(z) = 1 - \frac{2}{z} \ge 0, \quad f''(z) = \frac{2}{z^2} \ge 0 \quad \text{if } z \ge 2.$$

Since $f(z)$ is monotone increasing for $z > 2$ and tends to infinity as $z \to \infty$, there is $\mu \ge 2$ such that $f(z) \ge 0$ for $z \ge \mu$. For $z \in [1, \mu]$ we have

$$sF_\varepsilon(s) = z - 1 \le \mu - 1.$$

□

**Theorem 4.7** *Let in the conditions of* Theorem 4.6 *condition* (4.57) *be substituted by condition* (4.70). *Then the weak solution of problem* (4.54) *is unique.*

*Proof* We adapt the proof of Theorem 4.6. Let $u_1$, $u_2$ be two different solutions of problem (4.1). Set $u = u_1 - u_2$. Following the proof of Theorem 4.6 we arrive at the relation

$$\int_{\Omega_\varepsilon} G_\varepsilon(u(x,\tau))dx + \int_{Q_{\varepsilon,\tau}} \left(\sum_{i=1}^{n} a_{1i}(|D_i u_1|^{p_i-2} D_i u_1 - |D_i u_2|^{p_i-2} D_i u_2)\frac{D_i u}{u^2}\right) dz = I_1 + I_2 \tag{4.72}$$

with

$$I_1 = -\int_{Q\varepsilon} \sum_{i=1}^{n} (a_{1i} - a_{2i})|D_i u_2|^{p_i-2} D_i u_2 \frac{D_i u}{u^2}\, dz,$$

$$I_2 = -\int_{Q\varepsilon} (d(z,u_1) - d(z,u_2)) F_\varepsilon(u)\, dz.$$

The difference between this case and the one studied in Theorem 4.6 is that now the term $I_2$ has no definite sign. By Proposition 4.1

$$|I_2| \le C\int_{Q\varepsilon} u F_\varepsilon(u)\, dz = C\int_0^t \left(\int_{\Omega\cap(\varepsilon\le u\le\mu\varepsilon)} \cdots + \int_{\Omega\cap(\mu\varepsilon\le u)} \cdots\right) dt \equiv I_{21} + I_{22},$$

whence

$$I_{21} \le C\int_0^t \left(\int_{\Omega\cap(\varepsilon\le u\le\mu\varepsilon)} G_\varepsilon(u)dx\right) dt \le C\int_0^t \left(\int_{\Omega_\varepsilon} G_\varepsilon(u)dx\right) dt, \qquad I_{22} \le C|\Omega|T.$$

Let us introduce the function

$$Y(t) = \int_{\Omega\varepsilon} G_\varepsilon(u)\, dx.$$

Substituting the above inequalities into (4.72) and taking into account (4.35), we find that the function $Y(t)$ satisfies the Gronwall-type inequality

$$Y(t) \le C\left(\int_0^t Y(s)ds + \sum_{i=1}^{n}\int_{Q_\varepsilon} \left(|D_i u_1|^{p_i} + |D_i u_2|^{p_i}\right) dz + 1\right).$$

It follows that $Y(t) \le K$, which contradicts condition (4.63). □

**Corollary 4.1** (Comparison principle) *Let $u, v \in \mathbf{W}(Q_T)$ be two weak solutions of problem* (4.54) *such that $u(x,0) \le v(x,0)$ a.e. in $\Omega$. If the coefficients and the exponents of nonlinearity satisfy the conditions of* Theorem 4.6, *or* Theorem 4.7, *then $u \le v$ a.e. in $Q_T$.*

## 4.5 Global Regularity for a Class of Model Equations

Let us consider the following simplified version of problem (4.1):

$$\begin{cases} u_t - \sum_{i=1}^{n} \dfrac{d}{dx_i}\left(a_i(z)|D_i u|^{p_i(z)-2} D_i u\right) + c(z)|u|^{\sigma(z)-2}u = f(z) \quad \text{in } Q_T, \\ u = 0 \text{ on } \Gamma, \quad u(x,0) = u_0(x) \text{ in } \Omega. \end{cases} \tag{4.73}$$

We want to trace how the regularity of weak solutions depends on the regularity of the data, especially on the properties of the exponents $p_i(z)$ and $\sigma(z)$. Let us accept the notation

$$\begin{aligned} &\lambda_p(t) = \sum_{i=1}^{n} \max_{\overline{\Omega}} |p_{it}(x,t)|, && \lambda_a(t) = \sum_{i=1}^{n} \max_{\overline{\Omega}} |a_{it}(x,t)|, \\ &\lambda_\sigma(t) = \max_{\overline{\Omega}} |\sigma_t(x,t)|, && \lambda_c(t) = \max_{\overline{\Omega}} |c_t(x,t)|. \end{aligned} \tag{4.74}$$

**Theorem 4.8** *Let us assume that*

(a) *$a_i$, $p_i$ satisfy the conditions of* Theorem 4.1,
(b) *$p_{it}(z) \le 0$ for a.a. $z \in Q_T$,*
(c) *$\sigma(z) \in C(\overline{Q}_T)$, $c(z)$ measurable and bounded in $Q_T$, $\sigma_t(z)$ exists a.e. in $Q_T$ and*

$$\sigma_t(z) \le 0, \quad 0 \le c_0 \le c(z) \quad \text{a.e. in } Q_T,$$

(d) *$\lambda_p(t)$, $\lambda_\sigma(t)$, $\lambda_a(t)$, $\lambda_c(t) \in L^1(0,T)$ and*

$$\int_0^T \left(\lambda_p(t) + \lambda_a(t) + \lambda_\sigma(t) + \lambda_c(t)\right) dt = K < \infty.$$

(e) *$u_0 \in L^{\sigma(x,0)}(\Omega)$, $D_i u_0 \in L^{p_i(x,0)}(\Omega)$, $f \in L^2(Q_T)$.*

*Then the weak solution of problem* (4.73) *satisfies the estimate*

$$\sup_{t \in (0,T)} \int_\Omega \left[\sum_{i=1}^{n} a_i |D_i u\ |^{p_i} + c|u|^\sigma\right] dx + \int_{Q_T} |u_t|^2\, dz$$

$$+\int_{Q_T}\left(\sum_{i=1}^{n} a_i|D_i u|^{p_i}|\ln|D_i u||p_{it}| + c|u|^{\sigma}|\ln|u||\sigma_t|\right)dz$$
$$\leq C\left(\int_{\Omega}\left(\sum_{i=1}^{n}|D_i u_0|^{p_i(x,0)} + |u_0|^{\sigma(x,0)}\right)dx + \|f\|^2_{2,Q_T} + 1\right) \tag{4.75}$$

*with an absolute constant* $C = C(p^{\pm},\ \sigma^{\pm},\ q, T,\ K)$.

*Proof* Under the conditions of Theorem 4.8 the weak solution of problem (4.73) is unique and can be obtained as the limit of the sequence of Galerkin's approximations

$$u^{(m)} = \sum_{k=1}^{m} u_k^{(m)}(t)\psi_k(x), \quad \psi_k(x) \in H_0^s(\Omega),$$

where the functions $u_k^{(m)}(t)$ are solutions of problem (4.14) and $u^{(m)}(x,0) \to u_0$ in $\mathbf{V}_0(\Omega)$. Therefore, to prove Theorem 4.8 it suffices to derive estimate (4.75) for the approximate solutions $u^{(m)}$.

For the sake of simplicity, throughout the proof we use the notation $u$ for the approximate solution $u^{(m)}$. Fix some $m$, multiply relations (4.14) by $u_{k,t}$ and take the sum over $k = 1, \ldots, m$. This gives the equality

$$\|u_t\|^2_{2,\Omega} + \sum_{i=1}^{n}\int_{\Omega}\left(a_i|D_i u\ |^{p_i-2} D_i u\ D_i u_t\right)dx$$
$$+\int_{\Omega} c|u|^{\sigma-2}u\,u_t\,dx = \int_{\Omega} f\,u_t\,dx. \tag{4.76}$$

We will use the easily verified formulas

$$a_i|D_i u|^{p_i-2} D_i u\ D_i u_t = \frac{\partial}{\partial t}\left(a_i\frac{|D_i u\ |^{p_i}}{p_i}\right)$$
$$+ a_i|D_i u\ |^{p_i}\left(\frac{1}{p_i^2} - \frac{\ln|D_i u\ |^{p_i}}{p_i}\right)p_{it} - a_{it}\frac{|D_i u\ |^{p_i}}{p_i},$$

$$c|u|^{\sigma-2}u\,u_t = \frac{\partial}{\partial t}\left(c\frac{|u|^{\sigma}}{\sigma}\right) + c|u|^{\sigma}\left(\frac{1}{\sigma^2} - \frac{\ln|u|}{\sigma}\right)\sigma_t - c_t\frac{|u|^{\sigma}}{\sigma}.$$

Substituting them into (4.76) we rewrite it in the form

$$\|u_t\|^2_{2,\Omega} + Y'(t) = I_1 + \sum_{i=1}^{n} I_{2,i} + I_3, \tag{4.77}$$

where

$$Y(t)=\int_{\Omega}\left(\sum_{i=1}^{n}a_i\frac{|D_iu\,|^{p_i}}{p_i}+c\frac{|u|^{\sigma}}{\sigma}\right)dx,\quad I_1=-\int_{\Omega}f\,u_t dx, \tag{4.78}$$

$$I_{2,i}=\int_{\Omega}\left(-a_i|D_iu\,|^{p_i}\left(\frac{1}{p_i^2}-\frac{\ln|D_iu\,|^{p_i}}{p_i}\right)p_{it}+a_{it}\frac{|D_iu\,|^{p_i}}{p_i}\right)dx, \tag{4.79}$$

$$I_3=\int_{\Omega}\left(-c|u\,|^{\sigma}\left(\frac{1}{\sigma^2}-\frac{\ln|u\,|^{\sigma}}{\sigma}\right)\sigma_t+c_t\frac{|u\,|^{\sigma}}{\sigma}\right)dx. \tag{4.80}$$

The term $I_1$ can be immediately estimated:

$$|I_1|\le\frac{\delta}{2}\|u_t\|_{2,\Omega}^2+\frac{1}{2\delta}\|f\|_{2,\Omega}^2,\quad \delta\in(0,1).$$

Let us estimate $I_{2,i}$, $i=1,2,\ldots,n$. First,

$$\left|\int_{\Omega}a_i|D_iu\,|^{p_i}\frac{p_{it}}{p_i^2}dx\right|\le\max_{\overline{\Omega}}|p_{it}|\int_{\Omega}a_i|D_iu\,|^{p_i}\frac{1}{p_i^2}dx=\lambda_p(t)\int_{\Omega}a_i|D_iu\,|^{p_i}\frac{1}{p_i^2}dx.$$

Let us represent $\Omega=\Omega^+(t)\cup\Omega^-(t)$,

$$\Omega^+(t)=\{x\in\Omega\,:\,|D_iu(x,t)|\ge 1\},\qquad \Omega^-(t)=\{z\in\Omega\,:\,|D_iu(x,t)|<1\},$$

and denote

$$J_i^{\pm}(t)=\int_{\Omega^{\pm}(t)}a_i|D_iu\,|^{p_i}\left(\frac{\ln|D_iu|^{p_i}}{p_i}\right)p_{it}\,dx.$$

Since $p_{it}\le 0$ and $\ln|D_iu|\ge 0$ on $\Omega^+(t)$,

$$J_i^+(t)=-\int_{\Omega^+(t)}a_i|p_{it}||D_iu|^{p_i}\ln|D_iu|\,dx.$$

On the other hand,

$$J_i^-(t)\le a_1\,|\Omega|\max_{\overline{\Omega}}|p_{it}|\max_{\tau\in[0,1]}|\tau^{p_i}\ln\tau|\le C\lambda_p(t).$$

Gathering these formulas, we have: for every $t\in[0,T]$

$$\begin{aligned}J_i^+(t)+J_i^+(t)&\le-\int_{\Omega}a_i|p_{it}||D_iu|^{p_i}|\ln|D_iu||\,dx+2J_i^-(t)\\&\le-\int_{\Omega}a_i|p_{it}||D_iu|^{p_i}|\ln|D_iu||\,dx+C\lambda_p(t).\end{aligned}$$

Finally,

$$\int_\Omega a_{it} \frac{|D_i u|^{p_i}}{p_i} \, dx \le C\lambda_a(t) \int_\Omega \frac{|D_i u|^{p_i}}{p_i} \, dx \le C\lambda_a(t) Y(t).$$

It follows that

$$\sum_{i=1}^n I_{2,i} \le -\sum_{i=1}^n \int_\Omega a_i |p_{it}| |D_i u|^{p_i} |\ln |D_i u|| \, dx + C\left(\lambda_p(t) + \lambda_a(t)\right) Y(t) + C\lambda_p(t).$$

The term $I_3$ is estimated likewise. Splitting the domain $\Omega$ into the sets where $|u| > 1$ or $|u| \le 1$, and using the assumption $\sigma_t(z) \le 0$, we find that

$$\begin{aligned}
\int_\Omega c|u|^\sigma \frac{\ln |u|^\sigma}{\sigma} \sigma_t \, dx &= -\int_{\Omega \cap (|u|>1)} c|u|^\sigma |\ln |u|| |\sigma_t| \, dx \\
&\quad + \int_{\Omega \cap (|u| \le 1)} c|u|^\sigma |\ln |u|| |\sigma_t| dx \\
&\le -\int_\Omega c|u|^\sigma |\ln |u|| |\sigma_t| \, dx + C\lambda_\sigma(t),
\end{aligned}$$

while

$$\left| \int_\Omega c|u|^\sigma \frac{\sigma_t}{\sigma^2} dx \right| + \left| \int_\Omega c_t \frac{|u|^\sigma}{\sigma} dx \right| \le C\left(\lambda_c(t) + \lambda_\sigma(t)\right) Y(t).$$

Hence,

$$I_3 \le -\int_\Omega c|u|^\sigma |\ln |u|| |\sigma_t| \, dx + C\left(\lambda_c(t) + \lambda_\sigma(t)\right) Y(t) + C\lambda_\sigma(t).$$

Thus, the function $Y(t)$ satisfies the differential inequality

$$\begin{aligned}
&Y'(t) + \|u_t\|_{2,\Omega}^2 + \int_\Omega c|u|^\sigma |\ln |u|| \, |\sigma_t| \, dx \\
&\quad + \sum_{i=1}^n \int_\Omega |p_{it}| a_i |D_i u|^{p_i} |\ln |D_i u|| \, dx \\
&\le C\left(\Lambda(t) Y(t) + \lambda(t) - \|f\|_{2,\Omega}^2(t) + 1\right)
\end{aligned} \tag{4.81}$$

with

$$\Lambda(t) = \lambda_p(t) + \lambda_a(t) + \lambda_c(t) + \lambda_\sigma(t), \qquad \lambda(t) = \lambda_\sigma(t) + \lambda_p(t).$$

Omitting the nonnegative terms on the left-hand side, we estimate $Y(t)$ using Gronwall's lemma. Estimate (4.75) follows then after integration of (4.81) over the interval $(0, t)$. □

*Remark 4.3* Let us assume that in the conditions of Theorem 4.8 $\sigma \equiv \sigma(x)$, $p_i \equiv p_i(x)$, $a_{it} \le 0$, $c_t \le 0$ and $f = 0$ in $Q_T$. In this special case equalities (4.78)–(4.80) yield the relations

$$I_1 = 0, \quad I_{2,i} = \int_\Omega \frac{a_{it}}{p_i} |D_i u|^{p_i} \, dx \le 0, \quad I_3 = \int_\Omega \frac{c_t}{\sigma} |u|^\sigma \, dx \le 0$$

and equality (4.77) transforms into the inequality

$$Y'(t) \le \|u_t\|^2_{2,\Omega} + Y'(t) \le 0.$$

Integration of this inequality in $t$ leads to the estimate

$$Y(t) + \int_0^t \int_\Omega u_t^2 \, dz \le Y(0). \tag{4.82}$$

*Remark 4.4* The assertion of Theorem 4.8 remains true for the solutions of the equation

$$u_t - \sum_{i=1}^n \left[ \frac{d}{dx_i} \left( a_i(z) |D_i u|^{p_i(z)-2} D_i u \right) + c_i(z) |u|^{\sigma_i(z)-2} u \right] = f(z),$$

provided that the functions $c_i(z)$, $\sigma_i(z)$ satisfy the conditions of Theorem 4.8.

*Remark 4.5* The problem

$$\begin{cases} u_t = \Delta_p u + f \quad \text{in } Q_T, \\ u = 0 \text{ on } \Gamma_T, \qquad u(x,0) = u_0(x) \text{ in } \Omega, \end{cases} \tag{4.83}$$

with constant $p \in (1, \infty)$ is a partial case of problem (4.73). It was proved in [71, Lemma 2.1] that the solutions of (4.83) satisfy the estimate

$$\|u_t\|^2_{L^2(Q_T)} + \|u\|^p_{L^\infty(0,T;W^{1,p}(\Omega))} \le C \left( \|f\|^2_{L^2(Q_T)} + \|u_0\|^p_{W^{1,p}(\Omega)} \right), \tag{4.84}$$

which is contained in (4.75) if $p = const$. Moreover, if in the conditions of Theorem 4.8 the coefficients $a_i$, $c$ and the exponents $p_i$ and $\sigma$ are variable but independent of $t$, then estimate (4.84) is true as well for the solutions of problem (4.73).

## 4.6 Singular Perturbation of the $p(x,t)$-Laplace Operator

A solution of problem (4.1) can be obtained as the limit of the sequence of solutions of the perturbed equations. Let us assume that the exponents $p_i(z)$ are measurable, take values in an interval $[p^-, p^+] \subset (1, \infty)$ and satisfy the log-continuity condition (1.31). For the sake of simplicity we assume here that $b_i = d \equiv 0$. Set $\mathbf{s} = (s_1, \dots, s_n) \in \mathbb{R}^n$ and define the operator

$$\mathbf{A}(z, \mathbf{s}) = (A_1(z, s_1), \dots, A_n(z, s_n)) : Q \times \mathbb{R}^n \mapsto \mathbb{R}^n,$$
$$A_i(z, r) = |r|^{p_i(z)-2} r \text{ for all } r \in \mathbb{R} \text{ and a.a. } z \in Q.$$

We want to construct a solution of the problem

$$\begin{cases} u_t = \operatorname{div}\,(\mathbf{A}(z, \nabla u)) & \text{in } Q, \\ u = 0 \text{ on } \Gamma, \quad u(x,0) = u_0 \text{ in } \Omega \end{cases} \tag{4.85}$$

approximating the nonlinear operator $\mathbf{A}$ with variable nonlinearity by singularly perturbed operators with constant nonlinearity exponents defined on a function space smaller than the natural energy space $\mathbf{W}(Q)$. Let us introduce the family of the perturbed operators

$$\mathbf{A}_\varepsilon(z, \nabla u) := \varepsilon\, |\nabla u|^{p^+-2} \nabla u + \mathbf{A}(z, \nabla u), \quad \varepsilon > 0.$$

It is straightforward to check that $\mathbf{A}_\varepsilon(z, \nabla u)$ is defined for every $\varepsilon > 0$ for the elements of the Banach space

$$\mathbf{W}_+(Q) := L^2(Q) \cap L^{p^+}(0, T;\, W_0^{1,p^+}(\Omega)), \qquad \|u\|_{\mathbf{W}_+(Q)} = \|u\|_{2,Q} + \|\nabla u\|_{p^+,Q}.$$

For every $\varepsilon > 0$ $\mathbf{A}_\varepsilon(\cdot)$ possesses the properties[1]

$$\begin{array}{ll} \text{(a)} & \mathbf{A}_\varepsilon(z, \nabla u) : Q \times \mathbf{W}_+(Q) \mapsto \mathbf{W}'_+(Q), \quad \forall\, u,\, \phi \in \mathbf{W}_+(Q), \\ \text{(b)} & \mathbf{A}_\varepsilon(z, \nabla u) : Q \times \mathbf{W}_+(Q) \mapsto \mathbf{W}'_+(Q) \text{ is bounded}, \\ \text{(c)} & \mathbf{A}_\varepsilon \text{ is monotone}, \\ \text{(d)} & \mathbf{A}_\varepsilon(z, \nabla u) \text{ is hemi-continuous, that is, the mapping} \\ & \mathbb{R} \ni \mu \mapsto \displaystyle\int_Q \mathbf{A}_\varepsilon(z, \nabla(u + \mu v)) \cdot \nabla w \, dz \in \mathbb{R} \\ & \text{is continuous for all } u, v, w \in \mathbf{W}_+(Q). \end{array} \tag{4.86}$$

Let us consider the sequence of problems

$$\begin{cases} u_t = \operatorname{div}\,(\mathbf{A}_\varepsilon(z, \nabla u)) \text{ in } Q, \quad \varepsilon > 0, \\ u = 0 \text{ on } \Gamma, \quad u(x,0) = u_0(x) \in L^2(\Omega). \end{cases} \tag{4.87}$$

[1] such operators are sometimes called “Leray-Lions operators”.

**Definition 4.2** A function $u$ is called weak solution of problem (4.87) if $u \in L^{\infty}(0,T;\,L^2(\Omega)) \cap \mathbf{W}_+(Q)$, $\partial_t u \in \mathbf{W}'_+(Q)$, and for every test-function $\zeta \in \mathbf{W}_+(Q)$ with $\zeta_t \in \mathbf{W}'_+(Q)$

$$\int_Q \left(\zeta\, u_t + \mathbf{A}_\varepsilon(z, \nabla u)\cdot \nabla \zeta\right) dz = 0, \tag{4.88}$$

and $u(x,t) \to u_0(x)$ as $t \to 0$ weakly in $L^2(\Omega)$.

**Theorem 4.9** ([197], Chap. 2, Sect. 1, Theorem 1.2) *For every $\varepsilon > 0$ and $u_0 \in L^2(\Omega)$ problem* (4.87) *has a unique weak solution in the sense of* Definition 4.2, *and for every $t_1, t_2 \in [0,T]$*

$$\frac{1}{2}\int_\Omega u_\varepsilon^2\, dx\Big|_{t=t_1}^{t=t_2} + \varepsilon \int_Q |\nabla u_\varepsilon|^{p^+}\, dz + \sum_i \int_Q |D_i u_\varepsilon|^{p_i(z)}\, dz = 0. \tag{4.89}$$

*The function $u_\varepsilon : [0,T] \to L^2(\Omega)$ is continuous after possible redefining on a set of zero measure.*

It remains to show that the family of solutions of the perturbed problems (4.87) with $\varepsilon > 0$ contains a subsequence which converges to a solution of problem (4.85). Due to the uniform estimate (4.89) there exist $u \in \mathbf{W}(Q)$ and $\mathbf{A}^* \in \mathbf{W}'(Q)$, $B_i \in L^{p_i(\cdot)}(Q)$, such that (up to a subsequence)

$$u_\varepsilon \rightharpoonup u \text{ in } L^2(Q) \text{ and a.e. in } Q, \quad D_i u_\varepsilon \rightharpoonup B_i \text{ in } L^{p_i(\cdot)}(Q), \tag{4.90}$$

**Lemma 4.7** *There exists a subsequence $\{u_\varepsilon\}$ such that*

$$\begin{aligned} &\mathbf{A}_\varepsilon(z, \nabla u_\varepsilon) \rightharpoonup \mathbf{A}^* = (B_1, \dots, B_n) \text{ with some } B_i \in L^{p'_i(\cdot)}(Q) \\ &\text{and } \ \partial_t u_\varepsilon \rightharpoonup \partial_t u \text{ in } \mathbf{W}'(Q). \end{aligned}$$

*Proof* Equality (4.89) allows us to choose a subsequence such that $u_\varepsilon \rightharpoonup u$ in $L^2(Q)$, $D_i u_\varepsilon \rightharpoonup B_i$ in $L^{p_i(\cdot)}(Q)$ and $\varepsilon^{\frac{1}{p^+}}\, |\nabla u_\varepsilon|$ are uniformly bounded in $L^{p^+}(Q)$. Observe that

$$\varepsilon\, |\nabla u_\varepsilon|^{p^+-2}\nabla u_\varepsilon \rightharpoonup 0 \text{ in } L^{(p^+)'}(Q). \tag{4.91}$$

Indeed: by virtue of (4.89) with $t_1 = 0$, $t_2 = T$, for every $\phi \in \mathbf{W}_+(Q)$

$$\varepsilon \left| \int_Q |\nabla u_\varepsilon|^{p^+-2}\nabla u_\varepsilon \cdot \nabla\phi\, dz\right| \le \varepsilon^{\frac{1}{p^+}} \left(\varepsilon \int_Q |\nabla u_\varepsilon|^{p^+}\, dz\right)^{1-\frac{1}{p^+}} \|\nabla \phi\|_{p^+,Q} \to 0$$

as $\varepsilon \to 0$. Let $\phi \in \mathbf{W}(Q)$ and $\phi_k \in \mathbf{W}_+(Q)$ be such that $\|\phi_k - \phi\|_{\mathbf{W}(Q)} \to 0$ as $k \to \infty$. Accept the notation

$$\int_Q \mathbf{A}_\varepsilon(z, \nabla u_\varepsilon)\cdot\nabla\phi_k\,dz = \underbrace{\varepsilon\int_Q |\nabla u_\varepsilon|^{p^+-2}\nabla u_\varepsilon\cdot\nabla\phi_k\,dz}_{=J_1} + \sum_i \underbrace{\int_Q |D_i u_\varepsilon|^{p_i(z)-2}D_i u_\varepsilon\cdot D_i\phi_k\,dz}_{=J_{2,i}}.$$

According to (4.91) $J_1 \to 0$ as $\varepsilon \to 0$, while $J_{2,i} \to \int_Q B_i D_i\phi_k\,dz$ with $B_i \in L^{p_i'(\cdot)}(Q)$. Passing in (4.88) with $\zeta = \phi_k$ to the limit as $\varepsilon \to 0$ and then letting $k \to \infty$ we obtain: $\int_Q \mathbf{A}^*\cdot\nabla\phi_k\,dz \to \int_Q \mathbf{A}^*\cdot\nabla\phi\,dz$ for every $\phi \in \mathbf{W}(Q)$. By virtue of (4.88) and (4.91) $\|\partial_t u_\varepsilon\|_{\mathbf{W}'_+(Q)} \leq C$ uniformly with respect to $\varepsilon$, which means that $\partial_t u_\varepsilon \rightharpoonup \partial_t u$ in $\mathbf{W}'_+(Q)$ (up to a subsequence). Moreover, $\partial_t u \in \mathbf{W}'(Q)$ because

$$\int_Q \phi_k\partial_t u\,dz = -\int_Q \mathbf{A}^*\cdot\nabla\phi_k\,dz \to -\int_Q \mathbf{A}^*\cdot\nabla\phi\,dz \quad \text{as } k \to \infty.$$

□

**Corollary 4.2** *$\varepsilon\nabla u_\varepsilon \rightharpoonup 0$ in $\mathbf{W}_+(Q)$: for every $\phi \in \mathbf{W}_+(Q)$*

$$\varepsilon\left|\int_Q |\nabla\phi|^{p^+-2}\nabla\phi\cdot\nabla u_\varepsilon\,dz\right| \leq \varepsilon^{1-\frac{1}{p^+}}\left(\varepsilon\int_Q|\nabla u_\varepsilon|^{p^+}\,dz\right)^{\frac{1}{p^+}}\|\nabla\phi\|_{p^+,Q} \to 0$$

*as $\varepsilon \to 0$.*

**Corollary 4.3** *By* Lemma 4.5 $u \in C([0,T]; L^2(\Omega))$. *Since*

$$\int_Q \left(u_t\phi + \mathbf{A}^*\cdot\nabla u\right)dz = 0 \qquad \forall\phi \in \mathbf{W}(Q),$$

*it follows from* Lemma 1.20 *that*

$$\frac{1}{2}\int_\Omega u^2(x,t)\,dx\Big|_{t=0}^{t=T} + \int_Q \mathbf{A}^*\cdot\nabla u\,dz = 0. \tag{4.92}$$

To identify the limit $\mathbf{A}^*$ we follow the proof of Lemma 4.6. Taking in (4.88) $\zeta = u_\varepsilon$,

$$\frac{1}{2}\int_\Omega u_\varepsilon^2\,dx\Big|_{t=0}^{t=T} + \int_Q \mathbf{A}_\varepsilon(z,\nabla u_\varepsilon)\cdot\nabla u_\varepsilon\,dz = 0,$$

and using the monotonicity of $\mathbf{A}_\varepsilon$ we find that for every $\phi \in C^\infty([0,T]; C_0^\infty(\Omega))$

$$
-\int_Q \mathbf{A}_\varepsilon(z,\nabla u_\varepsilon)\cdot\nabla\phi\,dz-\int_Q \mathbf{A}_\varepsilon(z,\nabla\phi)\cdot\nabla(u_\varepsilon-\phi)\,dz-\frac{1}{2}\int_\Omega u_\varepsilon^2\,dx\Big|_{t=0}^{t=T}\geq 0.
$$

Letting $\varepsilon\to 0$ and using Lemma 4.7 together with Corollary 4.2 we arrive at the inequality

$$
-\int_Q \mathbf{A}^*\cdot\nabla\phi\,dz-\int_Q \mathbf{A}(z,\nabla\phi)\cdot\nabla(u-\phi)\,dz+\frac{1}{2}\int_\Omega u_0^2\,dx\geq\frac{1}{2}\int_\Omega u^2(x,T)\,dx.
$$

Plugging (4.92) we then have:

$$
\forall\,\phi\in C^\infty([0,T];C_0^\infty(\Omega))\qquad\int_Q\big(\mathbf{A}^*-\mathbf{A}(z,\nabla\phi)\big)\cdot\nabla(u-\phi)\,dz\geq 0.
$$

Since the smooth functions are dense in $\mathbf{W}(Q)$, the same is true for every $\phi\in\mathbf{W}(Q)$, which allows us to take $\phi=u\pm\delta\,\psi$ with an arbitrary $\psi\in\mathbf{W}(Q)$, $\delta\in\mathbb{R}$. Simplifying and letting $\delta\to 0$ we arrive at the conclusion that $\mathbf{A}^*=\mathbf{A}(z,\nabla u)$ a.e. in $Q$.

**Theorem 4.10** *Let $p_i(z)\in C_{\log}(\overline{Q})$ take values in the interval $[p^-,p^+]\subset(1,\infty)$. Then for every $u_0\in L^2(\Omega)$ problem (4.85) has at least one weak solution $u\in\mathbf{W}(Q)$ such that $u_t\in\mathbf{W}'(Q)$, $u:[0,T]\mapsto L^2(\Omega)$ is weakly continuous, and for every $\zeta\in\mathbf{W}(Q)$ with $\zeta_t\in\mathbf{W}'(Q)$*

$$
\int_Q(\zeta\,u_t+\mathbf{A}(z,\nabla u)\cdot\nabla\zeta)\,dz=0.
$$

## 4.7 The Cauchy Problem

Let us consider the Cauchy problem for the $p(x)$-Laplace equation

$$
\begin{cases} u_t-\operatorname{div}\mathscr{A}(z,\nabla u)=f(z) & \text{in } S_T=\mathbb{R}^n\times(0,T],\\ u(x,0)=u_0(x)\in L^2(\mathbb{R}^n), \end{cases}\tag{4.93}
$$

where $\mathscr{A}(z,\nabla u)=a(z)|\nabla u|^{p(x)-2}\nabla u$ and $z=(x,t)$ denotes the point of $S_T$. It is assumed that

$$
\begin{cases} a(z)\text{ is a measurable function},\\ 0<a_-\leq a(z)\leq a_+<\infty\text{ for a.e. } z\in S_T \end{cases}\tag{4.94}
$$

with some constants $a_\pm$,

$$
\begin{cases} p:\mathbb{R}^n\mapsto\mathbb{R}\text{ is a measurable function with}\\ p^-=\operatorname*{ess\,inf}_{\mathbb{R}^n}p(x)>1,\quad p^+=\operatorname*{ess\,sup}_{\mathbb{R}^n}p(x)<\infty. \end{cases}\tag{4.95}
$$

Unlike previous existence results, now we do not claim continuity of the exponent $p(x)$ and introduce the spaces $L^{p(\cdot)}$, $W_0^{1,p(\cdot)}$ in a different way. Let $\Omega \subseteq \mathbb{R}^n$ be a domain, not necessarily bounded. We will distinguish between two cases: either $\Omega = B_k = \{x \in \mathbb{R}^n : |x| < k\}$ with $k \in \mathbb{N}$, or $\Omega \equiv \mathbb{R}^n$. By $C_0^\infty(\Omega)$ we denote the space of infinitely differentiable functions with compact support in $\Omega$. Set $Q_T = \Omega \times (0, T)$ and define the space $W$ as the closure of $C^\infty([0, T]; C_0^\infty(\Omega))$ with respect to the norm

$$\|u\|_W = \|u\|_{2,Q_T} + \|\nabla u\|_{p(\cdot),Q_T}, \tag{4.96}$$

where $\|\cdot\|_{p(\cdot),Q_T}$ denotes the Luxemburg norm

$$\|h\|_{p(\cdot),Q_T} = \inf\left\{\lambda > 0 : \int_{Q_T} |h/\lambda|^{p(x)}\,dxdt < \infty\right\} \tag{4.97}$$

on the space of functions

$$L^{p(\cdot)}(Q_T) = \left\{u(z) \text{ is measurable in } Q_T : \int_{Q_T} |u(z)|^{p(x)}\,dxdt < \infty\right\}.$$

The dual space to $W$ is denoted by $W'$. This is the space of linear functionals over $W$,

$$\Phi \in W' \quad \Leftrightarrow \quad \begin{cases} \exists \phi_0 \in L^2(Q_T),\ \phi_i \in L^{p'(x)}(Q_T),\ i = 1, \dots, n, \\ \forall v \in W \quad (v, \Phi)_{2,Q_T} = (v, \phi_0)_{2,Q_T} + \sum_{i=1}^n (D_i v, \phi_i)_{2,Q_T}, \end{cases}$$

endowed with the norm

$$\|\Phi\|_{W'} = \sup\left\{(\Phi, u)_{2,Q_T} : \|u\|_W = 1\right\}.$$

**Definition 4.3** A function $u : S_T \mapsto \mathbb{R}$ is called weak solution of problem (4.93) if

1. $u \in W$, $u_t \in W'$,
2. for every $\phi \in C^1([0, T]; C_0^1(\mathbb{R}^n))$

$$\int_{S_T} (u_t\phi + \mathscr{A}(z, \nabla u) \cdot \nabla\phi - f\phi)\,dxdt = 0, \tag{4.98}$$

3. $u \in C_w^0([0, T]; L^2(\mathbb{R}^n))$, in particular, for every $\phi(x) \in C_0^1(\mathbb{R}^n)$

$$\int_{\mathbb{R}^n} \phi(x)(u - u_0)\,dz \to 0 \quad \text{as } t \to 0.$$

**Theorem 4.11** *Let conditions* (4.94) *and* (4.95) *be fulfilled. Then for every* $u_0 \in L^2(\mathbb{R}^n)$ *and* $f \in L^2(S_T)$ *problem* (4.93) *has at least one weak solution in the sense of* Definition 4.3. *The solution satisfies the energy estimate*

$$\frac{1}{2} \operatorname{ess} \sup_{(0,T)} \|u(t)\|^2_{2,\mathbb{R}^n} + \int_{S_T} |\nabla u|^{p(x)}\, dz \leq C \left( \|u_0\|^2_{2,\mathbb{R}^n} + \|f\|_{2,S_T} \right) \tag{4.99}$$

*with a constant* $C$ *depending on* $a_-$ *and* $T$.

Boundedness and uniqueness of weak solutions, as well as better regularity properties, are established under additional restrictions on the data. Let us denote by $V$ the closure of $C_0^\infty(\mathbb{R}^n)$ with respect to the norm

$$\|u\|_V = \|u\|_{2,\mathbb{R}^n} + \|\nabla u\|_{p(\cdot),\mathbb{R}^n}.$$

**Theorem 4.12** *Let the conditions of* Theorem 4.11 *be fulfilled.*

1. *If* $\sup_{S_T} |f| \leq C_f$ *and* $\sup_{\mathbb{R}^n} |u_0| \leq M$ *with finite positive constants* $C_f$ *and* $M$, *then the solution of problem* (4.93) *satisfies the estimate*
$$|u| \leq (1+M) e^{\frac{C_f}{1+M} T} \quad \text{a.e. in } S_T. \tag{4.100}$$
2. *If* $p(x) \geq 2$ *a.e. in* $\mathbb{R}^n$, *then problem* (4.93) *cannot have more than one bounded solution.*
3. *If* $u_0 \in V$ *and* supp $u_0 \subseteq B_R(0)$ *with a finite* $R > 0$, $f \in L^2(S_T)$ *and* $|a_t| \leq a_T = const$, *then* $u_t \in L^2(S_T)$, $|\nabla u|^{p(x)} \in L^\infty(0,T; L^1(\mathbb{R}^n))$ *and*
$$\|u_t\|^2_{2,S_T} + \operatorname{ess} \sup_{(0,T)} \int_{\mathbb{R}^n} |\nabla u|^{p(x)}\, dx \leq C$$
*with a finite constant* $C$.

## 4.7.1 Steklov's Means

Denote by $f_h$ the Steklov mean of the function $f \in L^{p(x)}(Q_T)$:

$$f_h = \frac{1}{h} \int_t^{t+h} f(x,\tau)\, d\tau.$$

**Proposition 4.2** *The operator* $S_h : f \to f_h$ *maps* $L^{p(x)}(Q_T)$ *into* $L^{p(x)}(Q_{T-h})$.

*Proof* By virtue of Hölder's inequality, for an arbitrary given $\varepsilon \in (0,T)$, every $h \in (0,\varepsilon)$ and a.e. $x \in \Omega$

$$
\begin{aligned}
|f_h|^{p(x)} &= \frac{1}{h^{p(x)}} \left| \int_0^h f(x, t+\tau)\, d\tau \right|^{p(x)} \\
&\le \frac{1}{h^{p(x)}} \left( \int_0^h |f(x, t+\tau)|^{p(x)}\, d\tau \right) \left( \int_0^h d\tau \right)^{\frac{p(x)}{p'(x)}} \\
&= \frac{1}{h} \int_0^h |f(x, t+\tau)|^{p(x)}\, d\tau = \left( |f|^{p(x)} \right)_h .
\end{aligned} \tag{4.101}
$$

Applying (4.101) we immediately obtain

$$
\begin{aligned}
\int_{Q_{T-\varepsilon}} |f_h|^{p(x)}\, dxdt &= \int_{Q_{T-\varepsilon}} \frac{1}{h^{p(x)}} \left| \int_0^h f(x, t+\tau)\, d\tau \right|^{p(x)} dxdt \\
&\le \frac{1}{h} \int_{Q_{T-\varepsilon}} \left( \int_0^h |f(x, t+\tau)|^{p(x)}\, d\tau \right) dxdt \\
&\le \frac{1}{h} \int_0^h \left( \int_{Q_T} |f|^{p(x)}\, dxdt \right) d\tau = \int_{Q_T} |f|^{p(x)}\, dxdt.
\end{aligned}
$$

□

**Corollary 4.4** *In the conditions of* Proposition 4.2

$$
\begin{aligned}
&\frac{1}{h^{p(x)}} \left| \int_0^h (f(x, t+\tau) - f(x, t))\, d\tau \right|^{p(x)} \\
&\qquad \le \frac{1}{h} \int_0^h |f(x, t+\tau) - f(x, t)|^{p(x)}\, d\tau.
\end{aligned} \tag{4.102}
$$

**Proposition 4.3** *The translation operator is continuous: for every fixed $\varepsilon > 0$ and $f \in L^{p(x)}(Q_T)$*

$$
\|f(x, t+h) - f(x, t)\|_{p(x), Q_{T-\varepsilon}} \to 0 \quad as\ h \to 0.
$$

*Proof* We consider separately the cases $|\Omega| < \infty$ and $|\Omega| = \infty$. Let us assume first that $|\Omega| < \infty$. Take $f \in L^{p(x)}(Q_T)$ and an arbitrary $\varepsilon > 0$. Since $C^0(\overline{Q}_T)$ is dense in $L^{p(x)}(Q_T)$, there exists $\phi \in C^0(\overline{Q}_T)$ such that $\|f - \phi\|_{p(x), Q_T} < \varepsilon$. For $\varepsilon \le 1$ this means that

$$
\int_{Q_T} |f - \phi|^{p(x)}\, dz < \varepsilon^{p^-}.
$$

Since $\phi$ is uniformly continuous in $\overline{Q}_T$, there exists $\delta > 0$ such that

$$
\forall x \in \Omega,\ \ t, \tau \in [0, T]\ \ |t - \tau| < \delta \ \Rightarrow\ \ |\phi(x, t) - \phi(x, \tau)| < \frac{\varepsilon}{(1 + 2|\Omega|T)^{\frac{1}{p^-}}}.
$$

Then

$$\int_{Q_{T-h}} |f(x,t+h)-f(x,t)|^{p(x)}\,dxdt \le \int_{Q_{T-h}} |f(x,t+h)-\phi(x,t+h)|^{p(x)}\,dxdt$$
$$+\int_{Q_{T-h}} |\phi(x,t+h)-\phi(x,t)|^{p(x)}\,dxdt + \int_{Q_{T-h}} |f(x,t)-\phi(x,t)|^{p(x)}\,dxdt$$
$$\le 2\varepsilon^{p^-} + \frac{\varepsilon^{p^-}|\Omega|T}{1+2|\Omega|T} \le 3\varepsilon^{p^-}.$$

Let now $|\Omega| = \infty$. Since $f \in L^{p(x)}(Q_T)$, for every $\varepsilon > 0$ one can find $R > 0$ such that

$$\int_0^T \int_{\mathbb{R}^n \setminus B_R(0)} |f(x,t+h)-f(x,t)|^{p(x)} dz < \varepsilon.$$

Set $Q_{T-h,R} = B_R(0) \times (0, T-h)$. Since

$$\begin{aligned}\int_{Q_{T-h}} |f(x,t+h)-f(x,t)|^{p(x)} dz &= \int_{Q_{T-h}\setminus Q_{T-h,R}} |f(x,t+h)-f(x,t)|^{p(x)} dz \\ &\quad + \int_{Q_{T-h,R}} |f(x,t+h)-f(x,t)|^{p(x)} dz \\ &< \varepsilon + \int_{Q_{T-h,R}} |f(x,t+h)-f(x,t)|^{p(x)} dz,\end{aligned}$$

the claim follows as in the case $|\Omega| < \infty$. □

*Remark 4.6* Proposition 4.3 is false if $p$ depends on $t$. It is known that in the case $p = p(x,t)$ the translation operator is unbounded unless $p = const$—see [115, Proposition 3.6.1]

**Proposition 4.4** *If $f \in W(Q_T)$, then $\|f_h - f\|_{W(Q_{T-h})} \to 0$ as $h \to 0$.*

*Proof* It is sufficient to check that for every $f \in L^{p(x)}(Q_T)$ $\|f_h - f\|_{p(x),Q_{T-\varepsilon}} \to 0$ as $h \to 0$. By (4.102)

$$\begin{aligned}\int_{Q_{T-\varepsilon}} |f_h - f|^{p(x)}\,dxdt &= \int_{Q_{T-\varepsilon}} \left|\frac{1}{h}\int_0^h (f(x,t+\tau)-f(x,t))\,d\tau\right|^{p(x)} dxdt \\ &\le \frac{1}{h}\int_{Q_{T-\varepsilon}} \left(\int_0^h |f(x,t+\tau)-f(x,t)|^{p(x)}\,d\tau\right) dxdt \\ &= \frac{1}{h}\int_0^h \left(\int_{Q_{T-\varepsilon}} |f(x,t+h)-f(x,t)|^{p(x)} dxdt\right) d\tau \\ &\equiv \frac{1}{h}\int_0^h F(\tau)\,d\tau.\end{aligned}$$

By Proposition 4.3 $F(\tau) \to 0$ as $\tau \to 0+$. For every $\varepsilon > 0$ there is $h(\varepsilon)$ such that $F(\tau) < \varepsilon$ if $\tau < h(\varepsilon)$. It follows that

$$\forall\, h < h(\varepsilon) \qquad \frac{1}{h}\int_0^h F(\tau)\, d\tau < \varepsilon,$$

whence the assertion. □

**Proposition 4.5** *If $u \in W(Q_T)$ and $u_t \in W'(Q_T)$, then $(u_t)_h = (u_h)_t$ and $(u_h)_t \rightharpoonup u_t$ in $W'(Q_{T-\varepsilon})$ for every $\varepsilon > 0$.*

*Proof* We make use of the easily verified property of Steklov's means:

$$\int_{Q_{T-h}} f_h g\, dz = \int_h^T \int_\Omega f g_{-h}\, dz \qquad \forall\, f \in L^p(Q_T),\ g \in L^{p'}(Q_T).$$

By the definition, for every $\Phi \in W(Q_T)$

$$\begin{aligned}
\langle (u_t)_h, \Phi\rangle &= \int_{Q_{T-h}} \left(\frac{1}{h}\int_0^h u_t(x,t+\tau)\Phi(x,t)\, d\tau\right) dz \\
&= \frac{1}{h}\int_0^h \left(\int_{Q_{T-h}} \left(\phi_0(x,t+\tau)\Phi + \sum \phi_i(x,t+\tau) D_i\phi\right) dz\right) d\tau \\
&= \int_{Q_{T-h}} \left(\phi_{0h}\Phi + \sum \phi_{ih} D_i \Phi\right) dz \\
&= \int_h^T \int_\Omega \left(\phi_0 \Phi_{-h} + \sum \phi_i D_i \Phi_{-h}\right) dz \\
&= \langle u_t, \Phi\rangle + \int_h^T \int_\Omega \left(\phi_0(\Phi_{-h} - \Phi) + \sum \phi_i D_i(\Phi_{-h} - \Phi)\right) dz \\
&\equiv \langle u_t, \Phi\rangle + J(h).
\end{aligned}$$

Using Hölder's inequality and applying Proposition 4.4 we find that

$$|J(h)| \le \|u_t\|_{W'} \|\Phi_{-h} - \Phi\|_{W(\Omega\times(h,T))} \to 0 \quad \text{as}\, h \to 0.$$

To prove the second claim, let us take $\Phi \in C^\infty(0,T; C_0^\infty(\Omega)) \cap W(Q_T)$ such that $\Phi(x,t) \equiv 0$ for $t < h$ and $t > T - h$. Then

$$\begin{aligned}
\int_{Q_T} (u_t)_h \Phi\, dz &= \int_{Q_T} u_t \Phi_{-h}\, dz = -\int_{Q_T} u(\Phi_{-h})_t\, dz = -\int_{Q_T} u(\Phi_t)_{-h}\, dz \\
&= -\int_{Q_T} u_h \Phi_t\, dz = -\int_{Q_T} u(\Phi_t)_{-h}\, dz = \int_{Q_T} (u_h)_t \Phi\, dz.
\end{aligned}$$

□

**Proposition 4.6** *If $u, v \in W(Q_T)$ and $u_t, v_t \in W'(Q_T)$, then for a.e. $t_1, t_2 \in (0, T)$*

$$\int_{t_1}^{t_2}\int_\Omega u_t v\,dxdt + \int_{t_1}^{t_2}\int_\Omega u v_t\,dxdt = \int_\Omega uv\,dx\Big|_{t=t_1}^{t=t_2}.$$

*Proof* Let $v_h$ and $w_h$ be Steklov's means of $v$ and $w$. Consider the function $v_h w_h \chi_k$ with $k \in \mathbb{N}$, $h > 0$ and $\chi_k$ defined in (1.39). For every $t_1 < t_2$

$$0 = \int_Q (v_h w_h \chi_k)_t\,dz \equiv \int_Q (v_h w_h)_t \chi_k\,dz - k\int_{\theta-\frac{1}{k}}^{\theta}\int_\Omega v_h w_h\,dz\Big|_{\theta=t_1}^{\theta=t_2}$$

and the conclusion follows as in the proof of Lemma 1.20. □

**Proposition 4.7** *If $u \in W(Q_T)$ and $u_t \in W'(Q_T)$, then $u \in C^0([0, T-\varepsilon]; L^2(\Omega))$ for every $\varepsilon \in (0, T)$.*

*Proof* We adapt the arguments used in the proof of Lemma 4.5. Since $u_h \in L^2(Q_T)$ and $(u_h)_t \in L^2(Q_{T-\varepsilon})$, it follows from [197, Chap. 1, Lemma 1.2] that $u_h \in C^0([0, T-\varepsilon]; L^2(\Omega))$ after possible redefining on a set of zero measure in $(0, T-\varepsilon)$. Thus, for every $h_1, h_2 \in (0, T-\varepsilon)$

$$\|u_{h_1} - u_{h_2}\|^2_{2,\Omega}(t) = 2\int_{Q_t}(u_{h_1} - u_{h_2})(u_{h_1} - u_{h_2})_t\,dz + \|u_{0h_1} - u_{0h_2}\|^2_{2,\Omega}.$$

By Proposition 4.4 $u_h \to u$ in $W(Q_{T-\varepsilon})$ as $h \to 0$, i.e., $\{u_h\}$ is a Cauchy sequence in $W(Q_{T-\varepsilon})$. By Proposition 4.5 $(u_h)_t \rightharpoonup u_t$ in $W'(Q_{T-\varepsilon})$ and, thus, $(u_h)_t$ are bounded in the norm of $W'(Q_{T-\varepsilon})$. Notice also that $(u_0)_h \equiv u_0$ and $\|u_{0h_2} - u_{0h_1}\|_{2,\Omega} = 0$. For every $t \in (0, T-\varepsilon)$

$$\begin{aligned}\|u_{h_1} - u_{h_2}\|^2_{2,\Omega}(t) &= 2\int_{Q_t}(u_{h_1} - u_{h_2})(u_{h_1} - u_{h_2})_t\,dz\\ &\leq 2\|(u_{h_1} - u_{h_2})_t\|_{W'}\|u_{h_1} - u_{h_2}\|_W \to 0 \quad \text{as } h_1, h_2 \to 0.\end{aligned}$$

This means that $\{u_h\}$ is a Cauchy sequence in $C^0([0, T-\varepsilon]; L^2(\Omega))$ and there is a function $\widetilde{u} \in C^0([0, T-\varepsilon]; L^2(\Omega))$ such that $u_h \to \widetilde{u}$. Since $u_h \to u \in W(Q_{T-\varepsilon})$, it necessary that $u = \widetilde{u}$ after possible redefining on a set of zero measure in $(0, T-\varepsilon)$. □

### 4.7.2 Problems in Bounded Cylinders

Let us consider the sequence of problems posed on bounded cylinders $Q_{T,k} = B_k \times (0, T]$, $k \in \mathbb{N}$,

$$\begin{cases} u_t - \operatorname{div}\mathscr{A}(z, \nabla u) = f \quad \text{in } Q_{T,k},\\ u = 0 \text{ on } \partial B_k \times (0, T], \quad u(x, 0) = u_0(x) \in L^2(B_k), \end{cases} \tag{4.103}$$

where for the initial data we take the restriction of $u_0$ to $B_k$, $f$ is the restriction of $f$ to $Q_{T,k}$. The natural energy space for problem (4.103) is defined by (4.96)–(4.97). The solution is understood is the sense of Definition 4.3 with the obvious changes: $u$ is a weak solution of problem (4.103) if $u \in C^0([0,T]; L^2(B_k)) \cap W(Q_{T,k})$, $u_t \in W'(Q_{T,k})$, for every test-function $\phi \in C^1([0,T]; C_0^1(B_k))$

$$\int_{Q_{T,k}} (u_t\phi + \mathscr{A}(z, \nabla u)\cdot\nabla\phi - f\phi)\, dz = 0 \tag{4.104}$$

and $(\phi, u - u_0)_{2,B_k} \to 0$ as $t \to 0$ for every $\phi(x) \in C_0^1(B_k)$.

**Theorem 4.13** *Let conditions* (4.94) *and* (4.95) *be fulfilled. For every*

$$u_0 \in L^2(\mathbb{R}^n), \quad f \in L^2(S_T), \quad k \in \mathbb{N},$$

*problem* (4.103) *has a unique weak solution* $u \equiv u_k$.

Let $\{\psi_i\}$ be the orthonormal basis of $L^2(B_k)$ composed of the eigenfunctions of problem (1.30) in $B_k$. Accept the notation

$$\mathscr{P}_N = \operatorname{span}\{\psi_1, \ldots, \psi_N\} \subset L^{p^+}(0,T; H_0^s(B_k)) \subset W(Q_{T,k}).$$

**Lemma 4.8** *The space* $L^{p^+}(0,T; H_0^s(B_k))$ *is dense in* $W(Q_{T,k})$.

*Proof* Let $u \in W(Q_{T,k})$. By the definition there is a sequence $\{u_s\}$ such that $u_s \in C^\infty(0,T; C_0^\infty(B_k))$ and $u_s \to u$ in $W(Q_{T,k})$. The assertion follows because $C^\infty(0,T; C_0^\infty(B_k)) \subset L^{p^+}(0,T; H_0^s(B_k))$. □

A solution of problem (4.103) will be obtained as the limit of the sequence

$$u^{(N)} = \sum_{i=1}^{N} \psi_i(x) d_{i,N}(t) \in \mathscr{P}_N$$

with the coefficients $d_{i,N}(t)$ to be defined. Substituting $u^{(N)}$ into Eq. (4.103), multiplying by $\psi_i$ and integrating over $B_k$ we obtain the system of ODEs for the coefficients $\mathbf{d}_N = \{d_{1,N}(t), \ldots, d_{N,N}(t)\}$:

$$\begin{cases} \mathbf{d}'_N(t) = \mathscr{F}(t, \mathbf{d}_N(t)), & t > 0 \\ d_{i,N}(0) = (u_0, \psi_i)_{2,B_k}, & i = 1, \ldots, N, \end{cases} \tag{4.105}$$

$$\mathscr{F}_i(t, \mathbf{d}_N(t)) = -\int_{B_k} \mathscr{A}(z, \nabla u^{(N)})\cdot\nabla\psi_i\, dx + \int_{B_k} f\,\psi_i\, dx.$$

Since $\mathscr{F}(t, \mathbf{d}_N)$ is continuous with respect to $d_{j,N}$, it follows from Peano's theorem that system (4.105) has a solution on an interval $[0, T_N)$.

**Lemma 4.9** *For every $N$ the function $u^{(N)}$ satisfies the estimate*

$$\operatorname{ess}\sup_{(0,T)} \|u^{(N)}(t)\|^2_{2,B_k} + \int_{Q_{T,k}} |\nabla u^{(N)}|^{p(x)}\,dz \le C\left(\|u_0\|^2_{2,\mathbb{R}^n} + \|f\|^2_{2,S_T}\right) \tag{4.106}$$

*where the constant $C$ depends on and $a_-$ and $T$, but is independent of $k$ and $N$.*

*Proof* Set $y(t) = \|u^{(N)}(t)\|^2_{2,Q_{T,k}\cap(0,t]}$. Multiplying the $i$th equation of system (4.105) by $d_{i,N}(t)$ and summing up the results we obtain the inequality

$$\frac{1}{2}y'(t) + a_-\int_{B_k} |\nabla u^{(N)}|^p\,dx \le \int_{B_k} |f||u^{(N)}|\,dx \le \sqrt{y(t)}\|f\|_{2,B_k}, \tag{4.107}$$

whence

$$\sqrt{y(t)} \le \sqrt{y(0)} + \int_0^t \|f(\cdot,s)\|_{2,B_k} ds.$$

Substituting this inequality into (4.107), integrating over the interval $(0,t)$ applying Hölder's and Young's inequalities we conclude that

$$\begin{aligned}
&\|u^{(N)}(t)\|^2_{2,B_k} + 2a_-\int_{Q_{T,k}} |\nabla u^{(N)}|^p\,dz \\
&\quad\le \|u_0\|^2_{2,B_k} + 2\left(\|u_0\|_{2,B_k} + \int_0^T \|f(\cdot,s)\|_{2,B_k} ds\right)\int_0^T \|f(\cdot,s)\|_{2,B_k} ds \\
&\quad\le C(T)\left(\|u_0\|^2_{2,B_k} + \|f\|^2_{2,Q_{T,k}}\right) \\
&\quad\le C(T)\left(\|u_0\|^2_{2,\mathbb{R}^n} + \|f\|^2_{2,S_T}\right).
\end{aligned}$$

□

**Corollary 4.5** *Estimates* (4.106) *are independent of $N$, which allows one to continue each of $u^{(N)}$ to the maximal existence interval* $[0,T]$.

**Lemma 4.10** *There is an independent of $k$ and $N$ constant $C$ such that*

$$\|(u^{(N)})_t\|_{L^{(p^+)'}(0,T;H^{-s}(B_k))} \le C. \tag{4.108}$$

We omit the proof because it is the literal repetition of the proof of Lemma 4.4.

**Lemma 4.11** *The sequence $\{u^{(N)}\}$ is precompact in $L^{\mu}(Q_{T,k})$ with some $\mu > 1$.*

*Proof* Estimates (4.106) and (4.108) and the inclusion $W \subset L^{p^-}(0,T;W_0^{1,p^-}(B_k))$ yield the independent of $k$, $N$ estimates

$$\|u^{(N)}\|_{L^{p^-}(0,T;W_0^{1,p^-}(B_k))} \le C. \tag{4.109}$$

The inclusion $W_0^{1,p^-}(B_k) \subset L^r(B_k)$ with $r < np^-/(n-p^-)$ is compact and $L^r(B_k) \subset H^{-s}(B_k)$. The assertion follows now from Lemmas 4.10 and 2.6. □

Due to estimates (4.106) and Lemmas 4.10, 4.11 we may find functions $\chi_k \in L^{p'(\cdot)}(Q_{T,k})$, $u_k \in L^{\mu}(Q_{T,k}) \cap W(Q_{T,k})$, $U \in L^\infty(0,T;L^2(B_k))$ such that

$$\begin{aligned}
&u^{(N)} \to u_k \text{ in } L^{\mu}(Q_{T,k}) \text{ with some } \mu > 1 \text{ and a.e. in } Q_{T,k},\\
&u^{(N)} \to U \ast\text{-weakly in } L^\infty(0,T;L^2(B_k)),\\
&\mathscr{A}(z,\nabla u^{(N)}) \rightharpoonup \chi_k \text{ in } L^{p'}(Q_{T,k}),\\
&(u^{(N)})_t \rightharpoonup u_{k,t} \text{ in } L^{(p^+)'}(0,T;H^{-s}(B_k)).
\end{aligned} \tag{4.110}$$

**Lemma 4.12** *For every $k \in \mathbb{N}$ we have $u_{k,t} \in W'(Q_{T,k})$ and*

$$\|u_{k,t}\|_{W'(Q_{T,k})} \le C$$

*with an independent of $k$ constant $C$.*

*Proof* By construction

$$\int_{Q_{T,k}} (u^{(N)})_t \phi^{(N)}\,dz = -\int_{Q_{T,k}} \mathscr{A}(z,\nabla u^{(N)}) \cdot \nabla\phi^{(N)}\,dz + \int_{Q_{T,k}} f\,\phi^{(N)}\,dz. \tag{4.111}$$

Take some $N \in \mathbb{N}$, fix $j \le N$ and test (4.111) with a function $\phi \in \mathscr{P}_j$:

$$\int_{Q_{T,k}} \left(u_t^{(N)}\phi + \mathscr{A}(z,\nabla u^{(N)}) \cdot \nabla\phi\right) dz = \int_{Q_{T,k}} f\,\phi\,dz. \tag{4.112}$$

Passing to the limit as $N \to \infty$ we obtain

$$\forall \phi \in \mathscr{P}_j \qquad \int_{Q_{T,k}} \left(u_{k,t}\phi + \chi_k \cdot \nabla\phi\right) dz = \int_{Q_{T,k}} f\,\phi\,dz. \tag{4.113}$$

Since $L^{p^+}(0,T;H_0^s(B_k))$ is dense in $W(Q_{T,k})$, letting $j \to \infty$ we conclude that (4.113) holds for an arbitrary $\phi \in W(Q_{T,k})$. It follows then that for every $\phi \in W(Q_{T,k})$

$$\left|\int_{Q_{T,k}} u_{k,t}\phi\,dxdt\right| \le C\|\phi\|_{W(Q_{T,k})}$$

with the constant $C$ from (4.106). □

**Lemma 4.13** *For every $k \in \mathbb{N}$ and a.e. $t \in (0, T)$*

$$\frac{1}{2}\int_{B_k} u_k^2\,dx\Big|_{\tau=0}^{\tau=t} + \int_0^t\int_{B_k} \chi_k \cdot \nabla u_k\,dz = \int_{Q_{T,k}} f\,u_k\,dz. \tag{4.114}$$

*Proof* It is shown in the proof of Lemma 4.12 that

$$\int_{Q_{T,k}} \left(u_{k,t}u_k + \chi_k \cdot \nabla u_k - f\,u_k\right)\,dz = 0.$$

The assertion follows now from Proposition 4.6 with $u = v = u_k$. □

**Lemma 4.14** *For every $k \in \mathbb{N}$ $u_k \in C_w(0, T; L^2(B_k))$ and $(\eta(x),\ u_k - u_0)_{2,B_k} \to 0$ as $t \to 0$ for every $\eta \in C_0^\infty(B_k)$.*

*Proof* Let us take for the test-function $\phi = \eta(x)\theta(t)$ with $\theta(t) \in C^1[0, T]$ such that $\theta(0) = \theta(T) = 0$ and $\eta(x) \in C_0^1(B_k)$. Denote $F(t) = \displaystyle\int_{B_k} u_k\eta\,dx$. By (4.113)

$$\int_0^T \theta'(t)F(t)\,dt = \int_0^T \theta(t)\Psi(t)\,dt, \quad \Psi = \int_{B_k} (\chi_k \cdot \nabla\eta - f\eta)\,dx \in L^1(0, T).$$

It follows that $F(t) \in W^{1,1}(0, T)$, whence $F(t)$ is absolutely continuous on $(0, T)$ and the limits $F(0)$ and $F(T)$ are well-defined. □

**Lemma 4.15** $\chi_k = \mathscr{A}(z, \nabla u_k)$ *a.e. in $Q_{T,k}$*

*Proof* We apply the standard monotonicity argument. Recall that

$$(\mathscr{A}(z, \mathbf{s}) - \mathscr{A}(z, \mathbf{r})) \cdot (\mathbf{s} - \mathbf{r}) \geq 0 \text{ for a.e. } z \in Q_{T,k} \text{ and all } \mathbf{s},\ \mathbf{r} \in \mathbb{R}^n.$$

By construction

$$\frac{1}{2}\int_{B_k} (u^{(N)})^2\,dx\Big|_{t=0}^{t=T} + \int_{Q_{T,k}} \mathscr{A}(z, \nabla u^{(N)}) \cdot \nabla u^{(N)}\,dz = \int_{Q_{T,k}} f\,u_k\,dz.$$

For every $\phi \in \mathscr{P}_N$

$$\begin{aligned}
\mathscr{A}(z, \nabla u^{(N)}) \cdot \nabla u^{(N)} &= \mathscr{A}(z, \nabla u^{(N)}) \cdot \nabla(u^{(N)} - \phi) + \mathscr{A}(z, \nabla u^{(N)}) \cdot \nabla\phi \\
&= (\mathscr{A}(z, \nabla u^{(N)}) - \mathscr{A}(z, \nabla\phi) \cdot \nabla(u^{(N)} - \phi) \\
&\quad + \mathscr{A}(z, \nabla u^{(N)}) \cdot \nabla\phi + \mathscr{A}(z, \nabla\phi) \cdot \nabla(u^{(N)} - \phi)
\end{aligned}$$

whence, by monotonicity,

$$\int_{Q_{T,k}} \mathscr{A}(z, \nabla u^{(N)}) \cdot \nabla u^{(N)} \, dz \geq \int_{Q_{T,k}} \mathscr{A}(z, \nabla u^{(N)}) \cdot \nabla \phi \, dz + \int_{Q_{T,k}} \mathscr{A}(z, \nabla \phi) \cdot \nabla (u^{(N)} - \phi) \, dz.$$

Since $u^{(N)} \in \mathscr{P}_N$

$$\begin{aligned} 0 &= \frac{1}{2} \int_{B_k} (u^{(N)})^2 \, dx \Big|_{t=0}^{t=T} + \int_{Q_{T,k}} \mathscr{A}(z, \nabla u^{(N)}) \cdot \nabla u^{(N)} \, dz - \int_{Q_{T,k}} f u^{(N)} \, dz \\ &\geq \frac{1}{2} \int_{B_k} (u^{(N)})^2 \, dx \Big|_{t=0}^{t=T} + \int_{Q_{T,k}} \mathscr{A}(z, \nabla u^{(N)}) \cdot \nabla \phi \, dz \\ &+ \int_{Q_{T,k}} \mathscr{A}(z, \nabla \phi) \cdot \nabla (u^{(N)} - \phi) \, dz - \int_{Q_{T,k}} f u^{(N)} \, dz. \end{aligned}$$

Fix an arbitrary $\phi \in \mathscr{P}_K$, $K \leq N$. Notice that due to Proposition 4.7 equality (4.114) is true for every $t \in [0, T]$. Letting $N \to \infty$ and using (4.110) we obtain

$$\frac{1}{2} \int_{B_k} u_k^2 \, dx \Big|_{t=0}^{t=T} + \int_{Q_{T,k}} \chi_k \cdot \nabla \varphi \, dz + \int_{Q_{T,k}} \mathscr{A}(z, \nabla \phi) \cdot \nabla (u_k - \phi) \, dz - \int_{Q_{T,k}} f u_k \, dz \leq 0.$$

Writing (4.113) in the form

$$\frac{1}{2} \int_{B_k} u_k^2 \, dx \Big|_{t=0}^{t=T} - \int_{Q_{T,k}} f u_k \, dz = - \int_{Q_{T,k}} \chi_k \cdot \nabla u_k \, dz$$

and substituting the result into the previous inequality we find that

$$\int_{Q_{T,k}} (\chi_k - \mathscr{A}(z, \nabla \phi)) \cdot \nabla (u_k - \phi) \geq 0$$

for every $\phi \in \mathscr{P}_K$. Letting $K \to \infty$ we may take for $\phi$ an arbitrary element from the main function space $W$. Let us choose $\phi$ in the special way: $\phi = u_k + \lambda w$ with $\lambda \in \mathbb{R}$ and $w \in W$: letting $\lambda \to 0$ we have

$$\int_{Q_{T,k}} (\chi_k - \mathscr{A}(z, \nabla u_k)) \cdot \nabla w \geq 0,$$

which is only possible if $\chi_k = \mathscr{A}(z, \nabla u_k)$ a.e. in $Q_{T,k}$.

This completes the proof of existence of a weak solution of problem (4.103). Uniqueness of the weak solution is an immediate byproduct of monotonicity of the operator $\mathscr{A}$. Assume that there are two different solutions $u_k^{(i)}$ of problem (4.103) and set $w = u_k^{(1)} - u_k^{(2)}$. Choosing $u_k^{(i)}$ for the test-function in identities (4.104) for $u_k^{(i)}$ and gathering the results we have that for every $\tau \in (0, T]$

$$\frac{1}{2}\int_{B_k} w(\tau)^2 dx \leq -\int_0^\tau \int_{B_k} \left(\mathscr{A}(z, u_k^{(1)}) - \mathscr{A}(z, u_k^{(2)})\right) \cdot \nabla(u_k^{(1)} - u_k^{(2)})\, dz \leq 0,$$

whence $w = 0$ a.e. in $Q_{T,k}$. □

*Remark 4.7* Identity (4.104) holds true for the test-functions $\phi \in W$ with $\phi_t \in W'$.

**Lemma 4.16** (Improved regularity) *Let* $u_0 \in V$ *and* supp $u_0 \subseteq B_R(0)$ *with a finite* $R > 0$, $f \in L^2(S_T)$ *and* $|a_t| \leq a_T = const$. *Then for all* $k > R$

$$\|u_{k,t}\|^2_{2,Q_{T,k}} + \operatorname{ess}\sup_{(0,T)} \int_{B_k} |\nabla u_k|^{p(x)}\, dx \leq C \tag{4.115}$$

*with a constant* $C$ *depending on* $\|u_0\|_{2,\mathbb{R}^n}$, $\|\nabla u_0\|_{p(\cdot),\mathbb{R}^n}$ *and* $\|f\|_{2,S_T}$, $a_T$ *but independent of* $k$.

*Proof* To prove the lemma it suffices to show that (4.115) holds for the Galerkin approximations of the regularized problems $u^{(N)}$. Let us multiply the $i$th equation of system (3.18) by $d'_{i,N}(t)$, integrate over $B_k$ and sum up:

$$\begin{aligned}&\|u_t^{(N)}(t)\|^2_{2,B_k} + \frac{d}{dt}\left(\int_{B_k} \frac{a(z)}{p(x)}|\nabla u^{(N)}(t)|^{p(x)}\, dx\right)\\ &= \int_{B_k} \frac{a_t(z)}{p(x)}|\nabla u^{(N)}(t)|^{p(x)}\, dx + \int_{B_k} f\, u_t^{(N)}\, dx.\end{aligned}$$

Integrating in $t$ and applying Young's inequality we arrive at the estimate

$$\begin{aligned}&\|u_t^{(N)}\|^2_{2,Q_{T,k}} + \int_{B_k} \frac{a(z)}{p(x)}|\nabla u^{(N)}(t)|^{p(x)}\, dx \leq \frac{1}{2}\|f\|^2_{2,Q_{T,k}} + \frac{1}{2}\|u_t^{(N)}\|^2_{2,Q_{T,k}}\\ &+ \frac{a_+}{p^-}\int_{B_k} |\nabla u_0^{(N)}|^{p(x)}\, dx + \frac{a_T}{p_-}\int_0^t \int_{B_k} |\nabla u^{(N)}(t)|^{p(x)}\, dz.\end{aligned}$$

Set

$$Y_N(t) = \int_0^t \int_{B_k} \frac{a(z)}{p(x)}|\nabla u^{(N)}(t)|^{p(x)}\, dz.$$

The function $Y_N(t)$ satisfies the linear differential inequality

$$Y_N'(t) \le \frac{a_T}{a_-} Y_N(t) + \frac{1}{2}\|f\|^2_{2,Q_{T,k}} + \frac{a_+}{p^-}\int_{B_k} |\nabla u_0^{(N)}|^{p(x)}\, dx.$$

Integration of this inequality gives the estimate:

$$Y_N(t) \le \frac{a_-}{a_T}\left(e^{\frac{a_T}{a_-}T} - 1\right)\left(\frac{1}{2}\|f\|^2_{2,Q_{T,k}} + \frac{a_+}{p^-}\int_{B_k} |\nabla u_0^{(N)}|^{p(x)}\, dx\right). \tag{4.116}$$

The proof will be completed if we show that the right-hand side of (4.116) is independent of $N$. Since $u_0 \in V$ and $\operatorname{supp} u_0 \subseteq B_R(0)$ by assumption, there exists a sequence $u_{0,\delta} \in C_0^\infty(B_R(0))$ such that $u_{0,\delta} \to u_0$ in $V$. Take an arbitrary $\varepsilon > 0$ and choose $\delta(\varepsilon)$ such that for all $\delta \in (0, \delta(\varepsilon))$

$$\|u_{0,\delta} - u_0\|_V = \|u_{0,\delta} - u_0\|_{2,B_R(0)} + \|\nabla(u_{0,\delta} - u_0)\|_{p(\cdot),B_R(0)} < \varepsilon.$$

The set $\{\psi_j\}$ is dense in $L^2(B_R(0)) \cap W_0^{1,p^+}(B_R(0))$ and there exists a sequence of functions $u_0^{(N)} \in \mathscr{P}_N$ such that for all $N$ from some $N(\varepsilon)$ on

$$\|u_{0,\delta} - u_0^{(N)}\|_V \le \|u_{0,\delta} - u_0^{(N)}\|_{2,B_R(0)} + C\|\nabla(u_{0,\delta} - u_0^{(N)})\|_{p^+,B_R(0)} < \varepsilon.$$

It follows that for all sufficiently big $N$

$$\begin{aligned}
\|\nabla u_0^{(N)}\|_{p(\cdot),B_R(0)} &\le \|\nabla(u_0 - u_0^{(N)})\|_{p(\cdot),B_R(0)} + \|\nabla u_0\|_{p(\cdot),B_R(0)} \\
&\le \|\nabla u_0\|_{p(\cdot),B_R(0)} + \|u_{0,\delta} - u_0\|_V + \|u_{0,\delta} - u^{(N)}\|_{W_0^{1,p^+}(B_R(0))} \\
&\le 1 + \|\nabla u_0\|_{p(\cdot),B_R(0)}.
\end{aligned}$$

The uniform in $N$ estimate on the last term in (4.116) follows now from (1.6) if we claim that $k \ge R$. □

### 4.7.3 Unbounded Domain

Let $\{u_k\}$ be the sequence of solutions of the regularized problems (4.103). Define the sequence of functions extended to the whole $S_T$

$$w_k = \begin{cases} u_k & \text{in } Q_{T,k}, \\ 0 & \text{in } S_T \setminus Q_{T,k}, \end{cases} \qquad f_k = \begin{cases} f & \text{in } Q_{T,k}, \\ 0 & \text{in } S_T \setminus Q_{T,k}. \end{cases}$$

According to the uniform estimates of Lemmas 4.9 and 4.12

$$w_k \text{ are bounded in } L^\infty(0,T;L^2(\mathbb{R}^n)) \text{ and in } W,$$
$$\mathscr{A}(z,\nabla w_k) \text{ and } w_{k,t} \text{ are bounded in } W'.$$

It follows that there exist functions $w \in W$, $\chi \in W'$, $U \in L^2(\mathbb{R}^n)$ such that

$$\begin{aligned}
&w_k(T) \to U \text{ in } L^2(\mathbb{R}^n), \qquad w_k \rightharpoonup w \text{ in } W,\\
&w_k \rightharpoonup w\text{*-weak in } L^\infty(0,T;L^2(\mathbb{R}^n)),\\
&\mathscr{A}(z,\nabla w_k) \rightharpoonup \chi \text{ and } w_{k,t} \rightharpoonup w_t \text{ in } W'.
\end{aligned}$$

**Lemma 4.17** *The sequence $\{w_k\}$ contains a subsequence which converges in $C^0([0,T];L^2(\Omega))$ on every compact $\Omega \subset \mathbb{R}^n$.*

*Proof* By Proposition 4.7 for every $k$ there is a subsequence $\{w_{m_k}\}$ converging in $C^0([0,T];L^2(B_k))$. Choosing subsequences $\{w_{m_{k+1}}\} \subset \{w_{m_k}\}$, $k = 1, 2, \ldots$, we conclude that the diagonal sequence $\{w_{m_m}\}$ converges in $C^0([0,T];L^2(B_k))$ for every natural $k$. □

An immediate corollary from Lemma 4.17 is the equality $U = w(T)$. Take an arbitrary $\phi \in C^1(0,T;C_0^1(\mathbb{R}^n))$ and choose $k_0 \in \mathbb{N}$ so big that $\operatorname{supp} \phi(\cdot,t) \subset B_k$ for all $k \geq k_0$ and $t \in [0,T]$. Since $C^1([0,T];C_0^1(B_k)) \subset W(Q_{T,k})$ for all $k \geq k_0$, the extended functions $w_k$ satisfy the identity

$$\int_{S_T} \left( w_{k,t}\phi + \mathscr{A}(z,\nabla w_k)\cdot\nabla\phi + f_k\phi \right) dz = 0.$$

Letting $k \to \infty$ we obtain

$$\int_{S_T} (w_t\phi + \chi\cdot\nabla\phi - f\phi)\, dz = 0.$$

Since $C^1(0,T;C_0^1(\mathbb{R}^n))$ is dense in $W(Q_{T,k})$ for every $k$, the same is true for $\phi = w_k$, which gives the energy identity as $k \to \infty$:

$$\int_{S_T} (w_t w + \chi\cdot\nabla w - f w)\, dz = 0. \tag{4.117}$$

By virtue of Proposition 4.6 the energy equality holds:

$$\frac{1}{2}\int_{\mathbb{R}^n} w^2\, dx \bigg|_{t=0}^{t=T} + \int_{S_T} (\chi\cdot\nabla w - f w)\, dx dt = 0.$$

It is now standard to check that $\chi = \mathscr{A}(z,\nabla w)$ a.e. in $S_T$.

### 4.7.4 Boundedness and Uniqueness of Weak Solutions

In this section we give the proof of Theorem 4.12, which is split into three assertions.

**Lemma 4.18** (The maximum principle) *Let* $\sup_{S_T} |f| \leq C_f$ *and* $\sup_{\Omega} |u_0| \leq M_0$ *with finite positive constants* $C_f$ *and* $M_0$. *Then the solution of problem* (4.93) *satisfies the estimate*

$$|u| \leq M e^{\frac{C_f}{M} T} \quad \text{a.e. in } S_T \text{ with } M = 1 + M_0. \tag{4.118}$$

*Proof* It is sufficient to show that estimate (4.118) holds true for the solutions of the auxiliary problems (4.103) $u_k$. Set $v_k := \mathrm{e}^{-\lambda t} u_k$ with a constant $\lambda$ to be defined and write identity (4.104) in the form

$$\int_{Q_{T,k}} \mathrm{e}^{\lambda t} \left(v_{k,t}\phi + \mathscr{A}(z, \mathrm{e}^{\lambda t}\nabla v_k) \cdot \nabla \phi\right) dz = -\int_{Q_{T,k}} \left(\lambda \mathrm{e}^{\lambda t} v_k - f\right) \phi \, dz.$$

Now set

$$\lambda = \frac{C_f}{M}, \quad v_{k,M} = \max\{v_k - M, 0\}, \quad \phi = \mathrm{e}^{-\lambda t} v_{k,M}.$$

By virtue of Remark 4.7 $\phi$ is an admissible test-function. Notice that

$$v_{k,t} v_{k,M} = \frac{1}{2}\partial_t v_{k,M}^2, \qquad \mathscr{A}(z, \mathrm{e}^{\lambda t}\nabla v_k) \cdot \nabla v_{k,M} = \mathscr{A}(z, \mathrm{e}^{\lambda t}\nabla v_{k,M}) \cdot \nabla v_{k,M} \geq 0.$$

Then

$$\begin{aligned}
&\int_{Q_{T,k}} \left(v_{k,t} v_{k,M} + \mathscr{A}(z, \mathrm{e}^{\lambda t}\nabla v_k) \cdot \nabla v_{k,M}\right) dz \\
&\quad = \frac{1}{2}\int_{B_k} v_{k,M}^2(T)\, dx + \int_{Q_{T,k}} \mathscr{A}(z, e^{\lambda t}\nabla v_k) \cdot \nabla v_{k,M} \, dz := J
\end{aligned}$$

with

$$\begin{aligned}
J &= -\int_{Q_{T,k}} \left(\lambda \mathrm{e}^{\lambda t} v_k - f\right) \phi \, dz = -\int_{Q_{T,k} \cap \{v_k > M\}} \left(\lambda v_k - \mathrm{e}^{-\lambda t} f\right) v_{k,M} \, dz \\
&\leq -\int_{Q_{T,k} \cap (v_k > M)} \left(\lambda M - C_f\right) v_{k,M} \, dz \leq 0.
\end{aligned}$$

It follows that $v_k \leq M$ a.e. in $Q_{T,k}$. In the same way we check then that $-v_k \leq M$. Thus, $|v_k| \leq M$ and $|u_k| \leq M \mathrm{e}^{\frac{C_f}{M} T}$. □

**Lemma 4.19** (Uniqueness of bounded solutions) *If* $p(x) \geq 2$ *a.e. in* $\mathbb{R}^n$, *problem* (4.93) *cannot have more than one bounded solution.*

*Proof* Let $u_1, u_2$ be two weak solutions of problem (4.93). Assume that there exists a finite constant $M$ such that $|u_i| \leq M$ a.e. in $S_T$, set $w = u_1 - u_2$ and introduce the function

$$\psi(s) = \begin{cases} 1 & \text{if } |s| \leq R, \\ s - R + 1 & \text{if } R < |s| < R + 1, \\ 0 & \text{if } R + 1 \leq |s| \,. \end{cases}$$

It is easy to see that $|\psi'(s)| \leq 1$. Taking $\psi(|x|)u_i$ for the test-functions in the integral identities (4.98) for $u_i$ and subtracting the results we arrive at the relation

$$\frac{1}{2}\int_{\mathbf{R}^n} w^2(T)\psi(|x|)\,dx + \int_0^T \int_{\mathbb{R}^n} \psi(|x|)\,(\mathscr{A}(z,u_1) - \mathscr{A}(z,u_2)) \cdot \nabla w\,dz = I$$

with

$$I = -\int_0^T \int_{\mathbb{R}^n} w\,(\mathscr{A}(z,u_1) - \mathscr{A}(z,u_2)) \cdot \nabla\psi(|x|)\,dz.$$

This integral is estimated as follows:

$$\begin{aligned} |I| &\leq a_+ \int_0^T \int_{\mathbb{R}^n \setminus B_R(0)} \left(|\nabla u_1|^{p-1} + |\nabla u_2|^{p-1}\right)(|u_1| + |u_2|)\,dz \\ &\leq C\left(\int_0^T \int_{\mathbb{R}^n \setminus B_R(0)} (|\nabla u_1|^p + |\nabla u_2|^p)\,dz + \int_0^T \int_{\mathbb{R}^n \setminus B_R} (|u_1|^p + |u_2|^p)\,dz\right) \\ &\equiv I_1(R) + I_2(R), \qquad C = C(a_+, p_\pm). \end{aligned}$$

Due to estimate (4.99), it is necessary that $I_1(R) \to 0$ as $R \to \infty$. Let us consider the integral $I_2(R)$. Since $p(x) \geq 2$ a.e. in $\mathbb{R}^n$, and $|u_i| \leq M$ by assumption, applying (4.99) we obtain

$$I_2(R) \leq T \sup_{\mathbb{R}^n} M^{p(x)-2} \operatorname*{ess\,sup}_{(0,T)} \int_{\mathbb{R}^n \setminus B_R(0)} \left(|u_1|^2 + |u_2|^2\right)\,dx \to 0 \quad \text{as } R \to \infty.$$

□

The assertion of item (3) of Theorem 4.12 follows from Lemma 4.16 as $k \to \infty$.

## 4.8 Stationary Solutions

In this section we study solvability of the Dirichlet problem for the stationary counterpart of the equation

$$u_t - \sum_{i=1}^{n} \frac{d}{dx_i}\left(a_i(x,u)|D_i u|^{p_i(x)-2} D_i u\right) + c(x,u)|u|^{\sigma(x)-2}u = \Phi(x).$$

Let us consider the problem

$$\begin{cases} -\sum_i D_i \left(a_i(x,u)|D_i u|^{p_i(x)-2} D_i u\right) + c(x,u)|u|^{\sigma(x)-2}u = \Phi(x) \quad \text{in } \Omega, \\ u = 0 \text{ on } \partial\Omega \end{cases} \tag{4.119}$$

with

$$\Phi(x) = f(x) + \operatorname{div} \mathbf{G}(x), \quad \mathbf{G}(x) \in \mathbf{V}'(\Omega).$$
$$f \in L^{\sigma'(x)}(\Omega), \quad \sigma'(x) = \frac{\sigma(x)}{\sigma(x)-1}.$$

It is assumed that the data of problems (4.119) satisfy the following conditions.

1. The coefficients $a_i(x,r)$, and $c(x,r)$ are Carathéodory functions. Unless explicitly stated, we always assume that $a_i$ and $c$ satisfy the conditions

$$\begin{cases} \forall x \in \overline{\Omega},\ r \in \mathbb{R} \quad 0 < a_0 \le a_i(x,r) < \infty, \\ \qquad\qquad\qquad\quad 0 < c_0 \le c(x,r) < \infty \end{cases} \tag{4.120}$$

   with some positive constants $a_0$, $c_0$.
2. The functions $p_i(x)$ and $\sigma(x)$ are bounded in $\Omega$: it is assumed that there exist constants $p^- \in (1,n)$, $p^+ < \infty$, $\sigma^- > 1$, $\sigma^+ < \infty$ such that for all $x \in \overline{\Omega}$

$$\begin{cases} p_i(x) \in (p^-, p^+], \quad \inf_\Omega p_i(x) > p^-, \quad i = 1, \dots, n, \\ \sigma(x) \in (\sigma^-, \sigma^+]. \end{cases} \tag{4.121}$$

3. The functions $p_i(x)$, $\sigma(x)$ are continuous in $\overline{\Omega}$, $p_i(x)$ satisfy the log-continuity condition (1.21).

**Definition 4.4** A locally integrable function $u(x)$ is called *weak solution* of problem (4.119) if

1. $u \in \mathbf{V}_\sigma(\Omega) \equiv \left\{u(x) | \, u \in L^{\sigma(\cdot)}(\Omega) \cap W_0^{1,1}(\Omega),\ |D_i u|^{p_i(x)} \in L^1(\Omega)\right\}$,
2. for any test-function $\zeta \in \mathbf{V}_\sigma(\Omega)$ the integral identity holds:

$$\int_\Omega \left(\sum_{i=1}^{n} a_i |D_i u|^{p_i(x)-2} D_i u \, D_i \zeta + c|u|^{\sigma(x)-2} u\zeta - f\,\zeta - \mathbf{G}\,\nabla\,\zeta\right) dx = 0. \tag{4.122}$$

### 4.8.1 A Model Equation

We begin with the special situation when the coefficients $a_i$ and $c$ do not depend on $u$ and the proof of existence of weak solutions is fairly easy.

**Theorem 4.14** *Let conditions* (4.120)–(4.121) *be fulfilled,* $f \in L^{\sigma'(x)}(\Omega)$, $\mathbf{G} \in \mathbf{V}'_\sigma(\Omega)$. *Assume that* $a_i(x,s) \equiv A_i(x)$ *and* $c(x,s) \equiv C(x)$. *Then the problem*

$$\begin{cases} -\sum_{i=1}^{n} D_i\left(A_i(x)|D_i u|^{p_i(x)-2} D_i u\right) + C(x)|u|^{\sigma(x)-2}u = \Phi(x) \quad \text{in } \Omega, \\ u = 0 \text{ on } \Gamma \end{cases} \tag{4.123}$$

*has at least one weak solution.*

*Proof* Let us introduce the operator $\mathscr{L} : \mathbf{V}_\sigma(\Omega) \mapsto \mathbf{V}'_\sigma(\Omega)$

$$(\mathscr{L}u, \zeta) = \int_\Omega \left(\sum_i A_i(x)|D_i u|^{p_i(x)-2} D_i u D_i \zeta + C(x)|u|^{\sigma(x)-2}u\zeta\right) dx.$$

The mapping $\mathscr{L} : \mathbf{V}_\sigma(\Omega) \mapsto \mathbf{V}'_\sigma(\Omega)$ is continuous and monotone.

**Lemma 4.20** *The operator* $L : \mathbf{V}_\sigma(\Omega) \mapsto \mathbf{V}'_\sigma(\Omega)$ *is coercive:*

$$\forall u \in \mathbf{V}_\sigma(\Omega) \quad (\mathscr{L}u, u) \geq C \min\left\{\left(\frac{\|u\|_{\mathbf{V}_\sigma}}{n+1}\right)^{p^-}, \left(\frac{\|u\|_{\mathbf{V}_\sigma}}{n+1}\right)^{p^+}\right\}.$$

*Proof* Let

$$\lambda = \|u\|_{\mathbf{V}_\sigma}, \quad \lambda_i = \|D_i u\|_{p_i(x)}, \quad i = 1, \dots, n, \qquad \lambda_0 = \|u\|_{\sigma(x)}.$$

Then there is $i \in \{0, 1, \dots, n\}$ such that $\lambda_i \geq \dfrac{\lambda}{n+1}$ (we do not loose the generality by assuming that $i = 1$). It follows from (1.6) that

$$\begin{aligned}
(\mathscr{L}u, u) &\geq C\left(A_{\mathbf{p}(\cdot)}(\nabla u) + A_{\sigma(\cdot)}(u)\right) \\
&\geq C \int_\Omega |D_1 u|^{p_1(x)}\, dx \\
&\geq C \begin{cases} \lambda_1^{p^-} & \text{if } \lambda_1 > 1, \\ \lambda_1^{p^+} & \text{if } \lambda_1 \leq 1 \end{cases} \\
&\geq C \min\left\{\left(\frac{\lambda}{n+1}\right)^{p^+}, \left(\frac{\lambda}{n+1}\right)^{p^-}\right\}.
\end{aligned}$$

□

The space $\mathbf{V}_\sigma(\Omega)$ is separable, and the mapping $\mathscr{L} : \mathbf{V}_\sigma(\Omega) \mapsto \mathbf{V}'_\sigma(\Omega)$ is monotone, continuous and coercive. By the Browder-Minty theorem [80, Theorem 7.3.2] for every $\Phi \in \mathbf{V}'_\sigma(\Omega)$ the equation $\mathscr{L}u = \Phi$ has at least one weak solution $u \in \mathbf{V}_\sigma(\Omega)$. □

### 4.8.2 The General Case

**Theorem 4.15** *Let conditions* (4.120)–(4.121) *be fulfilled,* $f \in L^{\sigma'(x)}(\Omega)$, $\mathbf{G} \in \mathbf{V}'_\sigma(\Omega)$. *Then problem* (4.119) *has at least one weak solution satisfying the estimate*

$$\|u\|_{\mathbf{V}_\sigma(\Omega)} \leq K \tag{4.124}$$

*where the constant $K$ depends on $p^\pm$, $\sigma^\pm$, $n$, $\|f\|_{\sigma'(\cdot)}$, $\|\mathbf{G}\|_{\mathbf{V}'_\sigma}$.*

Define the operator $\mathscr{L}_\Omega : \mathbf{V}_\sigma(\Omega) \mapsto \mathbf{V}'_\sigma(\Omega)$

$$(\mathscr{L}_\Omega u, \zeta) = \int_\Omega \left( \sum_i a_i(x,u)|D_i u|^{p_i(x)-2} D_i u D_i \zeta + c(x,u)|u|^{\sigma(x)-2} u \zeta \right) dx.$$

Since the space $\mathbf{V}_\sigma(\Omega)$ is separable, there exists a fundamental system $\{\phi_k(x)\} \subset \mathbf{V}_\sigma(\Omega)$. Let us search Galerkin's approximations $u^{(N)} = \sum_{k=1}^N c_{kN}\phi_k(x)$ as the solutions of the system of equations

$$\mathscr{L}_\Omega(u^{(N)}, \phi_k) = \int_\Omega \Phi\, \phi_k \, dx, \qquad k = 1, 2, \ldots, N.$$

Let $\mathscr{P}_N$ be the linear span of the system $\{\phi_1, \ldots, \phi_N\}$. The norm in $\mathscr{P}_N$ is defined by

$$\|\eta\|_{\mathscr{P}_N} = \|\eta\|_{\mathbf{V}_\sigma}.$$

Considering $\mathscr{P}_N$ as the $N$-dimensional space of vectors $\mathbf{v} = (v_1, \ldots, v_N)$ with the elements $\mathbf{v} = \sum_{i=1}^N v_i \phi_i(x)$ we introduce the scalar product $(\mathbf{v}, \mathbf{w})_{\mathscr{P}_N} = \sum_{i=1}^N v_i w_i$ and the norm $|\mathbf{v}|_N = \sqrt{(\mathbf{v}, \mathbf{v})}$. Then

$$\forall \eta \in \mathscr{P}_N \qquad \mathscr{L}_\Omega(u^{(N)}\ \eta) = \int_\Omega \Phi\, \eta \, dx. \tag{4.125}$$

For every fixed $u^{(N)}$ the left hand-side of this equality is a linear functional on $\mathscr{P}_N$. By Riesz's Theorem there exists $g(u^{(N)}) \in \mathscr{P}_N$ such that

$$\forall \eta \in \mathscr{P}_N \qquad \mathscr{L}_\Omega(u^{(N)}, \eta) = \left( g(u^{(N)}) - \Phi, \eta \right)_{\mathscr{P}_N}.$$

**Lemma 4.21** *The operator $\mathscr{L}_\Omega$ is coercive:*

$$\forall\, v \in \mathbf{V}_\sigma(\Omega) \qquad (\mathscr{L}_\Omega v,\, v) \geq C\, \min\left\{ \left(\frac{\|v\|_{\mathbf{V}_\sigma}}{n+1}\right)^{p^-},\, \left(\frac{\|v\|_{\mathbf{V}_\sigma}}{n+1}\right)^{p^+} \right\}.$$

(See the proof of Lemma 4.20.)

**Lemma 4.22** *The operator $g(\cdot) : \mathscr{P}_N \mapsto \mathscr{P}_N$ is continuous.*

*Proof* Let $\{v_k\} \subset \mathscr{P}_N$ and $|v_k - v|_N \to 0$. It follows that $|v_k - v|^{\sigma(x)} + \sum_i |D_i(v_k - v)|^{p_i(x)} \to 0$ almost everywhere in $\Omega$. Given an arbitrary $\varepsilon > 0$ one may choose a set $\Omega_\varepsilon \subset \Omega$ such that

$$\begin{aligned}
&|\Omega \setminus \Omega_\varepsilon| < \varepsilon,\\
&\sup_{\Omega_\varepsilon} |\phi_k| + \sum_i \sup_{\Omega_\varepsilon} |\nabla\, \phi_k| \leq K_\varepsilon = const,\\
&|v_k - v|^{\sigma(x)} + \sum_i |D_i(v_k - v)|^{p_i(x)} \to 0 \quad \text{uniformly on } \Omega_\varepsilon.
\end{aligned}$$

Then $\forall\, \eta \in \mathscr{P}_N$

$$(g(v_k) - g(v), \eta)_N = \left(\mathscr{L}_{\Omega_\varepsilon}(v_k, \eta) - \mathscr{L}_{\Omega_\varepsilon}(v, \eta)\right) + \left(\mathscr{L}_{\Omega\setminus\Omega_\varepsilon}(v_k, \eta) - \mathscr{L}_{\Omega\setminus\Omega_\varepsilon}(v, \eta)\right).$$

By the property of absolute continuity of the integral

$$\left|\mathscr{L}_{\Omega\setminus\Omega_\varepsilon}(v_k, \eta) - \mathscr{L}_{\Omega\setminus\Omega_\varepsilon}(v, \eta)\right| \leq C\, \left(\|v_k\|_{\mathbf{V}_\sigma} + \|v\|_{\mathbf{V}_\sigma}\right)\, \|\eta\|_{\mathbf{V}_\sigma(\Omega\setminus\Omega_\varepsilon)} \to 0$$

as $\varepsilon \to 0$. On the other hand,

$$\left|\mathscr{L}_{\Omega_\varepsilon}(v_k, \eta) - \mathscr{L}_{\Omega_\varepsilon}(v, \eta)\right| \to 0 \quad \text{when } k \to \infty \text{ for every fixed } \varepsilon.$$

Thus, for every $\eta \in \mathscr{P}_N$ $(g(v_k) - g(v), \eta)_N \to 0$, which means that $g(v_k) \to g(v)$ weakly in $\mathscr{P}_N$. Since the dimension of $\mathscr{P}_N$ is finite, the weak convergence $g(v_k) \to g(v)$ implies the strong convergence: $|g(v_k) - g(v)|_N \to 0$ when $k \to \infty$. □

**Lemma 4.23** *For every $N \in \mathbb{N}$ the solution $u^{(N)}$ of problem* (4.125) *satisfies the estimate $\|u^{(N)}\|_{\mathbf{V}_\sigma} \leq K$ with a finite constant $K$ independent of $N$.*

*Proof* Let us take $\eta = u^{(N)}$ for the test-function in (4.125) and then apply Young's inequality: $\forall\, \varepsilon > 0$

$$
\begin{aligned}
a_0 A_{\mathbf{p}(\cdot)}\left(\nabla u^{(N)}\right) &+ c_0 A_{\sigma(\cdot)}\left(u^{(N)}\right) \\
&\leq \varepsilon\, A_{\mathbf{p}(\cdot)}\left(\nabla u^{(N)}\right) + C_1(\varepsilon) \sum_i A_{p_i'(\cdot)}(G_i) \\
&\quad + \varepsilon\, A_{\sigma(\cdot)}\left(u^{(N)}\right) + C_2(\varepsilon)\, A_{\sigma'(\cdot)}(f).
\end{aligned}
$$

According to (1.6), this inequality provides the estimate $\|u^{(N)}\|_{\mathbf{V}_\sigma(\Omega)} \leq K$ with a constant $K$ depending only on $p^{\pm}$, $\sigma^{\pm}$, $\|\mathbf{G}\|_{\mathbf{V}'_\sigma}$ and $\|f\|_{\sigma'(\cdot)}$. □

**Lemma 4.24** *For every $N \in \mathbb{N}$ the equation $g(u^{(N)}) = 0$ has at least one solution in $\mathscr{P}_N$.*

*Proof* Consider the family of operators

$$
g_\tau(v) = (1-\tau)v + \tau g(v), \qquad \tau \in [0, 1].
$$

By Lemma 4.21, for every $|v|_N > n + 1$

$$
(g_\tau(v), v)_N = (1-\tau)(v, v)_N + \tau \mathscr{L}_\Omega(v, v) \geq C(1-\tau)|v|_N^2 + \tau C\left(\frac{|v|_N}{n+1}\right)^{p^-} > 0.
$$

It follows that $(g_\tau(v), v)_N > 0$ for all $\tau \in [0, 1]$ and every $\|v\|_{\mathbf{V}_\sigma} = |v|_N$ sufficiently large, say, for $|v|_N = R_N \geq n + 1$. For $\tau = 0$ the equation $g_0(v) = 0$ has in the ball $B_N = \{v \in \mathscr{P}_N : |v|_N < R_N\}$ only the trivial solution, and for every $\tau \in [0, 1]$ the boundary of the ball $B_N$ does not contain any solution of the the equation $g_\tau(v) = 0$. According to Browder's fixed point theorem, the equation $g_1(v) \equiv g(v) = 0$ has at least one solution in the ball $B_N$. □

By Lemma 4.23, the sequence of Galerkin's approximations contains a subsequence $\{u^{(N)}\}$ possessing the following properties: there exist functions $u \in \mathbf{V}_\sigma(\Omega)$, $A_i(x) \in L^{p_i'(x)}(\Omega)$, $v \in L^{\sigma'(x)}(\Omega)$ such that

$$
\begin{aligned}
& u^{(N)} \rightharpoonup u \text{ in } \mathbf{V}_\sigma(\Omega), \\
& a_i\left(x, u^{(N)}\right) |D_i u^{(N)}|^{p_i(x)-2} D_i u^{(N)} \rightharpoonup A_i(x) \text{ in } L^{p_i'(x)}(\Omega), \\
& c\left(x, u^{(N)}\right) |u^{(N)}|^{\sigma(x)-2} u \rightharpoonup v(x) \text{ in } L^{\sigma'(x)}(\Omega), \\
& D_i u^{(N)} \rightharpoonup D_i u \text{ in } L^{p_i(x)}(\Omega), \\
& u^{(N)} \rightharpoonup u \text{ in } L^{\sigma(x)}(\Omega), \\
& u^{(N)} \to u \text{ a.e. in } \Omega.
\end{aligned}
\tag{4.126}
$$

These properties allow one to pass to the limit in (4.125) when $N \to \infty$, which gives

$$\int_\Omega \left[ \sum_i A_i(x)\, D_i\eta + \nu(x)\,\eta - \Phi\,\eta \right] dx = 0, \qquad \eta \in \mathbf{V}_\sigma(\Omega). \tag{4.127}$$

To complete the proof, we have to identify the limits $A_i(x)$ and $\nu(x)$.

**Lemma 4.25** *For almost all $x \in \Omega$*

$$\begin{cases} A_i(x) = a_i(x,u)|D_i u|^{p_i(x)-2} D_i u, \quad i = 1, \ldots, n, \\ \nu(x) = c(x,u). \end{cases} \tag{4.128}$$

*Proof* Let us denote $\widetilde{a}_i(x,u,D_i v) \equiv a_i(x,u)|D_i v|^{p_i(x)-2} D_i v$. According to (4.35), for every $\xi \in \mathbf{V}_\sigma(\Omega)$

$$\sum_i \int_\Omega \left( \widetilde{a}_i\left(x, u^{(N)}, D_i u^{(N)}\right) - \widetilde{a}_i\left(x, u^{(N)}, D_i\xi\right)\right) D_i\left(u^{(N)} - \xi\right) dx \ge 0, \tag{4.129}$$

Let in (4.125) $\eta = u^{(N)} - \xi$ with $\xi \in \mathscr{P}_N$. Combining (4.125) with (4.129), we obtain:

$$\int_\Omega \Big[ \left(\Phi - c\left(x, u^{(N)}\right) |u^{(N)}|^{\sigma(x)-2} u^{(N)}\right)\left(u^{(N)} - \xi\right) - \sum_i \widetilde{a}_i\left(x, u^{(N)}, D_i\xi\right) D_i\left(u^{(N)} - \xi\right)\Big] dx \ge 0.$$

Letting $N \to \infty$ and using (4.126), we have that

$$\int_\Omega \Big[ (\Phi - \nu)(u - \xi) - \sum_i \widetilde{a}_i(x,u,D_i\xi) D_i(u-\xi) \Big] dx \ge 0.$$

Adding this inequality to (4.127) with $\eta = u - \xi$, we have: $\forall \xi \in \mathbf{V}_\sigma(\Omega)$

$$\sum_i \int_\Omega \Big[ A_i(x) - \widetilde{a}_i(x,u,D_i\xi) \Big] D_i(u-\xi)\, dx \ge 0.$$

Since $\xi$ is arbitrary, we may take $\xi = u \pm \varepsilon\zeta$ with $\varepsilon > 0$, $\zeta \in \mathbf{V}_\sigma(\Omega)$. Simplifying and then letting $\varepsilon \to 0$, we conclude that

$$\forall \zeta \in \mathbf{V}_\sigma(\Omega) \qquad \sum_i \int_\Omega \Big[ A_i(x) - \widetilde{a}_i(x,u,D_i u) \Big] D_i\zeta\, dx = 0.$$

This gives the first equality of (4.128). By (4.125) and (4.127) we have then that $\forall \eta \in \mathbf{V}_\sigma(\Omega)$

$$\sum_i \int_\Omega \left[\widetilde{a}_i\left(x, u^{(N)}, D_i u^{(N)}\right) - \widetilde{a}_i\left(x, u, D_i u\right)\right] D_i \eta\, dx$$
$$+ \int_\Omega c\left(x, u^{(N)}\right)\left(|u^{(N)}|^{\sigma(x)-2} u^{(N)} - |u|^{\sigma(x)-2} u\right) \eta\, dx$$
$$= \int_\Omega \left[v(x) - c\left(x, u^{(N)}\right) |u|^{\sigma(x)-2} u\right] \eta\, dx.$$

By virtue of (4.126) we may pass to the limit when $N \to \infty$ in all three terms of this equality. The two terms on the left-hand side tend to zero, whence

$$\forall\, \eta \in \mathbf{V}_\sigma(\Omega) \qquad \int_\Omega \left[v(x) - c\,(x, u)\, |u|^{\sigma(x)-2} u\right] \eta\, dx = 0.$$

□

Estimate (4.124) is a byproduct of Lemma 4.23.

### 4.8.3 Bounded Weak Solutions

**Theorem 4.16** *Let in the conditions of* Theorem 4.15 $\|\Phi(x)\|_{\infty,\Omega} = K < \infty$ *and* $c_0 > 0$. *Then the weak solution of problem* (4.119) *satisfies the estimate*

$$\|u\|_{\infty,\Omega} \le \max\left\{1;\ \left(\frac{K}{c_0}\right)^{1/(\sigma^- - 1)}\right\}$$

*with the constant* $c_0$ *from condition* (4.120).

*Proof* Let us define the constant $\mu$ from the relation $K = c_0 \mu^{\sigma^- - 1}$ and then set $M_\varepsilon = \max\{1;\ \mu + \varepsilon\}$ with an arbitrary $\varepsilon > 0$. Set $\zeta_\varepsilon = \max\{u - M_\varepsilon, 0\}$ and choose this function for the test-function in the integral identity (4.122). Notice that

$$\zeta_\varepsilon = \begin{cases} 0, & \text{if } u \le M_\varepsilon, \\ u - M_\varepsilon & \text{if } u > M_\varepsilon, \end{cases} \qquad \nabla\zeta = \begin{cases} 0 & \text{if } u \le M_\varepsilon, \\ \nabla u & \text{if } u > M_\varepsilon. \end{cases}$$

Identity (4.122) becomes

$$\int_{\Omega\cap(u\ge M_\varepsilon)} \left(\sum_{i=1}^n a_i |\nabla u|^{p_i(x)} + c|u|^{\sigma(x)-2} u(u - M_\varepsilon) - (f + \operatorname{div}\mathbf{G})(u - M_\varepsilon)\right) dx = 0.$$

Applying (4.120) and using the definition of the constant $M_\varepsilon$, we obtain the inequality

$$\left(c_0 M_\varepsilon^{\sigma^- - 1} - K\right) \int_{\Omega\cap(u\ge M_\varepsilon)} (u - M_\varepsilon)\, dx \le 0.$$

It follows that $u \leq M_\varepsilon$ for every $\varepsilon > 0$. Likewise we establish that $-u \leq M_\varepsilon$, whence

$$\|u\|_{\infty,\Omega} \leq M_\varepsilon = \max\left\{1;\ \varepsilon + \left(\frac{K}{c_0}\right)^{1/(\sigma^- -1)}\right\} \quad \text{for every } \varepsilon > 0.$$

□

**Theorem 4.17** *Let in the conditions of* Theorem 4.15 $c_0 \geq 0$ *and*

$$\Phi(x) \equiv f(x) \in L^{(q-1)/q}(\Omega) \quad \textit{with} \quad q \in \left(1, \frac{np^-}{n-p^-}\right).$$

*Then the solutions of problem* (4.119) *satisfy the estimate* $\|u\|_{\infty,\Omega} \leq K$ *with a constant* $M$ *depending on* $p^\pm$, $\|f\|$, $1/a_0$ *and* $n$.

*Proof* Fix an arbitrary $k \in \mathbb{N}$ and take the function $\zeta(x) = \max\{0, u-k\}$ for the test-function in Definition 4.4. Denote $\Omega_k = \Omega \cap \{x \in \Omega : u(x) > k\}$ and notice that

$$\nabla\zeta(x) = \begin{cases} \nabla u & \text{for } u > k, \\ 0 & \text{for } u \leq k. \end{cases}$$

According to (4.4)

$$\begin{aligned} I_1 &\equiv a_0 \sum_i \int_{\Omega_k} |D_i u|^{p_i(x)}\, dx \\ &\leq \int_{\Omega_k} \left(\sum_i a_i(x,u)|D_i u|^{p_i(x)} + c(x,u)|u|^{\sigma(x)-2} u\,(u-k)\right) dx \\ &= \int_{\Omega_k} (u-k) f \leq \|u-k\|_{q,\Omega_k} \|f\|_{q/(q-1),\Omega} \equiv I_2. \end{aligned}$$

Not loosing generality we may assume that $|\Omega| \equiv \operatorname{meas}\Omega < 1$ and $A_{\mathbf{p}(x)}(\nabla u) < 1$. Using (1.6) we estimate $I_1$ as follows:

$$I_1 \geq a_0 \sum_i \|D_i u\|^{p^+}_{p_i(\cdot)\Omega_k} \geq (2|\Omega_k|)^{-\frac{p^+}{p^-}} \|\nabla u\|^{p^+}_{p^-,\Omega_k}.$$

On the other hand, applying the Sobolev embedding theorem we have

$$I_2 \leq c|\Omega_k|^{\frac{1}{q}-\frac{1}{p^-}+\frac{1}{n}} \|\nabla u\|_{p^-,\Omega_k} \|f\|_{q/(q-1),\Omega}.$$

Gathering these estimates we obtain the inequality

$$\|\nabla u\|_{p^-,\Omega_k}^{p^-} \le \left(\frac{c}{a_0}\right)^{\frac{p^-}{p^+-1}} |\Omega_k|^{1+\left(\frac{1}{q}+\frac{1}{n}\right)\frac{p^-}{p^+-1}} \|f\|_{q/(q-1),\Omega}^{\frac{p^-}{p^+-1}}.$$

The conclusion follows now from [Chap. 2, Lemma 5.3, 187]. □

## 4.8.4 Systems of Elliptic Equations

Let us consider the system of equations

$$\begin{cases} -\sum_j D_j\left(a_{ij}(x,\nabla\mathbf{u})\right) + \sum_j c_{ij}(x)|u^{(j)}|^{\sigma_{ij}(x)-2}u^{(j)} = f^{(i)}(x) \quad \text{in } \Omega, \\ \qquad i = 1,\dots,n, \\ \mathbf{u} = 0 \text{ on } \partial\Omega, \quad \mathbf{u} = (u^{(1)},\dots,u^{(n)}). \end{cases} \tag{4.130}$$

We assume that the functions $a_{ij}(x,\mathbf{V})$ have the form

$$a_{ij}(x,\mathbf{V}) = A_{ij}(x)|V_{ij}|^{p_{ij}(x)-2}V_{ij}, \quad i,j = 1,\dots,n,$$

with coefficients $A_{ij}(x)$ and $c_{ij}(x)$ satisfying the inequalities

$$\forall x \in \overline{\Omega} \quad 0 < a_0 \le A_{ij}(x) < \infty, \quad 0 < c_0 \le c_{ij}(x) < \infty. \tag{4.131}$$

It is assumed that

$$p_{ij}(x) \in C_{\log}(\overline{\Omega}), \qquad \sigma_i(x),\ \sigma_{ij}(x) \in C^0(\overline{\Omega}) \tag{4.132}$$

and there exist constants $p^- > 1$, $p^+ < \infty$, $\sigma^+ < \infty$, $\sigma^- > 1$ such that for all $x \in \overline{\Omega}$

$$\begin{cases} p_{ij}(x) \in [p^-, p^+], \\ \sigma_i(x),\ \sigma_{ij}(x) \in [\sigma^-, \sigma^+], \qquad i,j = 1,\dots,n, \\ \inf_{\Omega} p_{ij}(x) \ge p^- \in (1,n). \end{cases} \tag{4.133}$$

The following growth and monotonicity conditions are fulfilled:

$$
\begin{aligned}
&\forall\,(x,\mathbf{V}),\ (x,\mathbf{U})\in\Omega\times\mathbb{R}^{n^2},\ (\mathbf{s},\mathbf{r})\in\mathbb{R}^n\\
&\sum_{ij} a_{ij}(x,\mathbf{V})\cdot V_{ij}\ge a_0\sum_{ij}|V_{ij}|^{p_{ij}(x)},\quad a_0=const>0,\\
&\sum_{ij}\left(a_{ij}(x,\mathbf{V})-a_{ij}(x,\mathbf{U})\right)\cdot\left(V_{ij}-U_{ij}\right)\ge 0,\\
&\sum_{ij} c_{ij}(x)|s^{(j)}|^{\sigma_{ij}(x)-2}s^{(j)}s^{(i)}\ge c_0\sum_{i}|s^{(i)}|^{\sigma_i(x)},\quad c_0=const>0,\\
&\sum_{ik} c_{ik}(x)\left(|s^{(k)}|^{\sigma_{ik}(x)-2}s^{(k)}-|r^{(k)}|^{\sigma_{ik}(x)-2}r^{(k)}\right)(s^{(i)}-r^{(i)})\ge 0.
\end{aligned}
\tag{4.134}
$$

Let us introduce the function space

$$
\mathbf{V}(\Omega)=\left\{\mathbf{u}:\ u^{(i)}\in L^{\sigma_i(x)}(\Omega)\cap W_0^{1,1}(\Omega),\ D_ju^{(i)}\in L^{p_{ij}(x)}(\Omega),\ i,j=1,\dots,n\right\},
$$

$$
\|\mathbf{u}\|_{\mathbf{V}}=\sum_{i=1}^{n}\|u^{(i)}\|_{\sigma_i(\cdot)}+\sum_{ij}\|D_ju^{(i)}\|_{p_{ij}(\cdot)}.
$$

The spaces $\mathbf{V}(\Omega)$ defined in this way are reflexive and separable Banach spaces. Let us introduce the operator $L(\mathbf{s}):\ \mathbf{V}(\Omega)\mapsto\mathbf{V}'(\Omega)$:

$$
\forall\,\mathbf{h}\in\mathbf{V}(\Omega)
$$

$$
(L(\mathbf{u}),\mathbf{h})\equiv\int_\Omega\left(\sum_{ij}a_{ij}(x,\nabla\mathbf{u})\cdot D_jh^{(i)}+\sum_{ij}c_{ik}(x)|u^{(j)}|^{\sigma_{ij}(x)-2}u^{(j)}h^{(i)}\right)dx.
$$

**Definition 4.5** A vector-valued function $\mathbf{u}=(u^{(1)},\dots,u^{(n)})$ is a weak solution of problem(4.130) if $\mathbf{u}\in\mathbf{V}(\Omega)$ and for every test-function $\mathbf{h}\in\mathbf{V}(\Omega)$

$$
\begin{aligned}
&(L(\mathbf{u}),\mathbf{h})-(\mathbf{f},\mathbf{h})\\
&\quad\equiv\int_\Omega\left(\sum_{ij}a_{ij}(x,\nabla\mathbf{u})\cdot D_jh^{(i)}+\sum_{ik}c_{ik}|u^{(k)}|^{\sigma_{ik}-2}u^{(k)}h^{(i)}-\mathbf{hf}\right)dx=0.
\end{aligned}
\tag{4.135}
$$

The proof of existence of a weak solution of problem (4.130) is an adaptation of the proof given in the case of a single scalar equation.

**Lemma 4.26** *The mapping $L:\ \mathbf{V}(\Omega)\mapsto\mathbf{V}'(\Omega)$ is coercive:*

$$
(L(\mathbf{u}),\mathbf{u})\ge C\,\min\left\{\left(\frac{\|\mathbf{u}\|_{\mathbf{V}}}{n(n+1)}\right)^{p^+},\ \left(\frac{\|\mathbf{u}\|_{\mathbf{V}}}{n(n+1)}\right)^{p^-}\right\}.
$$

*Proof* Denote $\lambda = \|\mathbf{u}\|_{\mathbf{V}}$, $\lambda_{0i} = \|u^{(i)}\|_{\sigma_i(\cdot)}$, $\lambda_{jk} = \|D_k u^{(j)}\|_{p_{jk}(\cdot)}$. There exists $\lambda_{ij}$ such that $n(n+1)\lambda_{ij} \geq \lambda$ (we may assume that $i = j = 1$). Then

$$\begin{aligned}(L(\mathbf{u}), \mathbf{u}) &\geq \min\{a_0, c_0\} \left(\textstyle\sum_i A_{\sigma_i(\cdot)}\left(u^{(i)}\right) + \sum_{ij} A_{p_{ij}(\cdot)}\left(D_j u^{(i)}\right)\right) \\ &\geq C\, A_{p_{11}(\cdot)}\left(D_1 u^{(1)}\right)\end{aligned}$$

and the conclusion follows as in the proof of Lemma 4.20. □

**Lemma 4.27** *The mapping $L : \mathbf{V}(\Omega) \mapsto \mathbf{V}'(\Omega)$ is continuous and monotone.*

**Proof.** The monotonicity immediately follows from (4.134) and (4.35). The continuity is obvious. □

**Theorem 4.18** *Let the exponents $\sigma_{ij}(x)$, $\sigma_i(x)$, $p_{ij}(x)$ satisfy* (4.132) *and* (4.133). *Assume that conditions* (4.134) *are fulfilled. Then for every $\mathbf{f} = (f^{(1)}, \dots, f^{(n)})$ such that $f^{(i)} \in L^{\sigma_i'(x)}(\Omega)$, problem* (4.130) *has a weak solution $\mathbf{u} \in \mathbf{V}(\Omega)$ which satisfies the estimate*

$$\int_\Omega \left( a_0 \sum_{ij} |D_j u^{(i)}|^{p_{ij}(x)} + c_0 \sum_i |u^{(i)}|^{\sigma_i(x)} \right) dx \leq C \sum_i \int_\Omega |f^{(i)}|^{\sigma_i'(x)}\, dx. \tag{4.136}$$

*Proof* Since the Banach space $\mathbf{V}(\Omega)$ is reflexive and separable, and the mapping $L$ is continuous, monotone and coercive, the existence follows by the Browder-Minty theorem [80, Chap. 7]. Estimate (4.136) follows from the proof of Lemma 4.26. □

## 4.9 Remarks

The results of this chapter generalize the results published in papers [44, 45, 47, 49, 58]. These works deal with the energy solutions, that is, the weak solutions which can be taken for the test-functions in the corresponding integral identities. Such solutions possess a series of specific properties which will be discussed in detail in the rest of the monograph. In the proof of uniqueness we follow ideas of [30, 45, 99, 100] where similar arguments were applied to the study of elliptic equations with nonstandard growth conditions. The assertions of Theorems 4.1, 4.2 extend to the equations

$$u_t - \sum_i \frac{d}{dx_i}\left[a_i(z,u)|D_i u|^{p_i(z)-2} D_i u + b_i(z,u)\right] + \sum_i d_i(z,u) D_i u + d(z,u) = 0,$$

which reduce to (4.1) by means of the substitution

$$\widetilde{b}_i(z,u) \equiv b_i(z,u) + \int_0^u d_i(z,s)\, ds, \quad \widetilde{d}(z,u) \equiv d(z,u) - \sum_i \int_0^u D_i d_i(z,s)\, ds.$$

Theorem 4.1 remains true if in (4.4)

$$\begin{cases} h_d(z) = h_d^0(z) + \operatorname{div} H(z), & h_d^0 \in L^{\lambda'}(Q_T), \\ H(z) = (H_1, \dots, H_n), & H_i \in L^{p_i'(z)}(Q_T). \end{cases}$$

No substantial change is needed to adapt the proofs of the existence and uniqueness theorems to the equations

$$u_t - \sum_i \frac{d}{dx_i}\left[a_i(z,u)|\nabla u|^{p(z)-2}D_i u + b_i(z,u)\right] + d(z,u) = 0.$$

For the main function spaces one may take

$$\mathbf{V}_t(\Omega) = \left\{u(x)|\, u(x) \in L^2(\Omega) \cap W_0^{1,1}(\Omega), \quad |\nabla u(x)| \in L^{p(x,t)}(\Omega)\right\},$$

and

$$\begin{aligned} &\mathbf{W}(Q_T) = \left\{u : [0,T] \mapsto \mathbf{V}_t(\Omega)|\, u \in L^2(Q_T),\ |\nabla u| \in L^{p(z)}(Q_T),\ u = 0 \text{ on } \Gamma_T\right\}, \\ &\|u\|_{\mathbf{W}(Q_T)} = \|\nabla u\|_{p(\cdot),Q_T} + \|u\|_{2,Q_T}. \end{aligned}$$

The unique solvability of the homogeneous Dirichlet problem for parabolic equations and systems of the form

$$\partial_t u - \operatorname{div} \mathbf{A}(\nabla u) = f \quad \text{in } Q_T$$

with the monotone operator $\mathbf{A}$ of the $p(x,t)$ structure satisfying conditions (4.86) is proved in [116]. The exponent $p \in C_{\log}(Q_T)$ is assumed to satisfy the inequalities $1 < p^- \le p \le p^+ < \infty$. The solution $u = (u_1, \dots, u_n)$ is constructed as the limit of the sequence of Galerkin's approximations in the space

$$W(Q_T) = \{u \in X(Q_T) \,|\, \partial_t u \in X'(Q_T)\},$$

where

$$\begin{aligned} &\mathbf{V}_t(\Omega) = L^2(\Omega)^n \cap W_0^{1,p(t,\cdot)}(\Omega)^{n\times n}, \\ &X(Q_T) = \left\{u \in L^2(Q_T)^n |\, \nabla u \in L^{p(\cdot,\cdot)}(Q_T)^{n\times n},\ u(\cdot,t) \in \mathbf{V}_t(\Omega) \text{ for a.e. } t \in (0,T)\right\}, \end{aligned}$$

and $X'(Q_T)$ is the dual of $X(Q_T)$ with respect to the inner product in $L^2(\Omega)^n$. It is shown that the problem has a unique weak solution for every $u_0 \in L^2(\Omega)^n$ and $f \equiv \widetilde{f} - \operatorname{div} F \in X'(Q_T)$. The authors of [116] provide a compete characterization of the dual space $X'(Q_T)$. Existence of periodic solutions of the system of $(p(x), q(x))$-Laplace parabolic equations with nonlocal terms is proved in [132].

In [7, 8] the subdifferential calculus is used to prove solvability of the homogeneous Dirichlet problem for the evolution $p(x)$-Laplace equation. It is shown that this problem has a unique weak solution for every $u_0 \in L^2(\Omega)$, $f \in L^2(Q)$ and under the assumption that $p(x)$ is measurable in $\Omega$ with values in an interval $[p^-, p^+] \subset (1, \infty)$. The solution continuously depends on the data $u_0$ and $f$. We refer also to [6, 9] for an application of this technique to the study of doubly nonlinear evolution equations with variable nonlinearities.

The subdifferential approach is used in [242] to prove the existence theorem for the homogeneous Neumann boundary value problem for the equations

$$\partial_t u - \Delta_{p(\cdot)} u + |u|^{p(x)-2} u = f, \qquad \partial_t u - \Delta_{p(\cdot)} u + |u|^{p(x)-2} u - \omega u = f,\ \omega > 0,$$

with continuous $p(x) > 2$. The subdifferential $\partial\phi_{p(x)}$, associated with the operator

$$\langle A(u), v\rangle = \int_\Omega \left(|\nabla u|^{p(x)-2}\nabla u \cdot \nabla v + |u|^{p(x)-2} u\right) v\,dx : W^{1,p(\cdot)}(\Omega) \mapsto W^{-1,p'(\cdot)}(\Omega)$$

on the space $W^{1,p(\cdot)}(\Omega) = \left\{u \in L^{p(\cdot)}(\Omega) : |\nabla u|^{p(x)} \in L^1(\Omega)\right\}$, is defined by the function $\phi_{p(x)} : L^2(\Omega) \mapsto \mathbb{R} \cup \{\infty\}$

$$\phi_{p(x)}(u) = \begin{cases} \displaystyle\int_\Omega \frac{|\nabla u|^{p(x)} + |u|^{p(x)}}{p(x)}\,dx & \text{if } W^{1,p(\cdot)}(\Omega), \\ +\infty & \text{if } u \in L^2(\Omega) \setminus W^{1,p(\cdot)}(\Omega). \end{cases}$$

With the same method the authors of [242] prove solvability of systems of parabolic differential inclusions with the homogeneous Neumann boundary conditions:

$$\begin{cases} \partial_t u - \Delta_{p(x)} u + |u|^{p(x)-2} u \in F(u, v), \\ \partial_t v - \Delta_{q(x)} v + |v|^{q(x)-2} v \in G(u, v), \\ \frac{\partial u}{\partial n}(x,t) = \frac{\partial v}{\partial n}(x,t) = 0, \\ u(x,0) \in L^2(\Omega), \quad v(x,0) \in L^2(\Omega), \quad p(x) > 2,\ q(x) > 2. \end{cases}$$

The homogeneous Neumann problem for the equation

$$\partial_t u = \operatorname{div}\left(|\nabla u|^{p(x,t)-2}\nabla u + f\right), \qquad p \in C_{\log}(Q),$$

is studied in [142]. The solution is constructed as a limit of the sequence of Galerkin's approximations in the space $\mathbf{W}(Q_T)$ with

$$\mathbf{V}_t(\Omega) = \left\{u : u \in L^2(\Omega) \cap W^{1,1}(\Omega),\ |\nabla u|^{p(x,t)} \in L^1(\Omega),\ \int_\Omega u(x)\,dx = 0\right\}.$$

A system of reaction-diffusion equations

$$u_t = \Delta_{p(x)}u - 2\lambda w, \qquad w_t = \Delta w - (f - u), \qquad p(x) \in C_{\log}(\Omega),$$

endowed with the homogeneous Neumann boundary conditions and the initial conditions is studied in [153]. The unique bounded solution is constructed as the limit of a sequence of Galerkin's approximations.

Various systems of quasilinear and semilinear parabolic equations with variable exponents were considered in [68, 193, 199, 247]. The initial and boundary value problem for the fourth-order equation

$$\begin{cases} \partial_t u + \operatorname{div}\left(|\nabla \Delta u|^{p(x)-2}\nabla \Delta u\right) = f(x,u) \quad \text{in } Q_T, \\ u = \Delta u = 0 \text{ on } \Gamma_T, \qquad u(x,0) = u_0(x) \text{ in } \Omega \end{cases}$$

is studied in [150]. The existence theorem is proved under the growth and regularity assumptions $u_0 \in W_0^{1,p(x)}(\Omega)$, $|f(x,u)| \le M\,|u|^{m(x)} + a(x)$ with $M = const > 0$, $a(x) \ge 0$, $p(x)$, $m(x) \in C_{\log}(\Omega)$,

$$\max\left\{\frac{(m^+ + 1)n}{n + m^+}, 2 - \frac{1}{p^+}\right\} < p^- < p^+ < n, \quad a(x) \in L^{(p')^+}(\Omega).$$

The solution is constructed by means of time-discretization. The uniqueness is proved under the additional restrictions

$$p^- > 2, \quad u_0 \in H_0^1(\Omega) \cap W_0^{3,p(x)}(\Omega), \quad \left|\operatorname{div}(|\nabla \Delta u_0|^{p(x)-2}\nabla \Delta u_0)\right| \in H_0^1(\Omega).$$

A class of higher-order systems of parabolic equations with variable nonlinearity is studied in [86].

The Dirichlet problem for non-divergent parabolic equations

$$u_t = u \operatorname{div}\left(|\nabla u|^{p(x)-2}\nabla u\right), \qquad 2 < p^- \le p(x) < \infty,$$

with nonnegative initial data and $p(x) \in C^1(\overline{\Omega})$ was considered in [200].

The parabolic variational inequalities with variable exponents of nonlinearity were studied in [92, 202]. In [146] the homogeneous Dirichlet problem for the anisotropic Barenblatt-type equation

$$f(x, \partial_t u) - \sum_{i=1}^{n} D_i\left(|D_i u|^{p_i(x)-2} D_i u\right) = g$$

is considered. It is shown that the problem admits a solution if $f(x,s) : \Omega \times \mathbb{R} \mapsto \mathbb{R}$ is a nondecreasing in $s$ Carathéodory function satisfying a suitable growth condition, $p_i \ge \min\left\{1, \frac{2n}{n+2}\right\}$ are continuous bounded functions, and the data $g$, $u(x,0)$ satisfy certain regularity assumptions.

The problem of existence of periodic solutions of the elliptic-parabolic equation

$$b(x)\partial_t u - \sum_{i=1}^{n} D_i(a_i|D_i u|^{p_i(x)-2}D_i u) + a_0|u|^{q(x)-2}u = f \quad \text{in } \Omega \times \mathbb{R} \tag{4.137}$$

with the boundary conditions

$$u\big|_{\Gamma^{(0)}} = 0, \quad \sum_{i=1}^{n} a_i|D_i u|^{p_i(x)-2}D_i u \cdot \nu_i(x)\big|_{\Gamma^{(1)}} = 0, \quad \Gamma^{(0)} \cup \Gamma^{(1)} = \Gamma, \tag{4.138}$$

is studied in [84]. It is assumed that $b(x) \geq 0$, the coefficients $a_i$, $a_0$ are locally bounded and separated away from zero by a constant. The exponents $q(x)$, $p_i(x)$ are measurable, a.e. bounded, and satisfy the inequalities $q(x) > 2$, $p_i(x) \geq 2$. Problem (4.137)–(4.138) in a bounded cylinder $\Omega \times (0, T)$ with the initial condition $\sqrt{b(x)}u(x, 0) = u_0(x)$ in $\Omega$ is considered in [85].

The study of PDEs with variable nonlinearity is motivated by various applications where such equations appear in the most natural way. Equations of the type (4.1) and their stationary counterparts appear in the mathematical descriptions of motions of the non-newtonian fluids [41, 276], in particular, electro-rheological fluids which are characterized by their ability to change the mechanical properties under the influence of the exterior electro-magnetic field [125, 156, 223, 228]. Most of the known results concern the stationary models, see, e.g., [2, 3]. Some properties of solutions of the system of modified nonstationary Navier-Stokes equations describing electro-rheological fluids are studied in [4]. Another important application is the image processing where the anisotropy and nonlinearity of the diffusion operator and convection terms are used to underline the borders of the distorted image and to eliminate the noise [1, 23, 97, 153, 155, 167, 191, 217, 259, 260]. The preprocessing procedure for the segmentation of digital images proposed in [217] consisted in solving the parabolic problem

$$\partial_t u - \operatorname{div}(\rho(|\nabla u|)\nabla u) = 0 \quad \text{in } Q_T$$

with the initial datum $u(x, 0) = u_0(x) \in [0, 1]$ representing the grey-scale intensity of the given image. As a prototype of the function $\rho$ one may take $\rho(s) = (1+s^2)^{-1}$. Such a choice provides slow diffusion in the regions where $|\nabla u|$ is high and fast diffusion whereas $|\nabla u|$ is small. In the result, the high gradient regions are enhanced, and the small gradients regions become more uniform.

The procedure for image restoration studied in [97] leads to a nonlinear parabolic equation which combines linear growth at infinity and variable growth near zero. In this case $p^- = 1$ and the solution belongs to the space $L^2\left(0, T; BV(\Omega) \cap L^2(\Omega)\right)$, such a situation is not discussed here. We refer also to [21, 22] for a review of results on the solvability of evolution nonlinear diffusion equations in the $BV$ spaces.

### 4.9.1 Renormalized and Entropy Solutions

Let us consider the problem

$$\begin{cases} u_t - \operatorname{div}\left(|\nabla u|^{p(x)-2}\nabla u\right) = f \text{ in } Q_T, \\ u = 0 \text{ on } \Gamma_T, \quad u(x,0) = u_0(x) \text{in } \Omega \end{cases} \tag{4.139}$$

maintaining all underlying hypothesis on the problem data except for the regularity: it is now claimed that $f \in L^1(Q_T)$ and $u_0 \in L^1(\Omega)$.

Given $\gamma > 0$, introduce the function $T_\gamma(x) = \min\{|x|, \gamma\} \operatorname{sign} x$ and its renormalization $\phi_\gamma(y) = T_{\gamma+1}(y) - T_\gamma(y)$. Take any positive function $s(\cdot) \in C^\infty(\mathbb{R})$ such that

$$s(x) = \begin{cases} 1 & |x| \le 1, \\ \in [0,1] & \text{if } 1 \le |x| \le 2, \\ 0 & \text{if } |x| \ge 2 \end{cases}$$

and then set

$$s_k(x) = \begin{cases} 1 & \text{if } |x| \le k-1, \\ s(x-(k-1)\operatorname{sign}(x)) & \text{if } |x| \ge k-1, \end{cases} \quad S_k(y) = \int_0^y s_k(z)\,dz.$$

For every $k \ge 2$ the function $S_k$ satisfies the conditions

$$\begin{cases} S_k(x) = S_k(T_{k+1}(x)), \quad \|S_k'\|_{\infty,\mathbb{R}} \le \|s\|_{\infty,\mathbb{R}}, \\ \operatorname{supp} S_k' \subset [-k-1, k+1], \quad \operatorname{supp} S_k'' \subset [-k-1,-k)] \cup [k, k+1]. \end{cases}$$

By $W^{1,p(\cdot)}(\Omega)$ we denote the set of measurable functions

$$W^{1,p(\cdot)}(\Omega) = \left\{u \in L^{p(\cdot)}(\Omega) : |\nabla u|^{p(x)} \in L^1(\Omega)\right\}$$

equipped with the norm

$$\|u\|_{W^{1,p(\cdot)}(\Omega)} = \|u\|_{L^{p(\cdot)}(\Omega)} + \|\nabla u\|_{L^{p(\cdot)}(\Omega)}$$

and define $W_0^{1,p(\cdot)}(\Omega)$ as the closure of $C_0^\infty(\Omega)$ with respect to the norm of $W^{1,p(\cdot)}(\Omega)$. Introduce the function space

$$V = \left\{v : v \in L^{p^-}(0,T; W_0^{1,p(\cdot)}(\Omega)),\ |\nabla v| \in L^{p(\cdot)}(Q_T)\right\}, \quad \|v\|_V = \|\nabla v\|_{p(\cdot),Q_T}.$$

The elements of the dual space $V'$ to $V$ are represented as follows:

$$w \in V' \quad \Leftrightarrow \quad \begin{cases} \exists \Psi = \{\psi_1, \dots, \psi_n\},\ \psi_i \in L^{p'(\cdot)}(\Omega) \\ \langle\langle w, \xi \rangle\rangle_{V',V} = \int_{Q_T} \Psi \cdot \nabla\xi \, dxdt \quad \forall \xi \in V. \end{cases}$$

**Definition 4.6** (*Renormalized solution*) A function $u : Q_T \mapsto \mathbb{R}$ is called renormalized solution of problem (4.139) if

$$T_\gamma(u) \in L^{p^-}(0, T; W_0^{1,p(\cdot)}(\Omega)) \text{ for every } \gamma > 0,$$
$$\nabla T_\gamma(u) \in (L^{p(\cdot)}(Q_T))^n \text{ for every } \gamma > 0,$$
$$\lim_{\gamma \to \infty} \int_{\{\gamma \le |u| \le \gamma+1\}} |\nabla u|^{p(x)} \, dxdt = 0$$

and for every renormalization $S \in C^\infty(\mathbb{R})$ with $S' \in [-M, M]$ for some $M > 0$, and every test-function $\phi \in L^\infty(Q) \cap V$

$$\partial_t S(u) - \operatorname{div}(S'(u)|\nabla u|^{p(x)-2}\nabla u) + S''(u)|\nabla u|^{p(x)} = f\, S'(u) \quad \text{in } \mathscr{D}'(Q_T)$$

(in the sense of distributions). The initial condition is fulfilled in the sense

$$S(u)|_{t=0} = S(u_0) \quad \text{a.e. in } \Omega.$$

**Theorem 4.19** (Theorem 3.1, [77]) *Let us assume that $p \in C^0(\overline{\Omega})$ with $1 < p^+ \le p^+ < n$, $f \in L^1(Q_T)$ and $u_0 \in L^1(\Omega)$. Then there exists at least one renormalized solution of problem (4.139) Moreover, if $p^- > 2 - \dfrac{1}{n+1}$, then $u \in L^{q^-}(0, T; W_0^{1,q(\cdot)}(\Omega))$ for every continuous exponent $q(x)$ satisfying the condition*

$$1 \le q(x) < \frac{n(p(x) - 1) + p(x)}{n+1} \quad \text{in } \overline{\Omega}.$$

The remormalized solution is unique, provided that the conditions of Theorem 4.19 are fulfilled—[77, Theorem 6.1].

Given $k > 0$, set

$$\theta_k(x) = \int_0^x T_k(y)\,dy, \qquad E = \left\{\phi \in V \cap L^\infty(Q_T) : \phi_t \in V' + L^1(Q_T)\right\}.$$

**Definition 4.7** (*Entropy solution*) A function $u : Q_T \mapsto \mathbb{R}$ is called entropy solution of problem (4.139) if

1. $T_k(u) \in V$ for every $k > 0$,
2. for all $k > 0$ and $\phi \in E$ the mapping $[0, T] \ni t \mapsto \int_\Omega \theta_k(u - \phi)(x, t)\,dx$ is a.e. equal to a continuous function,

3. for every $k > 0$ and $\phi \in E$

$$\int_{\Omega} (\theta_k(u-\phi)(T) - \theta_k(u_0-\phi)(0))\,dx + \langle \phi_t, T_k(u-\phi)\rangle$$
$$+ \int_{Q_T} |\nabla u|^{p(x)-2}\nabla u \cdot \nabla T_k(u-\phi)dxdt \le \int_{Q_T} f\,T_k(u-\phi)\,dxdt,$$

where $\langle \cdot, \cdot \rangle$ denotes the duality pairing between $V' + L^1(Q_T)$ and $V \cap L^\infty(Q_T)$.

**Theorem 4.20** (Proposition 7.1, [77]) *A function $u(x,t)$ is a renormalized solution of problem* (4.139) *if and only if it is an entropy solution.*

For the proof of existence of renormalized solutions given in [77] only continuity of $p(x)$ is required. Under the stronger assumption $p(x) \in C_{\log}(\overline{\Omega})$ the same result was obtained in [271].

The issues of existence and uniqueness of renormalized solutions to anisotropic parabolic equations with constant nonlinearity and $L^1$ data are discussed in the series of papers [73–75]. Paper [76] deals with the anisotropic equations with $L^1$ data and variable exponents of nonlinearity. Existence of relaxed solutions in the sense of [226] for parabolic equations with anisotropic growth and $L^1$ data was discussed in [192, 194], solvability of doubly nonlinear parabolic equations with constant anisotropic growth is studied in [108, 234, 244].

### *4.9.2 Equation with Measurable Exponent $p(x,t)$*

Following [12–14], let us consider the problem

$$\begin{cases} u_t = \operatorname{div} A(x,t,\nabla u) \text{ in } Q_T, \\ u = 0 \text{ on } \Gamma_T, \qquad u(x,0) = u_0(x) \in L^2(\Omega), \end{cases} \tag{4.140}$$

where $A(x,t,s)$ is a Caratéodory function satisfying the following monotonicity, coercivity and growth conditions:

$$\begin{aligned} &(A(x,t,\xi) - A(x,t,\eta)) \cdot (\xi - \eta) \ge 0 \quad \forall \xi, \eta \in \mathbb{R}^n, \\ &A(x,t,\xi) \cdot \xi \ge C_0 |\xi|^{p(x,t)} - \psi_0(x,t), \\ &|A(x,t,\xi)|^{p'(x,t)} \le C_1 |\xi|^{p(x,t)} + \psi_1(x,t) \end{aligned} \tag{4.141}$$

with positive constants $C_0$, $C_1$, some functions $\psi_0$, $\psi_1 \in L^1(Q_T)$ and measurable in $Q_T$ exponent $p(x,t)$ such that $1 < p^- \le p(x,t) \le p^+ < \infty$. Define the space

$$W_0 = \left\{ u : u \in L^2(Q_T),\ u(\cdot,t) \in W_0^{1,1}(\Omega) \text{ for a.e. } t \in [0,T],\ |\nabla u|^p \in L^1(Q_T) \right\}$$

with the norm

$$\|u\|_{W_0} = \|u\|_{2,Q_T} + \|\nabla u\|_{p(\cdot,\cdot),Q_T}.$$

Denote

$$S_0 = \left\{\phi \in C^\infty(\overline{Q}_T),\ \phi = 0 \text{ on } \Gamma_T, \phi(x,T) = 0\right\}$$

and define $H_0$ as the closure of $C^\infty([0,T]; C_0^\infty(\overline{\Omega}))$ in the norm of $W_0$.

**Definition 4.8** A function $u$ is called $H$-solution of problem (4.140) if $u \in H_0$ and satisfies the identity

$$\int_{Q_T} \left(-u\,\phi_t + |\nabla u|^{p(x,t)-2}\nabla u \cdot \nabla\phi\right) dxdt = \int_\Omega u_0\phi\,dx \qquad \forall \phi \in S_0. \quad (4.142)$$

Set

$$S_1 = \left\{\phi \in W_0,\ \phi_t \in L^2(Q_T),\ \phi(x,T) = 0\right\}.$$

**Definition 4.9** A function $u \in W_0$ is called $W$-solution of problem (4.140) if identity (4.142) holds for every test-function $\phi \in S_1$.

**Theorem 4.21** (Theorem 1, [13]) *Under conditions* (4.141) *problem* (4.140) *has a* $W$*-solution and a* $H$*-solution.*

The proof of Theorem 4.21 relies on the theory of monotone operators. Let us outline the main steps of the proof of existence of $H$ solutions. The change of the sought function $v = e^{-t}u$ converts the equation in (4.140) into

$$v_t - \operatorname{div} A(x,t,e^t\nabla v) + v = 0 \qquad (4.143)$$

with the coercive on $W_0$ operator $\mathscr{B}v := -\operatorname{div} A(x,t,e^t\nabla v) + v$. Consider the operator $\mathscr{M} : H_0 \mapsto H_0'$ with the domain $D(\mathscr{M}) = S_0$ defined by the equality $\mathscr{M}\phi = -\phi_t$. The operator $\mathscr{M}$ is monotone on $D(\mathscr{M})$ because

$$\begin{aligned}\int_{Q_T} \mathscr{M}(\phi-\psi)\cdot(\phi-\psi)\,dxdt &= -\int_{Q_T}(\phi-\psi)_t(\phi-\psi)\,dxdt\\ &= \frac{1}{2}\int_\Omega (\phi-\psi)^2(x,0)\,dx \geq 0.\end{aligned}$$

Let $\widetilde{\mathscr{M}}$ be the maximal monotone linear extension of $\mathscr{M}$. The adjoint operator $\mathscr{L} = \widetilde{\mathscr{M}}^*$ is also maximal monotone. It is known that the equation $\mathscr{L}u + \mathscr{B}u = 0$ has a unique solution $u \in H_0$, that is,

$$\int_{Q_T}\left(-v\,\phi_t + A(x,t,e^t\nabla v)\cdot\nabla\phi + v\phi\right) dxdt = 0 \qquad \forall \phi \in S_0.$$

It follows that the homogeneous Dirichlet problem for Eq. (4.143) with the initial datum $u_0 \equiv 0$ has a solution in $H_0$. The next step is the construction of a solution to the problem with a nonzero initial datum.

**Lemma 4.28** *For every* $u_0 \in L^2(\Omega)$ *there exist* $u \in H_0$ *and* $\mathscr{F} \in L^{p'(x,t)}(Q_T)$ *such that*

$$-\int_{Q_T} u\,\phi_t\,dxdt = \int_\Omega u_0\,\phi(x,0)\,dx + \int_{Q_T} \mathscr{F}\cdot\phi\,dxdt \qquad \forall\phi \in S_0. \quad (4.144)$$

Let $\{u_\varepsilon\}$ be the sequence of solutions of problem (4.87) with the isotropic perturbed equations:

$$\mathbf{A}_\varepsilon(x,t,\xi) = |\xi|^{p(x,t)-2}\xi + \varepsilon\,|\xi|^{p^+-2}\xi.$$

Passing to a subsequence we can find $u$ and $\mathscr{F}$ such that

$$u_\varepsilon \rightharpoonup u \text{ in } H_0, \qquad \mathbf{A}_\varepsilon(x,t,\nabla u_\varepsilon) \rightharpoonup \mathscr{F} \text{ in } L^{p'(\cdot,\cdot)}(Q_T).$$

The solution of problem (4.87) satisfies the identity

$$-\int_{Q_T} u_\varepsilon\phi\,dxdt = \int_{Q_t} \mathbf{A}_\varepsilon\cdot\nabla u_\varepsilon\,dxdt + \int_\Omega u_0\phi(x,0)\,dx \quad \forall\phi \in S_0.$$

Arguing as in Sect. 4.6 we let $\varepsilon \to 0$ in the last identity and arrive at (4.144).

To conclude the proof of existence of $H$-solutions it suffices to solve the problem

$$\begin{cases} w_t - \operatorname{div} A(x,t,\mathrm{e}^t(\nabla w + \nabla u)) + (w+u) = \operatorname{div}\mathscr{F} & \text{in } Q_T, \\ w = 0 \text{ on the parabolic boundary of } Q_T \end{cases}$$

with the already constructed $u \in H_0$ and $\mathscr{F} \in L^{p'(\cdot,\cdot)}(Q_T)$. This problem is solvable because the operator $A(x,t,\xi+\nabla u) + \mathscr{F}$ is a Leray-Lions operator satisfying conditions (4.141). We do not present here the method of construction of $W$-solutions, which is more complicated.

In general, the $H$ and $W$ solutions do not possess the same properties as the energy solutions, such as the weak $L^2(\Omega)$ regularity in time or the energy equalities. Moreover, for $H$ and $W$ solutions even the inclusion $v \in L^\infty(0,T;L^2(\Omega))$ need not be true. However, these properties hold if the exponent $p$ is independent of $t$ (still without any assumption on the continuity of $p(x)$).

**Theorem 4.22** (Theorem 2, [13]) *Let the exponent* $p$ *be independent of* $t$. *Then problem* (4.140) *has a unique* $H$ *solution* ($W$ *solution*) *which is* $L^2(\Omega)$*-continuous on* $[0,T]$ *and satisfies the energy equality*

$$\frac{1}{2}\int_\Omega u^2(x,t)\,dx + \int_{Q_t} A(x,t,\nabla u)\cdot\nabla u\,dxdt = \frac{1}{2}\int_\Omega u_0^2(x)\,dx \qquad \forall t \in [0,T].$$

In paper [11] existence and uniqueness of energy solutions of problem (4.140) is proved under the assumption that $p : Q \mapsto [p^-, p^+] \subset (1, \infty)$ is measurable in $Q$ and satisfies the log-continuity condition only with respect to the variable $t$. Under stronger conditions on the data problem (4.140) with measurable $p(x)$ was solved in [7, 8].

In [134] problem (4.140) for the equation

$$u_t - \operatorname{div} a(x, t, u, \nabla u) + a_0(x, t, u, \nabla u) = f$$

is studied under the monotonicity, coercivity and growth conditions

$$\begin{aligned}
&|a(x, t, s, \xi)| + |a_0(x, t, s, \xi)| \le \alpha \left(C(x, t) + |s|^{p(x)-1} + |\xi|^{p(x)-1}\right),\\
&(a(x, t, s, \xi) - a(x, t, s, \eta)) \cdot (\xi - \eta) > 0,\\
&a(x, t, s, \xi) \cdot \xi + a_0(x, t, s, \xi)\, s \ge \beta |\xi|^{p(x)} + \gamma\, |s|^{p(x)}, \quad \alpha, \beta, \gamma = const > 0,
\end{aligned}$$

with $C(x, t) \in L^{p'(\cdot)}(Q_T)$ and the independent of $t$ exponent $p(x) \in [p^-, p^+] \subset (1, \infty)$ continuous in $\overline{\Omega}$. Let $W^{1,x}L^{p(x)}(Q_T)$ be the closure of the set $C^\infty(0, T; C_0^\infty(\Omega))$ with respect to the norm

$$\|u\| = \|u\|_{p(\cdot),\Omega} + \sum_{i=1}^{n} \|D_{x_i} u\|_{p(\cdot),\Omega}.$$

It is shown in [134] that for every $u_0 \in L^2(\Omega)$ and $f \in W^{-1,x}L^{q(x)}(Q_T)$ (the dual space to $W^{1,x}L^{p(x)}(Q_T)$ problem (4.1) has at least one weak solution in $L^\infty(0, T; L^2(\Omega)) \cap W^{1,x}L^{p(x)}(Q_T)$, which is constructed as the limit of a sequence of Galerkin's approximations. In paper [196] the time discretization was used to prove the existence of a solution to the homogeneous Dirichlet problem for the equation

$$u_t = \Delta_{p(x)} u + f(x, t, u).$$

It was assumed that $p(x) \in C^0(\overline{\Omega})$ and $p(x) > 2$, and that the source term $f$ has either the subcritical growth, $|f| \le C(\phi(x, t) + |u|^\alpha)$ with $\alpha < p^- - 1$, or the critical growth $\alpha \le p^- - 1$ but meas $\Omega$ is sufficiently small.

### *4.9.3 On the Regularity of Weak Solutions*

Let $u(x, t)$ be a local weak solution of the equation

$$u_t - \operatorname{div}\left(|\nabla u|^{p(x,t)-2} \nabla u\right) u = 0 \quad \text{in } Q = \Omega \times (0, T], \tag{4.145}$$

where $\Omega \subset \mathbb{R}^n$ is a bounded domain, $T < \infty$. The exponent

$$p(z) : Q \mapsto [p^-, p^+] \subset (1, \infty)$$

is assumed to satisfy the log-continuity condition (1.31). The local solution is understood in the following way.

**Definition 4.10** A locally integrable in $Q$ function $u$ is called local weak solution of Eq. (4.145) if

1. $u \in C_{loc}(0, T; L^2(\Omega))$, $|\nabla u(z)|^{p(z)} \in L^1(Q)$,
2. for every test-function $\phi \in C^\infty(0, T; C_0^\infty(\Omega))$ and every $0 < t_1 < t_2 < T$

$$\int_\Omega u\,\phi\,dx\Big|_{t=t_1}^{t=t_2} + \int_{t_1}^{t_2}\int_\Omega \left(-\phi_t u + |\nabla u|^{p(z)-2}\nabla u \cdot \nabla\phi\right) dz = 0.$$

It is known that the local weak solutions are locally bounded in $Q$, or even Hölder continuous.

**Theorem 4.23** ([264]) *If*

$$\frac{2n}{n+2} < p(z) < \infty, \qquad p \in C_{\log}(Q_T), \tag{4.146}$$

*then every local weak solution of* Eq. (4.145) *is locally bounded in $Q$. Moreover, if $p^- > 2$, then the local solution is locally Hölder-continuous in $Q$.*

Local boundedness of weak solutions to more general equations with $p(x)$-growth condition is established in [135]. We refer also to [10] for elliptic equations with variable nonlinearity and to [219] for parabolic equations with anisotropic constant growth.

Besides the natural regularity prompted by the equation, $|\nabla u|^{p(z)} \in L^1(Q)$, the gradients of local solutions of Eq. (4.145) possess the property of higher integrability.

**Theorem 4.24** ([61]) *Let $u(x, t)$ be a bounded weak solution of* Eq. (4.145) *with the exponent $p(z)$ satisfying the conditions*

$$2 \le p^- \le p(z) \le p^+ < \infty, \qquad p \in C_{\log}(Q), \quad (n \ge 2). \tag{4.147}$$

*Then there exists a positive constant $\varepsilon > 0$ such that in every subdomain $Q'$, $\overline{Q'} \subset Q$,*

$$\int_{Q'} |\nabla u(z)|^{p(z)+\varepsilon} dz < \infty.$$

*The constant $\varepsilon$ depends only on $p^\pm$, the constant in condition* (1.31), *and*

$$\sup_{(0,T)} \|u(t)\|_{2,\Omega}^2 + \int_Q |\nabla u|^{p(z)} dz = M.$$

In [280] this result is extended to the systems of equations of the type (4.145)

$$\partial_t u_i = \operatorname{div}_x A_i(x,t,\nabla u) \quad i = 1,\dots,N, \tag{4.148}$$

where $u = (u_1,\dots,u_N)$ is the vector-valued function and $\{\nabla u\}_{ij} = D_{x_j}u_i$ denotes the gradient matrix. It is assumed that

$$A_i(x,t,\xi) : \Omega \times (0,T) \times \mathbb{R}^{n\times N} \mapsto \mathbb{R}^n$$

are Carathéodory functions satisfying the growth conditions

$$\begin{aligned} &\sum_{i=1}^{N} A_i(x,t,\xi)\cdot\xi_i \geq C_0|\xi|^{p(x,t)} - h_0, \\ &|A_i(x,t,\xi)| \leq C_1|\xi|^{p(x,t)-1} + h_1, \quad i = 1,\dots,N, \end{aligned} \tag{4.149}$$

with positive constants $C_0$, $C_1$ and measurable functions $h_0$, $h_1$ such that

$$\||h_1|^{p'} + |h_0|\|_{q,Q} \leq C_2, \quad C_2 = const, \quad p' = \frac{p}{p-1}, \quad q > 1. \tag{4.150}$$

**Theorem 4.25** ([280]) *Let $u = (u_1,\dots,u_N)$ with $u_i \in L^2(Q)$, $|\nabla u|^p \in L^1(Q)$, be a solution of system* (4.148) *in the sense of distributions. Assume that conditions* (4.149) *and* (4.150) *are fulfilled and the exponent $p(x,t)$ satisfies conditions* (4.146). *Then there exists $\delta > 0$ such that*

$$\||\nabla u|^{p(\cdot,\cdot)+\delta}\|_{1,Q'} \leq C$$

*for every subdomain $Q' \subset Q$, separated away from the parabolic boundary of $Q$. The constant $\delta$ depends on $n$ and the structural constants in* (4.149)–(4.150), *while $C$ depends also on $N$, $\||\nabla u|^{p(\cdot,\cdot)}\|_{1,Q}$ and $Q'$.*

The same result is obtained in [82] for the local solutions of degenerate parabolic systems of the form

$$\partial_t u - \operatorname{div} a(z,\nabla u) = \operatorname{div}\left(|F|^{p(z)-2}F\right), \qquad u = (u_1,\dots,u_N),$$

under the growth and coercivity assumptions

$$a : Q_T \times \mathbb{R}^{Nn} \mapsto \mathbb{R}^{Nn}, \qquad a(z,\xi)\cdot\xi \geq \nu\,|\xi|^{p(z)}, \qquad |a(z,\xi)| \leq L(1+|\xi|)^{p(z)-1}$$

with the constant $0 < \nu \leq 1 \leq L$. It is assumed that $n \geq 2$, the exponent $p(z)$ satisfies conditions (4.146), and the right-hand side $F : Q_T \mapsto \mathbb{R}^{Nn}$ satisfies the inclusion $F \in L^{p(\cdot)(1+\sigma)}(Q_T,\mathbb{R}^{Nn})$ with some $\sigma > 0$. The authors discuss the possibility of extension of the result to the systems with the right-hand side $b(z, \nabla u)$ subject to the growth condition $|b(z, \xi)| \leq L(1 + |\xi|)^{p(z)-1}$.

Local Hölder continuity of the gradients of solutions to the parabolic systems

$$\partial_t u - \operatorname{div} \left( a(z)|\nabla u|^{p(z)-2} \nabla u \right) = 0$$

is established in [83]. It is assumed that $a : Q_T \mapsto \mathbb{R}$ is Hölder continuous and separated away from zero and infinity, and that $p(z)$ satisfies (4.146) and is Hölder-continuous. In [266] the local Hölder continuity of the gradient is proved for the solutions of the scalar equation

$$\partial_t u - \operatorname{div} \left( (A(z)\nabla u \cdot \nabla u)^{\frac{p(z)-2}{2}} \cdot A(z)\nabla u \right) = \operatorname{div} \left( |f|^{p(z)-2} f \right)$$

with the exponent $p(z)$ satisfying (4.146) and Hölder-continuous in $Q_T$. The matrix $A$ is strictly positive definite and bounded, $\lambda^{-1}|\xi|^2 \leq A\xi \cdot \xi \leq \lambda|\xi|^2$ for some $\lambda \geq 1$, the components of the vector $f$ and the entries of the matrix $A(z)$ are assumed to be Hölder-continuous in $Q_T$.

It is proved in [258] that local solutions $u \in C_{loc}(0, T; L^2_{loc}(\Omega)) \cap W^{1,p(x)}_{loc}(Q_T) \cap L^\infty(Q_T)$ of the equation

$$\partial_t u - \operatorname{div} A(z, u, \nabla u) = B(z, u, \nabla u), \qquad z = (x, t) \in Q_T,$$

satisfy the Harnack inequality. This fact is proved under the assumptions that the functions $A : Q_T \times \mathbb{R}^{n+1} \mapsto \mathbb{R}^n$, $B : Q_T \times \mathbb{R}^{n+1} \mapsto \mathbb{R}$ satisfy the conditions

$$A(z, u, \nabla u) \cdot \nabla u \geq C_0|\nabla u|^{p(x)} - C_1,$$
$$|A(z, u, \nabla u)| + |B(z, u, \nabla u)| \leq C_2|\nabla u|^{p(x)-1} + C_3$$

with positive constants $C_i$ and a measurable exponent $p(x) \in C_{\log}(\Omega)$, $p(x) \in [p^-, p^+] \subset (2, n)$.

Stationary counterparts of Eq. (4.1) were studied by many authors. Not pretending to offer a complete account of the known results, we refer here to the review paper [157] and the papers [3, 28, 44, 45, 123] for an overview of the results published before 2011. Results on the solvability of anisotropic elliptic equations and problems with nonlinear boundary conditions can be found in [88–91, 106, 208]. Eigenvalue problems for elliptic equations with non-power nonlinearity in Orlicz-Sobolev spaces were studied in [206, 207]. The influence of the geometry of the problem domain on the solvability of elliptic problems with anisotropic nonlinearity is discussed in [123, 131, 255], see also [158]. Solvability of elliptic equations and systems with $p(u)$ and $p(x, u(x))$-Laplace operators and properties of solutions are studied in [19, 20].

# Chapter 5
# Space Localization of Energy Solutions

## 5.1 Introduction

Let us consider the evolution $p$-Laplace equation

$$u_t = \operatorname{div}\left(|\nabla u|^{p-2}\nabla u\right), \qquad (x,t) \in Q_T = \Omega \times (0,T], \tag{5.1}$$

with constant exponent of nonlinearity $p \in (1,2) \cup (2,\infty)$. Equations of the type (5.1) appear in numerous physical contexts, one of the most natural of them is the mathematical modeling of diffusion processes. According to the second Fick's law, the concentration $u$ obeys the equation $u_t = \operatorname{div}(K\nabla u)$ with a given diffusivity $K$. Assume that $K$ nonlinearly depends on the rate of change of the concentration, $|\nabla u|$, and is proportional to $|\nabla u|^{p-2}$ with a known constant $p \neq 2$. If $p > 2$, the diffusion is slow in the regions where $|\nabla u| \approx 0$. Conversely, for $p \in (1,2)$ the diffusion is fast wherever $|\nabla u| \approx 0$ and slow if $|\nabla u|$ is high.

By convention, we will say that the ranges $p \in (1,2)$ and $p > 2$ correspond to *fast* or *slow* diffusion. As in the case of the porous medium equation, the solutions of the diffusion equation (5.1) with $p \neq 2$ possess certain localization properties impossible for the solutions of the linear diffusion equation with $p = 2$. The solutions of the latter obey the strong maximum principle which prevents them from attaining the maximum and minimum values in the interior of the problem domain. As distinguished from this property, for $p \neq 2$ the solutions of the Dirichlet problem for Eq. (5.1) may be localized in space and time. More precisely, the following alternative holds: if $u$ is a solution of the homogeneous Dirichlet problem for Eq. (5.1), then

(a) $$1 < p < 2 \quad \Rightarrow \quad \exists\, T_1 : \ u(x,t) \equiv 0 \text{ for all } x \in \Omega \text{ and } t \geq T_1,$$

(b) $$\left.\begin{array}{r} p > 2 \\ u_0 \equiv 0 \text{ in } B_r(x_0) \subset \Omega \end{array}\right\} \quad \Rightarrow \quad \begin{array}{l} \exists\, t_*(x_0) : \ u \equiv 0 \text{ in } B_{s(t)}(x_0) \\ \text{for } t \in [0, t_*(x_0)] \text{ with some } 0 < s(t) \leq r, \end{array}$$



S. Antontsev and S. Shmarev, *Evolution PDEs with Nonstandard Growth Conditions*, Atlantis Studies in Differential Equations 4,
DOI 10.2991/978-94-6239-112-3_5

where $B_s(x_0) = \{x \in \mathbb{R}^n : |x - x_0| < s\}$. These properties complement each other: the former is called *extinction in a finite time*, the latter is usually referred to as *finite speed of propagation* of disturbances from the data. If $p > 2$ and the support of the initial function $u_0$ is compact in $\Omega$, then the support of the solution is expanding with time at a finite speed but eventually covers the whole of $\Omega$.

We begin with the study of the space localization of weak solutions of the isotropic diffusion-absorption equation

$$\begin{cases} u_t - \operatorname{div}\left(a(z,u)|\nabla u|^{p(z)-2}\nabla u\right) + c(z,u)|u|^{\sigma(z)-2}u = f \quad \text{in } Q, \\ u(x,0) = u_0(x) \text{ in } \Omega, \quad c(z,u) \geq 0 \text{ in } Q. \end{cases} \tag{5.2}$$

The exponents $p(z)$ and $\sigma(z)$ are measurable in $Q$ functions satisfying the conditions

$$p(z) : Q \mapsto [p^-, p^+], \qquad \sigma(z) : Q \mapsto [\sigma^-, \sigma^+]$$

with given finite constants $1 < p^- \leq p^+$, $1 < \sigma^- \leq \sigma^+$. For the solutions of Eq. (5.2), the property of finite speed of propagation of disturbances from the initial data may be caused either by slow diffusion, i.e., by the condition $p(z) \geq p^- > 2$, or by a suitable balance between the rates of diffusion and absorption, which can be expressed in terms of relations between the exponents $p(z)$ and $\sigma(z)$. In the latter case, the presence of the absorption term is necessary: $c(z,u) \geq c_0 > 0$ in $Q$.

Surprisingly, the space localization properties intrinsic for the solutions of *anisotropic* equations are conspicuously contrast to the properties typical for the solutions of the equation of isotropic diffusion. For the solutions of anisotropic equations the alternative "finite speed of propagation/vanishing in a finite time" is no longer valid. It turns out that the solutions may be localized in space even in the absence of the absorption terms, in this case the effect is caused by the anisotropy of the diffusion and depends on relations between the diffusion rates in different directions. Let us consider the anisotropic problem

$$\begin{cases} u_t - \sum_i D_i\left(a_i(z,u)|D_i u|^{p_i(x)-2}D_i u\right) + c(z,u)|u|^{\sigma(x)-2}u = f \quad \text{in } Q, \\ u = 0 \text{ on } \Gamma, \qquad u(x,0) = u_0(x) \text{ in } \Omega \end{cases} \tag{5.3}$$

with measurable and bounded in $\Omega$ exponents $p_i(x) > 1$ and $\sigma(x) > 1$. It is assumed that $a_i$ and $c$ are bounded, $a_i$ are strictly positive and $c$ is nonnegative. The study of the space localization of solutions of (5.3) in performed in two steps. We consider first the special case of the model anisotropic equation with constant exponents $p_i$

$$\partial_t u - \sum_{i=1}^n D_i\left(a_i|D_i u|^{p_i-2}D_i\right) = f. \tag{5.4}$$

It turns out that if the exponents $p_i$ satisfy the relation

$$\frac{1}{n-1} \geq \frac{1}{n-1}\sum_{i=2}^{n}\frac{1}{p_i} > \frac{1}{p_1},$$

then the disturbances from the data propagate in the direction $x_1$ only on a finite distance $L$, which can be estimated through the problem data. Moreover, we show that $L = 0$ under additional restrictions on the rate of vanishing of $u_0(x)$ and $f(x,t)$ near the boundaries of their supports. This means that the disturbances do not propagate in the direction $x_1$ and the solution possesses the property of infinite waiting time. An analogue of this property is established for the stationary solutions of Eq. (5.4) with variable exponents $p_i(x)$.

## 5.2 Localization via Diffusion-Absorption Balance

Let us consider the problem

$$\begin{cases} u_t - \operatorname{div}\left(a(z,u)|\nabla u|^{p(z)-2}\nabla u\right) + c(z,u)|u|^{\sigma(z)-2}u = f & \text{in } Q, \\ u(x,0) = u_0(x) \text{ in } \Omega \end{cases} \tag{5.5}$$

with measurable in $Q$ exponents $p(z)$, $\sigma(z)$ subject to the conditions

$$p(z) \subseteq (p^-, p^+) \subset (1,\infty), \quad \sigma(z) \subseteq (\sigma^-, \sigma^+) \subset (1,\infty) \tag{5.6}$$

with given finite constants $\sigma^{\pm}$, $p^{\pm}$. The coefficients $a$ and $c$ may depend on $z = (x.t)$, $u(z)$ and obey the following conditions: $a(z,u)$, $c(z,u)$ are Carathéodory functions, defined for $(z,r) \in \overline{Q} \times \mathbb{R}$,

$$\forall\,(z,r) \in \overline{Q} \times \mathbb{R} \quad \begin{array}{l} 0 < a_0 \leq a(z,r) \leq a_1 < \infty, \\ 0 \leq c_0 \leq c(z,u) \leq c_1 < \infty, \quad a_0, a_1, c_0, c_1 = const. \end{array} \tag{5.7}$$

Let us denote

$$\begin{array}{ll} B_\rho(x_0) = \{x \in \Omega \,|\, |x - x_0| < \rho\}, & S_\rho(x_0) = \partial B_\rho(x_0), \\ Q_\rho(x_0) = B_\rho(x_0) \times (0,T), & \Gamma_\rho(x_0) = \partial B_\rho(x_0) \times (0,T). \end{array}$$

We study the property of finite speed of propagation of disturbances for the local weak solutions of Eq. (5.5) in the cylinder $Q_{\rho_0} = B_{\rho_0} \times (0,T)$, $\overline{B}_{\rho_0} \subset \Omega$, regardless of the boundary conditions on $\partial\Omega$. We assume that for the solution under study the total energy in $Q_{\rho_0}$ is finite:

$$\operatorname*{ess\,sup}_{(0,T)} \|u(\cdot,\tau)\|^2_{L^2(B_{\rho_0})} + \int_{Q_{\rho_0}(x_0)} \left(|\nabla u^{\;p(z)} + c_0|u|^{\sigma(z)}\right) dz = M < \infty. \tag{5.8}$$

**Definition 5.1** A measurable function $u(x,t)$ is called local weak solution of Eq. (5.5) if for every $Q_\rho(x_0) \subset Q$

1. $u \in L^\infty(0,T; L^2(B_\rho(x_0)))$, $c(z)|u|^{\sigma(z)}$, $|\nabla u|^{p(z)} \in L^1(Q_\rho(x_0))$, $u_t \in \mathbf{W}'(Q_\rho(x_0))$,
2. for every test-function $\zeta$ such that $\zeta \in \mathbf{W}(Q_\rho(x_0))$, $\zeta_t \in \mathbf{W}'(Q_\rho(x_0))$, the integral identity holds:

$$\int_\Omega u\,\zeta\,dx\Big|_0^T + \int_Q \Big[-u\,\zeta_t + a|\nabla u|^{p(z)-2}\nabla u\cdot\nabla\zeta + c|u|^{\sigma(z)-2}u\,\zeta\Big]dz = \int_Q f\,\zeta\,dz.$$

Obviously, the weak solution of problem (5.5) $u \in \mathbf{W}(Q)$ (see Theorems 4.1 and 4.2) is a local weak solution in the sense of this definition.

### *5.2.1 The Basic Energy Relation*

**Theorem 5.1** *Let $p(z) \in C(\overline{Q})$ satisfy the oscillation condition*

$$\frac{p^+}{p^-} < 1 + \frac{2}{n} \quad \text{in } B_\rho(x_0) \times (0,T] \subseteq Q. \tag{5.9}$$

*If the local weak solution of* Eq. (5.5) *satisfies the energy estimate* (5.8), *then the following energy equality is fulfilled:*

$$\frac{1}{2}\int_{B_\rho} u^2(x,\tau)\,dx\Big|_{\tau=0}^{\tau=t} + \int_0^t \int_{B_\rho} \Big(a|\nabla u|^{p(z)} + c\,|u|^{\sigma(z)} - fu\Big)\,dxdt$$
$$= -\int_0^t \int_{S_\rho(x_0)} u\,a|\nabla u|^{p(z)-2}\nabla u\cdot\nu\,dS\,dt \equiv I(\rho,t), \tag{5.10}$$

*where $\nu$ is the unit outer normal to $S_\rho(x_0)$.*

*Proof* Let us denote $\omega = B_\rho \times (0,T)$. By Hölder's inequality (1.9)

$$\int_0^\rho |I(r,t)|\,dr \le 2a_1\,\|u\|_{p(\cdot,\cdot),\omega}\,\|\nabla u\|_{p(\cdot,\cdot),\omega},$$

and the inclusion $I(r,t) \in L^1(0,\rho)$ follows from Lemma 1.32, estimate (5.8) and (1.6). Let us denote $r = |x - x_0|$ and introduce the functions

$$\zeta_k(r) = \begin{cases} 0 & \text{if } r \geq \rho, \\ k(r-\rho) & \text{if } r \in [\rho - 1/k, \rho], \\ 1 & \text{if } r < \rho - 1/k, \end{cases} \qquad k \in \mathbb{N}.$$

Choosing $\eta = u\,\zeta_k(r)$ for the test-function in Definition 5.1 we arrive at the equality

$$\begin{aligned}\frac{1}{2}\int_{B_\rho(x_0)} u^2\zeta_k(r)\,dx\Big|_{t=0}^{t=T} &+ \int_0^t\int_{B_\rho(x_0)} \left(a|\nabla u|^p\zeta_k(r) + c|u|^\sigma\zeta_k(r)\right)\,dz \\ &= -\int_0^t\int_{B_\rho(x_0)} au\,|\nabla u|^{p-2}\nabla u\cdot\nabla\zeta_k(r)\,dz \\ &= -\int_0^t\int_{B_\rho(x_0)} au\;\zeta_k'(r)|\nabla u|^{p-2}\nabla u\cdot\nabla r\,dz \\ &= -k\int_0^t\int_{\rho-1/k<|x-x_0|<\rho} au\,\zeta_k'(r)\,|\nabla u|^{p-2}\nabla u\cdot\frac{x}{r}\,dz.\end{aligned}$$

By the Lebesgue dominated convergence theorem, every term of the first one of this chain of equalities has a limit as $k\to\infty$. This means that the last expression also has a limit as as $k\to\infty$:

$$\begin{aligned}\lim_{k\to\infty} k\int_0^t\int_{\rho-1/k<|x-x_0|<\rho} & au\,|\nabla u|^{p-2}\nabla u\cdot\mathbf{e}_r\,dz \\ &= \int_0^t\int_{S_\rho(x_0)} au\,|\nabla u|^{p-2}D_r u\,dSdt,\end{aligned}$$

where $\mathbf{e}_r$ denotes the unit vector of the $r$-axis in the spherical coordinate system. □

### 5.2.2 Finite Speed of Propagation of Disturbances

Given a local weak solution of problem (5.5), we introduce the energy functions

$$E(\rho,t) = \int_0^t\int_{B_\rho} |\nabla u|^{p(z)}dz, \qquad b(\rho,t) = \|u(\cdot,t)\|^2_{2,B_\rho}. \tag{5.11}$$

Let us denote

$$\overline{b}(\rho,t) = \operatorname*{ess\,sup}_{\tau\leq t}\|u(\cdot,\tau)\|^2_{2,B_\rho}.$$

Notice that there exist

$$
\begin{aligned}
E_\rho &= \int_0^t \int_{S_\rho} |\nabla u|^{p(z)}\, dxdt \in L^1(0,\rho_0),\\
E_t &= \int_{B_\rho} |\nabla u|^{p(z)}\, dx \in L^1(0,T), \qquad (5.12)\\
b_\rho(\rho,t) &= \|u(\cdot,t)\|^2_{2,S_\rho} \in L^1(0,\rho_0).
\end{aligned}
$$

**Theorem 5.2** *Let conditions* (5.6) *and* (5.7) *and the conditions of* Theorem 5.1 *be fulfilled. Assume that*

$$c_0 = 0, \qquad 2 < p^- \le p(x,t), \tag{5.13}$$

*and*

$$f = 0\, in\, Q_{\rho_0} = B_{\rho_0} \times (0,T), \quad u_0 = 0\, in\, B_{\rho_0}.$$

*If $p(z)$ satisfy the oscillation condition*

$$\frac{p^+}{p^-} < 1 + \frac{2}{n}\left(1 - \frac{1}{p^-}\right), \tag{5.14}$$

*then every local weak solution $u(z)$ of* Eq. (5.5) *in $Q_{\rho_0}$, satisfying* (5.8), *possesses the property of finite speed of propagation:*

$$u(x,t) = 0\, in\, B_{\rho(t)}(x_0)\, with\, 0 \le t < T$$

*and $\rho(t)$ given by the formula*

$$\rho^{1+\nu}(t) = \max\left\{0,\ \rho_0^{1+\nu} - Ct^\alpha M^{1-\beta}\right\}, \tag{5.15}$$

*with positive constants $C$, $\nu$, and $\alpha$, $\beta$ depending on the constants in conditions* (5.6) *and* (5.7). *The exponents $\alpha$, $\beta$, $\nu$ are given by formulas* (5.23) *below.*

*Remark 5.1* For small $t$ the function $\rho(t)$ defined by (5.15) is strictly positive and the set $B_{\rho(t)}(x_0)$ is nonempty.

*Remark 5.2* The oscillation condition (5.14) is stronger than (5.9). Since $p(z)$ is continuous, condition (5.14) is surely fulfilled for the sufficiently small $\rho_0$ and $T$.

*Proof* (Theorem 5.2) Let us transform the energy equality (5.10) into a nonlinear differential inequality for the energy function $E + b$. Applying Hölder's inequality (1.9) and using (1.6) we estimate

$$
\begin{aligned}
|I(\rho,t)| &\le C \int_{\Gamma_{\rho,t}(x_0)} |\nabla u|^{p(z)-1}|u|\,dz \le 2C \||\nabla u|^{p-1}\|_{p'(\cdot),\Gamma_{\rho,t}}\|u\|_{p(\cdot),\Gamma_{\rho,t}} \\
&\le 2C' \max\left\{E_\rho^{1-\frac{1}{p^+}},\ E_\rho^{1-\frac{1}{p^-}}\right\}\|u\|_{p^+,\Gamma_{\rho,t}},
\end{aligned}
\tag{5.16}
$$

$\Gamma_{\rho,t} = \partial B_\rho(x_0) \times (0,t)$, and transform (5.10) into

$$
\frac{1}{2}b(\rho,t) + a_0 E(\rho,t) \le 2C' \max\left\{E_\rho^{1-\frac{1}{p^+}},\ E_\rho^{1-\frac{1}{p^-}}\right\}\|u\|_{p^+,\Gamma_{\rho,t}}.
$$

Notice that since this inequality is true for almost every $t \in (0,T)$ and the right-hand side is nondecreasing as a function of $t$ the first term on the left-hand can be substituted by $\overline{b}/2$:

$$
\frac{1}{4}\overline{b} + a_0 E \le 2C' \max\left\{E_\rho^{1-\frac{1}{p^-}},\ E_\rho^{1-\frac{1}{p^-}}\right\}\|u\|_{p^+,\Gamma_{\rho,t}}.
\tag{5.17}
$$

Using the trace-interpolation inequality (1.62) with $q = p^+$, $p = p^-$, $\gamma = r = 2$ we find that

$$
\begin{aligned}
\|u\|_{p^+,\Gamma_{\rho,t}}^{p^+} &= \int_0^t \|u\|_{p^+,S_\rho}^{p^+}\,dt \\
&\le C\int_0^t \left(\|\nabla u\|_{p^-,B_\rho} + \rho^{-\delta}\|u\|_{2,B_\rho}\right)^{p^+\theta}\|u\|_{2,B_\rho}^{p^+(1-\theta)}\,dt \\
&\le C\max\left\{1,\rho^{-p^+\delta\theta}\right\}\int_0^t \left(\|\nabla u\|_{p^-,B_\rho}^{p^-} + b^{\frac{p^-}{2}}\right)^{\theta\frac{p^+}{p^-}}\overline{b}^{p^+\frac{1-\theta}{2}}\,dt \\
&\le C\max\left\{1,\rho^{-p^+\delta\theta}\right\}\overline{b}^{p^+\frac{1-\theta}{2}}\int_0^t \left(\|\nabla u\|_{p^-,B_\rho}^{p^-} + \overline{b}^{\frac{p^-}{2}-1}b\right)^{\theta\frac{p^+}{p^-}}\,dt \\
&\le C(1+M)^{\frac{p^-}{2}-1}\max\left\{1,\rho^{-p^+\delta\theta}\right\}\overline{b}^{p^+\frac{1-\theta}{2}}\int_0^t \left(\|\nabla u\|_{p^-,B_\rho}^{p^-} + \overline{b}\right)^{\theta\frac{p^+}{p^-}}\,dt
\end{aligned}
\tag{5.18}
$$

with the exponents

$$
\theta = \frac{p^-}{p^+}\frac{n(p^+-2)+2}{n(p^--2)+2p^-} < 1, \quad \delta = \frac{n(p^--2)+2p^-}{2p^-} > 1.
\tag{5.19}
$$

By assumption (5.14) $\theta p^+ < p^-$. Applying Hölder's inequality, from (5.18) we obtain

$$\begin{aligned}
\|u\|_{p^+,\Gamma_{\rho,t}} &\leq C(1+M)^{\frac{p^-}{2}-1} t^{\frac{1}{p^+}-\frac{\theta}{p^-}} \max\left\{1, \rho^{-\delta\theta}\right\} \\
&\quad \times \left(\int_0^t \left(\|\nabla u\|^{p^-}_{p^-,B_\rho} + b\right) dt\right)^{\frac{\theta}{p^-}} \overline{b}^{\frac{1-\theta}{2}} \\
&\leq C(1+M)^{\frac{p^-}{2}-1} t^{\frac{1}{p^+}-\frac{\theta}{p^-}} \max\left\{1, \rho^{-\delta\theta}\right\} \left(\|\nabla u\|^{p^-}_{p^-,Q_\rho} + t\,\overline{b}\right)^{\frac{\theta}{p^-}} \overline{b}^{\frac{1-\theta}{2}} \\
&\leq C(1+M)^{\frac{p^-}{2}-1} (1+t)^{\frac{\theta}{p^-}} t^{\frac{1}{p^+}-\frac{\theta}{p^-}} \max\left\{1, \rho^{-\delta\theta}\right\} \\
&\quad \times \left(\|\nabla u\|^{p^-}_{p^-,Q_\rho} + \overline{b}\right)^{\frac{\theta}{p^-}+\frac{1-\theta}{2}}.
\end{aligned}$$

Once again using Hölder's inequality (1.9) we estimate

$$\begin{aligned}
\int_{Q_\rho} |\nabla u|^{p^-} &\leq 2\|1\|_{\frac{p(\cdot)}{p(\cdot)-p^-},Q_\rho} \||\nabla u|^{p^-}\|_{\frac{p(\cdot)}{p^-},Q_\rho} \\
&\leq 2\|1\|_{\frac{p(\cdot)}{p(\cdot)-p^-},Q_\rho} \max\left\{\int_{Q_\rho} |\nabla u|^{p(z)}\,dz, \left(\int_{Q_\rho} |\nabla u|^{p(z)}\,dz\right)^{\frac{p^+}{p^-}}\right\} \\
&\leq 2\|1\|_{\frac{p(\cdot)}{p(\cdot)-p^-},Q_\rho} \max\left\{E, E^{\frac{p^-}{p^+}}\right\}.
\end{aligned}$$

If we now claim that $\rho_0 \leq 1$ and $T \leq 1$ are so small that $M \leq 1$, $E(\rho_0, T) + \overline{b}(\rho_0, T) \leq 1$ and $|Q_{\rho_0}| < 1$, the previous inequality takes the form

$$\int_{Q_\rho} |\nabla u|^{p^-}\,dz \leq C\,E^{\frac{p^-}{p^+}}. \tag{5.20}$$

Gathering it with (5.16) and substituting the result into (5.10) we arrive at the inequality

$$\begin{aligned}
\frac{1}{4}\overline{b} + a_0 E &\leq C\psi(\rho,t) \max\left\{E_\rho^{1-\frac{1}{p^+}}, E_\rho^{1-\frac{1}{p^-}}\right\} \left(E^{\frac{p^-}{p^+}} + \overline{b}^{1-\frac{p^-}{p^+}}\overline{b}^{\frac{p^-}{p^+}}\right)^{\frac{1-\theta}{2}+\frac{\theta}{p^-}} \\
&\leq C\phi(\rho,t) \max\left\{E_\rho^{1-\frac{1}{p^+}}, E_\rho^{1-\frac{1}{p^-}}\right\} (E+\overline{b})^{\frac{1}{\lambda}}
\end{aligned} \tag{5.21}$$

with

$$\phi(\rho,t) = (1+t)^{\frac{\theta}{p^-}} t^{\frac{1}{p^+}-\frac{\theta}{p^-}} \max\left\{1, \rho^{-\delta\theta}\right\}, \quad \frac{1}{\lambda} = \frac{p^-}{p^+}\left(\frac{1-\theta}{2} + \frac{\theta}{p^-}\right) \in (0,1),$$

and an absolute constant $C$. Applying to the right-hand side Young's inequality with the exponent $\frac{1}{\lambda}$ and simplifying, we reduce (5 21) to the inequality

$$E + b \le E + \overline{b} \le C'\phi(\rho, t) \max\left\{ E_\rho^{\lambda'\left(1-\frac{1}{p^+}\right)},\ E_\rho^{\lambda'\left(1-\frac{1}{p^-}\right)} \right\}, \quad \frac{1}{\lambda'} + \frac{1}{\lambda} = 1,$$

which can be written in the form

$$E_\rho \ge \begin{cases} (C'\phi(\rho, t))^{-\mu^+}(E + b)^{\mu^+} & \text{if } E_\rho \le 1, \\ (C'\phi(\rho, t))^{-\mu^-}(E + b)^{\mu^-} & \text{if } E_\rho \ge 1 \end{cases} \qquad \frac{1}{\mu^\pm} = \lambda'\left(1 - \frac{1}{p^\pm}\right).$$

Recall that $\rho_0$ and $T$ are already chosen so small that $E + b \le 1$ in $Q_{\rho_0}$ and continue the last inequality as follows:

$$\begin{aligned} (E + b)_\rho \ge E_\rho &\ge \min\{(C'\phi)^{-\mu^-}, (C'\phi)^{-\mu^-}\}(E + b)^{\mu^-} \\ &= C'' t^{-\mu^-\left(\frac{1}{p^+} - \frac{\theta}{p^-}\right)} \rho^{\delta\theta\mu^-}(E + b)^{\mu^-} \\ &= C'' t^{-\alpha}\rho^{\nu}(E + b)^{\beta} \end{aligned} \tag{5.22}$$

with the exponents

$$\begin{aligned} \beta &= \left(1 - \frac{p^-}{p^+}\left(\frac{1-\theta}{2} + \frac{\theta}{p^-}\right)\right)\frac{p^-}{p^- - 1} \in (0, 1), \\ \alpha &= \beta\left(\frac{1}{p^+} - \frac{\theta}{p^-}\right) > 0, \\ \nu &= \delta\theta\left(1 - \frac{1-\theta}{2} - \frac{\theta}{p^-}\right)\frac{p^-}{p^- - 1} > 0. \end{aligned} \tag{5.23}$$

Under the conditions (5.13) and (5.14) $\beta \in (0, 1)$, $\nu \in (0, 1)$ and $\alpha > 0$.

If $E(\rho_0, t) + b(\rho_0, t) = 0$, then $E + b \equiv 0$ in $Q_{\rho_0}$ because $E + b$ is a nondecreasing function of $\rho$. Let us assume that $E(\rho_0, t) + b(\rho_0, t) \not\equiv 0$. By continuity $E + \rho \ne 0$ for $\rho$ close to $\rho_0$, which allows one to divide both parts of (5.22) by $(E + b)^\beta$ and integrate the resulting inequality over the interval $(\rho, \rho_0)$ where $E + b$ remains positive:

$$M^{1-\beta} \ge (E + b)^{1-\beta}(\rho_0, t) \ge (E + b)^{1-\beta}(\rho, t) + C'' t^{-\alpha}\frac{1-\beta}{1+\nu}\left(\rho_0^{1+\nu} - \rho^{1+\nu}\right).$$

It is necessary that $E(\rho, t) + b(\rho, t) \equiv 0$ for $0 \le \rho \le \rho(t)$ with $\rho(t)$ defined from the equality

$$0 \le (E + b)^{1-\beta}(\rho(t), t) \le M^{1-\beta} - C'' t^{-\alpha}\frac{1-\beta}{1+\nu}\left(\rho_0^{1+\nu} - \rho^{1+\nu}(t)\right) = 0.$$

□

We proceed to study the effect of space localization in solutions of the isotropic Eq. (5.5) in the case $c_0 > 0$ when the influence of the absorption term becomes crucial.

**Theorem 5.3** *Let in* Theorem 5.2 *condition* (5.13) *be substituted by the following one:*

$$c_0 > 0, \quad 1 < \sigma^+ < p^- \leq p(x,t), \quad p^- \leq 2. \tag{5.24}$$

*Assume that there exists* $r \in (\sigma^+, p^-)$ *such that*

$$\frac{p^-}{p^+}\left(\frac{\theta}{p^-} + \frac{1-\theta}{r}\right) > \frac{1}{p^-} \quad \text{with} \quad \theta = \frac{p^-}{p^+}\frac{n(p^+-r)+r}{n(p^- - r)+p^-} \in (0,1). \tag{5.25}$$

*If the exponent* $p(x,t)$ *satisfies the oscillation conditions* (5.14) *and*

$$\frac{p^+}{p^-} - 1 < \begin{cases} \dfrac{1}{n}\left(p^+ - 1\right) & \text{if } p^- \geq n \\ \min\left\{\dfrac{p^+-1}{n-p^-}, \dfrac{1}{n}\left(p^+-1\right)\right\} & \text{if } p^- < n, \end{cases} \tag{5.26}$$

*then every local weak solution* $u(z)$ *of Eq.* (5.5) *in* $Q_{\rho_0}(x_0)$, *satisfying* (5.8), *possesses the property of finite speed of propagation:*

$$u(x,t) = 0 \text{ in } B_{\rho(t)}(x_0) \text{ with } 0 \leq t \leq t_* < T,$$

*where* $\rho(t)$ *is given by the formula*

$$\rho^{1+\beta}(t) = \max\left\{0, \rho_0^{1+\beta} - Ct^{\lambda}M^{1-\nu}\right\}, \tag{5.27}$$

*with positive constants* $C$, $\nu$, *and* $\lambda$, $\beta$ *given in* (5.32) *and depending only on the constants in conditions* (5.6) *and* (5.7).

*Remark 5.3* In the special case $p = const$ conditions (5.25) and (5.26) are immediate consequences of (5.24).

*Proof* The proof is an adaptation of the proof of Theorem 5.2. Let $u$ be a local energy solution of Eq. (5.5). Denote

$$D(\rho,t) = \int_{Q_\rho(x_0)} |u|^{\sigma(z)}\,dz.$$

We assume that $\rho_0 \leq 1$, $T \leq 1$ are so small that $|Q_{\rho_0}| \leq 1$ and $M \leq 1$. By Definition 5.1 $D(\rho,t) \in L^1(Q_{\rho_0})$, whence for a.e. $\rho \in (0,\rho_0)$ there exists

$$D_\rho = \int_0^t \int_{S_\rho} |u|^{\sigma(z)}\,dz \in L^1(0,\rho_0).$$

The energy functions $E$, $D$ and $b$ satisfy the energy relation (5.10), which leads to the inequality (cf. with (5.17))

$$\frac{1}{4}\overline{b} + a_0 E + c_0 D \leq 2C \max\left\{ E_\rho^{1-\frac{1}{p^+}},\ E_\rho^{1-\frac{1}{p^-}} \right\} \|u\|_{p^+,\Gamma_{\rho,t}} \tag{5.28}$$

with positive constants $a_0$ and $c_0$. To estimate $\|u\|_{p^+,\Gamma_{\rho,t}}$ we make use of the trace-interpolation inequality (1.62) with $q = p^+$, $p = p^-$, $\gamma = \sigma^+$ and some $r \in (\sigma^+, p^-)$:

$$\begin{aligned} \|u\|_{p^+,\Gamma_{\rho,t}}^{p^+} &\leq C \int_0^t \left( \|\nabla u\|_{p^-,B_\rho} + \rho^{-\delta} \|u|_{\sigma^+,B_\rho} \right)^{p^+\theta} \|u\|_{r,B_\rho}^{p^+(1-\theta)}\, dt \\ &\leq C\rho^{-p^+\delta\theta} \int_0^t \left( \|\nabla u\|_{p^-,B_\rho} + \|u\|_{\sigma^+,B_\rho} \right)^{\theta p^+} \|u\|_{r,B_\rho}^{p^+(1-\theta)}\, dt, \end{aligned} \tag{5.29}$$

where

$$\delta = 1 - \frac{n}{p^-} + \frac{1}{\sigma^+}, \qquad \theta = \frac{n(p^+ - r) + r}{n(p^- - r) + p^-}\frac{p^-}{p^+} \in (0, 1).$$

For every $r \in (\sigma^+, p^-)$ the inequality holds

$$\|u\|_{r,B_\rho} \leq \|u\|_{2,B_\rho}^{\frac{2}{r}\frac{r-\sigma^-}{2-\sigma^-}} \|u\|_{\sigma^+,B_\rho}^{\frac{\sigma^+}{r}\frac{2-r}{2-\sigma^+}}. \tag{5.30}$$

Indeed: by Hölder's inequality with $\varepsilon = \sigma^+ \dfrac{2-r}{2-\sigma^+}$

$$\begin{aligned} \|u\|_{r,B_\rho} &= \left( \int_{B_\rho} |u|^{r-\varepsilon} |u|^\varepsilon dx \right)^{\frac{1}{r}} \\ &\leq \|u\|_{2,B_\rho}^{\frac{r-\varepsilon}{2}} \left( \int_{B_\rho} |u|^{\frac{2\varepsilon}{2-r+\varepsilon}} dx \right)^{1-\frac{r-\varepsilon}{2}} = \|u\|_{2,B_\rho}^{\frac{2}{r}\frac{r-\sigma^+}{2-\sigma^+}} \|u\|_{\sigma^+,B_\rho}^{\frac{\sigma^+}{r}\frac{2-r}{2-\sigma^+}}. \end{aligned}$$

Plugging (5.30) into (5.29) we obtain

$$\begin{aligned} \|u\|_{p^+,\Gamma_{\rho,t}}^{p^+} \leq C\rho^{-\delta\theta p^+} \int_0^t & \left( \|\nabla u\|_{p^-,B_\rho}^{p^-} + \|u\|_{\sigma^+,B_\rho}^{\sigma^+\left(\frac{p^-}{\sigma^+}-1\right)} \|u\|_{\sigma^+,B_\rho}^{\sigma^+} \right)^{\theta\frac{p^+}{p^-}} \\ & \times \|u\|_{2,B_\rho}^{\frac{2p^+}{r}\frac{r-\sigma^+}{2-\sigma^+}(1-\theta)} \|u|_{\sigma^+,B_\rho}^{p^+\frac{\sigma^+}{r}\frac{2-r}{2-\sigma^+}(1-\theta)}\, dt. \end{aligned}$$

The assumptions $\overline{b} \le 1$, $\rho_0 \le 1$ provides the inequality

$$\|u\|_{\sigma^+,B_\rho} \le |B_\rho|^{\frac{1}{\sigma^+}-\frac{1}{2}} \|u\|_{2,B_\rho} \le C\,\overline{b}^{\frac{1}{2}} \le C,$$

while the assumption $E \le 1$ gives

$$\|\nabla u\|^{p^-}_{p^-,B_\rho} \le C \max\left\{E^{\frac{p^-}{p^+}},\ E\right\} \le C E^{\frac{p^-}{p^+}}.$$

Applying Hölder's inequality we now find that

$$\begin{aligned}\|u\|_{p^+,\Gamma_{\rho,t}} &\le C\rho^{-\delta\theta} t^{\frac{1}{p^+}-\gamma} \left(\|\nabla u\|^{p^-}_{p^-,Q_\rho} + \|u\|^{\sigma^+}_{\sigma^+,Q_\rho} + \overline{b}\right)^{\gamma+\frac{1-\theta}{r}\frac{r-\sigma^+}{2-\sigma^+}} \\ &\le C\rho^{-\delta\theta} t^{\frac{1}{p^+}-\gamma} \left(E^{\frac{p^-}{p^+}} + D + \overline{b}\right)^{\gamma+\frac{1-\theta}{r}\frac{r-\sigma^+}{2-\sigma^+}} \\ &\le C\rho^{-\delta\theta} t^{\frac{1}{p^+}-\gamma} \left(E + D + \overline{b}\right)^{\frac{p^-}{p^+}\left(\gamma+\frac{1-\theta}{r}\frac{r-\sigma^+}{2-\sigma^+}\right)}\end{aligned}$$

with the exponent

$$\gamma = \frac{\theta}{p^-} + \frac{2-r}{2-\sigma^+}\frac{1-\theta}{r}.$$

The energy inequality (5.28) takes the form

$$\overline{b} + E + D \le C\rho^{-\delta\theta} t^{\frac{1}{p^+}-\gamma} \max\left\{E_\rho^{1-\frac{1}{p^+}},\ E_\rho^{1-\frac{1}{p^-}}\right\} (E + D + \overline{b})^{\kappa} \tag{5.31}$$

with

$$\kappa = \frac{p^-}{p^+}\left(\frac{\theta}{p^-} + \frac{1-\theta}{r}\right).$$

Assumptions on the oscillation of $p(z)$ yield the inequality $\dfrac{1}{p^-} < \kappa < 1$. Indeed: the first inequality is equivalent to (5.25), the second one is immediate because

$$2 \ge p^- > r > \sigma^+ > 1 \quad \Rightarrow \quad \frac{p^-}{p^+}\left(\frac{\theta}{p^-} + \frac{1-\theta}{r}\right) \le \frac{\theta}{p^-} + \frac{1-\theta}{r} < \frac{1}{r} < 1.$$

Repeating the arguments we used to obtain the ordinary differential inequality (5.22) from the energy relation (5.21), we derive from (5.31) the differential inequality for the energy $E + D + b$:

$$(b + D + E)^{\nu} \le C t^{\lambda} \rho^{-\beta} (E + D + \rho)_\rho$$

with

$$\nu = (1-\kappa)(p^-)', \quad \lambda = (p^-)'\left(\frac{1}{p^+} - \gamma\right), \quad \beta = \delta\theta(p^-)'. \tag{5.32}$$

The assertion follows now exactly as in the proof of Theorem 5.2. □

*Remark 5.4* The assertions of Theorems 5.2 and 5.3 can be read in a different way. Assume that supp $u_0 \subset B_r(x_0) \subset \Omega$ and supp $f \subset Q_{r,x_0}$ with some $r > 0$. Take an arbitrary point $x' \in \partial B_r(x_0)$ and the ball $B_{\rho_0}(x'')$ touching $\partial B_r(x_0)$ at the point $x'$, of radius $\rho_0$ chosen in the proof of Theorem 5.2 (or Theorem 5.3). By Theorem 5.2 (or 5.3) $u(x,t) = 0$ in $B_{\rho(t)}(x'') \times (0, T_*]$ with some $T_*$ depending only on the total energy $M$ in the whole of the problem domain $Q$ and the exponents of nonlinearity. Rolling the point $x'$ over $\partial B_r(x_0)$ and repeating these arguments we find that $u(x,t) = 0$ in the annular set $\left(B_{r+\rho(t)}(x_0) \setminus B_r(x_0)\right) \times (0, T_*)$. It follows that supp $u(x,t)$ is contained in an expanding cylinder $B_{s(t)}(x_0) \times (0, T^*)$, where $s(t) > r$ is a monotone increasing function of $t$ and $T^*$ is the moment when the expanding ball $B_{s(t)}(x_0)$ touches $\partial\Omega$.

### 5.2.3 The Waiting Time Property

Let us now assume that

$$u_0(x) \equiv 0 \quad x \in B_{\rho_0}(x_0), \quad f(z,t) \equiv 0 \quad \text{in } Q_{\rho_0}(x_0), \tag{5.33}$$

and

$$\|u_0\|^2_{2,B_\rho(x_0)} + \|f\|^2_{2,Q_\rho(x_0)} \le \varepsilon(\rho - \rho_0)_+^{1/(1-\nu)} \tag{5.34}$$

for all $\rho \in [\rho_0, \rho_1]$ with some positive constant $\varepsilon$ and the exponent $\nu$ defined by (5.23) in the case of Theorem 5.2, or (5.32) in the case of Theorem 5.3. Assumption (5.34) means that the functions $u_0(x)$ and $f(z)$ are sufficiently "flat" near the boundaries of their supports.

**Theorem 5.4** *Let the conditions of* Theorem 5.1 *hold true and either the conditions of* Theorem 5.2, *or the conditions of* Theorem 5.3 *are fulfilled. There exist constants* $\varepsilon_*$ *and* $t_* \in (0, T]$ *such that if* $u_0$ *and* $f$ *satisfy condition* (5.34) *with* $\varepsilon \in (0, \varepsilon_*)$, *then every weak local solution* $u(z)$ *of* Eq. (5.5) *in* $Q_{\rho_1}(x_0)$ *possesses the waiting time property:*

$$u(x,t) = 0 \text{ in } B_{\rho_0}(x_0) \times (0, t^*).$$

*Proof* The proof is the same in both cases. For the sake of definiteness we assume that the conditions of Theorem 5.2 are fulfilled. The literal repetition of the derivation of inequality (5.22) leads to the nonhomogeneous ordinary differential inequality for the energy function $E + b$:

$$(E+b)^{\nu} \leq Ct^{\lambda}\rho^{-\beta}(E+b)_{\rho}+\varepsilon\,[\rho-\rho_0]_{+}^{\frac{\nu}{1-\nu}}, \quad \rho \in (\rho_0, \rho_1), \quad t \in (0, T). \quad (5.35)$$

This inequality coincides with inequality (3.23) already studied in the proof of Theorem 3.2: the required property follows by comparison of the energy $E + b$ with the majorant function. □

## 5.3 Localization Caused by Anisotropic Diffusion

Let us consider the Dirichlet problem for the degenerate anisotropic parabolic equation

$$\begin{cases} u_t - \sum_{i=1}^{n} D_i\left(a_i(z,u)|D_iu|^{p_i-2}D_iu\right) = f(z) \quad \text{in } Q, \\ u = 0 \text{ on } \Gamma, \qquad u(x,0) = u_0(x) \text{ in } \Omega, \end{cases} \quad (5.36)$$

where $\Omega \subset \mathbb{R}^n$ is a domain with Lipschitz-continuous boundary $\partial\Omega$, $\Gamma = \partial\Omega \times (0, T)$, $Q = \Omega \times (0, T)$, $z = (x, t)$. We assume that $p_i = const$ and

$$p_i \in (1, \infty), \quad i = 1, \ldots, n. \quad (5.37)$$

The weak solution of problem (5.36) is understood in the sense of Definition 4.1.

**Definition 5.2** We say that

(a) the weak solution of problem (5.36) is localized in the direction $x_i$ if

$$\exists s > 0 : \; u(x,t) \equiv 0 \text{ in } Q \cap \{x_i > s\};$$

(b) the weak solution possesses the waiting time property on the plane $x_i = s$ if

$$u_0(x) \equiv 0 \text{ in } \Omega \cap \{x_i > s\} \;\; \Rightarrow \;\; u(x,t) \equiv 0 \text{ in } \Omega \cap \{x_i > s\} \text{ for } t \in [0, t_*].$$

The coefficients $a_i$ are functions satisfying the following conditions:

$$\begin{aligned} &a_i(z,s) \text{ are Carathéodory functions,} \\ &a_0 \leq a_i(z,s) \leq A_0, \qquad a_0, A_0 = const > 0. \end{aligned} \quad (5.38)$$

By Theorem 4.1, under these assumptions for every $u_0 \in L^2(\Omega)$, $f \in L^2(Q)$ problem (5.36) has a weak solution $u \in \mathbf{W}(Q)$ which satisfies the estimate

$$\|u\|_{\mathbf{W}(Q)} \leq C \left\{\|u_0\|_{2,\Omega} + \|f\|_{2,Q}\right\}, \qquad C \equiv C(n, p_i, |\Omega|).$$

Let us introduce the notations

$$\begin{aligned} &\Omega(s) = \Omega \cap \{x_1 > s\}, \quad Q(s,t) = \Omega(s) \times (0,t), \\ &\omega(s) = \Omega \cap \{x_1 = s\}, \quad x' = (x_2, \ldots, x_n), \qquad x = (x_1, x'), \end{aligned}$$

and define the energy functions

$$\begin{aligned} E(s,t) &= \sum_{i=1}^{n} \int_{Q(s,t)} |D_i u|^{p_i}\, dxdt, \\ b(s,\tau) &= \int_{\Omega(s)} |u|^2(x,\tau)\, dx, \\ \overline{b}(s,t) &= \sup_{\tau \in (0,t)} b(s,\tau). \end{aligned}$$

We may always assume that

$$\{x_1 : x \in \Omega\} = (0,a) \quad \text{with some finite}\, a > 0.$$

By virtue of the definition of $\mathbf{W}(Q)$ the function $E(s)$ is differentiable for a.e. $s \in (0,a)$ and

$$E_s = -\sum_{i=1}^{n} \int_0^t \int_{\omega(s)} |D_i u|^{p_i}\, dx'dt.$$

**Lemma 5.1** *Let the conditions of* Theorem 4.1 *be fulfilled. For every solution* $u \in \mathbf{W}(Q)$ *of problem* (5.36) *and a.e.* $s \in (0,a)$ *such that* $\Omega \cap \{x_1 = s\} \neq \emptyset$

$$\begin{aligned} &\frac{1}{2}\int_{\Omega(s)} u^2(x,\tau)dx\Big|_{\tau=0}^{\tau=t} + \sum_{i=1}^{n} \int_{Q(s,t)} a_i(z,u)|D_i u|^{p_i}\, dxdt \\ &\quad + \int_0^t dt \int_{\omega(s)} a_1(z,u)\, u\, |D_1 u|^{p_1-2} D_1 u\, dx' = \int_{Q(s,t)} u\, f\, dxdt. \end{aligned} \tag{5.39}$$

*Proof* The energy solution can be taken for test–function in identity (4.10). Let us set

$$\phi_k(x_1, x', s) = \begin{cases} 1 & \text{for } x_1 > s + \frac{1}{k}, \\ k(x_1 - s) & \text{for } x_1 \in \left[s, s + \frac{1}{k}\right], \\ 0 & \text{for } x_1 < s, \qquad k \in \mathbb{N}, \end{cases}$$

and choose $u(x,t)\phi_k(x_1,x',s)$ for the test-function in the integral identity (4.10). The resulting identity has the form

$$\begin{aligned}\sum_{j=1}^{4} I_j(k,s) \equiv \sum_{i=1}^{n} \int_{Q(s,t)} a_i(z,u)|D_i u|^{p_i}\,\phi_k\,dxdt \\ +k\int_0^t dt\int_{\Omega(s+1/k)\setminus\Omega(s)} a_1(z,u)\,u\,|D_1u|^{p_1-2}D_1u\,dx \\ +\frac{1}{2}\int_{\Omega(s)}\phi_k u^2\,dx\Big|_{\tau=0}^{\tau=t} - \int_{Q(s,t)} u\,f\,\phi_k\,dxdt = 0.\end{aligned} \tag{5.40}$$

By the definition of $\mathbf{W}(Q)$

$$|D_i u|^{p_i}\,\phi_k,\ u\phi_k \in L^1(Q), \qquad u^2\phi_k \in L^1(\Omega) \quad \text{for a.e. } t \in (0,T),$$

which allows us to pass to the limit as $k \to \infty$ in $I_1$, $I_3$ and $I_4$:

$$\begin{aligned}\lim_{k\to\infty} I_1 &= \sum_{i=1}^{n}\int_{Q(s,t)} a_i(z,u)\,|D_iu|^{p_i}\,dxdt,\\ \lim_{k\to\infty} I_3 &= \frac{1}{2}\int_{\Omega(s)} u^2\,dx - \frac{1}{2}\int_{\Omega(s)} u_0^2\,dx,\\ \lim_{k\to\infty} I_4 &= \int_{Q(s,t)} u\,f\,dxdt.\end{aligned}$$

By virtue of (5.40) $I_2$ is bounded uniformly with respect to $k$, provided that so are the integrals $I_1$, $I_3$ and $I_4$. Writing $I_2$ in the form

$$I_2 = k\int_s^{s+1/k}\left(\int_0^t\left(\int_{\omega(s)} a_i(z,u)\,u\,|D_1u|^{p_1-2}D_1u\,dx'\right)dt\right)dx_1$$

and applying the Lebesgue theorem we conclude that there exists

$$\lim_{k\to\infty} I_2(k,s) = \int_0^t dt\int_{\omega(s)} a_1(z,u)\,u\,|D_1u|^{p_1-2}D_1u\,dx'.$$

Equality (5.40) transforms into (5.39) as $k \to \infty$. □

*Remark 5.5* Let

$$\Omega \subset \Omega_a \equiv \{x \in \mathbb{R}^n : x_i \in (0,a),\ i = 1,\dots,n\}$$

and let $u \in \mathbf{W}(Q)$ be a solution of problem (5.36). Set

$$u^*(x,t) = \begin{cases} u(x,t) & \text{in } Q, \\ 0 & \text{in } Q^{(a)} \equiv (\Omega_a \setminus \Omega) \times (0,T). \end{cases}$$

The continued function $u^*(x,t)$ belongs to $\mathbf{W}(Q^{(a)})$ and formally satisfies identity (5.39) in the cylinder $Q^{(a)}$. If $u^*(x,t)$ is localized in space, so is the function $u(x,t)$, which allows one to study the localization properties of the energy solutions of problem (5.36) formulated in the domain $\Omega \equiv \{x \in \mathbb{R}^n : x_i \in (0,a)\}$.

## 5.3.1 Differential Inequality for the Energy Function

We want to transform the energy relation (5.39) into the differential inequality for the energy function $E(s,t)$. Let us denote

$$J(s,t) = -\int_0^t dt \int_{\omega(s)} a_1\, u\, |D_1 u|^{p_1-2} D_1 u\, dx'$$

and write (5.39) in the form

$$\begin{aligned} &\frac{1}{2}\int_{\Omega(s)} u^2(x,\tau)dx\Big|_{\tau=0}^{\tau=t} \\ &+\sum_{i=1}^{n}\int_{Q(s,t)} a_i |D_i u|^{p_i}\, dxdt = J(s,t) + \int_{Q(s,t)} u\, f\, dxdt. \end{aligned} \tag{5.41}$$

We assume that $\Omega = \{x \in \mathbb{R}^n : \ x_i \in (0,a)\}$ (see Remark 5.5). Fix an arbitrary $\theta \in (0,1)$ and set

$$\lambda = 1 + 2\frac{p_1 - 1}{p_1}. \tag{5.42}$$

The following representation holds:

$$\begin{aligned} |u(s,x',t)| = |u|^{\theta}|u|^{1-\theta} &\le |u|^{\theta}\left(\int_s^a |D_1((u^2)^{\lambda/2})|dx\right)^{\frac{1-\theta}{\lambda}} \\ &= (\lambda/2)^{\frac{1-\theta}{\lambda}} |u|^{\theta}\left(\int_s^a (u^2)^{\frac{\lambda}{2}-1}|D_1(u^2)|dx\right)^{\frac{1-\theta}{\lambda}} \\ &\le \lambda^{\frac{1-\theta}{\lambda}} |u|^{\theta}\left(\int_s^a |D_1 u|^{p_1}dx\right)^{\frac{1-\theta}{p_1\lambda}}\left(\int_s^a u^2 dx\right)^{\frac{(1-\theta)(\lambda-1)}{2\lambda}}. \end{aligned} \tag{5.43}$$

Applying Hölder's inequality we find that

$$
\begin{aligned}
|J| &\le A_0 \int_0^t \int_{\omega(s)} |D_1 u|^{p_1 - 1} |u| \, dx' dt \\
&\le \left( \int_0^t \int_{\omega(s)} |D_1 u|^{p_1} \, dx' dt \right)^{\frac{p_1 - 1}{p_1}} \left( \int_0^t \int_{\omega(s)} |u|^{p_1} \, dx' dt \right)^{\frac{1}{p_1}} \\
&\le (-E_s(s,t))^{\frac{p_1-1}{p_1}} \cdot I^{\frac{1}{p_1}}.
\end{aligned}
\tag{5.44}
$$

Applying Hölder's inequality once again, we estimate $I$ as follows:

$$
\begin{aligned}
I &\le \int_0^t \int_{\omega(s)} |u|^{\theta p_1} \left( \int_s^a |D_1 u|^{p_1} \, dx_1 \right)^{\frac{1-\theta}{\lambda}} \left( \int_s^a u^2 \, dx_1 \right)^{\frac{(1-\theta)(p_1 - 1)}{\lambda p_1}} dx' dt \\
&\le \int_0^t \left[ \left( \int_{\omega(s)} \left( |u|^{\theta p_1} \cdot \left( \int_s^a |D_1 u|^{p_1} \, dx_1 \right)^{\frac{1-\theta}{\lambda}} \right)^{\frac{1}{\sigma}} dx' \right)^{\sigma} \right. \\
&\qquad \left. \times \left( \int_{\Omega(s)} u^2 dx \right)^{1-\sigma} \right] dt
\end{aligned}
$$

with the exponent

$$
\sigma = 1 - \frac{(1-\theta)(p_1 - 1)}{\lambda p_1} \equiv 1 - \frac{(1-\theta)(\lambda - 1)}{2\lambda} \in (0,1). \tag{5.45}
$$

Further,

$$
\begin{aligned}
I &\le \overline{b}^{1-\sigma}(s,t) \int_0^t \left[ \left( \int_{\omega(s)} |u|^{\frac{\theta p_1}{\sigma}} \cdot \left( \int_s^a |D_1 u|^{p_1} \, dx_1 \right)^{\frac{1-\theta}{\sigma\lambda}} dx' \right)^{\sigma} \right] dt \\
&\le \overline{b}^{1-\sigma}(s,t) \int_0^t \left[ \left( \int_{\omega(s)} |u|^{\frac{p_1 \theta}{\mu}} \, dx' \right)^{\mu} \left( \int_{\Omega(s)} |D_1 u|^{p_1} \, dx \right)^{\frac{1-\theta}{\lambda}} \right] dt
\end{aligned}
\tag{5.46}
$$

with the exponent

$$
\mu = \sigma - \frac{1-\theta}{\lambda} \equiv 1 - \frac{(1-\theta)(\lambda + 1)}{2\lambda} > 0.
$$

Let us apply Lemma 1.24 to the function $u(x_1, x', t)$ considered as a function of the variables $x' \in \omega(s)$. Let us claim that

$$
\frac{1}{q} = \frac{1}{n-1} \sum_{i=2}^{n} \frac{1}{p_i} \le \frac{1}{n-1}. \tag{5.47}
$$

(Notice that since $p_i > 1$, this claim is automatically fulfilled if $n = 2$.) For every $v \in \mathbf{V}(\omega(s))$ and every $r \in [1, \infty)$

$$\|v\|_{r,\omega(s)} \le C \left( \prod_{i=2}^{n} \|D_i v\|_{p_i,\omega(s)} \right)^{\frac{1}{n-1}}. \tag{5.48}$$

If $p_1\theta \ge \mu$, inequality (5.48) yields the estimate

$$\int_{\omega(s)} |u|^{\frac{p_1\theta}{\mu}}\, dx' \le C \left( \sum_{i=2}^{n} \int_{\omega(s)} |D_i u|^{p_i}\, dx' \right)^{\frac{p_1\theta}{\mu}\frac{1}{q}},$$

otherwise $\frac{p_1\theta}{\mu} \in (0, 1)$ and we have

$$\int_{\omega(s)} |u|^{\frac{p_1\theta}{\mu}}\, dx' \le C(a, n)\, \|u\|_{1,\omega(s)}^{\frac{p_1\theta}{\mu}} \le C' \left( \sum_{i=2}^{n} \int_{\omega(s)} |D_i u|^{p_i}\, dx' \right)^{\frac{p_1\theta}{\mu}\frac{1}{q}}.$$

Plugging these estimates to (5.46), we finally have:

$$I \le C\,\overline{b}^{1-\sigma}(s, t) \int_0^t (-E_{st})^{\frac{p_1}{q}\theta} (E_t)^{\frac{1-\theta}{\lambda}}\, dt.$$

Recall that all previous arguments hold true for every $\theta \in (0, 1)$. Let us now choose $\theta$ from the condition

$$\theta\, \frac{p_1}{q} + \frac{1-\theta}{\lambda} = 1. \tag{5.49}$$

Equation (5.49) always has a solution in the interval $(0, 1)$, provided that $p_1 > q$. Indeed: let

$$\Phi(\theta) = \theta\, \frac{p_1}{q} + \frac{1-\theta}{\lambda} - 1;$$

this function is continuous in $(0, 1)$,

$$\Phi(0) = \frac{1}{\lambda} - 1 < 0 \quad \text{and} \quad \Phi(1) = \frac{p_1}{q} - 1 > 0,$$

which is why $\Phi(\theta)$ has a root in the interval $(0, 1)$.

It follows now by Hölder's inequality that

$$I \le C\,(\overline{b}(s, t) + E(s, t))^{1-\sigma+\frac{1-\theta}{\lambda}} \cdot (-E_s)^{\frac{p_1}{q}\theta}.$$

Gathering this estimate with (5.44) we arrive at the estimate

$$|J| \leq C\,(\overline{b}(s,t)+E(s,t))^{\frac{1}{p_1}-\frac{\sigma}{p_1}+\frac{1-\theta}{p_1\lambda}}\,(-E_s(s,t))^{\frac{p_1-1}{p_1}+\frac{\theta}{q}}. \tag{5.50}$$

**Lemma 5.2** *Let the exponents $p_i$ satisfy conditions* (5.37) *and*

$$\frac{1}{n-1} \geq \frac{1}{n-1}\sum_{i=2}^{n}\frac{1}{p_i} = \frac{1}{q} > \frac{1}{p_1}. \tag{5.51}$$

*Then*

$$|J(s,t)| \leq \frac{1}{2}(\overline{b}+E) + C_1\,(-E_s)^{\alpha} \tag{5.52}$$

*with an absolute constant $C_1$ and the exponent*

$$\alpha = \frac{\frac{p_1-1}{p_1}+\frac{\theta}{q}}{\frac{p_1-1}{p_1}+\frac{\sigma}{p_1}-\frac{1-\theta}{p_1\lambda}} > 1 \tag{5.53}$$

*with $\theta$, $\sigma$, $\lambda$ defined in* (5.49), (5.45) *and* (5.42).

*Proof* By virtue of (5.51) conditions (5.47) and (5.49) are fulfilled, whence (5.50). The claim $\alpha > 1$ is equivalent to the inequality

$$0 < \frac{p_1\theta}{q} - \sigma + \frac{1-\theta}{\lambda} = 1-\sigma = \frac{(1-\theta)(\lambda-1)}{2\lambda}.$$

□

Plugging (5.52) into (5.41), we obtain the differential inequality for the energy function $E(s,t)$:

$$\overline{b} + E - C(-E_s)^{\alpha} \leq K\int_{Q(s,t)} u\,f\,dxdt + K\int_{\Omega(s)} u_0^2\,dx \tag{5.54}$$

with the exponent $\alpha > 1$ from (5.53) and absolute constants $C$, $K$.

### 5.3.2 Directional Stable Localization

The properties of directional localization of the solution follow from the properties of the functions satisfying the ordinary differential inequality (5.54).

**Theorem 5.5** *Let the conditions of* Lemma 5.2 *be fulfilled. If*

$$\begin{cases} u_0(x) \equiv 0 \text{ in } \Omega(s_0) \equiv \Omega \cap \{s_0 < x_1 < a\}, \\ f \equiv 0 \text{ in } Q(s_0, T) \equiv \Omega(s_0) \times (0, T), \end{cases}$$

*and*

$$M = \|u_0\|_{2,\Omega}^2 + \|f\|_{2,Q}^2$$

*is sufficiently small, then every solution $u(x,t)$ of problem* (5.36) *is localized in the direction $x_1$: there exists $s' \geq s_0$ such that*

$$u(x,t) \equiv 0 \text{ in } Q(s', T) \equiv \Omega(s') \times [0, T].$$

*The value of $s'$ is independent of $T$.*

*Proof* By virtue of (5.54) the energy function $E(s,t)$ satisfies the conditions

$$\begin{cases} CE^{\frac{1}{\alpha}} + E_s \leq 0 & \text{for } s > s_0, \\ 0 \leq E(s,t) \leq M & \text{for all } (s,t) \in (0,a) \times (0,T), \\ E_s \leq 0 \end{cases}$$

with the exponent $\alpha > 1$ defined in (5.53). The straightforward integration leads to the inequality

$$E^{1-\frac{1}{\alpha}}(s,t) = E^{1-\frac{1}{\alpha}}(s_0,t) - \frac{C\alpha}{\alpha - 1}(s - s_0) \leq M^{1-\frac{1}{\alpha}} - \frac{C\alpha}{\alpha - 1}(s - s_0).$$

It follows that for all $t \in [0, T]$

$$E(s', t) = 0 \quad \text{for } s' = s_0 + \frac{\alpha - 1}{C\alpha} M^{1-\frac{1}{\alpha}}.$$

Let us now claim that $s' < a$. Due to the energy estimate (4.12) this condition is always fulfilled if $M$ is sufficiently small. □

### 5.3.3 Infinite Waiting Time

**Theorem 5.6** *Let the conditions of* Theorem 5.5 *be fulfilled and, in addition, let us assume that the integral*

$$K(s) \equiv \int_s^{s_0} \frac{\Phi(z)}{(z - s_0)^{\alpha/(\alpha-1)}} dz, \quad \Phi(s) = \|u_0\|_{2,\Omega(s)}^{\frac{2}{\alpha}} + \|f\|_{L^1(0,T;L^2(\Omega(s)))}^{\frac{2}{\alpha}},$$

*is convergent. Then there exists a constant* $\varepsilon > 0$ *such that for*

$$M = \|u_0\|^2_{2,\Omega} + \|f\|^2_{2,Q} < \varepsilon,$$

*every solution of problem* (5.36) $u \in \mathbf{W}(Q)$ *possesses the property of infinite waiting time on the plane* $\Omega \cap \{x_1 = s_0\}$*:*

$$u(x,t) \equiv 0 \quad in\ Q \cap \{x_1 \geq s_0\}\ for\ all\ t \geq 0.$$

*Proof* The energy function $E$ satisfies the differential inequality (5.54). Let us estimate the first term on the right–hand side of this inequality in the following way: by Hölder's and Young's inequalities

$$\begin{aligned} 2\int_{Q(s,t)} u\, f\, dxdt &\leq 2\int_0^t \|u(\cdot,t)\|_{2,\Omega}\|f(\cdot,t)\|_{2,\Omega}\, dt \\ &\leq 2\sqrt{\overline{b}(s,t)}\int_0^t \|f(\cdot,t)\|_{2,\Omega}\, dt \\ &\leq \frac{1}{2}\overline{b}(s,t) + 2\,\|f\|^2_{L^1(0,T;L^2(\Omega(s)))}, \end{aligned}$$

which transforms (5.54) into the form

$$E \leq C\,(-E_s)^\alpha + K_1(\|f\|^2_{L^1(0,T;L^2(\Omega(s)))} + \|u_0\|^2_{2,\Omega}).$$

Rising both sides to the power $1/\alpha$, we finally arrive at the following inequality: for $s < s_0$

$$\begin{cases} E^{\frac{1}{\alpha}} + E_s \leq F(s), & F(s) \equiv K_2\left(\|f\|^{\frac{2}{\alpha}}_{L^1(0,T;L^2(\Omega(s)))} + \|u_0\|^{\frac{2}{\alpha}}_{2,\Omega(s)}\right), \\ 0 \leq E \leq M, & E_s \leq 0, \end{cases}$$

with a constant $K_2$ independent of $f$, $u_0$ and $E$. □

The assertion of Theorem 5.6 is a byproduct of the following technical result.

**Lemma 5.3** *If* $F(s) \equiv 0$ *for* $s \geq s_0$, *the integral*

$$K(s_0) = \left(\frac{\alpha}{\alpha-1}\right)^{\frac{\alpha}{\alpha-1}} \int_0^{s_0} \frac{F(\tau)}{(s_0-\tau)^{\frac{\alpha}{\alpha-1}}}\, d\tau$$

*is convergent and*

$$G(s_0) \equiv M - \left(\frac{\alpha-1}{\alpha}s_0\right)^{\frac{\alpha}{\alpha-1}} + K(s_0) \leq 0,$$

*then* $E(s) \equiv 0$ *in* $Q(s,t)$ *for all* $s \geq s_0$ *and* $t \in (0,T)$.

*Proof* Let $S(s)$ be the nonnegative solution of the problem

$$S'(s) + S^{\frac{1}{\alpha}}(s) = 0 \text{ for } s \in (0, s_0), \qquad S(0) = S_0 > 0$$

with $S_0$ to be defined. The function $S(s)$ is given by the explicit formula

$$S(s) = \max\left\{0, \left(S_0^{1-\frac{1}{\alpha}} - \frac{\alpha - 1}{\alpha} s\right)^{\frac{\alpha}{\alpha-1}}\right\}.$$

Let us fix the initial value $S(0)$ by the condition

$$s_0 = \frac{\alpha}{\alpha - 1} S_0^{1-\frac{1}{\alpha}},$$

i.e., $S(s_0) = 0$. Let us consider the function $W(s) \equiv E(s) - S(s)$ which satisfies on the interval $(0, s_0)$ the inequality

$$W'(s) + \frac{1}{\alpha}\int_0^1 \frac{d\lambda}{(\lambda E(s) + (1-\lambda)S(s))^{1-\frac{1}{\alpha}}} W(s) \le F(s).$$

Multiplying this inequality by

$$\exp\left(\frac{1}{\alpha}\int_0^t\int_0^1 \frac{d\lambda ds}{(\lambda E(s) + (1-\lambda)S(s))^{1-\frac{1}{\alpha}}}\right)$$

and then integrating over the interval $(0, s_0)$, we transform it to the form

$$\begin{aligned} E(s_0) &\le E(0) - S(0) \\ &+ \int_0^{s_0} F(s) \exp\left(\frac{1}{\alpha}\int_0^s\int_0^1 \frac{d\lambda\, dz}{(\lambda E(z) + (1-\lambda)S(z))^{1-\frac{1}{\alpha}}}\right) ds. \end{aligned}$$

Let us notice that by virtue of the equation for $S(t)$

$$\begin{aligned} &\frac{1}{\alpha}\int_0^s\int_0^1 \frac{d\lambda\, dz}{(\lambda E(z) + (1-\lambda)S(z))^{1-\frac{1}{\alpha}}} \\ &\quad\le \int_0^s \frac{dz}{S^{1-\frac{1}{\alpha}}(z)}\int_0^1 \frac{\frac{1}{\alpha}d\lambda}{\lambda^{1-\frac{1}{\alpha}}} = \int_0^s \frac{S'(z)}{S(z)}dz = \ln\frac{S(s)}{S(0)}. \end{aligned}$$

It follows that

$$E(s_0) \le E(0) - S(0)\left(1 - \int_0^{s_0} \frac{F(\tau)}{S(\tau)}d\tau\right) \le G(s_0) \le 0.$$

Since $E(s) \geq 0$ for all $s > 0$, it is necessary that $E(s_0) = 0$. Considering now the differential inequality for $E(s)$ on the interval $(s_0, s)$,

$$\begin{cases} E'(s) + E^{\frac{1}{\alpha}}(s) \leq 0 & \text{for } s > s_0, \\ E(s_0) = 0, \qquad E(s) \geq 0, \end{cases}$$

we conclude that these conditions are satisfied only if $E(s) = 0$ for $x_1 \geq s_0$, that is, if $\|u\|_{2,\Omega(s)}(t) \equiv 0$. □

## 5.4 Localized Stationary Solutions

In this section we study the properties of directional localization of stationary solutions to anisotropic diffusion equations of the type (5.3) with variable exponents $p_i(x)$.

Let the conditions of Theorems 4.15 and 4.16 be fulfilled and $u(x)$ be a bounded weak solution of the problem

$$\begin{cases} -\sum_{i=1}^{n} D_i \left( a_i(x,u)|D_i u|^{p_i(x)-2} D_i u \right) + c(x,u)|u|^{\sigma(x)-2} u = f \quad \text{in } \Omega, \\ u = 0 \text{ on } \partial\Omega. \end{cases} \tag{5.55}$$

The domain $\Omega$ is assumed to satisfy the following conditions:

$$\begin{cases} \exists\, L > 0 : \ \{x_1 : x \in \Omega\} = (0, L), \\ \forall s \in (0, L) \text{ the set } \omega(x_1) = \Omega \cap \{x_1 = s\} \text{ is a simple-connected} \\ \text{domain in } \mathbb{R}^{n-1},\ \partial\omega(t) \text{ is Lipschitz–continuous} \\ \exists \kappa \in (0,1) : \kappa\,\lambda(s) \leq \operatorname{diam}\omega(s) \leq \lambda(s), \quad \lambda(0) \geq 0, \quad \lambda(L) \geq 0, \\ \text{where } \lambda(t) \text{ is a given continuous function.} \end{cases} \tag{5.56}$$

Unless specially indicated, we assume that $L < \infty$. The solution $u \in \mathbf{V}_\sigma(\Omega)$ is understood in the sense of Definition 4.4, the coefficients and exponents are subject to conditions (4.120) and (1.31).

Set

$$\phi_k(x,s) = \begin{cases} 1 & \text{for } x_1 > s + \frac{1}{k}, \\ k(x_1 - s) & \text{for } x \in \left[s, s + \frac{1}{k}\right], \\ 0 & \text{for } x_1 < s, \qquad k \in \mathbb{N}, \end{cases}$$

and choose the function $u(x)\phi_k(x,s)$ for the test-function in the integral identity (4.122). The resulting identity has the form

$$\sum_{j=1}^{4} I_j(k,s) \equiv \sum_i \int\limits_{\Omega\cap\{x_1>s+1/k\}} a_i|D_i u|^{p_i}\,\phi_k\,dx$$
$$+k \int\limits_{\Omega\cap\{s<x_1<s+1/k\}} a_1\,u\,|D_1 u|^{p_1-2} D_1 u\,dxdt$$
$$+\int\limits_{\Omega\cap\{x_1>s\}} c(x)|u|^{\sigma(x)}\phi_k\,dx - \int\limits_{\Omega\cap\{x_1>s\}} f\,u\,\phi_k\,dx = 0. \tag{5.57}$$

The inclusion $u \in \mathbf{V}(\Omega)$ yields the inclusions

$$a_i|D_i u|^{p_i}\,\phi_k,\; c(x)|u|^{\sigma(x)}\phi_k,\; f\,u\,\phi_k \in L^1(\Omega),$$

which allow one to pass to the limit when $k \to \infty$ in $I_1$, $I_3$ and $I_4$:

$$\lim_{k\to\infty} I_1 = \sum_i \int_{\Omega\cap\{x_1>s\}} a_i|D_i u|^{p_i}\,dxdt,$$
$$\lim_{k\to\infty} I_3 = \int_{\Omega\cap\{x_1>s\}} c(x)|u|^{\sigma(x)}\,dx, \quad \lim_{k\to\infty} I_4 = -\int_{\Omega\cap\{x_1>s\}} f\,u\,dx.$$

Now notice that by virtue of (5.57) $I_2$ is bounded uniformly with respect to $k$, provided that so are the integrals $I_1$, $I_2$, $I_3$. Writing $I_2$ in the form

$$I_2 = k\int_s^{s+1/k} dx_1 \int_{\omega(x_1)} a_i\,u\,|D_i u|^{p_i-2} D_i u\,dx'$$

and applying the Lebesgue theorem we conclude that there exists

$$\lim_{k\to\infty} I_2(k,s) = \int_{\omega(s)} a_i\,u\,|D_i u|^{p_i-2} D_i u\,dx'.$$

Let us assume that

$$f(x) = 0 \quad \text{a.e. in } \Omega \cap \{x : x_1 > l\} \text{ with some } l \in (0, L). \tag{5.58}$$

Then the energy relation takes on the form

$$\forall s > l \quad \sum_i \int_s^L dx_1 \int_{\omega(x_1)} a_i|D_i u|^{p_i}\,dx' + \int_s^L \int_{\omega(s)} c\,|u|^{\sigma(x)}\,dx'$$
$$= -\int_{\omega(s)} a_1\,u\,|D_1 u|^{p_1-2} D_1 u\,dx'. \tag{5.59}$$

Let us introduce the function

$$J \equiv A_1 \int_{\omega(s)} |D_1 u|^{p_1(x)-1} |u| \, dx' \geq \left| \int_{\omega(s)} a_1 \, u \, |D_1 u|^{p_1-2} D_1 u \, dx' \right|.$$

By Hölder's inequality

$$J \leq A_1 \left\| |D_1 u|^{p_1(x)-1} \right\|_{\frac{p_1(\cdot)}{p_1(\cdot)-1}, \omega(s)} \|u\|_{p_1(\cdot),\omega(s)}.$$

By virtue of (1.6)

$$\begin{aligned} \left\| |D_1 u|^{p_1(x)-1} \right\|_{\frac{p_1(\cdot)}{p_1(\cdot)-1}, \omega(s)} &\leq \max \left\{ A_{p_1(\cdot),\omega(s)}^{\frac{p_1^- -1}{p_1^-}}(D_1 u), \ A_{p_1(\cdot),\omega(s)}^{\frac{p_1^+ -1}{p_1^+}}(D_1 u) \right\} \\ &= A_{p_1(\cdot),\omega(s)}^{\gamma_1(s)}(D_1 u) \end{aligned} \tag{5.60}$$

with the exponent

$$\gamma_1(s) = \begin{cases} \dfrac{p_1^- - 1}{p_1^-} & \text{if } A_{p_1(\cdot),\omega(s)}(D_1 u) < 1, \\ \dfrac{p_1^+ - 1}{p_1^+} & \text{otherwise.} \end{cases}$$

The second factor can be estimated by means of Lemma 1.14:

$$\|u\|_{p_1(\cdot),\omega(s)} \leq C \max \left\{ \lambda^{1+\frac{n-1}{\beta}-\frac{n-1}{p_1^-}}(s), \ \lambda^{1+\frac{n-1}{\beta}-\frac{n-1}{p_1^+}}(s) \right\} \|\widetilde{\nabla} u\|_{\beta,\omega(s)} \tag{5.61}$$

with $\beta = const > 0$ such that $\dfrac{1}{p_1(x)} > \dfrac{1}{\beta} - \dfrac{1}{n-1}$ in $\Omega$ if $\beta < n-1$. By $\widetilde{\nabla}$ we denote the gradient with respect to the variables $(x_2, \ldots, x_n)$ on the plane $x_1 = const$. Let us claim that $\beta = \min\limits_{j \neq 1} p_j^-$. Then

$$\begin{aligned} \|\widetilde{\nabla} u\|_{\beta,\omega(s)} &\leq 2 \sum_{j \neq 1} \|1\|^{\frac{1}{\beta}}_{\frac{p_j(\cdot)}{p_j(\cdot)-\beta},\omega(s)} \| |D_j u|^{\beta} \|^{\frac{1}{\beta}}_{\frac{p_j(\cdot)}{\beta},\omega(s)} \\ &\leq 2 \sum_{j \neq 1} \max \left\{ \lambda(s)^{\frac{n-1}{\beta}-\frac{n-1}{p_j^-}}, \ \lambda(s)^{\frac{n-1}{\beta}-\frac{n-1}{p_j^+}} \right\} \\ &\qquad \times \max \left\{ A_{p_j(\cdot),\omega(s)}^{\frac{1}{p_j^-}}(D_j u), \ A_{p_j(\cdot),\omega(s)}^{\frac{1}{p_j^+}}(D_j u) \right\}. \end{aligned}$$

Notice that for every $j = 2, \dots, n$

$$\max\left\{\lambda(s)^{\frac{n-1}{\beta}-\frac{n-1}{p_j^-}},\ \lambda(s)^{\frac{n-1}{\beta}-\frac{n-1}{p_j^+}}\right\} \le \max\left\{\lambda(s)^{\frac{n-1}{\beta}-\frac{n-1}{p^-}},\ \lambda(s)^{\frac{n-1}{\beta}-\frac{n-1}{p^-}}\right\}.$$

Fix an arbitrary $s \in (l, L)$ and choose $k \in \{2, \dots, n\}$ such that

$$A_{p_k(\cdot),\omega(s)}(D_k u) \ge A_{p_j(\cdot),\omega(s)}(D_j u) \quad \text{for all } j \ge 2.$$

There are two possibilities: either $A_{p_k(\cdot),\omega(s)}(D_k u) < 1$, or $A_{p_k(\cdot),\omega(s)}(D_k u) \ge 1$. In the former case the inequality

$$\sum_{j\neq 2} A^{\frac{1}{p_j^-}}_{p_j(\cdot),\omega(s)}(D_j u) \le n A^{\frac{1}{p^+}}_{p_k(\cdot),\omega(s)}(D_k u) \le n\left(\sum_{j\neq 2} A_{p_j(\cdot),\omega(s)}(D_j u)\right)^{\frac{1}{q^+}}$$

holds with $q^+ = \max_{j\neq 1} p_j^+$, otherwise

$$\sum_{j\neq 2} A^{\frac{1}{p^-}}_{p_k(\cdot),\omega(s)}(D_k u) \le n A^{\frac{1}{p^-}}_{p_k(\cdot),\omega(s)}(D_k u) \le n\left(\sum_{j\neq 2} A_{p_j(\cdot),\omega(s)}(D_j u)\right)^{\frac{1}{q^-}}$$

with $q^- = \max_{j\neq 1} p_j^-$. Thus, $J \le K\,\rho(s)\, n\, A^{\theta(s)}_{\mathbf{p}(\cdot),\omega(s)}(\nabla u)$ with the exponent and the coefficient

$$\theta(s) = \gamma_1(s) + \tau_j(s), \qquad \tau_j(s) = \begin{cases} \dfrac{1}{q^+} & \text{if } \max\limits_{j\ge 2} A_{p_j(\cdot),\omega(s)}(D_j u) < 1, \\ \dfrac{1}{q^-} & \text{otherwise,} \end{cases}$$

$$\rho(s) = \max\left\{\lambda(s)^{1+2\frac{n-1}{\beta}-2\frac{n-1}{p^-}},\ \lambda(s)^{1+2\frac{n-1}{\beta}-2\frac{n-1}{p^+}}\right\},$$

and with an absolute constant $K$ independent of $\lambda(s)$ and $u(x)$. Introducing the energy function

$$\Phi(s) \equiv \int_s^L dx_1 \int_{\omega(s)} A_{\mathbf{p}(\cdot),\omega(s)}(\nabla u)\,dx'$$

we rewrite (5.59) in the form

$$\begin{cases} \forall s > l \qquad \Phi^{\frac{1}{\theta(s)}}(s) + \psi(s)\Phi'(s) \le 0, \quad s \in (l, L), \\ \Phi(L) = 0, \qquad \Phi'(s) \le 0, \qquad \psi(s) = \left((K\,n\,a_0^{-1}\rho(s))\right)^{\frac{1}{\theta(s)}}. \end{cases} \tag{5.62}$$

**Lemma 5.4** *Let the conditions of* Theorems 4.15 *and* 4.16 *be fulfilled, the exponents* $p_i(x)$ *satisfy the oscillation conditions*

$$\frac{1}{p_1(x)} > \frac{1}{\min_{j\neq 1} p_j^-} - \frac{1}{n-1} \quad \text{if } \min_{j\neq 1} p_j^- < n-1 \tag{5.63}$$

*and the function* $f(x)$ *satisfies condition* (5.58). *Then the energy function* $\Phi(s)$ *satisfies the ordinary differential inequality* (5.62).

Let the energy function $\Phi(s)$ be uniformly bounded in the interval $(l, L)$ by a finite constant $M$. Then inequality (5.62) can be written in the form

$$\nu \Phi^{\mu} + \psi(s)\Phi'(s) \leq 0, \quad \mu = \inf_{(l,L)} \frac{1}{\theta(s)}, \quad \nu = \inf_{(l,L)} M^{\frac{1}{\theta(s)}-\mu}. \tag{5.64}$$

Let us claim that $\mu < 1$. This is true if $\theta(s) > 1$, which is guaranteed by the condition $p_1^- > \max_{j\neq 1} p_j^+$.

**Lemma 5.5** *Let the conditions of* Lemma 5.4 *be fulfilled. Assume that the exponents* $p_i(x)$ *satisfy the conditions*

$$\begin{cases} 0 < \dfrac{1}{p_1^+} \leq \dfrac{1}{p_1^-} < \dfrac{1}{p_2^+} & \text{if } n = 2, \\ \dfrac{1}{\min\limits_{j\neq 1} p_j^-} - \dfrac{1}{n-1} < \dfrac{1}{p_1^+} \leq \dfrac{1}{p_1^-} < \sum\limits_{j\neq 1} \dfrac{1}{\max_{j\neq 1} p_j^+} & \text{if } n \geq 3. \end{cases}$$

*Then there exists* $\varepsilon_* > 0$ *such that every solution* $u(x)$ *of problem* (5.55) *with the energy satisfying the inequality* $\Phi(l) \leq \varepsilon_*$ *vanishes on an interval* $(s_0, L) \subset (l, L)$.

*Proof* Let us integrate inequality (5.64) in the limits $(l, s)$,

$$\Phi^{1-\mu}(s) \leq \varepsilon_*^{1-\mu} - \nu(1-\mu) \int_l^s \frac{dt}{M^{1/\theta(t)}\psi(t)}, \tag{5.65}$$

and then take

$$\varepsilon_* < \left( \nu(1-\mu) \int_l^L \frac{dt}{M^{1/\theta(t)}\psi(t)} \right)^{1/(1-\mu)}.$$

The nonnegative and monotone decreasing function $\Phi(s)$ vanishes at a point $s_0 \in (l, L)$, and $\Phi(s) \equiv 0$ for all $s \in [s_0, L]$. □

**Theorem 5.7** *Let the conditions of* Lemma 5.5 *be fulfilled. Then there exists* $\varepsilon_* > 0$ *such that every weak solution of problem* (5.55) *with the energy* $\Phi(s)$ *satisfying the inequality* $\Phi(l) \leq \varepsilon_*$ *is localized in the direction* $x_1$: $u(x) = 0$ *a.e. in* $\Omega \cap \{x_1 > s_0\}$ *with some* $s_0 \in (l, L)$.

*Proof* Since the energy function $\Phi(s)$ satisfies the ordinary differential inequality (5.64), it follows from Lemma 5.5 that $\Phi(s) = 0$ for all $s > s_0$, provided that $\Phi(s)$ is sufficiently small. □

*Remark 5.6* The assertion of Theorem 5.7 holds true independently of $\varepsilon_*$, provided that

$$\int_l^L \frac{dt}{\psi(t)} = \infty.$$

The condition of divergence yields a condition on the rate of vanishing of the function $\psi(s)$ as $s \to L$, which is a restriction on the the shape of the problem domain $\Omega$ near the point $x_1 = L$.

## 5.5 Remarks

The contents of this chapter relies on the results published in [44, 45, 47, 50, 51]. Our approach to the study of space localization is based on a modification of the method of local energy estimates [36]—see the comments to Chap. 3. The properties of finite and infinite speed of propagation in nonnegative solutions of the doubly nonlinear equation with anisotropic diffusion and convection

$$\partial_t(u^m) - \left(|u_x|^{p-2}u_x\right)_x - \left(|u_y|^{q-2}u_y\right)_y \pm |u|^{\alpha-1}u_x = 0$$

with constant $p, q > 1$, $m, \alpha > 0$ are studied in [179] by means of comparison with suitable sub and super-solutions. We refer also to [36, Chap. 3], [35, 113, 180] for the ranges of the constant parameters $p, m, \gamma$ which provide the effect of finite speed of propagation in solutions of isotropic equations of the type

$$\partial_t(|u|^{m-1}u) - \Delta_p u + c|u|^{\gamma-2}u = 0.$$

The proof of the strong maximum principle and a description of the null sets for nonnegative continuous solutions of the anisotropic equations

$$\partial_t u - \sum_{i=1}^{n} D_i\left(|D_i u|^{p_i-2}D_i u\right) = f(u)$$

with constant $p_i$ are given in [256] under the assumption $f(s) \sim s^{p^- -1}$ for $s \sim 0$. The description of the null sets is based on comparison with barrier functions.

### 5.5.1 Equations of General Form

The results of this chapter refer to model Eqs. (5.5) and (5.36). A revision of the proofs shows that all assertions of Sect. 5.2 remain true for the local energy solutions of the equation

$$\partial_t u - \operatorname{div}(\mathscr{A}(z,u,\nabla u)) + \mathscr{C}(z,u) = 0, \tag{5.66}$$

provided that

1. $\mathscr{A}$ and $\mathscr{C}$ are Carathéodory functions satisfying the conditions

$$\begin{aligned} &\text{(a)} \quad \mathscr{A}(z,r,\xi)\cdot\xi \geq a_0|\xi|^{p(z)} \quad \forall z \in Q_{\rho_0},\ r\in\mathbb{R},\ \xi\in\mathbb{R}^n,\\ &\text{(b)} \quad |\mathscr{A}(z,r,\xi)| \leq a_1|\xi|^{p(z)-1},\\ &\text{(c)} \quad \mathscr{C}(z,r)r \geq c_0|r|^{\sigma(z)} - f(z)r \quad \forall z\in Q_{\rho_0},\ r\in\mathbb{R} \end{aligned} \tag{5.67}$$

   with the exponents $p(z), \sigma(z) : Q \mapsto (1,\infty)$, $p(z) \in C(Q_{\rho_0}(x_0))$, a given function $f \in L^2(Q_{\rho_0}(x_0))$ and positive constants $a_0, a_1, c_0$,
2. the solution $u \in L^\infty(0,T;L^2(B_\rho(x_0)))$, $|\nabla u|^{p(z)} \in L^1(Q_\rho(x_0))$, $u_t \in \mathbf{W}'(Q_\rho(x_0))$, satisfies the energy identity

$$\int_\Omega u\,\zeta\,dx\Big|_{t=0}^{t=T} + \int_Q \Big[-u\,\zeta_t + \mathscr{A}(z,u,\nabla u)\cdot\nabla\zeta + \mathscr{C}(z,u)\zeta\Big]\,dz = 0 \tag{5.68}$$

   with the test-function $\zeta$ such that $\zeta \in \mathbf{W}(Q_\rho(x_0))$, $\zeta_t \in \mathbf{W}'(Q_\rho(x_0))$.

Choosing $u(z)$ for the test-function and using the structure conditions (5.67) we arrive at the inequality (cf. with inequality (5.10))

$$\begin{aligned} \frac{1}{2}\int_{B_\rho} u^2(x,\tau)\,dx\Big|_{\tau=0}^{\tau=t} &+ \int_0^t\int_{B_{\rho_0}} \Big(a_0|\nabla u|^{p(z)} + c_0\,|u|^{\sigma(z)} - fu\Big)\,dxdt\\ &\leq a_1\int_0^t\int_{S_\rho(x_0)} |u||\nabla u|^{p(z)-1}\,dS\,dt, \end{aligned}$$

which leads to the differential inequality for the energy function $E+b$ in the case $c_0=0$, or $E+D+b$ if $c_0>0$.

Notice that unlike the existence results, the monotonicity of $\mathscr{A}$ is not required for deriving the localization properties.

The results of Sect. 5.3 extend to the anisotropic counterpart of Eq. (5.66) with constant exponents $p_i > 1$. It suffices to assume the fulfillment of the anisotropic growth conditions

$$\mathscr{A}(z,r,\xi)\cdot\xi \geq a_0\sum_{i=1}^{n}|\xi_i|^{p_i}, \quad |\mathscr{A}(z,r,\xi)| \leq a_1\sum_{i=1}^{n}|\xi_i|^{p_i-1}, \quad a_0, a_1 = const > 0,$$

condition (5.67) (c) with a constant $c_0 \geq 0$, and the inclusion $f \in L^2(Q) \cap L^1(0,T;L^2(\Omega))$.

### 5.5.2 On the Cauchy Problem

A combination of the existence results of Chap. 4 and results on the finite speed of propagation in this chapter allows one to prove that for every finite $T$ the Cauchy problem for Eq. (5.2) has an energy solution, provided that the data $u_0$ and $f$ are compactly supported in a ball $B_r(0)$ and the cylinder $Q_{r,T} = B_r(0)\times(0,T]$. Denote $S_T = \mathbb{R}^n \times (0,T]$ and consider the problem

$$\begin{aligned}&\mathscr{L}u \equiv u_t - \operatorname{div}\left(|\nabla u|^{p(z)-2}\nabla u\right) + c|u|^{\sigma(z)-2}u - f = 0 \quad \text{in } S_T,\\ &u(x,0) = u_0 \text{ in } \mathbb{R}^n\end{aligned} \tag{5.69}$$

with the exponents $p(z)$ and $\sigma(z)$ satisfying the conditions Theorem 4.1 in every cylinder $Q_{R,T}$. Let $v_R$ be the solution of the auxiliary Dirichlet problem

$$\begin{cases}\mathscr{L}v_R = 0 \text{ in } Q_{R,T} \text{ with } R = 1+r \text{ and } T \leq 1,\\ v_R = 0 \text{ for } |x| = R,\ t \in [0,T],\\ v_R(x,t) = u_0 \text{ in } B_R(0).\end{cases} \tag{5.70}$$

Assume that the data of problem (5.70) satisfy the conditions of Theorem 5.2 (or Theorem 5.3). The solution $v_R$ can be taken for the test-function in identity (4.10) for Eq. (5.70), which gives an independent of $R$ and $T$ estimate on the total energy $M$ (estimate (5.8)). The disturbances from the data propagate with finite speed and $\operatorname{supp} v_R \subset B_{r+\rho(t)}(0)$ for every $t \in [0,T]$, with the monotone increasing function $\rho(t)$ given in Theorem 5.2 (or Theorem 5.3). Moreover, $\rho(t)$ is independent of $R$—see Remark 5.4. Now set $R = 1 + (r + \rho(T))$ and define the function

$$u(x,t) = \begin{cases} v_R & \text{in } Q_{R,T},\\ 0 & \text{in } S_T \setminus Q_{R,T}.\end{cases}$$

The function $u(x,t)$ is a solution of the Cauchy problem (5.69) in the following sense:

1. $u \in \mathbf{W}(Q_{d,T}(0)) \cap L^{\infty}(Q_{d,T}(0))$, $u_t \in \mathbf{W}'(Q_{d,T}(0))$ for every $d > R$,
2. $\forall \phi \in \mathbf{W}(Q_{r,T}(0))$, $\phi_t \in \mathbf{W}'(Q_{r,T}(0))$ and $d > \max\{R, r\}$,

$$\int_{Q_{d,T}(0)} \left( u_t \phi + |\nabla u|^{p-2} \nabla u \cdot \nabla \phi + c|u|^{\sigma-2} u \phi \right) dz = 0,$$

3. $u(x,t) \to u_0$ as $t \to 0$ weakly in $L^2(B_d(0))$.

### *5.5.3 Localized Unbounded Solutions: An Example*

The influence of anisotropy on the behavior of solutions can be illustrated by the following simple example. Let us consider the Dirichlet problem for the equation of anisotropic diffusion

$$\begin{cases} u_t = u_{xx} + \left(|u_y|^{q-2} u_y\right)_y & \text{in } Q = \Omega \times (0, T], \\ u(x, y, 0) = u_0(x, y) \text{ in } \Omega, \\ u = 0 \text{ as } x = 0,\ x = a,\ y = b,\ u = \psi(x, t) \text{ for } y = 0 \end{cases} \tag{5.71}$$

with $\Omega = (0, a) \times (0, b)$, $T < \infty$, a constant exponent $q > 2$ and a given function $\psi(x, t)$. Let us construct a solution using separation of variables: $u = X(x, t) \cdot Y(y)$. Substituting this function into the equation and dividing by $|X|^{q-2} X \cdot Y$ we find that

$$\frac{X_t - X_{xx}}{|X|^{q-2} X} = \frac{(|Y'|^{q-2} Y')'}{Y} \quad \text{in } Q \cap \{(x, y, t) : \ X \cdot Y \neq 0\}.$$

Since the left-hand side is independent of $y$ and the right-hand side is independent of $x$ and $t$, it is necessary that there is a constant $\mu$ such that

$$\begin{cases} X_t - X_{xx} = \mu |X|^{q-2} X & \text{in } (0, a) \times (0, T), \\ X(0, t) = X(a, t) = 0, & X(x, 0) = X_0(x) \text{ in } (0, a) \end{cases} \tag{5.72}$$

and

$$(|Y'|^{q-2} Y')' = \mu Y \quad \text{in } (0, b), \qquad Y(b) = 0,\ Y(0) = const.$$

Let us fix an arbitrary $\mu > 0$ and a smooth $X_0(x) > 0$ in $(0, a)$. Under this choice of the data, for the sufficiently small $T$ problem (5.72) has a weak solution, strictly positive in $(0, a) \times (0, T)$, which may blow-up in a finite time (see, e.g., [222, Chap. 2] or Sect. 7.1 below). On the other hand, for every $L \in (0, b)$ the problem for $Y(y)$ admits a solution

$$Y_L(y) = K(L-y)_+^{\alpha} \quad \text{with}\, \alpha = \frac{q}{q-2} > 1, \;\; K = \frac{\mu(q-2)}{2(q-1)\alpha^{q-1}},$$

which is zero for all $y \in [L, b]$. Since the flux is continuous on the line $y = L$,

$$\lim_{y\to L^+} |Y'|^{q-2}Y' = \lim_{y\to L^-} |Y'|^{q-2}Y' = 0,$$

it is easy to check that the formal solution $u = X(x,t)Y(y)$ satisfies Eq. (5.71) in the sense of distributions: $u(x,y,t)$ satisfies the integral identity (4.10) for every smooth test-function $\phi(x,y,t)$, vanishing on the lateral boundaries of $Q$.

Another example of an exact solution is due to [256]. If

$$n-1 > p^-, \qquad \frac{p^+}{p^-} > \frac{n-2}{n-1-p^-}, \qquad n \geq 3,$$

then one may choose a constant $C$ in such a way that the function

$$\mathscr{U}(x) = \frac{C|x_n|^{\frac{p^+}{p^+-1}}}{\left(\sum_{i\neq n} x_i^{\frac{p^-}{p^--1}}\right)^{\frac{p^+-1}{p^+-p^-}}}$$

is a stationary solution of the anisotropic diffusion equation

$$\partial_t u = \sum_{i\neq n} D_i\left(|D_i u|^{p^--2}D_i u\right) + D_n\left(|D_n u|^{p^+-2}D_n u\right)$$

in $S = \{x \in \mathbb{R}^n : x_i > 0, \; i \neq n\} \times (0, T]$. This solution is localized in the direction $x_n$.

# Chapter 6
# Extinction in a Finite Time and the Large Time Behavior

## 6.1 Introduction

This chapter continues the study of the propagation properties of solutions of the Dirichlet problem for the anisotropic parabolic equations with variable nonlinearity

$$\begin{cases} u_t - \sum_{i=1}^{n} D_i \left( a_i |D_i u|^{p_i(z)-2} D_i u \right) + c|u|^{\sigma(z)-2} u = f(z) & \text{in } Q, \\ u = 0 \text{ on } \Gamma, \qquad u(x,0) = u_0(x) \text{ in } \Omega. \end{cases} \tag{6.1}$$

The coefficients $a_i$ and $c$ may depend on the independent variables and the solution $u$. We are interested now in the property of localization in time, that is, in the possibility of complete vanishing of solutions in a finite time. For the model equation (6.1) with constant exponents $p_i = p$ and $\sigma$ such an effect takes place if $\sigma + 1 < p$—see, e.g., [36, Chap. 2]. The situation is more complicated if the diffusion is anisotropic and variable. Let us illustrate the further results by the example of the model equation with two independent space variables:

$$\begin{cases} u_t = \left( |u_x|^{p(z)-2} u_x \right)_x + \left( |u_y|^{q(z)-2} u_y \right)_y + c_0 |u|^{\sigma(z)-2} u + f(z) & \text{in } Q, \\ u = 0 \text{ on } \Gamma, \\ u(x,0) = u_0(x) \text{ in } \Omega = (0,a) \times (0,a). \end{cases} \tag{6.2}$$

1. *Vanishing in a finite time.* Let $f(z) \equiv 0$ for all $t \geq t_f$. Then every (energy) solution of problem (6.2) vanishes in a finite time $t_* \geq t_f$ if

$$\frac{1}{\nu^+(t)} \equiv \frac{1}{\sigma^+(t)} + \frac{1}{2} \left( \frac{1}{p^+(t)} + \frac{1}{q^+(t)} \right) > 1 \quad \text{in } (0, T)$$



S. Antontsev and S. Shmarev, *Evolution PDEs with Nonstandard Growth Conditions*, Atlantis Studies in Differential Equations 4,
DOI 10.2991/978-94-6239-112-3_6

and the oscillation of the exponents of nonlinearity in $\Omega$ is appropriately small: for every $t \in (0, T)$

$$\frac{1}{\nu^-(t)} \equiv \frac{1}{\sigma^-(t)} + \frac{1}{2}\left(\frac{1}{p^-(t)} + \frac{1}{q^-(t)}\right) \le \frac{3}{2}.$$

The functions $\sigma^{\pm}(t)$, $p^{\pm}(t)$ are defined in (1.1).

2. *Vanishing at a prescribed moment.* Let us additionally assume that in the above conditions

$$\int_\Omega |f(x,y,t)|^{\sigma(x,y,t)}\,dxdy \begin{cases} \equiv 0 & \text{as } t \ge t_f, \\ \le C\,(t_f - t)^\mu & \text{for } t \in (0, t_f), \end{cases} \qquad C,\ \mu = const > 0,$$

with a suitably big exponent $\mu$. Then every (energy) solution of problem (6.1) vanishes at the instant $t = t_f$, provided that $C$ and $\|u_0\|_{2,\Omega}$ are sufficiently small.

3. *Vanishing of solutions of eventually linear equations.* Although the effect of vanishing in a finite time is never displayed by the solutions of the linear parabolic equations, it may happen that Eq. (6.1) with variable nonlinearity transforms into the linear one as $t \to \infty$ and nonetheless possesses solutions localized in time. A condition sufficient for such an effect can be formulated as a restriction on the rate of vanishing of $\nu^+(t) - 1$ as $t \to \infty$: if $f \equiv 0$ and $\|u_0\|_{2,\Omega} \le 1$, then every solution of problem (6.2) vanishes at a finite moment, provided that

$$\int_0^\infty \|u_0\|_{2,\Omega}^{2(\nu^+(t)-1)}\,dt = \infty \qquad \text{and} \qquad \int_0^\infty \frac{dt}{e^{t(1-\nu^+(t))}} < \infty.$$

Notice that these conditions are surely fulfilled if $\nu^+(t) \le \nu_0 < 1$.

4. *Large time behavior.* In case that the sufficient conditions of time localization are not fulfilled, we study the behavior of the norm $\|u(\cdot,t)\|_{2,\Omega}$ as $t \to \infty$. In dependence on the properties of the data, we establish the conditions of power or exponential decreasing of the $L^2(\Omega)$-norm of the solution.

The property of extinction in a finite time can be observed in solutions of equations of anisotropic diffusion even in the absence of absorption. Moreover, is turns out that for equations of anisotropic diffusion the effects of finite time extinction and directional space localization studied in Chap. 5 can be displayed simultaneously. For example, let $u(x,y,t)$ be an energy solution of the problem with two independent space variables and constant exponents $p$ and $q$:

$$\begin{cases} u_t = \left(|u_x|^{p-2}u_x\right)_x + \left(|u_y|^{q-2}u_y\right)_y & \text{in } Q, \\ u = 0 \text{ on } \Gamma, \\ u(x,0) = u_0(x) \text{ in } \Omega = (0,a)\times(0,a). \end{cases} \tag{6.3}$$

Let $u_0(x, y) \equiv 0$ in $\Omega \cap \{x > x_0\}$. Gathering the results of Chap. 5 and the assertions given below we find that every energy solution of this problem:

1. is localized in the direction $x$ if
$$1 < q < p;$$
2. vanishes in a finite time if
$$\frac{1}{p} + \frac{1}{q} > 1;$$
3. has the infinite waiting time on the plane $\{x = x_0\}$ if $1 < q < p$ and
$$F(x) \equiv \int_0^a |u_0(x, y)|^2 \, dy \le \varepsilon \max\left\{0, (x_0 - x)^\beta\right\} \quad \text{for } x < x_0$$
with a suitably small constant $\varepsilon > 0$ and $\beta \equiv \beta(p, q) > 1$;
4. in the borderline case
$$\frac{1}{p} + \frac{1}{q} = 1$$
the solution decreases exponentially as $t \to \infty$:
$$\|u(\cdot, t)\|_{2,\Omega}^2 \le \mathrm{e}^{-Ct} \|u_0\|_{2,\Omega}^2, \quad C = const > 0.$$

One easily sees that for certain $p$ and $q$ the properties of directional space localization and extinction in a finite time take place simultaneously. The same is true in the case $n \ge 3$, but the relations between the exponents $p_i$ that allow the simultaneous presence of these localization effects become more complicated and depend on the space dimension.

## 6.2 The Energy Relations

It is assumed that the coefficients $a_i(z, u)$, $c(z, u)$ depend on $z = (x, t)$ and $u(z)$, and are Carathédory functions such that

$$\forall\,(z, r) \in \overline{Q} \times \mathbb{R} \quad \begin{array}{l} 0 < a_0 \le a_i(z, r) \le A_0 < \infty, \\ 0 \le c_0 \le c(z, r) \le A_0 \end{array} \qquad a_0,\ c_0,\ A_0 = const. \tag{6.4}$$

The exponents $p_i(\cdot)$, $\sigma(\cdot)$ are defined on $Q$ and satisfy the conditions:

$$\begin{cases} 1 < p^- \leq p_i^-(t) \leq p_i(z) \leq p_i^+(t) \leq p^+ < \infty, \\ 1 < \sigma^- \leq \sigma^-(t) \leq \sigma(z) \leq \sigma^+(t) \leq \sigma^+ < \infty \end{cases} \tag{6.5}$$

with $p_i^{\pm}(t)$, $p^{\pm}$ and $\sigma^{\pm}(t)$, $\sigma^{\pm}$ introduced in accordance with convention (1.1). The solution of problem (6.1) is understood in the sense of Definition 4.1.

The constants $p^-$, $p^+$ are chosen so that $1 < p^- \leq p_i^-(t) \leq p_i^+(t) \leq p^+ < \infty$ for every $i = 1, \ldots, n$. The existence of an energy solution follows from Theorem 4.1. Let us recall the notation $Q_t = \Omega \times (0, t)$.

**Lemma 6.1** *Let conditions* (1.31) *and* (6.4) *be fulfilled. Every solution* $u \in \mathbf{W}(Q)$ *of problem* (6.1) *satisfies the energy identity: for every* $t \in [0, T]$

$$\frac{1}{2}\int_{\Omega} u^2(x,\tau)dx\bigg|_{\tau=0}^{\tau=t} + \sum_{i=1}^{n}\int_{Q_t} a_i(z,u)|D_i u|^{p_i(z)}\,dz + \int_{Q_t} c(z,u)|u|^{\sigma(z)}\,dz = \int_{Q_t} u\, f\, dz. \tag{6.6}$$

Identity (6.6) follows from Definition 4.1 if we take in (4.10) $\phi = u$ for the test-function, apply the formulas of integration by parts of Corollary 1.10 and use Lemma 4.5.

### *6.2.1 Estimate on the Total Energy*

The study of the energy functions $\Theta(t)$, $\Lambda(t)$, $\Lambda_\sigma(t)$ defined in (1.46) relies on the following uniform in $t$ estimates.

**Lemma 6.2** *Let conditions* (1.31), (6.4), (6.5) *be fulfilled,* $c_0 \geq 0$ *and* $u_0 \in L^2(\Omega)$.

1. *If* $f \in L^2(Q)$, *then every solution* $u \in \mathbf{W}(Q)$ *of problem* (6.1) *satisfies the estimate*

$$\frac{1}{4}\operatorname*{ess\,sup}_{(0,T)} \Theta(t) + a_0\int_0^T \Lambda(t)\,dt \leq \frac{1}{2}\Theta(0) + 4\left(\int_0^T \|f(\cdot,t)\|_{2,\Omega}\,dt\right)^2. \tag{6.7}$$

2. *If*

$$p^-(t) \geq \frac{2n}{n+2} \quad \textit{and} \quad \|f(\cdot,t)\|_{2,\Omega}^{(p^-(t))'} \in L^1(0,T),$$

*then*

$$\frac{1}{4}\operatorname*{ess\ sup}_{(0,T)} \Theta(t) + \frac{a_0}{2}\int_0^T \Lambda(t)\,dt$$
$$\le \frac{1}{2}\Theta(0) + C\int_0^T \left(\|f(\cdot,t)\|_{2,\Omega}^{(p^+)'(t)} + \|f(\cdot,t)\|_{2,\Omega}^{(p^-)'(t)}\right) dt \tag{6.8}$$

*with a constant* $C \equiv C(a_0, n, |\Omega|)$.

*Proof* 1. It follows from the energy relation (6.6) that

$$\frac{1}{2}\operatorname*{ess\ sup}_{t\in(0,T)} \int_{\Omega(0)} u^2(x,t)dx + a_0\sum_{i=1}^{n}\int_Q |D_i u|^{p_i(z)}\,dz + c_0\int_Q |u|^{\sigma(z)}\,dz$$
$$\le \int_Q |u|\,|f|\,dz + \frac{1}{2}\int_\Omega u_0^2\,dx. \tag{6.9}$$

Using Hölder's and Young's inequalities, we estimate

$$\begin{aligned}\int_Q u\,f\,dz &\le \int_0^T \|u(\cdot,t)\|_{2,\Omega}\|f(\cdot,t)\|_{2,\Omega}\,dt\\ &\le \operatorname*{ess\ sup}_{t\in(0,T)} \|u\|_{2,\Omega}\int_0^T \|f(\cdot,t)\|_{2,\Omega}\,dt\\ &\le \frac{1}{4}\operatorname*{ess\ sup}_{t\in(0,T)} \|u\|_{2,\Omega}^2 + 4\left(\int_0^T \|f(\cdot,t)\|_{2,\Omega}\,dt\right)^2,\end{aligned}$$

and (6.7) follows.

2. We apply Lemma 1.28 and Young's inequality:

$$\begin{aligned}\int_Q u\,f\,dz &\le \int_0^T \|u(\cdot,t)\|_{2,\Omega}\|f(\cdot,t)\|_{2,\Omega}\,dt\\ &\le C\int_0^T \max\left\{\Lambda^{\frac{1}{p^+(t)}}(t),\ \Lambda^{\frac{1}{p^-(t)}}(t)\right\}\|f(\cdot,t)\|_{2,\Omega}\,dt\\ &\le \frac{a_0}{2}\int_0^T \Lambda(t)\,dt + C'\int_0^T \max\left\{\|f(\cdot,t)\|_{2,\Omega}^{(p^+)'(t)},\ \|f(\cdot,t)\|_{2,\Omega}^{(p^-)'(t)}\right\} dt.\end{aligned}$$

□

**Lemma 6.3** *Let conditions* (1.31), (6.4) *and* (6.5) *be fulfilled. If* $c_0 > 0$ *and* $f \in L^{\sigma'(\cdot)}(Q)$, *then the solution of problem* (6.1) $u \in \mathbf{W}(Q) \cap L^{\sigma(\cdot)}(Q)$ *satisfies the estimate*

$$\frac{1}{2} \operatorname*{ess\ sup}_{t \in (0,T)} \Theta(t) + \frac{1}{2} \min\{a_0, c_0\} \int_0^T \Lambda_\sigma(t)\, dt \le \frac{1}{2} \Theta(0) + \left(\frac{2}{\min\{2;\ c_0\}}\right)^{\frac{1}{\sigma^- - 1}} \int_{Q_T} |f|^{\sigma'(z)}\, dz. \tag{6.10}$$

*Proof* The assertion follows from (6.9) and Young's inequality:

$$|u|\,|f| \le \frac{c_0}{2} |u|^{\sigma(z)} + \left(\frac{2}{c_0}\right)^{\frac{\sigma'(z)}{\sigma(z)}} |f|^{\sigma'(z)} \le \frac{c_0}{2} |u|^{\sigma(z)} + \left(\frac{2}{\min\{2;\ c_0\}}\right)^{\frac{1}{\sigma(z)-1}} |f|^{\sigma'(z)},$$

whence

$$\int_Q |u\, f|\, dz \le \frac{c_0}{2} \int_{Q_T} |u|^{\sigma(z)}\, dz + \left(\frac{2}{\min\{2;\ c_0\}}\right)^{\frac{1}{\sigma^- - 1}} \int_{Q_T} |f|^{\sigma'(z)}\, dz.$$

□

## 6.3 Differential Inequality for the Energy Function

Let $u(z) \in \mathbf{W}(Q)$ be a solution of problem (6.1). By virtue of (6.6), for every $t,\ t + \Delta t \in [0, T]$ the solution satisfies the identity

$$\frac{1}{2} \int_\Omega u^2(z)\Big|_{\tau=t}^{\tau=t+\Delta t} dx + \int_t^{t+\Delta t} \int_\Omega \left( \sum_i a_i |D_i u|^{p_i(z)} + c\, |u|^{\sigma(z)} \right) dz = \int_t^{t+\Delta t} \int_\Omega f\, u\, dz. \tag{6.11}$$

**Lemma 6.4** *Let in the conditions of* Lemma 6.1 $c_0 > 0$. *Then the functions* $\Theta(t)$, $\Lambda(t)$ *satisfy the differential inequality*

$$\frac{1}{2}\Theta'(t) + \min\{a_0, c_0\}\, \Lambda_\sigma(t) \le \int_\Omega |f\, u|\, dx \quad \forall a.e.\ t \in (0, T). \tag{6.12}$$

*Proof* For every $t,\ t+\Delta t \in [0, T]$ equality (6.11) gives the inequality

$$\frac{1}{2|\Delta t|}\Theta(\tau)\Big|_{\tau=t}^{\tau=t+\Delta t} \le -\frac{1}{|\Delta t|}\min\{a_0,\ c_0\}\int_t^{t+\Delta t} \Lambda_\sigma(\tau)\,d\tau + \frac{1}{|\Delta t|}\int_t^{t+\Delta t}\int_\Omega |f\,u|\,dz. \tag{6.13}$$

For $u \in \mathbf{W}(Q_T)$

$$\int_\Omega |D_i u|^{p_i(z)}\,dx,\ \int_\Omega |u|^{\sigma(z)}\,dx,\ \int_\Omega |u\,f|\,dx \in L^1(0,T),$$

whence for a.e. $t \in (0, T)$ every term on the right-hand side of (6.13) has a limit as $\Delta t \to 0$. It follows that there exists a limit of the left-hand side as $\Delta t \to 0$, whence

$$\frac{1}{2}\Theta'(t) + \min\{a_0, c_0\}\int_\Omega\left(\sum_{i=1}^n |D_i u|^{p_i(z)} + |u|^{\sigma(z)}\right)dx \le \int_\Omega |f\,u|\,dx.$$

□

**Corollary 6.1** *Let in the conditions of* Lemma 6.4 $c_0 \ge 0$. *Then*

$$\frac{1}{2}\Theta'(t) + a_0\,\Lambda(t) \le \int_\Omega |f\,u|\,dx \quad \forall a.e.\, t \in (0,T).$$

Let us denote

$$\frac{1}{\nu^+(t)} = \frac{1}{\sigma^+(t)} + \frac{1}{p^+(t)}, \qquad \frac{1}{p^+(t)} \equiv \frac{1}{n}\sum_{i=1}^n \frac{1}{p_i^+(t)}.$$

**Lemma 6.5** *Let in the conditions of* Lemma 6.3

$$\frac{1}{\sigma^-(t)} + \frac{1}{p^-(t)} = \frac{1}{\nu^-(t)} \le \frac{n+1}{n} \quad \textit{for}\ \frac{1}{n} < \frac{1}{p^-(t)} \equiv \frac{1}{n}\sum_{i=1}^n \frac{1}{p_i^-(t)}. \tag{6.14}$$

*Then the energy solution* $u \in \mathbf{W}(Q)$ *of problem* (6.1) *for a.e.* $t \in (0, T)$ *satisfies the differential inequality*

$$\frac{1}{2}\Theta'(t) + K_1 L(t)\,\Theta^{\nu^+(t)}(t) \le K_2\int_\Omega |f|^{\sigma'(z)}\,dx \tag{6.15}$$

*with the constants* $C$,

$$M = \operatorname{ess\,sup} \Theta(t), \quad K_1 = \frac{1}{2C}\min\{a_0, c_0\}, \quad K_2 = \left(\frac{2}{\min\{2; c_0\}}\right)^{\frac{1}{\sigma^- - 1}},$$

*from* (1.52), *and the coefficient*

$$L(t) = \begin{cases} 1 & \text{if } M \le 1, \\ M^{\nu^-(t) - \nu^+(t)} & \text{otherwise.} \end{cases} \tag{6.16}$$

*Proof* By Young's inequality

$$\int_\Omega |f\,u|\,dx \le \frac{c_0}{2}\int_\Omega |u|^{\sigma(z)}\,dx + \left(\frac{2}{\min\{2; c_0\}}\right)^{\frac{1}{\sigma^- - 1}} \int_\Omega |f|^{\sigma'(z)}\,dx.$$

Combining this inequality with (6.12) we have

$$\frac{1}{2}\Theta'(t) + \frac{1}{2}\min\{a_0, c_0\}\,\Lambda_\sigma(t) \le \left(\frac{2}{\min\{2; c_0\}}\right)^{\frac{1}{\sigma^- - 1}} \int_\Omega |f|^{\sigma'(z)}\,dx.$$

By Lemma 1.27

$$\begin{aligned} C\Lambda_\sigma(t) &\ge \min\left\{\Theta^{\nu^+(t)}(t),\ \Theta^{\nu^-(t)}(t)\right\} \\ &= \min\left\{M^{\nu^+(t)}\left(\frac{\Theta(t)}{M}\right)^{\nu^+(t)},\ M^{\nu^-(t)}\left(\frac{\Theta(t)}{M}\right)^{\nu^-(t)}\right\} \\ &\ge \left(\frac{\Theta(t)}{M}\right)^{\nu^+(t)} \min\left\{M^{\nu^+(t)}, M^{\nu^-(t)}\right\} \\ &= \Theta^{\nu^+(t)}(t)\min\left\{1,\ M^{\nu^-(t)-\nu^+(t)}\right\} \ge L(t)\,\Theta^{\nu^+(t)}(t). \end{aligned}$$

□

*Remark 6.1* Estimates on the "total energy" $M$ are given in Lemmas 6.2, 6.3.

**Lemma 6.6** *Let* $c_0 \ge 0$ *and let* $\operatorname{ess\,sup} \Theta(t) = M$. *If*

$$\frac{2}{p^-(t)} \le 1 + \frac{2}{n},$$

*then the solution of problem* (6.1) *satisfies the differential inequality*

$$\frac{1}{2}\Theta'(t) + \frac{a_0}{C}L(t)\,\Theta^{\nu^+(t)}(t) \le F(t) \qquad \forall \text{ a.e. } t \in (0, T) \tag{6.17}$$

*with the exponent and the right-hand side*

$$\nu^+(t) = \frac{p^+(t)}{2} \geq \frac{p^-(t)}{2} \geq \max\left\{\frac{n}{n+2}, \frac{1}{2}\right\}, \quad F(t) = K_2 \|f(\cdot,t)\|_{2,\Omega}^{\frac{2\nu^+(t)}{2\nu^+(t)-1}},$$

*the coefficients* $K_1$, *the constant* $C$ *from* (1.52), $L(t)$ *from* Lemma 6.5 *and a finite constant* $K_2 \equiv K_2(K_1, M, C, a_0, p^\pm)$.

*Proof* By virtue of Corollary 6.1 and (1.55)

$$\Theta'(t) + \frac{2\,a_0}{C}\,L(t)\,\Theta^{\nu^+(t)} \leq 2\int_\Omega |f\,u|\,dx.$$

Applying Hölder's and Young's inequalities we estimate the right-hand side

$$\int_\Omega |u|\,|f|\,dx \leq \sqrt{\Theta(t)}\ \|f(\cdot,t)\|_{2,\Omega} \leq \frac{a_0}{2C}\,L(t)\Theta^{\nu^+(t)}(t) + K_2\,\|f(\cdot,t)\|_{2,\Omega}^{\frac{2\nu^+(t)}{2\nu^+(t)-1}},$$

whence the assertion. □

## 6.4 Vanishing in a Finite Time

### *6.4.1 Sufficient Conditions for the Finite Time Extinction*

**Theorem 6.1** *Let* $u_0 \in L^2(\Omega)$, $f \in L^2(Q) \cap L^{\sigma'(\cdot)}(Q)$ *and* $c_0 > 0$. *Assume that the exponents* $\sigma(z)$, $p_i(z)$ *satisfy the conditions of* Lemma 6.5. *If* $f \equiv 0$ *for* $t \geq t_f$, *and if*

$$\frac{1}{\nu^+(t)} = \frac{1}{\sigma^+(t)} + \frac{1}{n}\sum_{i=1}^{n}\frac{1}{p_i^+(t)} \geq \frac{1}{\nu_0} > 1, \quad \nu_0 = const, \tag{6.18}$$

*then there exists* $t_* \geq t_f$ *such that*

$$\|u(\cdot,t)\|_{2,\Omega} \equiv 0\, in\ \Omega\ forall\, t \geq t_*.$$

*The extinction moment* $t_*$ *depends on* $\|u_0\|_{2,\Omega}$, $\|f\|_{\sigma'(\cdot),Q}$, $n$, $p^+$, $p^-$, $\sup \nu(t)$ *and* $\inf \nu(t)$.

*Proof* By Lemma 6.5 the energy function $\Theta(t)$ satisfies the ordinary differential inequality

$$\frac{1}{2}\,\Theta'(t) + K_1 L(t)\Theta^{\nu^+(t)}(t) \leq 0 \quad \text{in}\,(t_f, T). \tag{6.19}$$

By virtue of (6.10) $\Theta(t) \le M$ with a constant $M$ depending only on the data. Let us introduce the new energy function $Z(t) \equiv \Theta(t)/M$ and write (6.19) in the form

$$Z'(t) + 2\,K_1 L(t) M^{\nu^+(t)-1} Z^{\nu^+(t)}(t) \le 0 \quad \text{in}\,(t_f, T), \qquad Z(t_f) \le 1. \tag{6.20}$$

The function $Z(t)$ is nonnegative and monotone decreasing. Let us denote

$$T^* = \sup\{t > 0:\ Z(t) > 0\}.$$

If $T^* = t_f$, the assertion is proven. Let $T^* > t_f$. Since $Z^{\nu^+(t)} \ge Z^{\nu_0}(t)$, inequality (6.20) leads to the differential inequality

$$Z'(t) + 2\,K_1 L(t) M^{\nu^+(t)-1} Z^{\nu_0}(t) \le 0 \quad \text{in}\,(t_f, T^*), \qquad Z(t_f) \le 1.$$

Integration of this inequality gives: for $t \in [t_f, T^*)$

$$\begin{aligned} Z^{1-\nu_0}(t) &\le 1 - 2\,K_1(1-\nu_0) \int_{t_f}^{t} \min\left\{1, M^{\nu^-(\tau)-\nu^+(\tau)}\right\} M^{\nu^+(\tau)-1}\,d\tau \\ &\le 1 - 2\,K_1(1-\nu_0)\max\{1, 1/M\}\,(t - t_f). \end{aligned}$$

The conclusion follows now because $Z(t) \ge 0$. □

The assertion of Theorem 6.1 remains true in the case $c_0 = 0$, but under different conditions on the nonlinearity exponents.

**Theorem 6.2** *Let $c_0 \ge 0$, $u_0 \in L^2(\Omega)$, and let $f$ satisfy the conditions of* Lemma 6.2. *Assume that the exponents $p_i(z)$ satisfy the conditions of* Lemma 6.6. *If $f \equiv 0$ for $t \ge t_f$ and*

$$1 + \frac{2}{n} \ge \frac{2}{p^-(t)} \ge \frac{1}{\nu^+(t)} = \frac{2}{p^+(t)} \ge \frac{1}{\nu_0} > 1, \tag{6.21}$$

*then there exists $t_* \ge t_f$ such that*

$$\|u(\cdot, t)\|_{2,\Omega} \equiv 0\, in\ \Omega\ for\ all\, t \ge t_*.$$

The proof is an imitation of the proof Theorem 6.1: the energy function $\Theta(t)$ satisfies the differential inequality (6.19) with the exponent $\nu^+(t)$ given in (6.21) and is globally bounded by virtue of (6.7).

### 6.4.2 On the Balance Between Slow and Fast Diffusion

Let us discuss the conditions on the exponents $p_i(z)$ which guarantee finite time vanishing of the solutions in the isotropic and anisotropic cases. For the anisotropic equation

$$u_t = \sum_i D_i \left(|D_i u|^{p_i(z)-2} D_i u\right)$$

these conditions are given in Theorem 6.2. It is convenient to interpret this equation as the model of a diffusion process,

$$u_t = \operatorname{div}\left(\mathbf{K}(z, \nabla u)\nabla u\right), \quad z = (x, t) \in Q_T,$$

with the diagonal diffusion matrix $\mathbf{K}$, $\mathbf{K}_{ii}(z, \nabla u) = |D_i u|^{p_i(z)-2}$, $i = 1, \ldots, n$. Given a point $z_0$, the diffusion rate in the direction $x_i$ depends on $p_i(z_0)$ and $D_i u(z_0)$. By convention, we say that at the point $z_0$ the diffusion in the direction $x_i$ is *slow* if $p_i(z_0) > 2$ and *fast* in the opposite case $1 < p_i(z_0) \leq 2$.

In the special case of the isotropic diffusion and constant exponents of nonlinearity, $\mathbf{K}_{ii}(z, \nabla u) = |\nabla u|^{p-2}$, $p = const$, the conditions of vanishing in a finite time reduce to the well–known condition

$$2 > p \geq \frac{2n}{n+2}. \tag{6.22}$$

The former inequality of (6.22) guarantees extinction in a finite time (see [36, Chap. 2, Sect. 2]), while the latter provides the inclusion $W_0^{1,p}(\Omega) \subset L^2(\Omega)$. Moreover, it is known that in the case of constant isotropic diffusion and better regularity of the data this condition can be relaxed and transformed to the inequalities $2 > p > 1$—see, e.g., [36, pp. 79–83]. We also refer to [42, 48] for the case of isotropic and variable nonlinearity.

Let us turn to the case of anisotropic diffusion of variable rate.

$\mathbf{n = 2}$. Condition (6.21) becomes

$$1 < p_1^+(t) < \frac{p_2^+(t)}{p_2^+(t) - 1}, \qquad 1 < p_2^+(t). \tag{6.23}$$

It follows that unlike the isotropic case, the solutions always extinct in a finite time, provided that the diffusion in the direction $x_1$ is fast in comparison with the diffusion in the direction $x_2$. In particular, even in the case of very slow diffusion in the direction $x_2$, i.e. when $p_2^+(t) \to \infty$, the localization effect still takes place if the diffusion in the direction $x_1$ is very fast.

Figure 6.1 illustrates the difference between the equations of anisotropic and isotropic diffusion in the case $n = 2$: the solutions of the former vanish in a finite time if the point $(p_1^+(t), p_2^+(t))$ belongs to the shadowed zone for all $t$, while in the

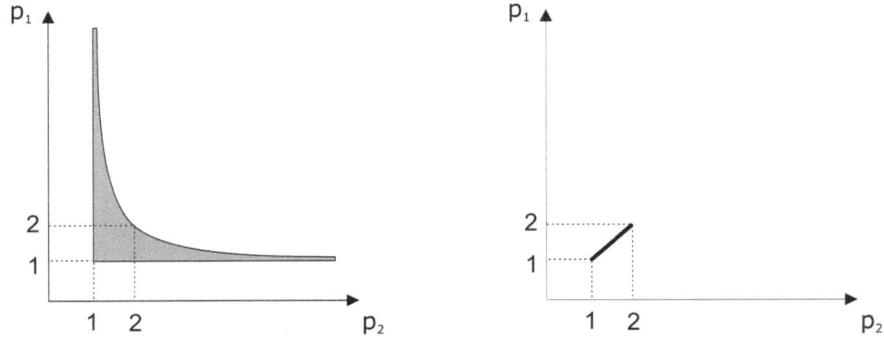


**Fig. 6.1** The admissible ranges of the exponents in the case $n = 2$

latter case the same effect takes place only if the point $(p_1(t), p_2(t))$ belongs to the open interval with the endpoints $(1, 1)$ and $(2, 2)$.

**n = 3**. In this case the solutions of the isotropic diffusion equation vanish in a finite time if

$$\frac{6}{5} < p^+(t) < 2.$$

Let us consider the situation when the diffusion is fast in the directions $x_1, x_2$. For the sake of simplicity we assume that $p_1^+(t) = p_2^+(t)$. The solutions extinct in a finite if

$$\frac{3}{2} < \frac{1}{p_1^+(t)} + \frac{1}{p_2^+(t)} + \frac{1}{p_3^+(t)},$$
$$1 < p_1^+(t) = p_2^+(t) < \frac{4p_3^+(t)}{3p_3^+(t) - 2} \searrow \frac{4}{3} \quad \text{as } p_3^+(t) \nearrow \infty.$$

It follows that if the diffusion in the direction $x_3$ is very slow, the solutions vanish in a finite time, provided that the diffusion in the directions $x_1, x_2$ is suitably fast.

In case of fast diffusion in the direction $x_1$, $(p_1(z) \in (1, 2))$, and slow diffusion of the same rate in the directions $x_2, x_3$, $(p_2(z) = p_3(z) > 2)$, we find that extinction takes place if

$$1 < p_1^+(t) < \frac{2p_2^+(t)}{3p_2^+(t) - 4} \quad \text{and} \quad p_2^+(t) = p_3^+(t) < 4.$$

In the limit case of fast diffusion in the direction $x_1$, $p_1^+(t) \searrow 1$, we arrive at the upper limit of the admissible rate of slow diffusion in the directions $x_2$, $x_3$: $p_2^+(t) = p_3^+(t) \nearrow 4$.

**n ≥ 3**. A similar analysis shows that in the case of higher space dimension the effect of finite time extinction in solutions of anisotropic equations takes place under stronger restrictions on the admissible rate of slow and fast diffusion in different directions.

## 6.5 Vanishing at a Prescribed Moment

**Theorem 6.3** *Let the conditions of* Theorem 6.1 *be fulfilled. Set*

$$S^{1-\nu_0}(t) = 2K_1(1-\nu_0)\int_t^{t_f} \min\{M^{\nu^+(\theta)-1};\ M^{\nu^-(\theta)-1}\}\,d\theta,$$
$$F(t) \equiv 2K_2\int_\Omega |f|^{\sigma'(z)}\,dx$$

*with the constants $K_1$, $K_2$ from the conditions of* Lemma 6.5 *and* $M = \operatorname{ess\,sup}\Theta(t)$. *If the integral*

$$K(t_f) = \int_0^{t_f} \frac{F(\tau)}{S(\tau)}\,d\tau$$

*is convergent and*

$$G(t_f) \equiv \|u_0\|^2_{2,\Omega} - S(0)(M - K(t_f)) \le 0,$$

*then*

$$\|u(\cdot,t)\|_{2,\Omega} \equiv 0 \textit{ in } \Omega \textit{ for all } t \ge t_f.$$

*Proof* By Lemma 6.5 the energy function $\Theta(t)$ satisfies the differential inequality

$$\Theta'(t) + 2K_1\,L(t)\,\Theta^{\nu^+(t)}(t) \le F(t) \quad \text{in } (0,T), \quad \Theta(0) = \|u_0\|^2_{2,\Omega}.$$

Following the proof of Theorem 6.1 we introduce the function $Z(t) \equiv \Theta(t)/M \le 1$. Since $Z \le 1$ and $\nu^+(t) \le \nu_0 < 1$, we have $Z^{\nu^+(t)} \ge Z^{\nu_0}$. The differential inequality for $Z$ can be written in the form

$$Z'(t) + 2K_1\,\mu(t)\,Z^{\nu_0}(t) \le \frac{1}{M}\,F(t) \quad \text{in } (0,T), \quad Z(0) = \frac{1}{M}\,\|u_0\|^2_{2,\Omega} \le 1,$$
$$\mu(t) = L(t)\,M^{\nu^+(t)-1} \equiv \min\{M^{\nu^+(t)-1};\ M^{\nu^-(t)-1}\}.$$

By $S(t)$ we denote the nonnegative solution of the problem

$$S'(t) + 2K_1\,\mu(t)\,S^{\nu_0}(t) = 0, \qquad S(0) = S_0 > 0,$$

which is given by the explicit formula

$$S(t) = \max\left\{0,\ \left(S(0)^{1-\nu_0} - 2K_1(1-\nu_0)\int_0^t \mu(t)\,d\tau\right)^{\frac{1}{1-\nu_0}}\right\}.$$

Let us fix the initial value $S(0)$ by the condition

$$\frac{S^{1-\nu_0}(0)}{2K_1(1-\nu_0)} = \int_0^{t_f} \mu(\tau)\,d\tau,$$

i.e., $S(t_f) = 0$. Notice that under this choice of $S(0)$ the function $S(t)$ can be written in the form

$$S^{1-\nu_0}(t) = 2K_1(1-\nu_0)\int_t^{t_f} \mu(\tau)\,d\tau \quad \text{if}\, t \le t_f.$$

Let us consider the function $W(t) \equiv Z(t) - S(t)$ which satisfies on the interval $(0, t_f)$ the inequality

$$W'(t) + 2K_1\nu_0\,\mu(t)\int_0^1 \frac{d\lambda}{(\lambda Z(t) + (1-\lambda)S(t))^{1-\nu_0}}\,W(t) \le \frac{1}{M}\,F(t).$$

Multiplying this inequality by

$$\exp\left(2K_1\nu_0\int_0^t \mu(\tau)\int_0^1 \frac{d\lambda d\tau}{(\lambda Z(\tau) + (1-\lambda)S(\tau))^{1-\nu_0}}\right)$$

and then integrating over the interval $(0, t_f)$, we arrive at the inequality

$$\begin{aligned} Z(t_f) &\le Z(0) - S(0) \\ &\quad + \frac{1}{M}\int_0^{t_f} F(\tau)\exp\left(2K_1\nu_0\int_0^\tau \mu(z)\int_0^1 \frac{d\lambda\,dz}{(\lambda Z(z) + (1-\lambda)S(z))^{1-\nu_0}}\right)d\tau. \end{aligned}$$

Notice that by virtue of the equation for $S(t)$

$$\begin{aligned} &2K_1\nu_0\int_0^\tau \mu(z)\int_0^1 \frac{d\lambda\,dz}{(\lambda Z(z) + (1-\lambda)S(z))^{1-\nu_0}} \\ &\quad\le 2K_1\int_0^\tau \frac{\mu(z)dz}{S^{1-\nu_0}(z)}\int_0^1 \frac{\nu_0\,d\lambda}{(1-\lambda)^{1-\nu_0}} = -\int_0^\tau \frac{S'(z)}{S(z)}\,dz = -\ln\frac{S(\tau)}{S(0)}. \end{aligned}$$

It follows that

$$Z(t_f) \le Z(0) - S(0)\left(1 - \frac{1}{M}\int_0^{t_f} \frac{F(\tau)}{S(\tau)}d\tau\right) \equiv \frac{1}{M}G(t_f) \le 0,$$

and since $Z(t) \ge 0$ for all $t > 0$, it is necessary that $Z(t_f) = 0$. Considering now the differential inequality for $Z(t)$ on the interval $(t_f, T)$,

$$\begin{cases} Z'(t) + 2K_1\,\mu(t)\,Z^{\nu_0}(t) \le 0 & \text{for } t \ge t_f, \\ Z(t_f) = 0, \qquad Z(t) \ge 0, \end{cases}$$

we conclude that these conditions are satisfied only if $Z(t) \equiv 0$ for $t \ge t_f$, that is, if $\|u\|_{2,\Omega}(t) \equiv 0$. □

*Remark 6.2* It is easy to see that the class of data $\{u_0, f\}$ for which the equation $G(t_f) = 0$ has a solution is nonempty. Let us assume, for example, that $u_0 \equiv 0$, $c_0 \ge 2$ and

$$\int_\Omega |f|^{\sigma'(z)}\,dx \le \varepsilon\,(t_f - t)^\mu \quad \text{for } t \le t_f \ \text{ with an exponent } \mu > \frac{\nu_0}{1-\nu_0}.$$

In this special case $K_2 = \left(\frac{2}{\min\{2;\,c_0\}}\right)^{\frac{1}{\sigma^- - 1}} = 1$ and, by virtue of Lemma 6.3,

$$M = \operatorname*{ess\,sup}_{t\in(0,T)} \Theta(t) \le 2\,K_2 \int_{Q_T} |f|^{\sigma'(z)}\,dz \le \frac{2\,\varepsilon\,t_f^{\mu+1}}{1+\mu} \le 1$$

for all sufficiently small $\varepsilon$. The condition $G(t_f) \le 0$ now reads $K(t_f) \le M$, that is,

$$K(t_f) \le \frac{2\,\varepsilon}{(2\,K_1(1-\nu_0))^{\frac{1}{1-\nu_0}}}\,\frac{t_f^{\mu+1-\frac{1}{1-\nu_0}}}{\mu + 1 - \frac{1}{1-\nu_0}} \le M$$

with the constants $K_1 = \frac{1}{2C}\min\{a_0, c_0\}$ independent of $\varepsilon$, $t_f$ and $M$. The last inequality is surely fulfilled if we claim that $\min\{a_0, c_0\}$ is suitably large.

The next theorem refers to the case $c_0 \ge 0$ and is a byproduct of Theorem 6.3.

**Theorem 6.4** *Let the conditions of* Theorem 6.2 *be fulfilled. If the integral* $K(t_f)$ *with*

$$F(t) \equiv 2K_2\,\|f(\cdot,t)\|_{2,\Omega}^{\frac{2\nu^+(t)}{2\nu^+(t)-1}}$$

*is convergent and* $G(t_f) \le 0$, *then*

$$\|u(\cdot,t)\|_{2,\Omega} \equiv 0 \textit{ in } \Omega \textit{ for all } t \ge t_f.$$

## 6.6 Large Time Behavior

According to Theorems 6.1, 6.2, every solution of problem (6.1) vanishes at a finite moment $t_*$, provided that $\nu^+(t) < 1$ and $f(x,t) \equiv 0$ from some $t_f$ on. Let us study the behavior of $\|u\|_{2,\Omega}(t)$ in the cases when at least one of these two conditions fails.

**Theorem 6.5** *Let the exponents $p_i(z)$ and $\sigma(z)$ satisfy the conditions of* Lemma 6.5 *if $c_0 > 0$, or* Lemma 6.6 *if $c_0 = 0$. Let $\Theta(t) \le M$ for all $t > 0$. Denote*

$$\frac{1}{\nu^+(t)} = \begin{cases} \dfrac{1}{\sigma^+(t)} + \dfrac{1}{n}\displaystyle\sum_{i=1}^{n} \dfrac{1}{p_i^+(t)} & \text{if } c_0 > 0, \\ \dfrac{2}{n}\displaystyle\sum_{i=1}^{n} \dfrac{1}{p_i^+(t)} & \text{if } c_0 = 0. \end{cases}$$

1. *If $\nu^+(t) \le 1$, then*

$$\|u\|^2_{2,\Omega}(t) \le \mathrm{e}^{-2K_1 \min\{1,1/M\} t} \left( \|u_0\|^2_{2,\Omega} + \int_0^t \Phi(\tau) \mathrm{e}^{2K_1 \min\{1,1/M\} \tau} \, d\tau \right)$$

   *with*

$$\Phi(t) = \begin{cases} 2K_2 \displaystyle\int_\Omega |f|^{\sigma'(z)} \, dx & \text{if } c_0 > 0, \\ 2K_2 \, \|f\|_{2,\Omega}^{\frac{2\nu^+(t)}{2\nu^+(t)-1}}(t) & \text{if } c_0 = 0 \end{cases}$$

   *and the constant $K_2$ from* (6.15) *if $c_0 > 0$, or from* (6.17) *if $c_0 = 0$.*
2. *If $\nu_0 \ge \nu^+(t) > 1$ and*

$$\Phi(t) \le f_0 (1+t)^{-\frac{\nu_0}{\nu_0 - 1}}$$

   *from some $t_0$ on, with a positive constant $f_0$, then there exists $C = const$ such that*

$$\|u\|^2_{2,\Omega}(t) \le C(1+t)^{-\frac{\nu_0}{\nu_0-1}} \quad \text{for } t \ge t_0.$$

*Remark 6.3* A byproduct of the first assertion of the theorem is that if $f \equiv 0$ for $t \ge t_*$ and $\nu^\pm(t) = 1$, then

$$\|u\|^2_{2,\Omega} \le \mathrm{e}^{-Ct} \|u_0\|^2_{2,\Omega} \quad \text{as } t \to \infty, \ C = const.$$

*Proof* 1. Since $\Theta(t) \le M$, we may apply (6.16) and continue inequalities (6.15), (6.17) as follows:

$$\Theta'(t) + 2K_1 \min\{1,\ 1/M\}\,\Theta(t) \le \Theta'(t) + 2K_1 L(t) M^{\nu^+(t)-1}\Theta(t)$$
$$\le \Theta'(t) + 2K_1 L(t) M^{\nu^+(t)} \left(\frac{\Theta(t)}{M}\right)^{\nu^+(t)} \le \Phi(t).$$

The conclusion now follows by Gronwall's inequality.

2. Without loss of generality we may assume that $t_0 = 0$. The function $\Theta(t)$ satisfies the differential inequality

$$\begin{aligned}
&\Theta'(t) + 2K_1(1+M)^{1-\nu_0}\Theta^{\nu_0}(t) \\
&\quad \le \Theta'(t) + 2K_1(1+M)^{\nu^+(t)-\nu_0}\Theta^{\nu_0}(t) \\
&\quad \le \Theta'(t) + 2K_1 L(t)(1+M)^{\nu^+(t)}\left(\frac{\Theta(t)}{1+M}\right)^{\nu^+(t)} \\
&\quad = \Theta'(t) + 2K_1 L(t)\Theta^{\nu^+(t)}(t) \le f_0(1+t)^{-\frac{\nu_0}{\nu_0-1}} \quad \text{for } t \ge 0.
\end{aligned}$$

Let $Y(t)$ be a solution of the equation

$$Y'(t) + 2K_1(1+M)^{1-\nu_0}Y^{\nu_0}(t) = f_*(1+t)^{-\frac{\nu_0}{\nu_0-1}}$$

with a constant $f_* > 0$ to be defined. A solution of this equation is given by the explicit formula

$$Y(t) = A(1+t)^{-\frac{1}{\nu_0-1}}$$

with the parameter $A$ chosen from the condition

$$G(A) \equiv -\frac{A}{\nu_0 - 1} + 2K_1(1+M)^{1-\nu_0}A^{\nu_0} - f_* = 0.$$

This algebraic equation has at least one solution $A^* > 0$ because $G(0) = -f_* < 0$, $G(\infty) = \infty$. Moreover, the solution $A^*$ is estimated from below

$$A^* = \left[\frac{(1+M)^{\nu_0-1}}{2K_1}\left(\frac{A^*}{\nu_0-1} + f_*\right)\right]^{\frac{1}{\nu_0}} \ge \left[\frac{(1+M)^{\nu_0-1}}{2K_1} f_*\right]^{\frac{1}{\nu_0}}.$$

If we claim that $f_* \ge \max\left\{f_0, 2K_1(1+M)^{1-\nu_0}\Theta^{\nu_0}(0)\right\}$, then

$$\Theta(0) \le Y(0) = A^* \quad \text{and} \quad f_* \ge f_0.$$

Let us introduce the function $W(t) = \Theta(t) - Y(t)$. The function $W(t)$ satisfies the linear differential inequality

$$W'(t) + D\,W(t) \le (f_0 - f_*)(1+t)^{-\frac{\nu_0}{\nu_0-1}} \le 0 \quad \text{for } t > 0, \qquad W(0) \le 0,$$

with the coefficient

$$\begin{aligned} D &= \frac{2K_1}{(1+M)^{\nu_0-1}} \frac{\Theta^{\nu_0}(t) - Y^{\nu_0}}{\Theta(t) - Y} \\ &\equiv \nu_0 \frac{2K_1}{(1+M)^{\nu_0-1}} \int_0^1 (\lambda\Theta(t) + (1-\lambda)Y(t))^{\nu_0-1}\, d\lambda \ge 0. \end{aligned}$$

It follows that $W(t) \le 0$ for all $t > 0$:

$$0 \le \Theta(t) \le Y(t) = A^*(1+t)^{-\frac{1}{\nu_0-1}}.$$

□

## 6.7 Limit Cases

Let us now turn to the situation when in inequalities (6.15), (6.17)

$$\nu^+(t) \nearrow 1 \quad \text{as } t \to \infty.$$

If this happens, the arguments based on comparison with solutions of equations with constant exponents of nonlinearity are no longer valid because the nonlinear differential inequality (6.15) eventually transforms into the linear one. We will rely on the following assertion.

**Lemma 6.7** *Let a nonnegative function $\Theta(t)$ satisfy the conditions*

$$\begin{cases} \Theta'(t) + C\,\Theta^{\mu(t)}(t) \le 0 \quad \textit{for a.e. } t \ge 0, \\ \Theta(t) \le \Theta(0) < \infty, \quad \Theta(0) > 0, \end{cases} \tag{6.24}$$

*with $\mu(t) \in (0,1)$ and $C = const > 0$. If the exponent $\mu(t)$ is monotone increasing, then $\Theta(t) \equiv 0$ for all $t \ge t_*$ with $t_*$ defined from the equality*

$$C \int_0^{t_*} \Theta^{\mu(s)-1}(0)\, ds = \int_0^\infty \frac{dz}{e^{z(1-\mu(z))}}.$$

*Proof* Let us consider the function $J(t) = \Theta(t)/\Theta(0)$, satisfying the conditions

$$\forall \text{ a.e. } t > 0 \quad J'(t) + C\Theta^{\mu(t)-1}(0)J^{\mu(t)} \leq 0, \quad J(0) = 1, \quad J'(t) \leq 0.$$

Introducing the new independent variable $\tau = C\int_0^t \Theta^{\mu(t)-1}(0)\,dt$, the new exponent $a(\tau) \equiv \mu(t)$ and the new thought functior $I(\tau) \equiv J(t)$, we find that the function $I(\tau)$ satisfies the conditions

$$\forall \text{ a.e. } \tau > 0 \quad I'(\tau) + I^{a(\tau)} \leq 0, \quad I(0) = 1, \quad I(\tau) \geq 0, \quad I'(\tau) \leq 0. \tag{6.25}$$

By monotonicity of $I(\tau)$ there is an interval $[0, \varepsilon)$ where $I(\tau) > 0$, otherwise $I(\tau) \equiv 0$ for all $\tau \geq 0$. Since $I(\tau) \leq 1$ and $a(\tau) \in (0, 1)$, then $I(\tau) \leq I^{a(\tau)}(\tau)$, which leads to the inequality

$$I'(\tau) + I(\tau) \leq I'(\tau) + I^{a(\tau)}(\tau) \leq 0. \tag{6.26}$$

This inequality yields $I(\tau) \leq e^{-\tau}$, whence

$$\tau \leq -\ln I(\tau) \quad \text{and} \quad a(\tau) \leq a\left(-\ln I(\tau)\right). \tag{6.27}$$

Combining this inequality with (6.25) we have that

$$I'(\tau) + I^{a(-\ln I(\tau))}(\tau) \leq 0. \tag{6.28}$$

The straightforward integration of this inequality over the interval $(0, \tau)$ gives:

$$\int_1^{I(\tau)} \frac{dr}{r^{a(-\ln r)}} \leq -\tau.$$

Introducing the new variable $z = -\ln r$, we rewrite it in the form

$$\int_0^{-\ln I(\tau)} \frac{dz}{e^{z(1-a(z))}} \geq \tau \qquad \text{for } \tau \geq 0.$$

According to the choice of $t_*$

$$\int_0^{-\ln I(\tau_*)} \frac{dz}{e^{z(1-a(z))}} \geq \tau_* = \int_0^{\infty} \frac{dz}{e^{z(1-a(z))}},$$

which is impossible unless $I(\tau_*) = \Theta(t_*)/\Theta(0) = 0$. □

**Theorem 6.6** *Let in the conditions of* Theorem 6.5 $f \equiv 0$ *and* $\Theta(t) \le M$. *Assume that* $\nu^+(t)$ *is monotone increasing and* $\nu^+(t) \nearrow 1$ *as* $t \to \infty$. *If the equation*

$$2K_1 \min\{1, 1/M\} \int_0^R \|u_0\|_{2,\Omega}^{2(\nu^+(s)-1)}\, ds = \int_0^\infty \frac{dz}{e^{z(1-\nu^+(z))}}$$

*has a root* $R = t_*$, *then the solution of problem* (6.1) $u \in \mathbf{W}(Q)$ *vanishes at the instant* $t = t_*$:

$$\|u(\cdot,t)\|_{2,\Omega} = 0 \text{ for } t \ge t_*.$$

*Proof* By Lemmas 6.5, 6.6 for every finite $T > 0$ the energy function $\Theta(t)$ satisfies the ordinary differential inequality (6.15) or (6.17), which can be written in the form

$$\Theta'(t) + 2K_1 \min\{1, 1/M\}\, \Theta^{\nu^+(t)}(t) \le 0.$$

The assertion follows then from Lemma 6.7. □

## 6.8 Remarks

The material of this chapter is based on the results published in [48, 54]. The underlying technique is the analysis of ordinary differential inequalities for a local energy function [36, Chap. 2]. This approach is applicable to solutions of equations more complicated than (6.1). To be precise, let us consider the problem

$$\begin{cases} \partial_t u - \operatorname{div}(\mathscr{A}(z,u,\nabla u)) + \mathscr{C}(z,u) = 0 & \text{in } Q = \Omega \times (0,T], \\ u = 0 \text{ on } \Gamma, \qquad u(x,0) = u_0 \in L^2(\Omega) \end{cases} \tag{6.29}$$

with $\mathscr{A}$ and $\mathscr{C}$ being Carathéodory functions subject to the structural assumptions

$$\begin{aligned} &(a) \quad \mathscr{A}(z,r,\xi)\cdot\xi \ge a_0 \sum_{i=1}^n |\xi_i|^{p_i(z)} \quad \forall z \in Q,\ r \in \mathbb{R},\ \xi \in \mathbb{R}^n, \\ &(b) \quad |\mathscr{A}(z,r,\xi)| \le a_1 \sum_{i=1}^n |\xi_i|^{p_i(z)-1}, \\ &(c) \quad \mathscr{C}(z,r)r \ge c_0|r|^{\sigma(z)} - f(z)r \quad \forall z \in Q,\ r \in \mathbb{R} \end{aligned} \tag{6.30}$$

with measurable exponents $p_i(z), \sigma(z) : Q \mapsto (1,\infty)$, $f \in L^2(Q)$, and positive constants $a_0, a_1, c_0$. We assume that $p_i \in C(Q)$. Let $u(z)$ be an energy solution of problem (6.29):

1. $u \in \mathbf{W}(Q)$, $u_t \in \mathbf{W}'(Q)$,
2. for every $\zeta \in \mathbf{W}(Q)$, $\zeta_t \in \mathbf{W}'(Q)$,

$$\int_{\Omega} u\,\zeta\,dx\Big|_t^{t+h} + \int_t^{t+h}\int_{\Omega}\Big[-u\,\zeta_t + \mathscr{A}(z,u,\nabla u)\cdot\nabla\zeta + \mathscr{C}(z,u)\zeta\Big]\,dxdt = 0.$$

By Corollary 1.10

$$\frac{1}{2h}\int_{\Omega} u^2\,dx\Big|_t^{t+h} + \frac{1}{h}\int_t^{t+h}\int_{\Omega}\Big[\mathscr{A}(z,u,\nabla u)\cdot\nabla u + \mathscr{C}(z,u)u\Big]\,dxdt = 0.$$

Letting $h \to 0$ (see the proof of Lemma 6.4) we conclude that for a.e. $t \in (0,T)$ the energy $y(t) = \|u(t)\|^2_{2,\Omega}$ satisfies the relation

$$\frac{1}{2}y'(t) + a_0\sum_{i=1}^{n}\int_{\Omega}|D_i u|^{p_i(x,t)}\,dx + c_0\int_{\Omega}|u|^{\sigma(x,t)}dx \le \int_{\Omega}|u|\,|f|\,dx,$$

which coincides with the basic energy inequality (6.12). This observation allows one to extend to the energy solutions of problem (6.29) all conclusions about the finite time extinction and the asymptotic behavior stated in Sects. 6.4–6.7.

A special case of problem (6.29) under the structural assumptions (6.30) (a), (c)

$$\mathscr{A}(z,r,\xi) = (|r|^{\delta} + d_0)|\xi|^{p(z)-2}\xi, \qquad \mathscr{C}(z,r) = c_0|r|^{\sigma-2}r - f(z)$$

with constant $\delta > 0$, $d_0 > 0$, $c_0 \ge 0$, $\sigma > 1$ is studied in [143, 144, 151, 152]. The property of finite time extinction is established for globally bounded solutions of equations satisfying condition (6.30) (b). Another special case

$$\mathscr{A}(z,r,\xi) = |\xi|^{p(z)-2}\xi, \qquad \mathscr{C}(z,r) = b(z)|r|^{q} - \gamma r$$

with constant $q > 0$, $\gamma > 0$ is considered in [221]. Sufficient conditions of finite extinction of solutions of the homogeneous Neumann problem for the evolution $p(z)$-Laplacian are obtained in [142].

Papers [7, 8] deal with the weak solutions of the problem

$$\begin{cases} u_t - \Delta_{p(x)}u = f \quad \text{in } Q, \\ u = 0 \text{ on } \Gamma, \quad u(x,0) = u_0(x) \text{ in } \Omega \end{cases}$$

with the exponent $p : \Omega \mapsto [p^-, p^+] \subset (1,\infty)$, $p \in C_{\log}(\Omega)$. The authors derive the decay rates of solutions and study the possibility of extinction in a finite time. There are established sufficient conditions for convergence of the sequence of solutions to the problems

$$\begin{cases} u_{nt} = \Delta_{p_n(x)} u_n + f_n & \text{in } Q, \\ u_n = 0 \text{ on } \Gamma, \quad u_n(x,0) = u_{0n}(x) \text{ in } \Omega \end{cases}$$

to some $u \in L^2(\Omega)$ under the assumptions that $u_{0n} \to u_0$ in $L^2(\Omega)$, $f_n \to f \in L^2(Q)$, and

$$p_n(x) = \begin{cases} q_n(x) \to \infty & \text{for } x \in D, \\ q(x) & \text{for } x \in \Omega \setminus \overline{D} \end{cases} \qquad \text{as } n \to \infty,$$

where $q(x) : \Omega \setminus \overline{D} \mapsto (1, \infty)$ and $q_n(x) : D \mapsto (1, \infty)$ are measurable functions. The limit function $u \in C([0,T]; L^2(\Omega)) \cap W^{1,2}_{\text{loc}}((0,T); L^2(\Omega))$ satisfies the equation $u_t - \Delta_{q(x)} u = f$ in the sense of distributions on $\Omega \setminus \overline{D}$, and is a solution of an evolutional quasi-variational inequality on $D$.

Paper [220] deals with a class of parabolic equations of Leray-Lions type. The questions of existence and uniqueness, the rates of decay, and the phenomenon of extinction in a finite time are studied in the situation when the initial datum does not satisfy the natural integrability assumptions.

Initial an boundary-value problems for equations of the $p(x)$-Laplace type can be interpreted as the abstract Cauchy problem for equations $u'(t) + Au(t) = f(x, u(t))$ or inclusions $u'(t) + Au(t) \in f(x, u(t))$, which allows one to study the asymptotic behavior of solutions using the theory of nonlinear semi-groups in Banach spaces, see [186, 203]. For the evolution equations of $p(x)$-Laplace type the natural function spaces are $L^2(\Omega)$ and $W^{1,p(x)}(\Omega)$. This approach was developed in the series of works [214, 236, 237, 238, 239, 240, 241]. For various evolution equations and inclusions of $p(x)$-Laplace structure, the questions the existence of global attractors, their fractal dimension and the continuity properties of flows were studied. The pullback attractors for non-autonomous equations generalizing the evolution $p(x)$-Laplacian are studied in [181].

# Chapter 7
# Blow-up in Equations with Variable Nonlinearity

A remarkable property of nonlinear evolution equations is the possibility of formation of singularities in solutions of problems with smooth input data, which leads to blow-up, or explosion, of initially regular solutions. In the last decades this feature has attracted attention of many researchers. The mathematical models based on evolution equations whose solutions may blow-up in a finite time, or when time goes to infinity, appear in various applications.

We derive sufficient conditions of finite time blow-up in solutions of equations with nonstandard growth conditions of the type (4.1). Given a solution $u(x,t)$, we say that $u$ blows-up at the instant $t^* < \infty$ if

$$\|u(\cdot,t)\|_{\infty,\Omega} \to \infty \text{ as } t \to t^* - .$$

The following two variants of Eq. (4.1) are studied:

- the semilinear equation with nonlinear sources

$$u_t = \Delta u + b(x,t)|u|^{\sigma(x,t)-2}u,$$

- the quasilinear anisotropic equation of the type reaction-diffusion with variable exponents of nonlinearity not depending on $t$

$$u_t = \sum_{i=1}^{n} D_i\left(a_i(x,t)|D_i u|^{p_i(x)-2}D_i u\right) + \sum_{i=1}^{n} b_i(x,t)|u|^{\sigma_i(x)-2}u.$$

The underlying conditions on the data of problem (4.1) guarantee the existence of an energy solution on a small time interval (see Chap. 4).

The effect of finite time blow-up in solutions of equations with time-dependent nonlinearities is studied in Chap. 10.



S. Antontsev and S. Shmarev, *Evolution PDEs with Nonstandard Growth Conditions*, Atlantis Studies in Differential Equations 4,
DOI 10.2991/978-94-6239-112-3_7

## 7.1 Semilinear Equation with Variable Source

Let us consider the semilinear problem

$$\begin{cases} u_t = \Delta u + b(x,t)|u|^{\sigma(x,t)-2}u & \text{in } Q, \\ u(x,0) = u_0(x) \geq 0 \text{ in } \Omega, & u = 0 \text{ on } \Gamma \end{cases} \tag{7.1}$$

with a continuous coefficient $b(x,t)$ and an exponent $\sigma(z) \in C^0(\overline{Q})$ satisfying the conditions

$$0 \leq b^- < b(x,t) \leq b^+ < \infty, \qquad \sigma(z) : Q \mapsto [\sigma^-, \sigma^+] \subset [2, \infty).$$

By Theorem 4.1, under these conditions for every $u_0 \in L^\infty(\Omega)$ problem (7.1) has a local in time bounded weak solution in the sense of Definition 4.1. Moreover, $u \geq 0$ a.e. in $Q_T$, provided that $u_0 \geq 0$ in $\Omega$, which allows us to consider the source terms of the form $b(x,t)u^{\sigma(z)-1}$. To study the possibility of the blow-up we apply the eigenfunction method of Kaplan [177]. Let $\lambda > 0$ and $\phi(x) \geq 0$ be the first eigenvalue and eigenfunction of the Dirichlet problem for the Laplace operator in $\Omega$:

$$-\Delta\phi = \lambda\phi \text{ in } \Omega, \qquad \phi = 0 \text{ on } \Gamma. \tag{7.2}$$

We normalize $\phi$ by the condition $\displaystyle\int_\Omega \phi(x)dx = 1$. Let us introduce the functions

$$\begin{aligned} \alpha(t) &= \left(\int_\Omega b^{\frac{1}{2-\sigma^-(t)}}(x,t)\phi(x)dx\right)^{2-\sigma^-(t)}, \\ \beta(t) &= \int_\Omega b(x,t)\phi(x)dx, \\ \mu(t) &= \int_\Omega u(x,t)\phi(x)dx, \\ \sigma^-(t) &= \min_{x\in\Omega} \sigma(x,t) \end{aligned} \tag{7.3}$$

and denote

$$A(t) = \alpha(t) - \frac{\lambda^{\sigma^-(t)-1}}{\sigma^-(t)-1}, \qquad B(t) = \beta(t) + \frac{\sigma^-(t)-2}{\sigma^-(t)-1}. \tag{7.4}$$

We will assume that

$$A^- = \min_{t\geq 0} A(t) > 0, \qquad B^+ = \max_{t\geq 0} B(t) < \infty. \tag{7.5}$$

**Theorem 7.1** *Let the data of problem* (7.1) *satisfy the conditions*

$$\begin{cases} 2 < \sigma^- \leq \sigma^-(t), \\ 0 < b^- \leq b(x,t) \leq b^+ < \infty, \end{cases} \tag{7.6}$$

$$\begin{cases} -\lambda\mu(0) + b^-\mu^{\sigma^- -1}(0) - b^+ > 0, \\ -\lambda + b^-(\sigma^- - 1)\mu^{\sigma^- -1}(0) > 0. \end{cases} \tag{7.7}$$

*Then every weak solution blows-up at a moment* $t^* \equiv t^*(\mu(0), \sigma^-, b^\pm) < \infty$.

*Remark 7.1* Conditions (7.7) are surely fulfilled for the sufficiently large $\mu(0)$.

**Theorem 7.2** *Let inequalities* (7.5) *be fulfilled,* $b^- > 0$ *and*

$$g(t, \mu(0)) = A^-\mu^{\sigma^-(t)-1}(0) - B^+ > 0 \quad \textit{for every}\, t \geq 0. \tag{7.8}$$

*Assume that one of the conditions is fulfilled:*

(i) *either*

$$\sigma^-(t) \geq \sigma^- > 2 \quad \textit{for all}\, t > 0, \tag{7.9}$$

(ii) *or*

$$\begin{cases} \mu(0) > 1, \quad \sigma^-(t) > 2\, \textit{for all finite}\, t > 0, \\ \sigma^-(t)\, \textit{is monotone decreasing and}\, \sigma^-(t) \to 2\, \textit{as}\, t \to \infty, \\ \displaystyle\int_1^\infty \mathrm{e}^{s(2-\sigma^-(s))}\, ds < \infty. \end{cases} \tag{7.10}$$

*Then every solution of problem* (7.1) *blows-up at a finite moment* $t^* < \infty$.

## 7.1.1 The Differential Inequality

Let $u(x,t)$ be a weak solution of the semilinear problem (7.1) in the cylinder $Q_T$ of a finite height $T$. By Definition 4.1, for every test-function $\phi(x) \in H_0^1(\Omega)$ and every $t, t+h < T$

$$\int_t^{t+h} \int_\Omega (u_t\phi + \nabla u \cdot \nabla\phi - d(x,t,u)\,\phi)\, dxdt = 0. \tag{7.11}$$

Let us choose the eigenfunction $\phi$ for the test-function in (7.11), divide the resulting equality by $h$, and let $h \to 0$. Applying the Lebesgue differentiation theorem we find that for a.e. $t < T$

$$\mu'(t) = \int_\Omega u_t \phi dx = -\int_\Omega \nabla u \cdot \nabla \phi dx + \int_\Omega b(x,t) u^{\sigma(x,t)-1} \phi dx$$
$$= -\lambda\mu + \int_\Omega b u^{\sigma(x,t)-1} \phi\, dx. \tag{7.12}$$

Using the representation

$$I = \int_\Omega b u^{\sigma(x,t)-1} \phi\, dx$$
$$= \int_{\Omega\cap(u\geq 1)} b u^{\sigma(x,t)-1}\phi\, dx + \int_{\Omega\cap(u<1)} b u^{\sigma(x,t)-1}\phi\, dx \tag{7.13}$$

we evaluate $I$ in the following way: since in the conditions of Theorems 7.1 and 7.2 (i) $\sigma(x,t) \geq \sigma^- > 2$, then

$$I \geq \int_{\Omega\cap(u\geq 1)} b u^{\sigma^- -1}\phi dx = \int_\Omega b u^{\sigma^- -1}\phi dx - \int_{\Omega\cap(u<1)} b u^{\sigma^- -1}\phi dx$$
$$\geq \int_\Omega b u^{\sigma^- -1}\phi dx - \int_\Omega b\phi dx.$$

Applying the inverse Hölder inequality

$$\int_\Omega |u||v| dx \geq \left(\int_\Omega |u|^q dx\right)^{\frac{1}{q}} \left(\int_\Omega |v|^{\frac{q}{q-1}} dx\right)^{\frac{q-1}{q}} \tag{7.14}$$

with the exponent $q = \frac{1}{\sigma^- -1} \in (0,1)$, we may estimate

$$\int_\Omega b u^{\sigma^- -1}\phi dx = \int_\Omega u^{\sigma^- -1}\phi^{\sigma^- -1} b\phi^{2-\sigma^-(t)} dx$$
$$\geq \left(\int_\Omega u\phi dx\right)^{\sigma^- -1} \left(\int_\Omega b^{\frac{1}{2-\sigma^-}}(x,t)\phi dx\right)^{2-\sigma^-} \tag{7.15}$$
$$= \alpha(t)\mu^{\sigma^- -1}(t).$$

Gathering these formulas we arrive at the ordinary differential inequality for the function $\mu(t)$:

$$\mu'(t) \geq -\lambda\mu(t) + \alpha(t)\mu^{\sigma^- -1}(t) - \beta(t) \equiv f(\mu(t)), \tag{7.16}$$

with the functions $\alpha(t)$, $\beta(t)$ defined in (7.3).

If condition (7.9) is substituted by the weaker condition (7.10), the same arguments lead to the inequality

$$\mu'(t) \geq -\lambda\mu(t) + \alpha(t)\mu^{\sigma^-(t)-1}(t) - \beta(t) \equiv F(t,\mu(t)). \tag{7.17}$$

Unlike (7.16), inequality (7.17) involves the variable exponent.

## 7.1.2 Analysis of the Differential Inequality

We are now in position to proof Theorems 7.1 and 7.2. Under conditions (7.6)

$$0 < b^- \leq \alpha(t) = \left(\int_\Omega b^{\frac{1}{2-\sigma^-}}(x,t)\phi(x)dx\right)^{2-\sigma^-},$$
$$0 < b^- \leq \beta(t) = \int_\Omega b(x,t)\phi(x)dx \leq b^+ < \infty,$$

and (7.16) yields the inequality with constant coefficients and exponents:

$$\mu'(t) \geq -\lambda\mu(t) + b^-\mu^{\sigma^- -1}(t) - b^+ \equiv f(\mu(t)), \quad \sigma^- = const > 2. \tag{7.18}$$

The function $f(s)$ is concave and attains its minimum at the point

$$s_* = \left(\frac{\lambda}{b^-(\sigma^- - 1)}\right)^{\frac{1}{\sigma^- -2}}.$$

Conditions (7.7) mean that $f(\mu(0)) > 0$, $f'_\mu(\mu(0)) > 0$ and inequality (7.18) guarantees that $\mu(t)$ is a strictly positive and increasing function of $t$ whence strict positivity of $f(\mu(t))$ for all $t \geq 0$. Dividing both parts of (7.18) by $f(\mu(t))$ and integrating we arrive at the inequality

$$J(\mu(t)) = \int_{\mu(0)}^{\mu(t)} \frac{ds}{f(s)} \geq t.$$

Since the integral $J(s)$ is convergent at $s = \infty$, this inequality is possible only if there exists $t^*$ such as $\mu(t) \to \infty$ as $t \to t^*$. This completes the proof of Theorem 7.1.

Under the conditions of Theorem 7.2 $\mu(t)$ satisfies the differential inequality (7.17). Applying Young's inequality

$$a\,b \leq \frac{1}{p}(\varepsilon a)^p + \frac{p-1}{p}\left(\frac{b}{\varepsilon}\right)^{\frac{p}{p-1}}$$

with $a = \mu$, $b = \lambda$, $\varepsilon = 1$, $p = \sigma^-(t) - 1$, we have:

$$\mu(t)\lambda \leq \frac{1}{\sigma^-(t)-1}(\lambda\mu(t))^{\sigma^-(t)-1} + \frac{\sigma^-(t)-2}{\sigma^-(t)-1}.$$

Plugging this inequality into (7.17), we obtain

$$\begin{aligned}\mu'(t) &\geq F(t, \mu(t)) \\ &\geq A(t)\mu^{\sigma^-(t)-1}(t) - B(t) \\ &\geq A^- \mu^{\sigma^-(t)-1}(t) - B^+ \equiv g(t, \mu(t))\end{aligned} \tag{7.19}$$

with the coefficients $A(t)$, $B(t)$, $A^-$, $B^+$ defined in (7.4) and (7.5). Since

$$g'_\mu(t, \mu) = A^- \left(\sigma^-(t) - 1\right) \mu^{\sigma^-(t)-1} > 0,$$

the function $g(t, \mu)$ is increasing as a function of $\mu$, whence, by virtue of the first condition in (7.8) and the inequality $\mu' \geq g(t, \mu)$, we have $g(t, \mu(t)) > g(t, \mu(0)) > 0$ for every $t > 0$. It follows that $\mu'(t) > 0$ for all $t > 0$. If condition (7.9) is fulfilled, the conclusion about the finite time blow-up of the solution $u$ follows exactly as in the proof of Theorem 7.1. Let us assume that condition (7.10) is fulfilled. Since $g(t, \mu)$ in increasing as a function of $\mu$, $\mu$ is an increasing function of $t$ and $g(t, \mu) \to \infty$ as $\mu \to \infty$, there exists $t'$ such that $g(t, \mu) \geq \frac{1}{2}A^- \mu^{\sigma^-(t)-1}(t)$ for all $t \geq t'$. Inequality (7.19) gives

$$\mu'(t) \geq \frac{1}{2}A^- \mu^{\sigma^-(t)-1}(t) \quad \text{for } t \geq t', \qquad \mu(t') \geq \mu(0) > 1.$$

Let us introduce the new independent variable $\theta = \frac{A^-}{2}(t - t')$ and denote $\gamma(\theta) = \sigma^-(t)$, $\nu(\theta) \equiv \mu(t)$. For the function $\nu(\theta)$ we have the conditions

$$\nu'(\theta) \geq \nu^{\gamma(\theta)-1}(\theta), \quad \nu(\theta) \geq 1 \quad \text{for } \theta > 0, \tag{7.20}$$

which yield the inequality $\nu'(\theta) \geq \nu(\theta)$. Integration of this inequality gives

$$\ln \nu(\theta) \geq \ln \left(\frac{\nu(\theta)}{\nu(0)}\right) \geq \theta,$$

and for the monotone decreasing function $\gamma(\theta)$ we have: $\gamma(\theta) \geq \gamma(\ln \nu(\theta))$. In the result we have the autonomous differential inequality for $\nu(\theta)$:

$$\nu'(\theta) \geq \nu^{\gamma(\ln \nu(\theta))-1}, \quad \nu(\theta) \geq 1 \quad \text{for } \theta \geq 0.$$

Integrating and changing the variable of integration we finally obtain the inequality

$$I(\ln \nu(\theta)) \equiv \int_0^{\ln \nu(\theta)} \frac{d\tau}{e^{\tau(\gamma(\tau)-2)}} \geq \int_{\nu(0)}^{\nu(\theta)} \frac{ds}{s^{\gamma(\ln s)} - 1} \geq \theta.$$

If $I(\infty) < \infty$, this inequality leads to a contradiction unless there exists a finite $\theta^*$ such that $v(\theta) \to \infty$ as $\theta \to \theta^*$. The proof of Theorem 7.2 is completed.

## 7.2 Equations with Vanishing Sources

Let us consider the case when the coefficient $b(x, t)$ in Eq. (7.1) is not necessarily strictly positive everywhere in the problem domain and is allowed to vanish on a set of zero measure. Recall the notation $\sigma^+(t) = \sup\{\sigma(x, t) | \, x \in \Omega\}$.

**Theorem 7.3** *Let us assume that* $\sigma(x, t)$, $b(x, t)$ *and* $\mu(0) = \displaystyle\int_\Omega u_0 \, \phi \, dx$ *satisfy the following conditions*:

$$\begin{gathered} \sigma(x, t) \geq \sigma^- > 2, \qquad \mu(0) > 1, \\ \inf\{b(x, t) | \, (x, t) \in \Omega \times (0, \infty)\} \geq b^- > 0. \end{gathered} \tag{7.21}$$

*Let us assume also that there exists a constant* $\delta > 0$ *and a function* $q(t)$ *such that*

$$-\lambda + DM > \delta, \quad q(t) \in (0, 1) \quad q(t)\,(\sigma^-(t) - 1) > 1,$$

*where*

$$D = 2^{1-\sigma^+} \min\{1, |\Omega|^{2-\sigma^+}\}, \qquad M = b^- \min_{t \geq 0} \left( \frac{|\Omega|^{\frac{q-1}{q}}}{\max_\Omega \left(\phi^{\sigma-2}\right)} \right).$$

*Then every solution of problem* (7.1) *blows-up in a finite time.*

**Theorem 7.4** *Let us assume that* $\sigma(x, t)$, $b(x, t)$ *and* $\mu(0)$ *satisfy the following conditions*:

$$\sigma(x, t) \geq \sigma^- > 2, \qquad \mu(0) > 1, \qquad b(x, t) \geq 0. \tag{7.22}$$

*Let us assume also that there exist constants* $\beta > 0$, $\gamma > 1$, $q \in (0, 1)$ *and* $\delta > 0$ *such that*

$$q(\sigma^- - 1) > 1, \qquad \int_\Omega \frac{dx}{b^{\frac{q\gamma}{1-q}}(x, t)} = B(t) \leq \beta, \tag{7.23}$$

*and*

$$-\lambda + DM' > \delta$$

*with*

$$D = 2^{1-\sigma^+} \min\{1, |\Omega|^{2-\sigma^+}\}, \qquad M' = \frac{|\Omega|^{\frac{(q-1)(\gamma-1)}{q\gamma}}}{\max_{t \geq 0} \left(\phi^{\sigma(x,t)-2}\right)}.$$

*Then every solution of problem* (7.1) *blows-up in a finite time.*

*Remark 7.2* The assertions of Theorems 7.3 and 7.4 remain true for the semilinear equations with various sources

$$u_t = \sum_{i=1}^{n} D_i \left(a_i(x) D_i u\right) + \sum_{i=1}^{K} b_i(x,t)|u|^{\sigma_i(x,t)-2}u, \qquad 1 \le K \le n,$$

provided that $b_i(x,t) \ge 0$ for all $i$, and there is at least one $j \in \{1, \dots, K\}$ such that $b_j(x,t)$ and $\sigma_j(x,t)$ satisfy the conditions of Theorems 7.3 and 7.4.

Multiplying Eq. (7.1) by $\phi$ and integrating over $\Omega$ we arrive at the relation

$$\begin{aligned}\mu'(t) &= \int_\Omega u_t \phi\, dx = -\int_\Omega \nabla u \cdot \nabla \phi\, dx + \int_\Omega b\, u^{\sigma-1} \phi\, dx \\ &= -\lambda\, \mu(t) + \int_\Omega b u^{\sigma-1} \phi\, dx \equiv -\lambda\, \mu(t) + I(t).\end{aligned} \tag{7.24}$$

To estimate $I(t)$ from below we use the inverse Hölder inequality

$$\int_\Omega |u||v| dx \ge \|u\|_{q,\Omega} \|v\|_{q',\Omega} \quad \text{with } q \in (0,1),\ q' = \frac{q}{q-1}. \tag{7.25}$$

Let us take $q(t)$ such that for all $t \ge 0$ $q(t) \in (0,1)$, $q(t)(\sigma(x,t)-1) > 1$ in $\Omega$, and estimate the integral $I(t)$ as follows:

$$I(t) = \int_\Omega (u\,\phi)^{\sigma-1} \left(b\,\phi^{2-\sigma}\right) dx \ge \left(\int_\Omega (u\phi)^{(\sigma-1)q} dx\right)^{\frac{1}{q}} \left(\int_\Omega (b\,\phi^{2-\sigma})^{\frac{q}{q-1}} dx\right)^{\frac{q-1}{q}}.$$

Using (1.6) and (1.9), we also have that for the same $q$

$$\begin{aligned}\mu(t) &= \int_\Omega u\,\phi\, dx \le C\, \|u\phi\|_{(\sigma(\cdot,t)-1)q,\Omega} \\ &\le C(t) \max\left\{ \left(\int_\Omega |u\,\phi|^{(\sigma-1)q} dx\right)^{\frac{1}{q(t)(\sigma^-(t)-1)}}, \left(\int_\Omega \dots\right)^{\frac{1}{q(t)(\sigma^+(t)-1)}} \right\}\end{aligned} \tag{7.26}$$

with

$$C(t) = 2\,\|1\|_{\frac{(\sigma-1)q}{(\sigma-1)q-1},\Omega} \le 2 \max\left\{1,\ |\Omega|^{\frac{\sigma^+-2}{\sigma^+-1}}\right\}.$$

Writing inequality (7.26) in the equivalent form

$$\min\left\{ \left(\frac{\mu(t)}{C(t)}\right)^{\sigma^-(t)-1}, \left(\frac{\mu(t)}{C(t)}\right)^{\sigma^+(t)-1} \right\} \le \left(\int_\Omega (u\,\phi)^{(\sigma-1)q} dx\right)^{\frac{1}{q}},$$

we obtain the needed estimate on $I(t)$:

$$I(t) \geq \left(\int_{\Omega} (u\,\phi)^{(\sigma-1)q} dx\right)^{\frac{1}{q}} \left(\int_{\Omega} (b\,\phi^{2-\sigma})^{\frac{q}{q-1}} dx\right)^{\frac{q-1}{q}}$$
$$\geq D \min\left\{\mu^{\sigma^-(t)-1}(t),\, \mu^{\sigma^+(t)-1}(t)\right\} \left(\int_{\Omega} (b\,\phi^{2-\sigma})^{\frac{q}{q-1}} dx\right)^{\frac{q-1}{q}}$$

with $D = 2^{1-\sigma^+} \min\{1,\ |\Omega|^{2-\sigma^+}\}$. In the case that $b(x,t)$ is separated away from zero [see (7.21)]

$$g(t) \equiv \left(\int_{\Omega} (b\,\phi^{2-\sigma})^{\frac{q}{q-1}} dx\right)^{\frac{q-1}{q}} \geq b^- \min_{t\geq 0} \left(\frac{|\Omega|^{\frac{q-1}{q}}}{\max_{x\in\Omega}\left(\phi^{\sigma(x,t)-2}\right)}\right) \equiv M, \tag{7.27}$$

and the differential inequality for $\mu(t)$ takes on the form

$$\mu'(t) \geq -\lambda\,\mu(t) + DM \min\left\{\mu^{\sigma^-(t)-1}(t),\ \mu^{\sigma^+(t)-1}(t)\right\}. \tag{7.28}$$

Let us specially notice that $D$ and $M$ are independent of $u$ and depend only on $\sigma(x,t)$ and $|\Omega|$.

Estimate (7.27) on $g(t)$ ceases to be true if $b^- = 0$. Let us estimate $g(t)$ from below in a different way: by Hölder's inequality with $\gamma > 1$

$$\frac{1}{g(t)} = \left(\int_{\Omega} |b\,\phi^{2-\sigma}|^{\frac{q}{q-1}} dx\right)^{\frac{1-q}{q}} \leq \|b^{\frac{q}{q-1}}\|^{\frac{1-q}{q}}_{\gamma,\Omega} \|\phi^{\frac{q(2-\sigma)}{q-1}}\|^{\frac{1-q}{q}}_{\gamma',\Omega}$$
$$\leq |\Omega|^{\frac{1-q}{q}\frac{\gamma-1}{\gamma}} \|b^{\frac{q}{q-1}}\|^{\frac{1-q}{q}}_{\gamma,\Omega} \max_{Q_T}\left(\phi^{\sigma(x,t)-2}\right) \tag{7.29}$$
$$= K\left(\int_{\Omega} \frac{dx}{b^{\frac{q\gamma}{1-q}}(x,t)}\right)^{\frac{1-q}{q\gamma}} \equiv K B(t), \qquad K \equiv K(|\Omega|, \phi, \sigma, \gamma).$$

Let the conditions of Theorem 7.3 be true. By virtue of (7.24) $\mu(t)$ is continuous. Writing (7.28) in the form

$$\frac{\mu'(t)}{\mu(t)} \geq -\lambda + DM \min\left\{\mu^{\sigma^-(t)-2}(t),\, \mu^{\sigma^+(t)-2}(t)\right\}, \tag{7.30}$$

integrating and using the continuity of $\mu(t)$, we conclude that for some $s > 0$

$$
\begin{aligned}
\mu(t) &\geq \mu_0 \exp\left(\int_0^t \left(-\lambda + DM \min\left\{\mu^{\sigma^-(s)-2}(s), \mu^{\sigma^+(s)-2}(s)\right\}\right) ds\right) \\
&\geq \mu_0 \exp\left(\int_0^t (-\lambda + DM)\, ds\right) \geq \mu_0 e^{\delta t} > \mu(0) \qquad \text{for } t \in (0, s]
\end{aligned}
$$

and, thus, $\mu(t) \geq \mu(0) > 1$ for all $t \in [0, s]$. Set

$$T^* = \sup\{t > 0 : \mu(t) < \infty\}.$$

Since $\mu(s) > 1$, this process can be continued as long as the function $\mu(t)$ remains bounded, which gives the estimate $\mu(t) > 1$ for all $t \in [0, T^*]$. Now we may transform (7.30) to the form

$$
\begin{aligned}
\mu'(t) &\geq \mu^{\sigma^-(t)-1}(t)\left(-\lambda\mu^{2-\sigma^-(t)}(t) + DM\right) \\
&\geq \mu^{\sigma^-(t)-1}(t)\,(-\lambda + DM) \qquad (7.31)\\
&\geq \delta\, \mu^{\sigma^-(t)-1}(t) \geq \delta\, \mu^{\sigma^- -1}(t).
\end{aligned}
$$

Integrating this inequality we find that $\mu(t)$ becomes infinite in a finite time: in the case that $\sigma^-(t) \geq \sigma^- = const > 2$ we have

$$\mu^{\sigma^- -2}(t) \geq \frac{\mu^{\sigma^- -2}(0)}{1 - \delta\,(\sigma^- - 2)\,\mu^{\sigma^- -2}(0)\, t} \to \infty \quad \text{as} \quad t \to t^* = \frac{\mu^{2-\sigma^-}(0)}{\delta\,(\sigma^- - 2)}.$$

Assume the conditions of Theorem 7.4. According to assumption (7.23) $B(t) \leq \beta < \infty$, which gives the needed estimate from below: $g(t) \geq \dfrac{1}{K\beta} \equiv M'$. This leads to the differential inequality (7.28) with the constant $M'$ instead of $M$, and the conclusion follows.

## 7.3 Various Extensions

### *7.3.1 Eventually Linear Equations*

Let us illustrate the assertion of Theorem 7.2 by the following example: assume that $u$ is a weak solution of problem (7.1) for the equation

$$u_t = \Delta u + u^{1+\varepsilon(t)}, \qquad (7.32)$$

where $\varepsilon(t)$ is a monotone decreasing positive function such that $\varepsilon(t) \to 0$ as $t \to \infty$. Equation (7.32) is semilinear in every cylinder of finite height, but it transforms into the linear heat equation as $t \to \infty$. We assume that the initial function $u_0$ is as large as required in Theorem 7.2. Moreover, increasing $\mu(0)$ we may guarantee that in (7.19) $g(t, \mu(t)) \geq \frac{1}{2}A^- \mu^{1+\varepsilon(t)}(t)$ for all $t \geq 0$, so that the sufficient condition of the finite time blow-up reduces to the following one:

$$\int_1^{\infty+} \frac{d\tau}{e^{\varepsilon(\tau)\tau}} < \infty.$$

The simplest convergence test shows that this condition is fulfilled if, say,

$$\varepsilon(\tau) = \alpha \frac{\ln \tau}{\tau} \quad \text{with any } \alpha > 1.$$

### 7.3.2 *Equations with Vanishing Sources*

In the conditions of Theorem 7.4, the coefficient $b(x, t)$ need not be separated away from zero and is allowed to vanish on a set of zero measure in $\Omega$ for every $t = const$. The admissible rate of vanishing of $b(x, t)$ is defined in terms of the exponent $\sigma(x, t)$: we claim that $\|1/b(x, t)\|_{\frac{q\gamma}{1-q}, \Omega} = B(t) < \infty$ uniformly in $t$. Let us consider the model problem

$$\begin{cases} u_t = \Delta u + |x|^\alpha u^{\sigma(x,t)-1} & \text{in } Q = B_1(0) \times (0, T], \\ u = 0 \text{ on } \Gamma, \qquad u(x, 0) = u_0(x) \text{ in } \Omega \end{cases}$$

with constant $\alpha \geq 0$. The domain $\Omega$ is the unit ball $B_1(0) = \{x : |x| < 1\}$. By Theorem 4.2, for every nonnegative $u_0 \in L^\infty(B_1(0))$ this problem has a local in time nonnegative weak solution. Following the proof of Theorem 7.4 one sees that the possibility of blow-up depends on the convergence of the integral

$$\int_\Omega \frac{dx}{b^{\frac{q\gamma}{1-q}}(x, t)} = C \int_0^1 s^{n-1-\alpha \frac{q\gamma}{1-q}} ds \quad \text{with some } 1 > q > \frac{1}{\sigma^- - 1} \text{ and } \gamma > 1.$$

Such constants exist and the integral is convergent if

$$\alpha < n\left(\sigma^- - 2\right).$$

### 7.3.3 *Regional Blow-up*

The study of the blow-up phenomenon can be confined to a subdomain of the problem domain $\Omega$. Let us take a subdomain $D \subset \Omega$, such that meas $D > 0$, $\partial D \in C^1$, and denote by $\phi$ and $\lambda$ be the first eigenfunction and the corresponding eigenvalue of the problem

$$-\Delta\phi = \lambda\phi \text{ in } D, \qquad \phi = 0 \text{ on } \partial D. \tag{7.33}$$

Let us introduce the function

$$\mu(t) = \int_D u\,\phi\,dx.$$

Given a solution $u \in \mathbf{W}(Q)$, we may formally consider the semilinear Eq. (7.1) (at least for small times) as the heat equation with the bounded free term $f(x,t) \equiv b(x,t)\,u^{\sigma(x,t)-1}$. It follows then from the classical parabolic theory that $u \in L^2(0,\theta; H_0^1(\omega))$ for every subdomain $\omega \subset \Omega$ with the sufficiently smooth boundary $\partial\omega$ and every $\theta < t^*$. This observation justifies the forthcoming arguments. Let us multiply the equation by the function $\phi$ and integrate over $D$:

$$\begin{aligned}\mu'(t) &= \int_D u_t\phi\,dx = \int_D u\Delta\phi\,dx - \int_{\partial D} u\,(\nabla\phi,\,\mathbf{n})dS + \int_D b\,u^{\sigma(x,t)}\,\phi\,dx\\ &= -\lambda\mu - \int_{\partial D} u\,(\nabla\phi,\,\mathbf{n})\,dS + \int_D b\,u^{\sigma(x,t)}\phi\,dx,\end{aligned}$$

where $\mathbf{n}$ denotes the outward normal to $\partial D$. Since $\phi > 0$ in $D$ and $\phi = 0$ on $\partial D$, it is necessary that $(\nabla\phi,\,\mathbf{n}) \le 0$ on $\partial D$, whence

$$-\int_{\partial D} u\,(\nabla\phi,\,\mathbf{n})\,dS \ge 0$$

for every nonnegative function $u$. The differential inequality for $\mu(t)$ takes on the form

$$\mu'(t) \ge -\lambda\mu + \int_D b\,u^{\sigma(x,t)}\,\phi\,dx$$

and its analysis follows the proof of Theorem 7.1.

### 7.3.4 Equations with Nonlocal Sources

Let us consider the problem

$$\begin{cases} u_t = \Delta u + f(x,t,u) & \text{in } Q_T,\\ u(x,0) = u_0(x) \text{ in } \Omega, & u = 0 \text{ on } \Gamma_T, \end{cases} \tag{7.34}$$

with the nonlocal source

$$f(x,t,u) = \sum_{k=1}^{N} b_k(x,t)u^{\sigma_k(x,t)-1} + \sum_{i=N+1}^{L} c_i(x,t)\int_\Omega d_i(s,t)u^{\sigma_i(s,t)-1}ds,$$

with $b_k \geq 0, c_i \geq 0, d_i \geq 0$ and integer $0 \leq N < n$ and $N+1 \leq L \leq n$. The further analysis is formal: we assume without proof that problem (7.34) has a nonnegative weak solution $u(z)$ understood in the sense of Definition 4.1. Multiplying (7.34) by $u$ and integrating over $\Omega$ we arrive at the relation [cf. with (7.12)]

$$\mu'(t) = -\lambda\mu + I_1 + I_2, \tag{7.35}$$

where

$$I_1 = \int_\Omega \left( \sum_{k=1}^{N} b_k(x,t) u^{\sigma_k(x,t)-1} \right) \phi dx,$$

$$I_2 = \int_\Omega \left( \sum_{i=N+1}^{L} c_i(x,t) \int_\Omega d_i(s,t) u^{\sigma_i(s,t)-1} ds \right) \phi dx.$$

$I_1$, $I_2$ are estimated from below in the following way [cf. with (7.13)–(7.15)]:

$$\begin{aligned}
I_1 &\geq \int_{\Omega\cap(u\geq 1)} \left( \sum_{k=1}^{N} b_k(x,t) u^{\sigma_k(x,t)-1} \right) \phi dx \geq \int_{\Omega\cap(u\geq 1)} \left( \sum_{k=1}^{N} b_k(x,t) u^{\sigma_k^-(x,t)-1} \right) \phi dx \\
&= \int_\Omega \left( \sum_{k=1}^{N} b_k(x,t) u^{\sigma_k^-(x,t)-1} \right) \phi dx - \int_{\Omega\cap(u<1)} \left( \sum_{k=1}^{N} b_k(x,t) u^{\sigma_k^-(x,t)-1} \right) \phi dx \\
&\geq \int_\Omega \left( \sum_{k=1}^{N} b_k(x,t) u^{\sigma_k^-(x,t)-1} \right) \phi(x) dx - \int_\Omega \left( \sum_{k=1}^{N} b_k(x,t) \right) \phi(x) dx \\
&\geq \sum_{k=1}^{N} \left( \int_\Omega b_k^{-\frac{1}{\sigma_k^-(x,t)-2}}(x,t)\phi(x)dx \right)^{-\frac{1}{\sigma_k^-(x,t)-2-}} \left( \int_\Omega u(x,t)\phi(x)dx \right)^{\sigma_k^-(x,t)-1} \\
&\quad - \int_\Omega \left( \sum_{k=1}^{N} b_k(x,t) \right) \phi dx = \sum_{k=1}^{N} \alpha_k(t) \mu^{\sigma_k^-(t)-1} - \beta(t),
\end{aligned}$$

with

$$\alpha_k(t) = \left( \int_\Omega b_k^{\frac{1}{2-\sigma_k^-(t)}}(x,t)\phi(x)dx \right)^{2-\sigma_k^-(t)}, \quad \beta(t) = \int_\Omega \left( \sum_{k=1}^{N} b_k(x,t) \right) \phi dx$$

for $k = 1, \ldots, N$,

$$I_2 \geq \sum_{i=N+1}^{L} \int_\Omega c_i \phi \left( \int_\Omega u\phi ds \right)^{\sigma_i^- - 1} \left( \int_\Omega d_i^{\frac{1}{2-\sigma_i^-(t)}} \phi^{\frac{\sigma_i - 1}{\sigma_i - 2}} ds \right)^{2-\sigma_i^-} dx$$

$$
\begin{aligned}
&- \sum_{i=N+1}^{L} \int_{\Omega} \left( c_i(x,t) \int_{\Omega} d_i(s,t) ds \right) \phi dx \\
&= \sum_{i=N+1}^{L} \mu^{\sigma_i^- - 1}(t) \theta_i(t) \int_{\Omega} c_i(x,t)\phi(x) dx - \beta(t) \\
&= \sum_{i=N+1}^{L} \alpha_i(t) \mu^{\sigma_i^- - 1} - \beta(t),
\end{aligned}
$$

with

$$
\begin{aligned}
\theta_i(t) &= \left( \int_{\Omega} d_i^{\frac{1}{2-\sigma_i^-}}(x,t) \phi^{\frac{\sigma_i - 1}{\sigma_i - 2}}(x) dx \right)^{2-\sigma_i^-} \\
\alpha_i(t) &= \theta_i(t) \int_{\Omega} c_i(x,t)\phi(x) dx, \\
\beta(t) &= \sum_{i=N+1}^{L} \int_{\Omega} \left( c_i(x,t) \int_{\Omega} d_i(s,t)\, ds \right) \phi(x) dx
\end{aligned}
$$

for $i = N+1, \ldots, L$. Gathering these formulas we arrive at the differential inequality

$$
\mu'(t) \geq -\lambda \mu + \sum_{i=1}^{L} \alpha_i(t) \mu^{\sigma_i^-(t)-1}(t)(t) - \beta(t), \tag{7.36}
$$

which can be studied exactly as (7.17).

### 7.3.5 Semilinear Equations of General Form

Without any change in the arguments one may consider the homogeneous Dirichlet problem for more general semilinear equations

$$
u_t = Lu + f(x,t,u), \tag{7.37}
$$

where $L$ denotes the linear elliptic operator

$$
Lu = \sum_{i,j=1}^{n} D_i \left( a_{ij}(x) D_j u \right) + \sum_{i=1}^{n} D_i \left( a_i(x) u \right) + a_0(x) u
$$

and the right-hand side $f$ has the form

$$f(x,t,u) = \sum_{k=1}^{N} b_k(x,t)u^{\sigma_k(x,t)-1} + \sum_{i=N}^{K} c_i(x,t) \int_{\Omega} d_i(s,t)u^{\sigma_i(s,t)-1}ds \quad (7.38)$$

with $b_k \geq 0$, $c_i \geq 0$, $d_i \geq 0$. Let us denote by $L^*$ the adjoint operator

$$\begin{cases} L^*w \equiv -D_{x_j}\left(a_{ij}(x)D_{x_i}w\right) - a_i(x)D_{x_i}w + a_0(x)w, \\ (u, L^*w)_\Omega = (w, Lu)_\Omega \quad \forall\, u,\ w \in W_0^{1,2}(\Omega), \end{cases}$$

and assume that there exist a nonnegative function $\phi \geq 0$ and a positive number $\lambda > 0$ such that

$$\forall\, \psi \in H_0^1(\Omega) \quad (\psi, L^*\phi)_{2,\Omega} = -\lambda(\psi, \phi)_{2,\Omega}. \quad (7.39)$$

Multiplying (7.37) by $u$ and integrating over $\Omega$ we arrive at inequality (7.36) for the function

$$\mu(t) = \int_{\Omega} u(x,t)\phi(x)dx.$$

The above arguments apply to study more general situations. Let us consider the problem

$$\begin{cases} u_t = Lu + f(x,t,u) \quad \text{in } Q_T, \\ B_k u = 0 \text{ on } \Gamma_T,\ k = 1, \dots, m, \\ u(x,0) = u_0(x) \text{ in } \Omega, \end{cases} \quad (7.40)$$

where $f$ is defined by (7.38), $L$ is a linear operator of order $2m$, with the coefficients independent of $t$, and such that the eigenvalue problem for the adjoint operator

$$L^*\phi = \lambda\phi \text{ in } x \in \Omega, \qquad B_k^*\phi = 0 \text{ on } \Gamma,\ k = 1, \dots, m$$

has a nonnegative solution $\phi$ with positive $\lambda$. Multiplying Eq. (7.40) by $\phi$ and integrating by parts in $\Omega$ we obtain an ordinary differential inequality for the function $\mu(t)$.

## 7.4 Equations with Anisotropic $p(x)$-Growth

Let $u(x,t)$ be a weak solution of the parabolic equation

$$\begin{cases} u_t = \sum_{i=1}^{n} D_i\left(a_i(x,t)|D_i u|^{p_i(x)-2}D_i u\right) + \sum_{i=1}^{K} b_i(x,t)|u|^{\sigma_i(x)-2}u \quad \text{in } Q_T, \\ u(x,0) = u_0(x) \text{ in } \Omega, \qquad u = 0 \text{ on } \Gamma_T, \quad 1 \leq K \leq n. \end{cases} \quad (7.41)$$

It is assumed that the coefficients $a_i, b_i$ and the exponents $p_i, \sigma_i$ satisfy the following conditions:

$$\begin{cases} \text{there exist constants } a_i^{\pm} \in (0, \infty),\ b_i^{\pm} \in [0, \infty) \text{ such that} \\ a_i(x,t) \in (a_i^-, a_i^+), \quad b_i(x,t) \in (b_i^-, b_i^+), \end{cases} \tag{7.42}$$

$$\begin{cases} \sigma_i(x) \in C(\overline{\Omega}),\ p_i \in C_{\log}(\overline{\Omega}), \\ \text{there exist constants } p_i^{\pm},\ \sigma_i^{\pm} \in (1, \infty) \text{ such that} \\ p_i(x) \in [p_i^-, p_i^+], \quad \sigma_i(x) \in [\sigma_i^-, \sigma_i^+]. \end{cases} \tag{7.43}$$

For the sake of brevity of notation, in the case that $1 \le K < n$ we simply set $b_i \equiv 0$ for $i = K+1, \dots, n$. Let us define the functions

$$f(t) = \frac{1}{2} \int_0^t \int_\Omega u^2(x,s)\,dx\,ds, \tag{7.44}$$

$$E(t) = \sum_{i=1}^n \int_\Omega \left( \frac{a_i(x,t)}{p_i(x)} |D_i u|^{p_i(x)} - \frac{b_i(x,t)}{\sigma_i(x)} |u|^{\sigma_i(x)} \right) dx \tag{7.45}$$

and assume that

$$E(0) = \sum_{i=1}^n \int_\Omega \left( \frac{a_i(x,0)}{p_i(x)} |D_i u_0|^{p_i(x)} - \frac{b_i(x,0)}{\sigma_i(x)} |u_0|^{\sigma_i(x)} \right) dx \le 0. \tag{7.46}$$

**Theorem 7.5** *Let conditions* (7.42), (7.43) *and* (7.46) *be fulfilled. Assume that* $|u_0|^{\sigma(x)} \in L^1(\Omega)$, $|D_i u_0|^{p_i(x)} \in L^1(\Omega)$,

$$a_{it}, b_{it} \in C^0(\overline{Q}_T), \qquad a_{it} \le 0, b_{it} \ge 0 \text{ in } Q_T.$$

*If there exists a constant* $\lambda > 0$ *such that*

$$\begin{cases} \dfrac{1}{p_i^+} \ge \lambda \ge \dfrac{1}{\sigma_i^-} \text{ for all } i = 1, \dots, K, \\ \dfrac{1}{p_j^+} \ge \lambda > \dfrac{1}{\sigma_j^-} \text{ and } b_j^- > 0 \text{ for at least one } j \in \{1, \dots, n\}, \end{cases} \tag{7.47}$$

*then every nonstationary solution of problem* (7.41) *blows-up in a finite time: there exists* $t^* < \infty$ *such that* $\|u(\cdot,t)\|_{\infty,\Omega} \nearrow \infty$ *as* $t \to t^*-$.

*Remark 7.3* In this chapter we are confined to the case when the exponents $p_i$ and $\sigma$ do not depend on $t$. Sufficient conditions for the blow-up of solutions of the equation

$$u_t = \sum_{i=1}^{n} D_i \left( a_i |D_i u|^{p_i(x,t)-2} D_i u \right) - c|u|^{\sigma(x,t)-2} u$$

are given in Corollary 10.2 in Chap. 10, where this effect is studied for the doubly degenerate equations.

*Remark 7.4* The assertion of Theorem 7.5 is applicable, for example, to the equations

$$\text{(a)} \quad u_t = \sum_{i=1}^{n} D_i(|D_i u|^{p_i(x)-2} D_i u) + |u|^{\sigma(x)-2} u, \qquad \text{(b)} \quad u_t = \Delta u + \sum_{i=1}^{K} |u|^{\sigma_i(x)-2} u.$$

For the former equation, the solutions blow-up in a finite if $p_i(x) > 1$ for all $i = 1, \dots, n$ and $p_j^+ < \sigma^-$ for some $j \in \{1, \dots, n\}$. For the latter equation the restrictions on the exponents read: $\sigma_i(x) \geq 2, \sigma_j^- > 2$ for at least one $j \in \{1, \dots, K\}$. To equation (b) Theorem 7.3 is applicable.

*Remark 7.5* A revision of the proof of Theorem 7.5 given below shows that the condition $b_i \geq 0$ for all $i = 1, \dots, K$ can be relaxed and the assertion holds true if condition (7.47) is substituted by the following one: there exists a constant $\lambda > 0$ such that

$$\begin{cases} \dfrac{1}{p_i^+} \geq \lambda \text{ and } \left( \lambda - \dfrac{1}{\sigma_i^-} \right) b_i(x,t) \geq 0 \text{ for all } i = 1, \dots, K, \\ \dfrac{1}{p_j^+} \geq \lambda > \dfrac{1}{\sigma_j^-} \text{ and } b_j^- > 0 \text{ for at least one } j \in \{1, \dots, n\}. \end{cases} \tag{7.48}$$

An example of such a situation is furnished by the equation with constant exponents of nonlinearity

$$\begin{cases} u_t = u_{xx} + u_{yy} + (|u_z|^{p_3-2} u_z) + u - |u|^{-1/2} u + |u| u, \\ p_1 = p_2 = 2, \quad p_3 \in (1, 2], \quad \sigma_1 = 2, \quad \sigma_2 = 3/2, \quad \sigma_3 = 3, \\ b_1 = 1, \quad b_2 = -1, \quad b_3 = 1. \end{cases}$$

In this case we may take $\lambda = 1/2$.

### 7.4.1 The Energy Relations

**Lemma 1** *Under the conditions of Theorem* 7.5 *the solution of problem* (7.41) *satisfies the estimate*

$$\frac{1}{2} \int_\Omega u^2(x,t)\, dx + \sum_{i=1}^{n} \int_0^t \int_\Omega \left( a_i |D_i u|^{p_i} - b_i |u|^{\sigma_i} \right) dx = \frac{1}{2} \int_\Omega u_0^2 dx \tag{7.49}$$

*Proof* Equality (7.49) follows from (4.10) if we take the solution $u$ for the test-function. □

The second energy relation follows from Remark 4.3 to Theorem 4.8: under the conditions of Theorem 7.5 the energy $E(t)$ satisfies estimate (4.82)

$$\forall \text{ a.e. } t > 0 \qquad E(t) + \int_0^t \int_\Omega u_t^2 dxdt \leq E(0). \tag{7.50}$$

Using the function $f(t)$ defined in (7.44) we may rewrite equality (7.49) in the form

$$\begin{aligned} f'(t) &\equiv \frac{1}{2} \int_\Omega u^2(x,t)dx \\ &= \frac{1}{2} \int_\Omega u_0^2\, dx + \sum_{i=1}^n \int_0^t \int_\Omega \left(-a_i\, |D_i u|^{p_i} + b_i\, |u|^{\sigma_i}\right) dxdt \geq 0. \end{aligned} \tag{7.51}$$

It follows that

$$f''(t) \equiv \int_\Omega u\, u_t dx = \sum_{i=1}^n \int_\Omega \left(-a_i\, |D_i u|^{p_i} + b_i\, |u|^{\sigma_i}\right) dx. \tag{7.52}$$

Multiplying (7.52) by a positive constant $\lambda$, adding to (7.50) and taking into account the assumption $E(0) \leq 0$, we arrive at the inequality

$$E(t) + \lambda \sum_{i=1}^n \int_\Omega \left(-a_i\, |D_i u|^{p_i} + b_i\, |u|^{\sigma_i}\right) dx + \int_0^t \int_\Omega u_t^2 dsdx \leq \lambda f''(t).$$

Using (7.45) we can write it in the form

$$\begin{aligned} &\sum_{i=1}^n \int_\Omega \left(\left(\frac{1}{p_i} - \lambda\right) a_i\, |D_i u|^{p_i} + \left(\lambda - \frac{1}{\sigma_i}\right) b_i\, |u|^{\sigma_i}\right) dx \\ &\qquad + \int_0^t \int_\Omega u_t^2 dx \leq \lambda f''(t). \end{aligned} \tag{7.53}$$

Assumption (7.47) allows one to choose $\lambda > 0$ in such a way that

$$\begin{aligned} \forall\, i = 1, \ldots, n \qquad & \frac{1}{\sigma_i^-} \leq \lambda \leq \frac{1}{p_i^+}, \\ \exists\, j \in \{1, \ldots, n\}: \qquad & \frac{1}{\sigma_j^-} < \lambda \leq \frac{1}{p_j^+}. \end{aligned}$$

Keeping on the left-hand side the term corresponding to the sub-index $j$ and dropping the others, we reduce the differential inequality (7.53) to the form

$$\left(\lambda - \frac{1}{\sigma_j^-}\right) b_j^- \int_\Omega |u|^{\sigma_j}\, dx + \int_0^t \int_\Omega u_t^2 dx \le \lambda f''(t). \tag{7.54}$$

## 7.4.2 Second-Order Differential Inequalities

To prove Theorem 7.5 we consider separately the cases $p_j^+ > 2$ and $1 < p_j^+ \le 2$. Let us denote by $t^*$ the time of existence of the solution $u$:

$$t^* = \sup\{t > 0 : \|u\|_{\infty,\Omega} < \infty \text{ for } t < t^*\}.$$

The theorem is proved if we show that $t^*$ is finite.

**Case 1**: $p_j^+ > 2$. Dropping the first nonnegative term on the left-hand side of (7.54), we have

$$0 < \int_0^t \int_\Omega u_t^2 dx \le \lambda f''(t). \tag{7.55}$$

Using Hölder's inequality and (7.55), we may write the following chain of relations:

$$\begin{aligned}
\left(f'(t) - f'(0)\right)^2 &= \left(\int_0^t \frac{d}{dt}\left(\frac{1}{2}\int_\Omega u^2 dx\right) dt\right)^2 = \left(\int_0^t \int_\Omega u u_t dx\right)^2 \\
&\le \left(\int_0^t \|u_t\|_{2,\Omega} \|u\|_{2,\Omega} dt\right)^2 \le \|u_t\|^2_{2,\Omega\times(0,t)} \|u\|^2_{2,\Omega\times(0,t)} \\
&\le \lambda f''(t) \int_0^t \int_\Omega u^2 dx dt = 2\lambda f''(t) f(t),
\end{aligned}$$

which leads to the second-order nonlinear ordinary differential inequality for $f(t)$:

$$\begin{aligned}
&\left(f'(t) - f'(0)\right)^2 \le 2\lambda f''(t) f(t) \le \frac{2}{p_+} f''(t) f(t) \qquad \text{for } t > 0, \\
&f(0) = 0, \qquad f'(t) > 0.
\end{aligned} \tag{7.56}$$

We want to prove that the function $f'(t)$ becomes unbounded at a finite instant. Let us show first that if $f'(t)$ exists for all $t > 0$, it is necessary that $f'(t) \nearrow \infty$ as $t \to \infty$. Assume the contrary: there exists a positive constant $L$ such that $0 \le f'(t) \le L$ for all $t > 0$. The function $f$ is strictly positive by definition, $f''$ is strictly positive and increasing by virtue of (7.55), while $f'$ is strictly positive and monotone increasing due to the Lagrange intermediate value theorem. Fix an arbitrary $\tau > 0$. By assumption $f(t) \le f(0) + Lt = Lt$, and by virtue of (7.56)

$$\frac{p^+}{2Lt} \le \frac{f''(t)}{\left(f'(t) - f'(0)\right)^2} \quad \text{in } (\tau, t).$$

A straightforward integration of this inequality over the interval $(\tau, t)$ leads to the contradiction:

$$L \geq f'(t) - f'(0) \geq \frac{f'(\tau) - f'(0)}{1 - \dfrac{p^+}{2L}(f'(\tau) - f'(0)) \ln \dfrac{t}{\tau}} \geq L + 1,$$

provided that

$$t \geq \tau \exp\left(\frac{2L}{p^+}\left(\frac{1}{f'(\tau) - f'(0)} - \frac{1}{L+1}\right)\right).$$

Thus, if $f'(t)$ exists for all $t > 0$, then $f'(t) \nearrow \infty$ as $t \to \infty$. It follows that there exists a moment $t_0$ and a constant $1 < \nu < \frac{p_j^+}{2}$ such that $(f'(t) - f'(0))^2 \geq \frac{2\nu}{p_j^+}(f'(t))^2$ for $t \geq t_0$. This observation allows one to continue (7.56) as follows:

$$\nu(f'(t))^2 \leq \frac{p_j^+}{2}\left(f'(t) - f'(0)\right)^2 \leq f''(t) f(t) \qquad \text{for } t \geq t_0.$$

Assuming that $f'(t)$ remains bounded for all finite $t$, we write this inequality in the form

$$\frac{\nu f'(t)}{f(t)} \leq \frac{f''(t)}{f'(t)}, \quad t > t_0,$$

and integrate over the interval $(t_0, t)$:

$$\frac{f'(t)}{f^\nu(t)} \geq K, \qquad K = \frac{f'(t_0)}{f^\nu(t_0)}. \tag{7.57}$$

Integrating (7.57) we obtain the estimate

$$f^{\nu-1}(t) \geq \frac{f^{\nu-1}(t_0)}{1 - (\nu - 1)\dfrac{f'(t_0)}{f(t_0)}(t - t_0)} \nearrow \infty \quad \text{as } t \nearrow T = t_0 + \frac{f(t_0)}{(\nu - 1) f'(t_0)}.$$

Substituting this estimate into (7.57) we conclude that $f'(t)$ becomes infinite as $t \to T-$.

We are now in position to show that $t^*$ (the blow-up moment) is finite. The assumption $t^* = \infty$ leads to the following contradiction: on one hand, $\|u(\cdot, t)\|_{\infty,\Omega} < \infty$ on every finite interval, on the other hand

$$\infty > |\Omega| \|u\|^2_{\infty,\Omega} \geq \int_\Omega u^2(x,t)dx \equiv 2f'(t) \nearrow \infty \quad \text{as } t \to T - .$$

This completes the proof of Theorem 7.5 in the case $p_j^+ > 2$.

**Case 2:** $1 < p_j^+ \leq 2$. Applying Hölder's inequality and inequality (1.6), we continue (7.54) as follows:

$$\min\left\{(f'(t))^{\frac{\sigma_j^+}{2}}, (f'(t))^{\frac{\sigma_j^-}{2}}\right\} = \min\left\{\|u\|_{2,\Omega}^{\sigma_j^+}, \|u\|_{2,\Omega}^{\sigma_j^-}\right\}$$
$$\leq C\int_\Omega |u|^{\sigma_j}\,dx \leq C'\,f''(t).$$

In our conditions on the data $f$, $f'$ and $f''$ are strictly positive for $t > 0$, and there exists some $t_0$ such that $f'(t) > 1$ for all $t \geq t_0$. It follows that

$$(f'(t))^{\frac{\sigma_j^-}{2}} \leq C'\,f''(t) \quad \text{for } t \geq t_0, \qquad f'(t) \geq f'(0) = \frac{1}{2}\|u_0\|_{2,\Omega}^2 > 0.$$

Integration of this inequality leads to the following one:

$$f'(t) \geq \left((f'(0))^{1-\frac{\sigma_j^-}{2}} - \frac{1}{C}\left(\frac{\sigma_j^-}{2} - 1\right)t\right)^{\frac{2}{2-\sigma_j^-}}, \qquad \sigma_j^- > 2.$$

The right-hand side of this formula becomes infinite at a finite moment $T$. This leads to the desired conclusion because $t^*$ cannot exceed $T$:

$$|\Omega|\,\|u\|_{\infty,\Omega}^2 \geq \int_\Omega u^2(x,t)\,dx \equiv 2\,f'(t) \to \infty \quad \text{as } t \to T-.$$

The proof of Theorem 7.5 is completed.

## 7.5 Remarks

There is a vast literature devoted to study the blow-up effect in solutions of evolution equations with constant nonlinearity. An exhaustive review of the existing methods and known results can be found in the papers and monographs [137–139, 209, 222, 233]. Following [137], one may group the most common questions related to the study of this phenomenon in the following way: given a nonlinear PDE, does it admit solutions that blow-up in finite time? in case of the affirmative answer, when does this happen? what is the structure of the blow-up set (a finite set of points, a set of nonzero measure, or the whole of the space domain)? what is the structure of the solution when time approaches the blow-up instant from below? Each of this issues requires specific methods of study. Numerous approaches have been proposed to answer these questions in the case of equations with constant nonlinearity, but most of them do not work in the case of an equation with variable growth conditions. Here we are confined to the cases of semi-linear equations, or equations with $p(x)$-growth. In both cases it is possible to derive sufficient conditions for the finite time blow-up. However, many of the questions related to the fine structure of the blow-up set, or a detailed description of the solution profile near the blow-up instant remain open.

The presentation follows the papers [52, 53], see also [60] on the blow-up of solutions of doubly nonlinear parabolic equations with variable growth conditions.

The first result on the blow-up in solutions of evolution equations with variable nonlinearity was obtained in the paper [218] for semi-linear parabolic and hyperbolic equations with local or nonlocal low-order terms

$$(a)\quad u_t = \Delta u + a(x)u^{q(x)}, \qquad (b)\quad u_t = \Delta u + a(x)\int_{\Omega} u^{q(y)}(y,t)dy \tag{7.58}$$

for $1 < q(x) < \infty$, $0 < a_0 \leq a(x) < \infty$. The proof was based on the Kaplan eigenfunction method [177]. Sufficient and necessary conditions for the finite time blow-up in solutions of the semilinear Eq. (7.58)(a) with $a \equiv 1$ are found in [130]. The equation is considered in the domain $\Omega \times (0, T)$. The authors show that in the case $\Omega = \mathbb{R}^n$ all solutions blow up in finite time if $1 < q^- \leq q^+ \leq 1 + \frac{2}{n}$, the equation admits global non-trivial solutions if $q^- > 1 + \frac{2}{n}$, whereas for $q^- < 1 + \frac{2}{n} < q^+$ there are both global in time solutions and solutions that blow-up in a finite time. It is shown that in the case of a bounded domain $\Omega$ the existence of global in time solutions depends on the size of $\Omega$. A lower bound for the blow-up moment in solutions of the Dirichlet problem for Eq. (7.58)(a) is derived in [67]. Conditions of global existence and finite time blow-up of solutions for systems of parabolic equations with variable nonlinearity were obtained in [144, 248]. Existence and nonexistence of global in time solutions for systems of semilinear equations of the type (7.58) were studied in [68]. A special case of blow-up of solutions of the semi-linear equation with variable exponent of nonlinearity and positive initial energy is addressed in [257, 263]. For the system of semilinear equations with non-power nonlinearity

$$u_t = \Delta u + e^{m(x)u+p(x)v}, \qquad v_t = \Delta v + e^{q(x)u+n(x)v}$$

conditions of the finite time blow-up and its rates were derived in [198].

The techniques of study of the finite time blow up or vanishing phenomena are applicable to equations and systems which do not belong to any of the already considered types. Let us present a few examples of such equations.

### *7.5.1 Higher-Order Equations*

The following class of the plate equations with memory is considered in [39]:

$$
\begin{aligned}
& u_{tt} + \Delta^2 u - \Delta_{\mathbf{p}(x,t)} u + \int_0^t g(t-s)\Delta u(s)ds - \varepsilon \Delta u_t + f(u) = 0 \text{ in } Q_T, \\
& u = \partial u/\partial \nu = 0, \ \Gamma_T = \partial\Omega \times (0,T), \\
& u(x,0) = u_0(x), \ u_t(x,0) = u_1(x), \ x \in \Omega,
\end{aligned} \tag{7.59}
$$

where $\varepsilon > 0$ is a constant, $\partial u/\partial \nu$ is the outer normal derivative and

$$
-\Delta_{\mathbf{p}(x,t)} u = -\sum_{i=1}^{n} D_i \left( |D_i u|^{p_i(x,t)-2} D_i u \right), \quad \mathbf{p} = (p_1, p_2, \dots, p_n),
$$

is the $\mathbf{p}(x,t)$-Laplace operator, $g \geq 0$ is a memory kernel which decays exponentially, $f(u)$ is a given nonlinear function. Problem (7.59) with the memory term

$$
\int_0^t g(t-s)\Delta u(s)ds
$$

can be regarded as a fourth-order viscoelastic plate equation with the low-order perturbation of $\mathbf{p}(x,t)$-Laplace type. Another interpretation is that of the equation of elastoplastic flow with memory effects. In the recent years, the plate equations with lower-order perturbation of $p$-Laplacian type with constant $p$ were intensively studied. Paper [18] is devoted to study problem (7.59) with $p = const$. Results on the local and global in time existence and uniqueness of weak solutions for problem (7.59) can be found in [39]. This is the first example of an equation that combines the memory term and a lower-order perturbation of $\mathbf{p}(x,t)$-Laplace type.

## 7.5.2 $p(x,t)$-Curl Systems Arising in Electromagnetism

A class of $p$-curl systems arising in electromagnetism was considered in [40]. Let $\Omega$ be a bounded open subset of $\mathbb{R}^3$ with Lipschitz boundary $\Gamma$, $Q_T = \Omega \times (0,T)$ and $\Sigma_T = \Gamma \times (0,T)$. The problem is described by the system

$$
\begin{aligned}
& \partial_t h + \nabla \times \left( |\nabla \times h|^{p-2} \nabla \times h \right) = f(h), \quad \nabla \cdot h = 0 \quad \text{in } Q_T, \\
& |\nabla \times h|^{p-2} \nabla \times h \times n = 0, \quad h \cdot n = 0 \quad \text{on } \Sigma_T, \\
& h(\cdot, 0) = h_0 \quad \text{in } \Omega,
\end{aligned} \tag{7.60}
$$

where $h$ is the unknown magnetic field, $\nabla\times = \text{curl}$, $\nabla\cdot = \text{div}$, $6/5 < p = const < \infty$. The nonlinear term $f(h)$ represents sources or sinks and satisfies the compatibility conditions $\nabla \cdot f(h) = 0$ in $Q_T$ and $f(h) \cdot n = 0$ on $\Sigma_T$. In [40], local and global in time solvability of the problem is proved in dependence on the values of $\sigma$ and $p$. Qualitative properties of solutions, such as the finite time blow-up (an explosion of the magnetic field), or the finite time extinction (disappearance of the magnetic field), were studied. The same methods are applicable to equations of the type (7.60) with variable $p(x, t)$.

### 7.5.3 Navier-Stokes Type Equations with Variable Structure

A modified Oberbeck-Boussinesq problem with a thermo-absorption term is considered in paper [31]. The problem consists in solving the system of equations

$$\begin{aligned}
&\operatorname{div} \mathbf{u} = 0 \quad \text{in} \quad Q_T = \Omega \times (0, T), \quad \Omega \subset \mathbf{R}^n, \ n \geq 2, \\
&\frac{\partial \mathbf{u}}{\partial t} + \mathbf{div}(\mathbf{u} \otimes \mathbf{u}) = \mathbf{f}(\theta) - \alpha \, |\mathbf{u}|^{\sigma(\theta)} \, \mathbf{u} - \nabla p + \nu \Delta \mathbf{u} \quad \text{in} \quad Q_T, \\
&\frac{\partial \theta}{\partial t} + \mathbf{u} \cdot \nabla \theta = \Delta \varphi(\theta) \quad \text{in} \quad Q_T,
\end{aligned} \tag{7.61}$$

endowed with the initial and boundary conditions

$$\begin{aligned}
\mathbf{u} &= \mathbf{u}_0 && \text{in} \quad \Omega \quad \text{for} \quad t = 0, \\
\mathbf{u} &= \mathbf{0} && \text{on} \quad \Gamma_T = \partial\Omega \times (0, T), \\
\theta &= \theta_0 && \text{in} \quad \Omega \quad \text{for} \quad t = 0, \\
\theta &= \theta_0 && \text{on} \quad \Gamma_T.
\end{aligned} \tag{7.62}$$

Here $\mathbf{u}$ is the velocity, $p$ is the pressure, $\theta$ is the absolute temperature, $\nu$ is the constant kinematic viscosity and $\varphi$ is a nonlinear function which usually expresses the thermal conductivity. In this model, the exponent $\sigma(\theta)$ is an unknown function because it depends on one of the components of the vector-valued solution $(\mathbf{u}, p, \theta)$. For a detailed discussion of the Oberbeck-Boussinesq model and pertinent results see, e.g., [41, 111, 112, 117, 140, 262] and references therein. Existence of a weak solution for any $n \geq 2$ and its uniqueness in the case $n = 2$ were proved for problem (7.61) and (7.62) under certain restrictions on $\sigma$, $\varphi$ and $f$. The phenomenon of finite time extinction and the rate of decay as $t \to \infty$, power-like and exponential, were also studied.

Paper [32] deals with the generalized Navier-Stokes equations

$$\begin{aligned}
&\operatorname{div} \mathbf{u} = 0, \\
&\frac{\partial \mathbf{u}}{\partial t} + \mathbf{div}(\mathbf{u} \otimes \mathbf{u}) = \mathbf{f} - \nabla p + \mathbf{div}\, \mathbf{S} \quad \text{in} \quad Q_T,
\end{aligned} \tag{7.63}$$

endowed with the initial and boundary conditions

$$\mathbf{u} = \mathbf{u}_0 \text{ in } \Omega \text{ for } t = 0, \qquad \mathbf{u} = \mathbf{0} \text{ on } \Gamma_T. \tag{7.64}$$

Here $\mathbf{u}$ denotes the velocity field, $p$ stands for the pressure divided by the constant density, and $\mathbf{f}$ is the external forces field. $\mathbf{S}$ is the deviatoric stress tensor which depends on $(\mathbf{x}, t)$ and on the strain rate tensor $\mathbf{D}$:

$$\mathbf{S} = \nu(D_{II})\mathbf{D}, \quad \mathbf{D} \equiv \mathbf{D}(\mathbf{u}) = \frac{1}{2}\left(\nabla\,\mathbf{u} + \nabla\,\mathbf{u}^T\right), \quad D_{II} = \frac{1}{2}|\mathbf{D}|^2,$$

being $D_{II}$ the second invariant of $\mathbf{D}$. Many constitutive laws have been proposed in the rheological literature to model diverse fluids under different flow conditions (see, e.g., [41, 111, 112, 117, 140, 262, 277]). It is assumed that a general deviatoric stress tensor $\mathbf{S}$ has a variable $q$-structure in the following sense:

(A) $\mathbf{S} : Q_T \times M^N_{\text{sym}} \to M^N_{\text{sym}}$ is a Carathéodory function;
(B) growth condition:
$|\mathbf{S}(\mathbf{x}, t, \mathbf{A})| \le C|\mathbf{A}|^{q(\mathbf{x},t)-1}$ for all $\mathbf{A}$ in $M^N_{\text{sym}}$ and for a.a. $(\mathbf{x}, t)$ in $Q_T$;
(C) coercivity condition:
$\mathbf{S}(\mathbf{x}, t, \mathbf{A}) : \mathbf{A} \ge C|\mathbf{A}|^{q(\mathbf{x},t)}$ for all $\mathbf{A}$ in $M^N_{\text{sym}}$ and for a.a. $(\mathbf{x}, t)$ in $Q_T$;
(D) monotonicity condition:
$(\mathbf{S}(\mathbf{x}, t, \mathbf{A}) - \mathbf{S}(\mathbf{x}, t, \mathbf{B})) : (\mathbf{A} - \mathbf{B}) > 0$ for all $\mathbf{A}$ and $\mathbf{B}$ in $M^N_{\text{sym}}$, with $\mathbf{A} \ne \mathbf{B}$, and for a.a. $(\mathbf{x}, t)$ in $Q_T$.

Here $C$ is a positive constant and $M^N_{\text{sym}}$ is the vector space of all symmetric $N \times N$ matrices, equipped with the inner product $\mathbf{A} : \mathbf{B}$ and the norm $|\mathbf{A}| = \sqrt{\mathbf{A} : \mathbf{A}}$. It is assumed that $q \in C_{\log}(Q_T)$ and satisfies the inequalities

$$1 < \alpha \le q(x, t) \le \beta < \infty \text{ for all } (x, t) \in Q_T.$$

Unlike the case when a fluid satisfies conditions (B)–(C) with constant exponent $q$, the fluids that satisfy conditions (B)–(C) with variable $q$ do not fall into any of the studied subclasses of non-Newtonian fluids, such as pseudo-plastic, or dilatant fluids. For example, such fluids may go back and forth from the consistency of a liquid to that of a gel with the response times of order of milliseconds. This is also the case of electro-rheological and magneto-rheological fluids whose rheological properties can be controlled through application of an electric or magnetic field. An example of such fluids are suspensions dispersed with some polymeric colloids which exhibit trembling shear behavior under the influence of electric or magnetic fields.

Results on the existence and qualitative properties of weak solution to problem (7.63)–(7.64), such as finite time of extinction and large time behavior, can be found in [32]. This study is based on methods and results from [36, 117, 216, 262, 277–279]. A deep insight into the mathematical theory of electro-rheological fluids is offered in the monograph [228].

# Chapter 8
# Equations with Double Isotropic Nonlinearity

In this chapter we study the Dirichlet problem for the class of equations with double variable nonlinearity

$$\begin{cases} u_t = \operatorname{div}\left(a(z,u)|u|^{\alpha(z)}|\nabla u|^{p(z)-2}\nabla u\right) + f(z) \quad z=(x,t)\in Q, \\ u(x,0) = u_0(x) \text{ in } \Omega, \quad u=0 \quad \text{on} \quad \Gamma = \partial\Omega\times(0,T]. \end{cases} \tag{8.1}$$

Unlike the already considered equations with nonstandard growth conditions, Eq. (8.1) is not monotone, which entails specific difficulties. Let us introduce the functions

$$\begin{aligned} \gamma(z) &= \frac{\alpha(z)}{p(z)-1}, \\ v(z) &= \int_0^u |s|^{\gamma(z)}\,ds = \frac{u|u|^{\gamma(z)}}{\gamma(z)+1}, \\ u(z) &= \Phi_0(z,v) = \frac{|v|^{\frac{-\gamma}{1+\gamma}}v}{(1+\gamma)^{\frac{1}{1+\gamma}}} \end{aligned} \tag{8.2}$$

and write problem (8.1) in the formally equivalent form

$$\begin{cases} \partial_t\Phi_0(z,v) = \operatorname{div}\left(b(z,v)|\nabla v + \mathscr{B}(v)|^{p(z)-2}(\nabla v + \mathscr{B}(v))\right) + f \quad \text{in } Q, \\ v(x,0) \equiv v_0(x) = \dfrac{u_0|u_0|^{\gamma(x,0)}}{1+\gamma(x,0)} \text{ in } \Omega, \quad v = 0 \text{ on } \Gamma \end{cases} \tag{8.3}$$

with

$$b(z,v) \equiv a(z,\Phi(z,v)), \qquad \mathscr{B}(v) = -\nabla\gamma\cdot\int_0^u |s|^{\gamma(z)}\ln|s|\,ds.$$

Problem (8.3) will be the subject of the further study.

 
S. Antontsev and S. Shmarev, *Evolution PDEs with Nonstandard Growth Conditions*, Atlantis Studies in Differential Equations 4,
DOI 10.2991/978-94-6239-112-3_8

A solution to problem (8.3) is obtained as the limit of solutions of the regularized problems for equations of the type studied in Chap. 4. The comparison principle and uniqueness are proved for the solutions subject to some additional restrictions, but under weaker assumptions on the data, their proofs are practically independent of the proof of the existence theorem. To be precise, the comparison principle and uniqueness are true for the weak solutions with $\partial_t \Phi_0(z, v) \in L^1(Q)$. In order to ensure that this class of solutions is nonempty, in the end of the chapter we show that the constructed solution belongs to this class, provided that the data of the problem satisfy some additional conditions.

Equations of this type with constant exponents $\alpha$ and $p$ arise in the mathematical modelling of various physical processes such as flows of incompressible turbulent fluids or gases in pipes, processes of filtration in porous media, or glaciology—see, [15, 93, 109, 110, 171] and the further references therein. The questions of existence and uniqueness of solutions to equations like (8.1) and (8.3) with constant exponents $\alpha$ and $p$ were studied by many authors—[109, 110, 119, 166, 188, 189, 120, 244]. Doubly nonlinear equations with the exponents depending only on $x$ were studied in [6, 87]. Solvability of the problem

$$\begin{cases} \partial_t(|u|^{m(x)-2}u) = \Delta_{p(x)}u & \text{in } \Omega \times (0, T], \\ |u(x,0)|^{m(x)-2}u(x,0) = v_0 \text{ in } \Omega, \end{cases}$$

under the homogeneous Dirichlet or Neumann boundary conditions is studied in [6]. The problem is considered as an abstract Cauchy problem in the variable exponent spaces under the assumptions that the functions $p(x), m(x) : \Omega \mapsto (1, \infty)$ are measurable and

$$p(x) \in C_{\log}(\Omega), \quad \operatorname*{ess\,inf}_{\Omega}(p^*(x) - m(x)) > 0, \quad p^*(x) = \frac{np(x)}{\max\{n - p(x), 0\}}.$$

Paper [87] deals with the class of equations

$$\partial_t(u + |u|^{m(x)-2}u) - \Delta_p u + |u|^{q(x)-2}u = f$$

with constant $p$ and independent of $t$ exponents $m(x)$ and $q(x)$.

We are especially interested in the cases when the variable exponents of nonlinearity may depend on $t$.

## 8.1 Assumptions and Results

The existence result is established for the problem (8.3) which is formally equivalent to problem (8.1). The coefficient $a(z, r)$ and the exponents on nonlinearity $p(z)$, $\alpha(z)$ satisfy the conditions:

1. $a(z, r)$ is a Carathéodory function, there exists positive constants $a^{\pm}$ such that

$$\forall z \in Q,\ r \in \mathbb{R} \qquad a^{-} \le a(z, r) \le a^{+} < \infty, \tag{8.4}$$

2. $\alpha(z)$, $p(z)$ are measurable and bounded in $Q$, there exist constants $\alpha^{\pm}$, $p^{\pm}$ such that

$$-1 < \alpha^{-} \le \alpha(z) \le \alpha^{+} < \infty, \quad 1 < p^{-} \le p(z) \le p^{+} < \infty, \quad \alpha^{-} + p^{-} > 1, \tag{8.5}$$

3. the exponent $\gamma(z) = \dfrac{\alpha(z)}{p(z) - 1}$ satisfies

$$|\nabla \gamma(z)|^{p(z)} \in L^{1}(Q), \quad \partial_t \gamma(z) \in L^{2}(Q). \tag{8.6}$$

**Definition 8.1** A function $v(z)$ is called weak solution of problem (8.3) if

1. $v \in \mathbf{W}(Q) \cap L^{\infty}(Q)$, $\partial_t \Phi_0(z, v) \in \mathbf{W}'(Q)$,
2. for every $\phi \in \mathbf{W}(Q)$

$$\int_Q \left(\phi\, \partial_t \Phi_0(z, v) + b(z, v)|\nabla v + \mathscr{B}(v)|^{p-2}(\nabla v + \mathscr{B}(v)) \cdot \nabla \phi - f\, \phi\right) dz = 0, \tag{8.7}$$

3. $\forall \phi(x) \in C_0^{\infty}(\Omega)$

$$\int_{\Omega} \Phi_0(z, v(z))\, \phi(x)\, dx \to \int_{\Omega} \Phi_0((x, 0), v_0(x))\, \phi(x)\, dx \quad \text{as } t \to 0.$$

The main existence result is given in the following theorem.

**Theorem 8.1** *Let conditions* (8.4)–(8.6) *be fulfilled. Assume that* $p \in C_{\log}(\overline{Q})$. *Then for every* $f \in L^{1}(0, T; L^{\infty}(\Omega))$, $u_0$, $v_0 \in L^{\infty}(\Omega)$ *problem* (8.3) *has at least one weak solution* $v(z)$ *in the sense of* Definition 8.1.

The uniqueness result is proved under the assumption that either $a(z, v) \equiv 1$, or $\alpha(z) \equiv 0$. In the latter case $\Phi_0(z, v) \equiv v$ so that the equation transforms into the evolutional $p(z)$-Laplacian equation.

**Theorem 8.2** *Let us assume that the data of problem* (8.3) *satisfy the conditions*

$$\begin{cases} a(z, u) \equiv 1, \quad \Phi_0(z, s) \in C^{1}(Q \times \mathbb{R}), \\ \Phi_0(z, s) \text{ is an increasing function of } s. \end{cases} \tag{8.8}$$

*Let conditions* (8.4) *and* (8.5) *be fulfilled. Then for every weak solutions* $v_1$, $v_2$ *with* $\partial_t \Phi_0(z, v_i) \in L^{1}(Q)$, *and a.e.* $t \in (0, T)$

$$\|\Phi_0(z, v_1(z)) - \Phi_0(z, v_2(z))\|_{1,\Omega} \\ \leq \|\Phi_0(x, 0, v_{01}) - \Phi_0(x, 0, v_{02})\|_{1,\Omega} + \|f_1 - f_2\|_{1,Q}. \tag{8.9}$$

**Theorem 8.3** *Let $v_1, v_2$ be two weak solutions of problem* (8.3) *with $\alpha(z) \equiv 0$ and the data $v_{0i}$, $f_i$. Let the coefficient $a(z, s)$ be Hölder-continuous with respect to $s$:*

$$|a(z, s) - a(z, r)| \leq C\,|s - r|^{\beta}, \quad C = const, \quad \beta \in (0, 1].$$

*If conditions* (8.4) *and* (8.5) *are fulfilled and $\partial_t u_i \in L^1(Q)$, then for a.e. $t \in (0, T)$*

$$\|v_1(x, t) - v_2(x, t)\|_{L^1(\Omega)} \leq \|v_{01} - v_{02}\|_{L^1(\Omega)} + \|f_1 - f_2\|_{L^1(Q)}. \tag{8.10}$$

The uniqueness is proved in a narrower class of functions than the existence, but since the proofs of Theorems 8.2 and 8.3 are practically independent of the proof of Theorem 8.1, the conditions on the exponents $\alpha(z)$, $p(z)$ are less restrictive.

## 8.2 Regularized Problems

A solution of problem (8.3) has the form $v = \Phi_0^{-1}(z, u)$ where $u$ is obtained as the limit of the sequence of solutions of the regularized problems

$$\begin{cases} u_t = \operatorname{div}\left(\mathscr{A}_{\varepsilon,K}(z, u)|\nabla u|^{p(z)-2}\nabla u\right) + f(z) & \text{in } Q, \\ u(x, 0) = u_0 \text{ in } \Omega, \quad u = 0 \text{ on } \Gamma \end{cases} \tag{8.11}$$

with the coefficient $\mathscr{A}_{\varepsilon,K}(z, u) = a(z, u)(\varepsilon + \min\{K, |u|\})^{\alpha(z)}$, depending on the given parameters $\varepsilon \in (0, 1)$, $K > 0$. For every $\varepsilon \in (0, 1)$ and $1 < K < \infty$ the coefficient $\mathscr{A}_{\varepsilon,K}(z, u)$ is separated away from zero and infinity, which allows us to treat problem (8.11) as the Dirichlet problem for the evolution $p(z)$-Laplacian studied in Chap. 4. The next assertion is an immediate byproduct of Theorems 4.1 and 4.3.

**Theorem 8.4** *For every $u_0 \in L^2(\Omega)$, $f \in L^2(Q)$, $\varepsilon > 0$, $K > 0$ problem* (8.11) *has at least one energy solution in the sense of* Definition 4.1 *$u \in C([0, T]; L^2(\Omega)) \cap \mathbf{W}(Q)$, $u_t \in \mathbf{W}'(Q)$. Moreover, if $u_0 \in L^\infty(\Omega)$, $f \in L^1(0, T; L^\infty(\Omega))$, then this solution belongs to $L^\infty(Q)$ and obeys the estimate*

$$\|u\|_{\infty,Q} \leq \|u_0\|_{\infty,\Omega} + \int_0^T \|f(\cdot, s)\|_{\infty,\Omega}\, ds \equiv K_0. \tag{8.12}$$

The solution is obtained as the limit of the sequence of Galerkin's approximations $u^{(m)} = \sum_i^m c_{i,m}(t)\psi_m(x)$, where $\{\psi_m\}$ is the system of eigenfunctions of problem (1.30). For every finite $m$ the system of ODEs for the coefficients $c_i(t)$, $i = 1, \dots, m$, can be solved on the interval $[0, T]$. The functions $u^{(m)}$ admit the uniform estimate

$$\|u^{(m)}\|_{L^\infty(0,T;L^2(\Omega))} + \int_Q |\nabla u^{(m)}|^{p(z)}\,dz \le C\left(\|u_0\|^2_{2,\Omega} + \|f\|^2_{2,Q}\right),$$

which allows one to pass to the limit in the sequence $\{u^{(m)}\}$. The justification of the limit passage as $m \to \infty$ is based on the monotonicity of the operator $|s|^{p-2}s : \mathbb{R}^n \mapsto \mathbb{R}^n$ and the formulas of integration by parts in $t$ for the elements of the spaces $\mathbf{W}(Q)$, $\mathbf{W}'(Q)$ given Lemma 1.20.

According to (8.12), for $u_0 \in L^\infty(\Omega)$, $f \in L^1(0, T; L^\infty(\Omega))$ and $K > K_0$ the solutions of the regularized problem (8.11) do not depend on the parameter $K$ and, in fact, are solutions of the problems

$$\begin{cases} \partial_t u_\varepsilon = \operatorname{div}\left(\mathscr{A}_\varepsilon(z, u_\varepsilon)|\nabla u_\varepsilon|^{p(z)-2}\nabla u_\varepsilon\right) + f(z) & \text{in } z \in Q, \\ u_\varepsilon(x,0) = u_0 \text{ in } \Omega, \quad u_\varepsilon = 0 \text{ on } \Gamma, \\ \mathscr{A}_\varepsilon(z, u_\varepsilon) = a(z, u_\varepsilon)\,(\varepsilon + |u_\varepsilon|)^{\gamma(z)(p(z)-1)}, \end{cases} \tag{8.13}$$

with $\varepsilon > 0$ and $\gamma(z) = \dfrac{\alpha(z)}{p(z)-1}$. For every $\phi \in \mathbf{W}(Q)$ the weak energy solutions $u_\varepsilon$ of problem (8.13) satisfy the equality

$$\int_Q \left[\phi\,\partial_t u_\varepsilon + \mathscr{A}_\varepsilon(z, u_\varepsilon)|\nabla u_\varepsilon|^{p(z)-2}\nabla u_\varepsilon \cdot \nabla\phi - f\phi\right] dz = 0. \tag{8.14}$$

### 8.2.1 A Priori Estimates

It is assumed that the conditions of Theorem 8.1 are fulfilled. Let us introduce the vector-valued function $\mathbf{G}_\varepsilon = \mathscr{A}_\varepsilon(z, u_\varepsilon)|\nabla u_\varepsilon|^{p(z)-2}\nabla u_\varepsilon$ and write Eq. (8.13) in the form

$$\partial_t u_\varepsilon = \operatorname{div}\mathbf{G}_\varepsilon + f(z). \tag{8.15}$$

Define the functions

$$v_\varepsilon(z) = \int_0^{u_\varepsilon(z)} (\varepsilon + |s|)^{\gamma(z)}\,ds, \qquad \gamma(z) = \frac{\alpha(z)}{p(z)-1} \ge \gamma^- > -1. \tag{8.16}$$

The following formulas hold:

$$
\begin{aligned}
&\text{(a)} \quad \nabla v_\varepsilon = (\varepsilon + |u_\varepsilon|)^\gamma \, \nabla u_\varepsilon + \nabla \gamma \int_0^{u_\varepsilon} (\varepsilon + |s|)^\gamma \, \ln(\varepsilon + s) \, ds, \\
&\text{(b)} \quad \partial_t v_\varepsilon = (\varepsilon + |u_\varepsilon|)^\gamma \, \partial_t u_\varepsilon + \gamma_t \int_0^{u_\varepsilon} (\varepsilon + |s|)^\gamma \, \ln(\varepsilon + s) \, ds.
\end{aligned}
\tag{8.17}
$$

**Lemma 8.1** *The functions* $\mathbf{G}_\varepsilon$ *satisfy the estimate*

$$
\int_Q |\mathbf{G}_\varepsilon|^{p'(z)} \, dz \le C, \qquad p'(z) = \frac{p(z)}{p(z) - 1}, \tag{8.18}
$$

*with an independent of* $\varepsilon$ *constant* $C = C(a^\pm, \ p^\pm, \ \alpha^\pm, \ K_0, \ \|f\|_{2,Q}, \ \|\nabla\gamma\|_{p(\cdot),Q})$.

*Proof* Since $u_\varepsilon \in \mathbf{W}(Q) \cap L^\infty(Q)$, then $v_\varepsilon \in \mathbf{W}(Q)$ for every $\varepsilon > 0$. The first inclusion immediately follows from (8.17)(a) and the estimates $|u_\varepsilon| \le K_0$, $|v_\varepsilon| \le \dfrac{1}{\gamma^- + 1}(1 + K_0)^{\gamma^+ + 1}$:

$$
|\nabla v_\varepsilon|^{p(z)} \le C(a^+, \varepsilon, K_0) \left[ |\nabla u_\varepsilon|^{p(z)} + |\nabla \gamma|^{p(z)} \right] \in L^1(Q).
$$

Choosing $v_\varepsilon \in \mathbf{W}(Q)$ for the test-function in (8.14) and applying (1.40) we find that $\forall$ a.a. $t_2, t_1 \in (0, T)$

$$
\begin{aligned}
\mu_\varepsilon(u_\varepsilon, v_\varepsilon) &+ \int_{t_1}^{t_2} \int_\Omega (\varepsilon + |u_\varepsilon|)^\gamma \mathbf{G}_\varepsilon \cdot \nabla u_\varepsilon \, dz \\
&= \int_{t_1}^{t_2} \int_\Omega f v_\varepsilon \, dz - \int_{t_1}^{t_2} \int_\Omega \mathbf{G}_\varepsilon \cdot \nabla \gamma \int_0^{u_\varepsilon} (\varepsilon + |s|)^\gamma \, \ln(\varepsilon + |s|) \, ds dz.
\end{aligned}
$$

It follows that

$$
\frac{1}{(1 + a^+)^{\frac{1}{p^+}}} \int_{t_1}^{t_2} \int_\Omega |\mathbf{G}_\varepsilon|^{p'(z)} \, dz \le C + M \int_{t_1}^{t_2} \int_\Omega |\mathbf{G}_\varepsilon| \, |\nabla \gamma| \, dz
$$

with constants $C$ and $M$ which depend on $\gamma^\pm$, $K_0$, $\|\gamma_t\|_{2,Q}$, but are independent of $\varepsilon$. The needed estimate follows after application of Young's inequality to the last term. □

**Lemma 8.2** *Under the foregoing assumptions* $\|\partial_t u_\varepsilon\|_{\mathbf{W}'(Q)} \le C$ *with an independent of* $\varepsilon$ *constant* $C$.

*Proof* Combining (8.14) with (8.18) we find that

$$\begin{aligned}\left|\int_Q \phi\,\partial_t u_\varepsilon\,dz\right| &\le \int_Q |\mathbf{G}_\varepsilon|\,|\nabla\phi|\,dz + \int_Q |f|\,|\phi|\,dz\\ &\le \|\mathbf{G}_\varepsilon\|_{p'(\cdot),Q}\|\nabla\dot{\varphi}\|_{p(\cdot),Q} + \|f\|_{2,Q}\,\|\phi\|_{2,Q}\\ &\le C\,\|\phi\|_{\mathbf{W}(Q)}\end{aligned}$$

with an independent of $\varepsilon$ constant $C$. □

**Lemma 8.3** *The sequence of solutions of problem* (8.13) *is relatively compact in* $L^s(Q)$ *with some* $1 < s < \infty$.

*Proof* Uniform boundedness of $u_\varepsilon$ and estimate (8.18) entail the inequality

$$\begin{aligned}\int_Q |\nabla(|u_\varepsilon|^{\nu-1}u_\varepsilon)|^{p^-}\,dz &\le C_1\int_Q (\varepsilon+|u_\varepsilon|)^{(\nu-1)p^-}\,|\nabla u_\varepsilon|^{p^-}\,dz\\ &\le C_2\int_Q (\varepsilon+|u_\varepsilon|)^{\gamma(z)\,p(z)}\,|\nabla u_\varepsilon|^{p(z)}\,dz\\ &\quad + C_3\int_Q (\varepsilon+|u_\varepsilon|)^{((\nu-1)-\gamma)\frac{p(z)p^-}{p(z)-p^-}}\,dz \le C_4,\end{aligned} \tag{8.19}$$

provided that $\nu \ge 1+\gamma^+$. The uniform estimates

$$\|\partial_t u_\varepsilon\|_{\mathbf{W}'(Q)} \le C, \qquad \|\nabla\left(|u_\varepsilon|^{\nu-1}u_\varepsilon\right)\|_{p^-,Q} \le C$$

and Lemma 2.6 yield relative compactness of the sequence $\{u_\varepsilon\}$ in $L^s(Q)$ with some $s \in (1,\infty)$. □

Gathering the above estimates we may find a function $u \in L^\infty(Q)\cap L^s(Q)$ and a subsequence of $\{u_\varepsilon\}$ (we conserve for this subsequence the same notation $u_\varepsilon$) such that

$$\begin{aligned}&\|u_\varepsilon\|_{\infty,Q} \le K_0 \quad K_0 = const \text{ independent of } \varepsilon,\\ &u_\varepsilon \to u \text{ in } L^s(Q),\ 1 < s < \infty,\\ &u_\varepsilon \to u \quad \text{a.e. in } Q,\\ &\partial_t u_\varepsilon \rightharpoonup \partial_t u \quad \text{in } \mathbf{W}'(Q).\end{aligned} \tag{8.20}$$

Since the mapping $u_\varepsilon \mapsto v_\varepsilon$ is monotone, there exists the inverse function $u_\varepsilon = \Phi_\varepsilon(z, v_\varepsilon)$. By Lemma 8.1, the uniform in $\varepsilon$ boundedness of $u_\varepsilon$ yields the inequality

$$|\nabla v_\varepsilon|^{p(z)} \le C(K_0)\left[|\mathbf{G}_\varepsilon|^{p'(z)} + |\nabla\gamma|^{p(z)}\right] \in L^1(Q),$$

whence $\|\nabla v_\varepsilon\|_{p(\cdot),Q} \leq C$ with an independent of $\varepsilon$ constant $C$. It follows that there exist a subsequence of $\{v_\varepsilon\}$ and a vector-valued function $\mathbf{V} = (V_1, \ldots, V_n)$, $V_i \in L^{p'(\cdot,\cdot)}(Q)$, such that

$$|\nabla v_\varepsilon|^{p(z)-2} D_i v_\varepsilon \rightharpoonup V_i \quad \text{in } L^{p'(\cdot)}(Q).$$

On the other hand, continuity of the mapping $u_\varepsilon \mapsto v_\varepsilon$ and the convergence $u_\varepsilon \to u$ give

$$\begin{aligned}
&|v_\varepsilon| \leq C(K_0),\\
&v_\varepsilon \to v \quad \text{in } L^q(Q) \text{ for some } q \in (1, \infty),\\
&v_\varepsilon \to v \quad \text{a.e. in } Q,\\
&D_i v_\varepsilon \rightharpoonup D_i v \text{ in } L^{p(z)}(Q),\\
&v = \int_0^u |s|^\gamma \, ds = \frac{u\,|u|^\gamma}{\gamma+1} \in \mathbf{W}(Q).
\end{aligned} \tag{8.21}$$

The next step is to check that the limit function $u = \lim\limits_{\varepsilon\to 0} \Phi_\varepsilon(z, v_\varepsilon)$ is a solution of problem (8.3).

## 8.3 Existence of Weak Solution

Let us write the regularized problems (8.13) as the problem for the functions $v_\varepsilon$ defined by (8.16):

$$\partial_t \Phi_\varepsilon(z, v_\varepsilon) = \operatorname{div}\Big(b(z, v_\varepsilon)|\nabla v_\varepsilon + \mathscr{B}(v_\varepsilon)|^{p(z)-2}(\nabla v_\varepsilon + \mathscr{B}(v_\varepsilon))\Big) + f \quad \text{in } Q, \tag{8.22}$$

$$v_\varepsilon = 0 \text{ on } \Gamma, \quad v_\varepsilon(x, 0) = v_0(x) \text{ in } \Omega,$$

where

$$\begin{aligned}
&\mathscr{B}(v_\varepsilon) = -\nabla\gamma \int_0^{u_\varepsilon} (\varepsilon + |s|)^\gamma \,\ln(\varepsilon + |s|)\, ds,\\
&b(z, v_\varepsilon) \equiv a(z, u_\varepsilon), \qquad u_\varepsilon = \Phi_\varepsilon(z, v_\varepsilon).
\end{aligned}$$

For every $\varepsilon > 0$ problem (8.13) has at least one bounded solution $u_\varepsilon = \Phi_\varepsilon(z, v_\varepsilon) \in \mathbf{W}(Q)$. The corresponding function $v_\varepsilon = \Phi_\varepsilon^{-1}(z, u_\varepsilon)$ is a solution of problem (8.22) in the sense of Definition 8.1. By virtue of (8.21) and (8.20) there exist functions $u \in L^2(Q)$ and $\Lambda \in (L^{p'(\cdot)}(Q))^n$ such that for a subsequence of solutions of problem (8.22) $\{v_\varepsilon\}$ (for this subsequence we conserve the same notation $\{v_\varepsilon\}$)

$$
\begin{aligned}
& v_\varepsilon \to v \equiv \frac{1}{\gamma+1}|u|^\gamma u, \\
& \Phi_\varepsilon(z, v_\varepsilon) \to \Phi_0(z, v) \equiv u, \\
& \mathscr{B}(v_\varepsilon) \to \mathscr{B}(v) = \frac{\nabla\gamma}{\gamma+1} v\,(1-(1+\gamma)\ln|v|) \\
& \qquad = \frac{\nabla\gamma}{\gamma+1}\left(\frac{u|u|^\gamma}{\gamma+1} - u|u|^\gamma \ln|u|\right) \quad \text{a.e. in } Q, \\
& b(z, v_\varepsilon)|\nabla v_\varepsilon + \mathscr{B}(v_\varepsilon)|^{p(z)-2}(\nabla v_\varepsilon + \mathscr{B}(v_\varepsilon)) \rightharpoonup \Lambda \quad \text{in } \left(L^{p'(\cdot)}(Q)\right)^n .
\end{aligned}
\tag{8.23}
$$

Letting $\varepsilon \to 0$ in (8.7) for $u_\varepsilon$, we find that for every $\phi \in \mathbf{W}(Q)$

$$
\int_Q (u_t\,\phi + \Lambda\,\nabla\phi - f\,\phi) = 0, \qquad u = \Phi_0(z, v). \tag{8.24}
$$

To complete the proof of existence amounts to check that for all $\phi \in \mathbf{W}(Q)$

$$
\int_Q \Lambda(z)\cdot\nabla\phi\,dz = \int_Q b(z, v)|\nabla v + \mathscr{B}(v)|^{p(z)-2}(\nabla v + \mathscr{B}(v))\cdot\nabla\phi\,dz.
$$

Let us define the function $F(\xi, \eta) = |\nabla\xi + \mathscr{B}(\eta)|^{p(z)-2}(\nabla\xi + \mathscr{B}(\eta))$ and make use of the representation

$$
\begin{aligned}
& F(v_\varepsilon, v_\varepsilon) - F(v, v) \\
& \quad \equiv |\nabla v_\varepsilon + \mathscr{B}(v_\varepsilon)|^{p(z)-2}(\nabla v_\varepsilon + \mathscr{B}(v_\varepsilon)) - |\nabla v + \mathscr{B}(v)|^{p(z)-2}(\nabla v + \mathscr{B}(v)) \\
& \quad = [F(v_\varepsilon, v_\varepsilon) - F(v_\varepsilon, v)] + [(F(v_\varepsilon, v)) - F(v, v)] \\
& \quad \equiv J_1^{(\varepsilon)} + J_2^{(\varepsilon)}.
\end{aligned}
$$

### 8.3.1 Step 1: $J_1^{(\varepsilon)} \to 0$ as $\varepsilon \to 0$.

Let us assume that $p(z) \geq 2$ at a point $z \in Q$. According to Proposition 1.8 with $a = \nabla v_\varepsilon + \mathscr{B}(v_\varepsilon)$, $b = \nabla v_\varepsilon + \mathscr{B}(v)$ and $p = p(z)$

$$
|J_1^{(\varepsilon)}| \leq C(p^+, p^-)\,|\mathscr{B}(v_\varepsilon) - \mathscr{B}(v)|\left(|\mathscr{B}(v_\varepsilon)|^{p(z)-1} + |\mathscr{B}(v)|^{p(z)-1} + |\nabla v_\varepsilon|^{p(z)-1}\right).
$$

If $p(z) \in (1, 2)$, we apply Proposition 1.9 to obtain

$$
\begin{aligned}
|J_1^{(\varepsilon)}| & \leq C(p^+, p^-)\,|\mathscr{B}(v_\varepsilon) - \mathscr{B}(v)|^{1-\alpha} \\
& \quad \times \left(|\mathscr{B}(v_\varepsilon)|^{p(z)-2+\alpha} + |\mathscr{B}(v)|^{p(z)-2+\alpha} + |\nabla v_\varepsilon|^{p(z)-2+\alpha}\right).
\end{aligned}
$$

Set

$$\begin{aligned}&q^+(z)=\max\{p(z),\,2\}, &&q^-(z)=\min\{p(z),\,2\},\\ &Q^+=Q\cap\{z:p(z)\geq 2\}, &&Q^-=Q\cap\{z:\,p(z)\in(1,2)\}.\end{aligned}$$

For every test-function $\phi\in\mathbf{W}(Q)$

$$\begin{aligned}\left|\int_Q J_1^{(\varepsilon)}\cdot\nabla\phi\,dz\right| &\leq\left|\int_{Q^+}\dots dz\right|+\left|\int_{Q^-}\dots dz\right|\\ &\leq C_1(p^\pm)\int_{Q^+}|\mathscr{B}(v_\varepsilon)-\mathscr{B}(v)|\\ &\quad\times\left(|\mathscr{B}(v_\varepsilon)|^{p(z)-1}+|\mathscr{B}(v)|^{p(z)-1}+|\nabla v_\varepsilon|^{p(z)-1}\right)dz\\ &\quad+C_2(p^\pm)\int_{Q^-}|\mathscr{B}(v_\varepsilon)-\mathscr{B}(v)|^{1-\beta}\\ &\quad\times\left(|\mathscr{B}(v_\varepsilon)|^{p(z)-2+\beta}+|\mathscr{B}(v)|^{p(z)-2+\beta}+|\nabla v_\varepsilon|^{p(z)-2+\beta}\right)dz\end{aligned}\tag{8.25}$$

with $2-p<\beta<1$. Applying Hölder's inequality we conclude that the right-hand side of this inequality tends to zero as $\varepsilon\to 0$.

### 8.3.2 Step 2: $J_2^{(\varepsilon)}\to F(v,v)$ as $\varepsilon\to 0$.

Taking $v_\varepsilon$ for the test-function in identity (8.7) for $u_\varepsilon$ and applying the integration-by-parts formula (1.40) we obtain the energy relation

$$\mu_\varepsilon(u_\varepsilon,v_\varepsilon)+\int_Q b(z,v_\varepsilon)F(v_\varepsilon,v_\varepsilon)\cdot\nabla v_\varepsilon=\int_Q f\,v_\varepsilon\,dz.\tag{8.26}$$

Recall that since $u_\varepsilon$ and $v_\varepsilon$ are uniformly bounded by a constant, $|\mu_\varepsilon|$ are also bounded uniformly with respect to $\varepsilon$.

**Lemma 8.4** *For every $\phi\in\mathbf{W}(Q)$*

$$\lim_{\varepsilon\to 0}\int_{Q_T}b(z,v_\varepsilon)F(v_\varepsilon,v)\cdot\nabla\phi\,dz=\int_{Q_T}b(z,v)F(v,v)\cdot\nabla\phi\,dz.\tag{8.27}$$

*Proof* By monotonicity

$$0\leq\int_Q b(z,v_\varepsilon)\left(F(v_\varepsilon,v)-F(\eta,v)\right)\cdot\nabla(v_\varepsilon-\eta))\,dz.\tag{8.28}$$

Subtracting (8.26) from (8.28), we get

$$
\begin{aligned}
0 &\le -\mu_\varepsilon(u_\varepsilon, v_\varepsilon) - \int_Q b(z, v_\varepsilon) F(v_\varepsilon, v_\varepsilon) \cdot \nabla v_\varepsilon \, dz \\
&\quad + \int_{Q_T} b(z, v_\varepsilon)\Big(F(v_\varepsilon, v) - F(\eta, v)\Big) \cdot \nabla(v_\varepsilon - \eta)\, dz + \int_Q f\, v_\varepsilon \, dz \qquad (8.29)\\
&\equiv -\mu_\varepsilon(u_\varepsilon, v_\varepsilon) + \sum_{i=1}^{5} I_i(\varepsilon),
\end{aligned}
$$

where

$$
\begin{aligned}
I_1(\varepsilon) &= \int_{Q_T} b(z, v_\varepsilon)(F(v_\varepsilon, v) - F(v_\varepsilon, v_\varepsilon)) \cdot \nabla v_\varepsilon \, dz,\\
I_2(\varepsilon) &= -\int_Q b(z, v_\varepsilon) F(\eta, v) \cdot \nabla v_\varepsilon \, dz,\\
I_3(\varepsilon) &= -\int_{Q_T} b(z, v_\varepsilon) F(v_\varepsilon, v) \cdot \nabla \eta \, dz,\\
I_4(\varepsilon) &= \int_{Q_T} b(z, v_\varepsilon) F(\eta, v) \cdot \nabla \eta \, dz,\\
I_5(\varepsilon) &= \int_Q f\, v_\varepsilon \, dz.
\end{aligned}
$$

Using the continuity of $b(z, s)$ with respect to $s$ and the weak convergence

$$
\begin{aligned}
&\nabla v_\varepsilon \rightharpoonup \nabla v,\\
&b(z, v_\varepsilon) F(v_\varepsilon, v_\varepsilon) = (b(z, v_\varepsilon) - b(z, v)) F(v_\varepsilon, v_\varepsilon) + b(z, v) F(v_\varepsilon, v_\varepsilon) \rightharpoonup \Lambda,
\end{aligned}
$$

it is easy to see that

$$
\begin{aligned}
&\lim\ I_2(\varepsilon) = -\int_Q b(z, v) F(\eta, v) \cdot \nabla v dz,\\
&\lim\ I_3(\varepsilon) = -\int_Q \Lambda \cdot \nabla \eta dz, \quad \lim I_4(\varepsilon) = \int_Q b(z, v) F(\eta, v) \cdot \nabla \eta dz.
\end{aligned}
$$

Let us consider the term $I_1(\varepsilon)$. Since

$$
\mathscr{B}(v_\varepsilon) \to \mathscr{B}(v) \text{ a.e. in } Q \quad \text{and} \quad \int_Q |\nabla v_\varepsilon|^{p(z)}\, dz \le C
$$

we find, repeating the proof of (8.25), that $|I_1| \to 0$ as $\varepsilon \to 0$. Letting $\varepsilon \to 0$ in (8.29) and using the inequality

$$\liminf_{\varepsilon \to 0} \mu_\varepsilon(u_\varepsilon, v_\varepsilon) \geq \mu(u, v) = \int_Q u_t \, v \, dz$$

with $\mu$ given in Corollary 1.11, we obtain the inequality

$$\begin{aligned} 0 \leq -\mu(u, v) &- \int_Q b(z, v) F(v, v) \cdot \nabla v \, dz \\ &- \int_Q \Lambda \cdot \nabla \eta \, dz + \int_Q b(z, v) F(\eta, v) \cdot \nabla \eta \, dz + \int_Q f \, v \, dz. \end{aligned} \tag{8.30}$$

Combining (8.30) with (1.40) we have

$$0 \leq \int_Q (\Lambda - b(z, v) F(\eta, v) \cdot \nabla(v - \eta)) \, dz. \tag{8.31}$$

Let us take $\eta = v - \lambda w$ with $\lambda = const$ and an arbitrary $w \in \mathbf{W}(Q)$. Under this choice of the test-function

$$0 \leq \lambda \int_Q (\Lambda - b(z, v) F(v - \lambda \, w, v)) \cdot \nabla w \, dz. \tag{8.32}$$

Simplifying and letting $\lambda \to 0$ we find that

$$0 \leq \int_Q (\Lambda - b(z, v) F(v, v)) \cdot \nabla w \, dz.$$

Since $w \in \mathbf{W}(Q)$ is arbitrary, it follows that $\Lambda = b(z, v) F(v, v)$. The proof of Theorem 8.1 is completed. □

## 8.4 Comparison Principle and Uniqueness for Strong Solutions

We consider separately the cases of constant and variable coefficient $a$. We will prove that the weak solution is unique in the class of functions

$$\mathscr{V}(Q) \equiv \left\{ v(z) : v \in \mathbf{W}(Q) \cap L^\infty(Q), \; \partial_t \Phi_0(z, v) \in L^1(Q) \cap \mathbf{W}'(Q) \right\}.$$

The solutions possessing this additional regularity are called strong solutions. In the final section of this chapter we will show that under additional assumption on the regularity of the data every weak solution is also a strong solution.

### 8.4.1 Proof of Theorem 8.2: The case $a \equiv 1$

Let $v_i \in \mathscr{V}(Q)$ be two bounded weak solutions of problem (8.3) with the data $\{f_i, v_{0i}\}$, $i = 1, 2$. Introduce the functions

$$w = \Phi_0(z, v_1) - \Phi_0(z, v_2), \quad v = v_1 - v_2,$$
$$F(s) = |\mathscr{B}(v) + \nabla v|^{p(z)-2} (\mathscr{B}(v) + \nabla v).$$

By (8.7), for every test-function $\phi \in \mathbf{W}(Q)$

$$\int_Q \left(\phi\, \partial_t w + (F(v_1) - F(v_2)) \cdot \nabla\phi\right) dz = \int_Q (f_1 - f_2)\, \phi\, dz. \tag{8.33}$$

Taking for the test-function $\phi_{k,\delta,\theta}$ defined in (1.43) and applying Lemma 1.22 we have that for a.e. $\theta \in (0, T)$ there exists the limit of the first term on the left-hand side of (8.33):

$$\int_Q \phi_{k,\delta,\theta} \partial_t w\, dz \to \int_\Omega |w|\, dx\Big|_{t=0}^{t=\theta} \quad \text{as } k \to \infty,\ \delta \to 0. \tag{8.34}$$

On the other hand, the rest of the terms in (8.33) are continuous functions of $\theta$ because of the property of absolute continuity of the integral. It follows that (8.34) is true for all $t \in [0, T]$. The second term on the left-hand side of (8.33) with $\phi(z) = \chi_{k,\theta}(t)\, T_\delta(v(z))$ is represented in the form

$$\begin{aligned} I_2 &= \int_Q (F(v_1) - F(v_2)) \cdot \nabla\phi\, dz = \int_Q \chi_{k,\theta} (F(v_1) - F(v_2))\, \nabla T_\delta(v)\, dz \\ &= \int_Q \chi_{k,\theta}\, T'_\delta(v)\, (F(v_1) - F(v_2))\, \nabla v\, dz. \end{aligned} \tag{8.35}$$

Let us denote

$$\zeta_i = \nabla v_i + \mathscr{B}(v_i), \quad i = 1, 2$$

so that

$$\nabla v_i = \zeta_i - \mathscr{B}(v_i), \qquad F(v_i) = |\zeta_i|^{p(z)-2} \zeta_i, \qquad \zeta_i = |F(v_i)|^{p'(z)-2} F(v_i)$$

[recall that $\mathscr{B}(s)$ is defined in (8.3)]. Passing to the limit as $k \to \infty$, for every fixed $\delta$ and $\theta$ we obtain the equality

$$
\begin{aligned}
\lim_{k\to\infty} I_2 &= \int_{Q_\theta} T'_\delta(v)\Big(F(v_1) - F(v_2)\Big)\cdot \nabla v\, dz \\
&= \int_{Q_\theta} T'_\delta(v)\Big(F(v_1) - F(v_2)\Big)(\zeta_1 - \zeta_2)\, dz \\
&\quad - \int_{Q_\theta} T'_\delta(v)\Big(F(v_1) - F(v_2)\Big)(\mathscr{B}(v_1) - \mathscr{B}(v_2))\, dz \\
&\equiv J_1(\delta) - J_2(\delta).
\end{aligned}
$$

Applying (4.35) we may write

$$
\begin{aligned}
J_1(\delta) &= \int_{Q_\theta} T'_\delta(v)\left(|\zeta_1|^{p(z)-2}\zeta_1 - |\zeta_2|^{p(z)-2}\zeta_2\right)(\zeta_1 - \zeta_2)\, dz \\
&\geq 2^{-(p^-)'} \int\limits_{Q_\theta\cap\{p(z)\geq 2\}} T'_\delta(v)\, |F(v_1) - F(v_2)|^{p(z)}\, dz \\
&\quad + (p^- - 1) \int\limits_{Q_\theta\cap\{p(z)\in(1,2)\}} T'_\delta(v)\, |F(v_1) - F(v_2)|^2 \\
&\qquad \times \left(|F(v_1)|^{p(z)} + |F(v_2)|^{p(z)}\right)^{\frac{p(z)-2}{p(z)}} dz.
\end{aligned} \tag{8.36}
$$

Next,

$$
\begin{aligned}
J_2(\delta) &\leq \int_{Q_\theta\cap\{z:\, p(z)\geq 2\}} T'_\delta(v)\, |F(v_1) - F(v_2)|\, |\mathscr{B}(v_1) - \mathscr{B}(v_2)|\, dz \\
&\quad + \int_{Q_\theta\cap\{z:\, 1<p(z)<2\}} \ldots \equiv \mathscr{K}^{(1)}(\delta) + \mathscr{K}^{(2)}(\delta).
\end{aligned}
$$

To estimate $J^{(1)}(\delta)$ we make use of the following elementary lemma.

**Lemma 8.5** *For every $p(z) \in [p^-, p^+] \subset (1, \infty)$ and $\varepsilon \in (0, 1)$*

$$
a\, b \leq \varepsilon\, a^{p'(z)} + \frac{\varepsilon^{1-p^+}}{p^-}\left(\frac{p^+}{p^+ - 1}\right)^{1-p^-} b^{p(z)} \qquad \forall a,\, b \geq 0.
$$

*Proof* The assertion follows from Young's inequality

$$
\begin{aligned}
a\, b &= (\varepsilon\, p'(z))^{\frac{1}{p'(z)}} a\, (\varepsilon\, p'(z))^{-\frac{1}{p'(z)}} b \\
&\leq \frac{1}{p'(z)}\left((\varepsilon\, p'(z))^{\frac{1}{p'(z)}} a\right)^{p'(z)} + \frac{1}{p(z)}\left((\varepsilon\, p'(z))^{-\frac{1}{p'(z)}} b\right)^{p(z)}
\end{aligned}
$$

and the inequalities

$$\varepsilon^{-\frac{p(z)}{p'(z)}} = \varepsilon^{-(p(z)-1)} = \mathrm{e}^{-(p(z)-1)\ln\varepsilon} \le \mathrm{e}^{-(p^+-1)\ln\varepsilon} = \varepsilon^{1-p^+},$$

$$\begin{aligned}(p'(z))^{-\frac{p(z)}{p'(z)}} &= \left(1-\frac{1}{p(z)}\right)^{p(z)-1} = \mathrm{e}^{(p(z)-1)\ln\left(1-\frac{1}{p(z)}\right)} \\ &\le \mathrm{e}^{(p^--1)\ln\left(1-\frac{1}{p^+}\right)} = \left(\frac{p^+}{p^+-1}\right)^{1-p^-}.\end{aligned}$$

□

Applying Lemma 8.5 we have:

$$\begin{aligned}\mathscr{K}^{(1)}(\delta) \equiv & \int\limits_{Q_\theta\cap\{p(z)\ge 2\}} \left(T_\delta'(v)\right)^{\frac{1}{p(z)}} |F(v_1)-F(v_2)| \\ & \left(T_\delta'(v)\right)^{\frac{1}{p(z)}} |\mathscr{B}(v_1)-\mathscr{B}(v_2)|\,dz \\ \le\, & \varepsilon \int\limits_{Q_\theta\cap\{p(z)\ge 2\}} T_\delta'(v)\,|F(v_1)-F(v_2)|^{p(z)}\,dz \\ & + C(\varepsilon,p^-,p^+) \int\limits_{Q_\theta\cap\{p'(z)\ge 2\}} T_\delta'(v)\,|\mathscr{B}(v_1)-\mathscr{B}(v_2)|^{p(z)}\,dz.\end{aligned} \tag{8.37}$$

By Young's inequality

$$\begin{aligned}\mathscr{K}^{(2)}(\delta) \equiv & \int\limits_{Q_\theta\cap\{p(z)<2\}} \left(\sqrt{T_\delta'(v)}|F(v_1)-F(v_2)|\left(|F(v_1)|^p+|F(v_2)|^p\right)^{\frac{p-2}{2p}}\right) \\ & \times \left(\sqrt{T_\delta'(v)}|\mathscr{B}(v_1)-\mathscr{B}(v_2)|\left(|F(v_1)|^p+|F(v_2)|^p\right)^{\frac{2-p}{2p}}\right) dz \\ \le\, & \varepsilon \int\limits_{Q_\theta\cap\{p(z)<2\}} T_\delta'(v)|F(v_1)-F(v_2)|^2\left(|F(v_1)|^p+|F(v_2)|^p\right)^{\frac{p-2}{p}}\,dz \\ & + \frac{1}{4\varepsilon} \int\limits_{Q_\theta\cap\{p(z)<2\}} T_\delta'(v)|\mathscr{B}(v_1)-\mathscr{B}(v_2)|^2\left(|F(v_1)|^p+|F(v_2)|^p\right)^{\frac{2-p}{p}}\,dz\end{aligned} \tag{8.38}$$

Gathering (8.36)–(8.38) we arrive at the inequality

$$\begin{aligned}J_1(\delta)-J_2(\delta) &\ge J_1(\delta)-\mathscr{K}^{(1)}(\delta)-\mathscr{K}^{(2)}(\delta) \\ &\ge (2^{-(p^-)'}-\varepsilon) \int\limits_{Q_\theta\cap\{p(z)\ge 2\}} T_\delta'(v)|F(v_1)-F(v_2)|^{p(z)}\,dz\end{aligned}$$

$$- C(\varepsilon, p^-, p^+) \int\limits_{Q_\theta \cap \{p(z) \geq 2\}} T'_\delta(v)\, |\mathscr{B}(v_1) - \mathscr{B}(v_2)|^{p(z)}\, dz$$

$$+ (p^- - 1 - \varepsilon) \int\limits_{Q_\theta \cap \{p(z) \in (1,2)\}} T'_\delta(v)\, |F(v_1) - F(v_2)|^2$$

$$\times \left(|F(v_1)|^{p(z)} + |F(v_2)|^{p(z)}\right)^{\frac{p(z)-2}{p(z)}} dz$$

$$- \frac{1}{4\varepsilon} \int\limits_{Q_\theta \cap \{p(z) < 2\}} T'_\delta(v) |\mathscr{B}(v_1) - \mathscr{B}(v_2)|^2 \left(|F(v_1)|^p + |F(v_2)|^p\right)^{\frac{2-p}{p}} dz.$$

For the sufficiently small $\varepsilon \equiv \varepsilon(p^-)$ we have

$$\begin{aligned}
\lim_{k\to\infty} I_2 \geq -C &\int_{Q_\theta \cap \{p(z) \geq 2\}} T'_\delta(v)\, |\mathscr{B}(v_1) - \mathscr{B}(v_2)|^{p(z)}\, dz \\
&- \frac{1}{4\varepsilon} \int_{Q_\theta \cap \{p(z) < 2\}} T'_\delta(v) |\mathscr{B}(v_1) - \mathscr{B}(v_2)|^2 \\
&\times \left(|F(v_1)|^p + |F(v_2)|^p\right)^{\frac{2-p}{p}} dz
\end{aligned} \tag{8.39}$$

with a positive constant $C \equiv C(p^{\pm})$. It remains to show that the right-hand side of the last inequality tends to zero as $\delta \to 0$. We will use the following Lemmas.

**Lemma 8.6** *For every* $\eta \in (0, 1)$

$$|\mathscr{B}(v_1) - \mathscr{B}(v_2)| \leq |\nabla\gamma(z)|\, |v_1 - v_2|^{1-\eta} \left| \int_{u_1}^{u_2} |s|^{\gamma(z)}\, |\ln|s||^{\frac{1}{\eta}}\, ds \right|^\eta .$$

*Proof* By the definition

$$\mathscr{B}(v) = -\nabla\gamma \cdot \int_0^u |s|^{\gamma(z)} \ln|s|\, ds, \qquad v = \int_0^u |s|^{\gamma(z)}\, ds.$$

Not loosing generality we may assume that $u_1 \geq u_2$ and, thus, $v_1 \geq v_2$. Then for every $\eta \in (0, 1)$

$$\begin{aligned}
|\mathscr{B}(v_1) - \mathscr{B}(v_2)| &= |\nabla\gamma(z)| \left| \int_{u_1}^{u_2} |s|^{\gamma(z)} \ln|s|\, ds \right| \\
&\leq |\nabla\gamma(z)| \int_{u_1}^{u_2} |s|^{\gamma(z)(1-\eta)} |s|^{\gamma(z)\eta} |\ln|s||\, ds
\end{aligned}$$

and the assertion follows by Young's inequality with $p = \dfrac{1}{1-\eta}$, $q = \dfrac{1}{\eta}$. □

**Lemma 8.7** *For every $p, r > 0$, $q \geq \dfrac{r}{2}$ and all $v \in \mathbb{R}$*

$$\frac{\delta^p |v|^r}{(\delta^2 + v^2)^q} \leq \varepsilon^{p+r-2q}.$$

*Proof* It suffices to notice that

$$\frac{\delta^p |v|^r}{(\delta^2 + v^2)^q} \leq \frac{\delta^p (\delta^2 + v^2)^{\frac{r}{2}}}{(\delta^2 + v^2)^q} = \frac{\delta^p}{(\delta^2 + v^2)^{q-\frac{r}{2}}}$$
$$= \left(\frac{\delta^2}{\delta^2 + v^2}\right)^{q-\frac{r}{2}} \delta^{p-2(q-\frac{r}{2})} \leq \delta^{p+r-2q}.$$

□

*Proof* (End of the proof of Theorem 8.2) Fix some $\eta \in (0, 1)$ and denote

$$K(u_1, u_2, \eta) = \left| \int_{u_1}^{u_2} |s|^{\gamma(z)} \, |\ln |s||^{\frac{1}{\eta}} \, ds \right|^{\eta}.$$

Since $u_i$ are bounded, so is $K(u_1, u_2, \eta)$. By virtue of Lemmas 8.6 and 8.7 the first term on the right-hand side of (8.39) is estimated as follows: for $\eta \in (0, 1 - 1/p^-)$

$$\begin{aligned}
&\int_{Q_\theta \cap \{p(z) \geq 2\}} T'_\delta(v) \, |\mathscr{B}(v_1) - \mathscr{B}(v_2)|^{p(z)} \, dz \\
&\quad \leq \int_{Q_\theta \cap \{p(z) \geq 2\}} \frac{\delta^2}{(\delta^2 + \eta^2)^{\frac{3}{2}}} |v|^{p(1-\eta)} |\nabla \gamma|^p K^p \, dz \\
&\quad \leq \int_{Q_\theta \cap \{p(z) \geq 2\}} \delta^{p(z)-1-p(z)\eta} |\nabla \gamma|^{p(z)} K^{p(z)} \, dz \\
&\quad \leq \delta^{p^- - 1 - p^- \eta} \int_{Q_\theta \cap \{p'(z) \geq 2\}} |\nabla \gamma|^{p(z)} K^{p(z)} \, dz \to 0 \quad \text{as } \delta \to 0
\end{aligned} \tag{8.40}$$

because $K^{p(z)}$ is uniformly bounded and $|\nabla \gamma(z)|^{p(z)} \in L^1(Q)$. The second term of (8.39) is estimated in a similar way:

$$\begin{aligned}
&\int_{Q_\theta \cap \{p(z) < 2\}} T'_\delta(v) |\mathscr{B}(v_1) - \mathscr{B}(v_2)|^2 \left(|F(v_1)|^p + |F(v_2)|^p\right)^{\frac{2-p}{p}} \, dz \\
&\leq \int_{Q_\theta \cap \{p(z) < 2\}} \frac{\delta^2}{(\delta^2 + \eta^2)^{\frac{3}{2}}} |v|^{2(1-\eta)} \left(|F(v_1)|^p + |F(v_2)|^p\right)^{\frac{2-p}{p}} |\nabla \gamma|^2 K^2 \, dz
\end{aligned}$$

$$
\begin{aligned}
&\le \delta^{1-2\eta} \int\limits_{Q_\theta \cap \{p(z)<2\}} \left(|F(v_1)|^p + |F(v_2)|^p\right)^{\frac{2-p}{p}} |\nabla\gamma|^2 K^2\, dz \\
&\le \delta^{1-2\eta} \int\limits_{Q_\theta \cap \{p(z)<2\}} \left[\frac{2}{p}|\nabla\gamma|^p K^p + \left(1-\frac{2}{p}\right)\left(|F(v_1)|^p + |F(v_2)|^p\right)\right] dz \to 0
\end{aligned}
\tag{8.41}
$$

as $\delta \to 0$, because $|F(v_i)|^{p'(z)} = |\nabla v_i + \mathscr{B}(v_i)|^{p(z)} \in L^1(Q)$. Plugging (8.40)–(8.41) to (8.39), we obtain, letting $\delta \to 0$ in (8.35): for a.e. $\theta \in (0, T)$

$$
\int_\Omega |w(x,\theta)|\, dx \le \int_\Omega |w(x,0)|\, dx + \int_Q \operatorname{sign} w\, (f_1 - f_2)\, dz,
$$

whence the assertion of Theorem 8.2. □

## 8.4.2 *Proof of Theorem 8.3: The case* $\alpha = 0$.

Let us consider the Dirichlet problem for the evolutional $p(z)$-Laplace equation

$$
\begin{cases}
v_t = \operatorname{div}\left(a(z,v)|\nabla v|^{p(z)-2}\nabla v\right) + f(z) & \text{in } Q, \\
v(x,0) = v_0(x) \text{ in } \Omega, \quad v = 0 \text{ on } \Gamma.
\end{cases}
\tag{8.42}
$$

This equation is the special case of Eq. (8.1): $\alpha = 0$. Given two weak solutions of problem (8.42) $v_1$ and $v_2$, we denote $v = v_1 - v_2$. Following the proof of Theorem 8.2 we see that to prove Theorem 8.3 amounts to show that $\lim\limits_{\delta\to 0}(I_1 + I_2) \ge 0$, where

$$
\begin{aligned}
I_1 &= \int_Q a(z,v_1)\, T'_\delta(v) \left(\left(|\nabla v_1|^{p-2}\nabla v_1 - |\nabla v_2|^{p-2}\nabla v_2\right)\cdot \nabla v\right) dz, \\
I_2 &= \int_Q T'_\delta(v)(a(z,v_1) - a(z,v_2))|\nabla v_2|^{p-2}\nabla v_2 \cdot \nabla v\, dz.
\end{aligned}
$$

**Proposition 8.1** ([98]) *Let* $1 < p < \infty$. *There is a constant* $C(p)$ *such that*

$$
\left(|\xi|^{p-2}\xi - |\zeta|^{p-2}\zeta\right)\cdot(\xi - \zeta) \ge C\,|\xi - \zeta|^2\,(|\xi| + |\zeta|)^{p-2}.
$$

*Proof* For $p \in (1, 2]$ the assertion is a byproduct of (4.35). Let $p \ge 2$. Take some $\xi, \zeta \in \mathbb{R}^n$ and assume, without loss of generality, that $|\xi| \ge |\zeta|$. Denote

$$
A = \left(|\xi|^{p-2}\xi - |\zeta|^{p-2}\zeta\right)\cdot(\xi - \zeta) = |\xi|^p + |\zeta|^p - \left\{|\xi|^{p-2} + |\zeta|^{p-2}\right\}(\xi, \zeta).
$$

Since $(\xi, \zeta) = \frac{1}{2}\left(|\xi|^2 + |\zeta|^2 - |\xi - \zeta|^2\right)$, we have

$$\begin{aligned} A &= \frac{1}{2}\left\{|\xi|^{p-2} + |\zeta|^{p-2}\right\}|\xi - \zeta|^2 + |\xi|^p + |\zeta|^p - \frac{1}{2}\left(|\xi|^{p-2} + |\zeta|^{p-2}\right)\left(|\xi|^2 + |\zeta|^2\right) \\ &= \frac{1}{2}\left\{|\xi|^{p-2} + |\zeta|^{p-2}\right\}|\xi - \zeta|^2 + \frac{1}{2}\left(|\xi|^{p-2} - |\zeta|^{p-2}\right)\left(|\xi|^2 - |\zeta|^2\right). \end{aligned}$$

For $p \geq 2$ the mapping $X \mapsto X^{p-2}$ is nondecreasing. The second term of the last inequality is then nonnegative and can be dropped. Moreover, $|\xi| \geq \frac{1}{2}(|\xi| + |\zeta|)$ by assumption. It follows that for $p \geq 2$

$$A \geq \frac{1}{2}\left\{|\xi|^{p-2} + |\zeta|^{p-2}\right\}|\xi - \zeta|^2 \geq \frac{1}{2}|\xi|^{p-2}|\xi - \zeta|^2 \geq \frac{1}{2^{p-1}}\left(|\xi| + |\zeta|\right)^{p-2}|\xi - \zeta|^2.$$

□

Applying Proposition 8.1 we have

$$\begin{aligned} I_1 &\geq C\int_{Q_\theta} \frac{\delta^2}{(\delta^2 + v^2)^{3/2}}\,|\nabla(v_1 - v_2)|^2\left(|\nabla v_1| + |\nabla v_2|\right)^{p-2} dz, \\ &= C\int_{Q_\theta \cap \{v \neq 0\}} \frac{\delta^2}{(\delta^2 + v^2)^{3/2}}\,|\nabla(v_1 - v_2)|^2\left(|\nabla v_1| + |\nabla v_2|\right)^{p-2} dz, \end{aligned}$$

$$\begin{aligned} I_2 &\leq C\int_{Q_\theta \cap \{v \neq 0\}} \frac{\delta^2}{(\delta^2 + v^2)^{3/2}}\,|v|^\beta\,|\nabla v_2|^{p-1}\,|\nabla(v_1 - v_2)|\,dz \\ &\leq C\int_{Q_\theta \cap \{v \neq 0\}} \frac{\delta^2}{(\delta^2 + v^2)^{3/2}}\,|v|^\beta\left(|\nabla v_1| + |\nabla v_2|\right)^{p-1}|\nabla(v_1 - v_2)|\,dz \\ &= C\int_{Q_\theta \cap \{v \neq 0\}} \left(\frac{\delta^2}{(\delta^2 + v^2)^{3/2}}\right)^{1/2} |\nabla(v_1 - v_2)|\left(|\nabla v_1| + |\nabla v_2|\right)^{\frac{p-2}{2}}\left(|\nabla v_1| + |\nabla v_2|\right)^{\frac{p}{2}} \\ &\quad\times\left(\frac{\delta^2}{(\delta^2 + v^2)^{3/2}}\right)^{1/2} |v|^\beta\,dz \\ &\leq C\,\varepsilon\int_{Q_\theta \cap \{v \neq 0\}} \frac{\delta^2}{(\delta^2 + v^2)^{3/2}}\,|\nabla(v_1 - v_2)|\left(|\nabla v_1| + |\nabla v_2|\right)^{p-2} dz \\ &\quad + \frac{C}{4\varepsilon}\int_{Q_\theta \cap \{v \neq 0\}} \frac{\delta^2\,|v|^{2\beta}}{(\delta^2 + v^2)^{3/2}}\left(|\nabla v_1| + |\nabla v_2|\right)^p dz. \end{aligned}$$

For all sufficiently small $\varepsilon$ these inequalities yield

$$I_1 + I_2 \geq -\frac{C}{4\varepsilon}\int_{Q_\theta \cap \{v \neq 0\}} \frac{\delta^2\,|v|^{2\beta}}{(\delta^2 + v^2)^{3/2}}\left(|\nabla v_1| + |\nabla v_2|\right)^p dz.$$

For $\beta > 1/2$ we have

$$I_1 + I_2 \geq -\frac{C\delta^{2\beta-1}}{4\varepsilon} \int_{Q_\theta \cap \{v \neq 0\}} (|\nabla v_1| + |\nabla v_2|)^p \, dz \to 0.$$

For $\beta = 1/2$ one has

$$\frac{\delta^2 \, |v|^{2\beta}}{(\delta^2 + v^2)^{3/2}} = \frac{\delta^2 \, |v|}{(\delta^2 + v^2)^{3/2}} \leq \frac{1}{2} \frac{\delta}{(\delta^2 + v^2)^{1/2}}$$

and

$$I_1 + I_2 \geq -\frac{C}{8\varepsilon} \int_{Q_\theta \cap \{v \neq 0\}} \frac{\delta}{(\delta^2 + v^2)^{1/2}} (|\nabla v_1| + |\nabla v_2|)^p \, dz \to 0$$

by the Lebesgue Theorem.

## 8.5 Existence of Solutions $u \in \mathscr{V}(Q)$: $L^1$-Estimate for $\partial_t \Phi(z, v)$

Let us check that problem (8.3) indeed admits solutions in $\mathscr{V}(Q)$, which means that the class of uniqueness is nonempty. The proof of integrability of $\partial_t \Phi_0(z, v) \equiv \partial_t u$ reduces to checking that for the finite-dimensional approximations $v_\varepsilon^{(m)}$ of solutions of the regularized problems (8.22) the norms $\|\partial_t \Phi_\varepsilon(z, v_\varepsilon^{(m)})\|_{1,Q}$ are bounded uniformly with respect to $\varepsilon$ and $m$.

The function $u_\varepsilon^{(m)} = \Phi_\varepsilon(z, v_\varepsilon^{(m)})$ is a weak solution of problem (8.22) with the data

$$u_0^{(m)} = \sum_{i=1}^{m} u_{0i} \psi_i(x) \to u_0, \qquad f^{(m)} = \sum_{i=1}^{m} f_i(t) \psi_i(x) \to f$$

and satisfies (8.14) with an arbitrary $\phi \in \mathbf{W}(Q)$. Let us fix some $\varepsilon > 0$, $m \in \mathbb{N}$, and introduce the function

$$V = \partial_t \Phi_\varepsilon(z, v_\varepsilon^{(m)}) \equiv \sum_{i=1}^{m} c'_{i,m,\varepsilon}(t) \psi_i(x).$$

Set $\Psi = \nabla v_\varepsilon^{(m)} + \mathscr{B}_\varepsilon(v_\varepsilon^{(m)})$, $\mathscr{F} = \left(|\Psi|^{p-2}\Psi\right)_t$. Differentiating Eq. (8.22) for $u_\varepsilon^{(m)}$ in $t$, we write the equation for $V$ in the form

$$V_t = \operatorname{div} \mathscr{F} + f_t^{(m)}. \tag{8.42}$$

This equation is fulfilled in the following sense: for every test-function $\phi \in \mathbf{W}(Q)$

$$\int_Q \left[\phi\, V_t + \mathscr{F} \cdot \nabla\phi - f_t^{(m)}\, \phi\right] dz = 0.$$

It is straightforward to check that

$$\begin{aligned} V &= (\Phi_\varepsilon)_v v_t + (\Phi_\varepsilon)_t, \\ \mathscr{F} &= (p-1)|\Psi|^{p-2}\Psi_t + |\Psi|^{p-2}\Psi\, \ln|\Psi|\, p_t, \\ \Psi_t &= \nabla v_t + (\mathscr{B}_\varepsilon)_v\, v_t + (\mathscr{B}_\varepsilon)_t. \end{aligned}$$

Combining these formulas we conclude that

$$\begin{aligned} v_t &= \frac{V - (\Phi_\varepsilon(z,v))_t}{(\Phi_\varepsilon(z,v))_v}, \\ \Psi_t &= \nabla\left(\frac{V - (\Phi_\varepsilon(z,v))_t}{(\Phi_\varepsilon(z,v))_v}\right) + (\mathscr{B}_\varepsilon(v))_v \frac{V - (\Phi_\varepsilon(z,v))_t}{(\Phi_\varepsilon(z,v))_v} + (\mathscr{B}_\varepsilon(v))_t, \\ \mathscr{F} &= (p-1)|\Psi|^{p-2}\left[\nabla\left(\frac{V - (\Phi_\varepsilon(z,v))_t}{(\Phi_\varepsilon(z,v))_v}\right)\right. \\ &\quad \left. + (\mathscr{B}_\varepsilon(v))_v \frac{V - (\Phi_\varepsilon(z,v))_t}{(\Phi_\varepsilon(z,v))_v} + (\mathscr{B}_\varepsilon(v))_t\right] + |\Psi|^{p-2}\Psi\, \ln|\Psi|\, p_t. \end{aligned}$$

Let us introduce the functions

$$h_\mu(\sigma) = \begin{cases} \dfrac{2}{\mu}\left(1 - \dfrac{|\sigma|}{\mu}\right) & \text{if } |\sigma| < \mu, \\ 0 & \text{if } |\sigma| \geq \mu, \end{cases}$$

$$H_\mu(\sigma) = \int_0^\sigma h_\mu(s) ds, \qquad \mathbf{H}_\mu(\sigma) = \int_0^\sigma \int_0^q h_\mu(s)\, ds dq.$$

According to the definition

$$\begin{cases} h_\mu(\sigma) \geq 0, & \lim\limits_{\mu\to 0} \sigma h_\mu(\sigma) = 0, \\ \left|H_\mu(\sigma)\right| \leq 1, & \lim\limits_{\mu\to 0} H_\mu(\sigma) = \operatorname{sign}\sigma, \quad \lim\limits_{\mu\to 0} \mathbf{H}_\mu(\sigma) = |\sigma|. \end{cases} \tag{8.43}$$

Multiplying (8.42) by $H_\mu(V)$ and integrating by parts in $t$, we arrive at the equality

$$\int_Q H_\mu(V)\, V_t\, dz = \int_Q \partial_t \mathbf{H}(V(z))\, dz$$

$$
\begin{aligned}
&= \int_\Omega \mathbf{H}_\mu(V(z))\,dx - \int_\Omega \mathbf{H}_\mu(V(x,0))\,dx \\
&= -\int_Q \mathscr{F}\,\nabla H_\mu(V)\,dz + \int_Q f_t^{(m)}\,H_\mu(V)\,dz \qquad (8.44)\\
&= -\int_Q \mathscr{F}\,h_\mu(V)\,\nabla V\,dz + \int_Q f_t^{(m)}\,H_\mu(V)\,dz.
\end{aligned}
$$

Let us consider the simple case: $p_t = 0$, $\gamma_t = 0$, $\Phi_\varepsilon \equiv \Phi_\varepsilon(x, v)$. In this case

$$
\begin{aligned}
\mathscr{F} &= (p-1)|\Psi|^{p-2}\left[\nabla\left(\frac{V}{(\Phi_\varepsilon)'_v}\right) + (\mathscr{B}_\varepsilon(v))'_v\frac{V}{(\Phi_\varepsilon)'_v}\right]\\
&= (p-1)\frac{|\Psi|^{p-2}}{(\Phi_\varepsilon)'_v}\left[\nabla V - \frac{V\,\nabla v}{(\Phi_\varepsilon)'_v} + V\,(\mathscr{B}_\varepsilon(v))'_v\right].
\end{aligned}
$$

Since

$$
\begin{aligned}
&(\Phi_\varepsilon)'_v = (\gamma+1)^{-\frac{\gamma}{\gamma+1}}v^{-\frac{\gamma}{\gamma+1}},\\
&(\mathscr{B}_\varepsilon(v))'_v = \nabla\gamma\cdot(\varepsilon + |\Phi_\varepsilon(x,v)|)^\gamma\,\ln(\varepsilon + |\Phi_\varepsilon(x,v)|)(\Phi_\varepsilon)'_v,
\end{aligned}
$$

the previous equality becomes

$$
\mathscr{F} = (p-1)\,|\Psi|^{p-2}\left[\frac{\nabla V}{(\Phi_\varepsilon)'_v} - \frac{V\,\nabla v}{((\Phi_\varepsilon)'_v)^2} + V\,\nabla\gamma\cdot(\varepsilon + |\Phi_\varepsilon|)^\gamma\,\ln(\varepsilon + |\Phi_\varepsilon|)\right].
$$

Let us write (8.44) in the form

$$
\int_\Omega \mathbf{H}_\mu(V)\,dx\Big|_{t=0}^{t=T} = I_1 + I_2 + I_3 + \int_Q f_t^{(m)}\,H_\mu(V)\,dz \qquad (8.45)
$$

with

$$
\begin{aligned}
I_1 &= -\int_Q (p-1)\,|\Psi|^{p-2}\,(\gamma+1)^{\frac{\gamma}{\gamma+1}}v^{\frac{\gamma}{\gamma+1}}\,|\nabla V|^2\,h_\mu(V)\,dz \le 0,\\
I_2 &= \int_Q (p-1)\,|\Psi|^{p-2}\,(\gamma+1)^{\frac{2\gamma}{\gamma+1}}v^{\frac{2\gamma}{\gamma+1}}\,\nabla V\cdot\nabla v\,\left(V\,h_\mu(V)\right)\,dz,\\
I_3 &= -\int_Q (p-1)\,|\Psi|^{p-2}\,(\nabla\gamma\cdot\nabla V)\,(\varepsilon + |\Phi_\varepsilon|)^\gamma\,\ln(\varepsilon + |\Phi_\varepsilon|)\,\left(V\,h_\mu(V)\right)\,dz.
\end{aligned}
$$

Dropping the nonpositive term $I_1$ on the right-hand side of (8.45), letting $\mu \to 0$ and using (8.43) we finally obtain:

$$
\int_\Omega |V(x,t)|\,dx \le \int_\Omega |V(x,0)|\,dx + \int_Q |f_t|\,dz, \qquad V = \partial_t\Phi_\varepsilon(x,v). \qquad (8.46)
$$

Since the right-hand side of this inequality is independent of $m$ and $\varepsilon$, the needed estimate follows by passing to the limit as $m \to \infty$ and $\varepsilon \to 0$. These arguments are summarized in the following assertion.

**Theorem 8.5** *Let the conditions of* Theorem 8.1 *be fulfilled and, additionally,* $f_t \in L^1(Q)$, $p \equiv p(x)$, $\alpha \equiv \alpha(x)$ *and*

$$\operatorname{div}\left(b((x,0), v_0)|\nabla v_0 + \mathscr{B}(v_0)|^{p(x)-2}(\nabla v_0 + \mathscr{B}(v_0))\right) \in L^1(\Omega).$$

*Then* $V = \partial_t \Phi(x, v) \in L^1(Q)$ *and inequality* (8.46) *is fulfilled.*

# Chapter 9
# Strong Solutions of Doubly Nonlinear Anisotropic Equations

This chapter continues the study of doubly nonlinear equations. We consider the anisotropic problem

$$\begin{aligned} &\frac{d}{dt}(|v|^{m(z)} \operatorname{sign} v) = \sum_{i=1}^{n} D_i\big(a_i(z)|D_i v|^{p_i(z)-2} D_i v\big) + b(z)|v|^{\sigma(z)-2} v + g \text{ in } Q,\\ &v = 0 \text{ on } \Gamma, \quad v(x,0) = v_0(x) \text{ in } \Omega. \end{aligned} \tag{9.1}$$

The total derivative with respect to $t$ is calculated according to the rule

$$\frac{d}{dt}\psi(v,m) = \psi_v(v,m)\, v_t + \psi_m(v,m)\, m_t.$$

Equations of the type (9.1) with constant exponents $m$ and $p_i$ arise in the mathematical modelling of various physical processes such as flows of incompressible turbulent fluids or gases in pipes, processes of filtration in porous media or glaciology [36, 109, 110]. The questions of existence, uniqueness and propagation of solutions to doubly degenerate equations of the form

$$\begin{aligned} &\text{(a)} \qquad u_t = \operatorname{div}\left(a\,|u|^{\alpha}|\nabla u|^{p-2}\nabla u\right) + f(x,t,u),\\ &\text{(b)} \qquad \partial_t\left(|u|^{\beta-1}u\right) = \operatorname{div}\left(a\,|\nabla u|^{p-2}\nabla u\right) + g(x,t,u) \end{aligned} \tag{9.2}$$

with constant exponents of nonlinearity have been studied by many authors. We refer here to papers [79, 95, 101, 109, 182, 183, 188, 234, 250] for the results concerning the boundary value problems in cylinder domains and to [103, 107, 165] for a discussion of the Cauchy problem. Stationary solutions of the anisotropic version of Eq. (9.2)(a) were studied in [43, 55].

Equation (9.1) contains, as partial cases, equations studied in the previous chapters. In the case $p_i \equiv 2$ Eq. (9.1) can be regarded as the porous mediumequation,

 
S. Antontsev and S. Shmarev, *Evolution PDEs with Nonstandard Growth Conditions*, Atlantis Studies in Differential Equations 4,
DOI 10.2991/978-94-6239-112-3_9

if $m \equiv 1$ it transforms into the anisotropic evolution $p(x,t)$-Laplace equation. The material of this chapter are based on results of the papers [29, 56, 57, 59]. We are specially interested in the situation when the equation is nonlinear with respect to the solution and the components of the gradient, and when the exponents of nonlinearity depend on $t$. A combination of these two peculiarities in the same anisotropic equation prevents us from using the arguments employed in the previous chapter.

The main result is the theorem of existence of bounded strong solutions of problem (9.1). These solutions belong to a Sobolev variable exponent space and possess the additional regularity property:

$$|v|^{m(z)-1}v_t^2 \in L^1(Q).$$

As a byproduct of the existence theorem we obtain a series of energy estimates for the constructed strong solution. These estimates are interesting in itself: they provide information on the regularity of the constructed solution and turn out to be the key tool in the study of the localization and blow-up properties given in the next chapter.

## 9.1 Existence of Strong Solutions

The notations $p_i^{\pm}(t)$, $p^{\pm}$, $m^{\pm}(t)$ and $m^{\pm}$ are explained in (1.1), while

$$p^+ = \max\{p_i^+, i = 1, \dots, n\}, \qquad p^- = \min\{p_i^-, i = 1, \dots, n\}.$$

For every fixed $t \in [0, T]$ we introduce the Banach space

$$\mathbf{V}_t(\Omega) = \left\{u(x) \in L^{m(\cdot,t)+1}(\Omega) \cap W_0^{1,1}(\Omega) : \ |D_i u(x)|^{p_i(x,t)} \in L^1(\Omega)\right\},$$
$$\|u\|_{\mathbf{V}_t(\Omega)} = \|u\|_{m(\cdot,t)+1,\Omega} + \sum_i \|D_i u\|_{p_i(\cdot,t),\Omega},$$

and denote by $\mathbf{V}_t'(\Omega)$ its dual. By $\mathbf{W}(Q)$ we denote the Banach space

$$\mathbf{W}(Q) = \Big\{|u|^{m(z)+1} \in L^1(Q) : \ |D_i u|^{p_i(z)} \in L^1(Q),$$
$$u(\cdot,t) \in \mathbf{V}_t(\Omega) \text{ for a.e. } t \in (0,T)\Big\},$$
$$\|u\|_{\mathbf{W}(Q)} = \sum_i \|D_i u\|_{p_i(\cdot,\cdot),Q} + \|u\|_{m(\cdot,\cdot)+1,Q}.$$

$\mathbf{W}'(Q)$ is the dual of $\mathbf{W}(Q)$ (the space of linear functionals over $\mathbf{W}(Q)$):

$$w \in \mathbf{W}'(Q) \quad \Leftrightarrow \quad \begin{cases} \exists w = (w_0, w_1, \dots, w_n), \ w_0 \in L^{(m(\cdot)+1)'}(Q), \quad w_i \in L^{p_i'(\cdot)}(Q), \\ \forall \phi \in \mathbf{W}(Q) \ \langle\langle w, \phi \rangle\rangle = \int_Q \Big( w_0 \phi + \sum_i w_i D_i \phi \Big) dz. \end{cases}$$

The norm in $\mathbf{W}'(Q)$ is defined by

$$\|v\|_{\mathbf{W}'(Q)} = \sup \left\{ \langle\langle v, \phi \rangle\rangle | \, \phi \in \mathbf{W}(Q), \ \|\phi\|_{\mathbf{W}(Q)} \le 1 \right\}.$$

Let

$$\mathbf{V}_+(\Omega) = \left\{ u(x) | \ u \in L^{m^+ + 1}(\Omega) \cap W_0^{1,1}(\Omega), \ |\nabla u| \in L^{p^+}(\Omega) \right\}.$$

Since $\mathbf{V}_+(\Omega)$ is separable, it is a span of a countable set of linearly independent functions $\{\psi_k\} \subset \mathbf{V}_+(\Omega)$. Notice certain abuse of notation: in the previous chapters the same notations were used for the function space with $m \equiv 1$.

It is convenient to introduce the functions

$$\Phi(v, m) = m(z) \int_0^v |s|^{m(z)-1} ds \equiv |v|^{m(z)} \operatorname{sign} v$$

$$f(z, v) = b(z)|v|^{\sigma(z)-2} v + g(z)$$

and to write problem (9.1) in the form

$$\begin{cases} \partial_t \Phi(v, m) = \sum_{i=1}^n D_i \Big( a_i(z) |D_i v|^{p_i(z)-2} D_i v \Big) + f(z, v) \text{ in } Q, \\ v(x, 0) = v_0(x) \text{ in } \Omega, \\ v = 0 \text{ on } \Gamma. \end{cases} \tag{9.3}$$

**Definition 9.1** A function $v(z)$ is called *strong solution* of problem (9.3) if

1. $v \in L^\infty(Q) \cap \mathbf{W}(Q)$, $\Phi_v'(v, m) v_t^2 \in L^1(Q)$ and $|D_i v|^{p_i(z)} \in L^\infty(0, T; L^1(\Omega))$,
2. for every test-function $\phi \in \mathbf{W}(Q)$

$$\int_Q \Big[ \phi \frac{d}{dt} \Phi(v, m) + \sum_{i=1}^n a_i |D_i v|^{p_i(z)-2} D_i v \cdot D_i \phi \Big] dz = \int_Q f(z, v) \phi dz, \tag{9.4}$$

3. for every $\psi(x) \in C_0^\infty(\Omega)$

$$\int_\Omega \psi(x) (\Phi(v(z), m(z)) - \Phi(v_0, m_0)) dx \to 0 \ \text{ as } t \to 0, \ m_0 = m(x, 0). \tag{9.5}$$

Let us assume that the coefficients of Eq. (9.3) and the exponents of nonlinearity are subject to the following conditions:

$$\begin{cases} m(z),\ \sigma(z),\ p_i(z) \in C^0(\overline{Q}_T), \\ p_i(z) \text{ satisfy the log-continuity condition (1.31)}, \\ \text{there exist finite positive constants } a^{\pm}, b^{\pm} p^{\pm}, m^{\pm}, \sigma^{\pm} \text{ such that} \\ 0 < a^- \leq a_i(z) \leq a^+, \quad -b^- \leq b(z) \leq b^+, \\ 1 < p^- \leq p_i(z) \leq p^+, \quad 1 < \sigma^- \leq \sigma(z) \leq \sigma^+, \\ 0 < m^- \leq m(z) \leq m^+, \end{cases} \tag{9.6}$$

$$m_t \in L^\infty(Q), \quad \exists\, \mu = const > 0: \ \ 0 \leq m_t \leq \mu \text{ a.e. in } Q, \tag{9.7}$$

$$\begin{cases} 0 \leq -p_{it}(z) \leq C_{pi}, & 0 \leq b(z)\sigma_t(z) \leq C_\sigma, \\ 0 \leq -a_{it}(z) \leq C_a, & 0 \leq b_t(z) \leq C_b \end{cases} \tag{9.8}$$

with nonnegative constants $C_a$, $C_b$, $C_\sigma$, $C_{pi}$. The main existence result is given in the following theorem.

**Theorem 9.1** *Let the exponents $m(z)$, $\sigma(z)$, $p_i(z)$ satisfy conditions (9.6)–(9.8). Let us assume that there exists a constant $\alpha > 0$ such that if $p^-(t) < n$ for some $t \in [0, T]$, then*

$$m^+(t) + 1 + \alpha < \frac{np^-(t)}{n - p^-(t)}, \qquad \text{where} \quad \frac{1}{p^-(t)} = \frac{1}{n}\sum_{i=1}^{n} \frac{1}{p_i^-(t)}. \tag{9.9}$$

*Assume that $m(z)$ satisfies the oscillation condition*

$$\frac{m^+ - 1}{2} < \min\left\{m^-, \left(1 - \frac{1}{p^-}\right)(m^- + 1)\right\} \text{ if } m^+ > 1 \text{ and } \sum_{i=1}^{n} \frac{1}{p_i^-} > 1. \tag{9.10}$$

*Then for every $v_0 \in L^\infty(\Omega) \cap \mathbf{V}_0(\Omega)$ and every $g \in L^\infty(Q)$ with $|g_t(z)|^{1+\frac{1}{m(z)}} \in L^1(Q)$ problem (9.3) has at least one strong solution in a cylinder $Q_{T^*}$,*

$$T^* = \sup\{\theta : \ \|v(t)\|_{\infty,\Omega} < \infty \ \ \forall t \in (0, \theta)\}.$$

*Moreover, for small $\tau$ the solution satisfies the estimate*

$$\|v(t)\|_{\infty,\Omega} \leq \left(1 + \|v_0\|_{\infty,\Omega}\right) e^{At}, \quad t \in [0, \tau], \tag{9.11}$$

*with a constant $A$ independent of $v$. If either $\sup_Q(\sigma(z) - m(z)) \leq 1$, or $b^+ \leq 0$, then estimate (9.11) remains true for all $T > 0$, the solution exists globally in time and $T^* = T$. In the case $m^- \geq 1$ the solution satisfies the inclusion $\Phi_v(v, m)v_t \in L^2(Q_{T^*})$.*

Let us illustrate the assertions of Theorem 9.1 by the example of the model equation with constant coefficients and one space variable:

$$\frac{d}{dt}\left(|v|^{m(z)-1}v\right) = \left(|v_x|^{p(z)-2}v_x\right)_x. \tag{9.12}$$

- $m > 0$ and $p > 1$ are constants. In this simplest case the conditions of the theorem reduce to the conditions on the regularity of the initial datum.
- $m \equiv m(x) > 0$ and $p \equiv p(x) > 1$. A strong solution exists if $m$ and $p$ are log-continuous in $\Omega$.
- $m \equiv m(x) > 0$ and $p \equiv p(x,t) > 1$: it is required that, in addition to the above conditions, $p_t \in [0, C_p]$ with a positive constant $C_p$.
- $m \equiv m(x,t) > 0$ and $p \equiv p(x,t) > 1$. We have to add to the above conditions the restriction $m_t(x,t) \in [0, \mu]$.
- For $n > 1$ conditions (9.9) and (9.10) confine the assertion of Theorem 9.1 to the equations with slow diffusion. Nonetheless, in this case Eq. (9.12) still admits weak solutions (see Theorem 8.1).

The same is true for Eq. (9.12) with lower-order terms,

$$\frac{d}{dt}\left(|v|^{m(z)-1}v\right) = \left(|v_x|^{p(z)-2}v_x\right)_x + b\,|v|^{\sigma(z)-2}v + g(z), \quad b = const,$$

but the results are confined to a small time interval and the possibility of continuation of the solution to the arbitrary interval $(0, T)$ depends on the relations between $m(z)$, $p(z)$, $\sigma(z)$, and on the signs of $\sigma_t(z)$ and $b$. The solution exists globally in time if either $\sup_Q(\sigma(z) - m(z)) < 1$, or $b = const < 0$. Both conditions are well-known for the equations with constant exponents of nonlinearity—see, e.g., [36, 52, 267, 272].

## 9.2 Regularized Problem

A solution of problem (9.3) is constructed as the limit of the sequence of solutions of the regularized problems

$$\begin{cases} \dfrac{d}{dt}\Phi_\varepsilon(v,m) = \operatorname{div}\mathscr{F}(\nabla v, z) + f_K(v,z) \text{ in } Q, \\ v(x,0) = v_0(x) \text{ in } \Omega, \quad v = 0 \text{ on } \Gamma, \end{cases} \tag{9.13}$$

where $\mathscr{F}(\nabla v, z) = \{\mathscr{F}_1, \ldots, \mathscr{F}_n\}$, $\mathscr{F}_i(\nabla v, z) = a_i(z)|D_i v|^{p_i(z)-2}D_i v$, and

$$\Phi_\varepsilon(v,m) = \int_0^{v(z)} A_\varepsilon(s, m(z))ds, \quad A_\varepsilon(s,m) = \varepsilon(1 + l|s|^{l-1}) + m(z)|s|^{m(z)-1}$$

with the constant exponent $l = \max\{1, m^+\}$ and a given parameter $\varepsilon > 0$. For every $K > 1$ the function $f_K$ is defined by the equality

$$f_K(w,z) = g(z) + b(z) \times \begin{cases} (\min\{|w|, K\})^{\sigma(z)-m(z)-1} \ |w|^{m(z)-1}w & \text{if } \sigma(z) \geq m(z)+1, \\ |w|^{\sigma(z)-2}w & \text{otherwise.} \end{cases} \tag{9.14}$$

**Theorem 9.2** *Let $v_0 \in L^{l+1}(\Omega) \cap \mathbf{V}_0(\Omega)$. For every $\varepsilon > 0$, $K > 1$ problem* (9.13) *has at least one solution $v$ such that*

1. $v \in \mathbf{W}(Q) \cap L^{l+1}(Q)$,
2. $\dfrac{d}{dt}\Phi_\varepsilon(v,m) \in L^{1+\frac{1}{l}}(Q) \cap \mathbf{W}'(Q)$,
3. *for every test-function $\phi \in L^{l+1}(Q) \cap \mathbf{W}(Q)$*

$$\int_Q \left[\phi \frac{d}{dt}\Phi_\varepsilon(v,m) + \mathscr{F}(\nabla v, z)\cdot\nabla\phi\right]dz = \int_Q f_K(v,z)\,\phi\,dz, \tag{9.15}$$

4. *for every $\phi \in C_0^\infty(\Omega)$*

$$\int_\Omega \phi(x)(\Phi_\varepsilon(v,m) - \Phi_\varepsilon(v_0(x), m_0(x)))dx \to 0 \quad \text{as}\, t \to 0, \quad m_0 = m(x,0).$$

### 9.2.1 Galerkin's Approximations

Let us take a linearly independent system $\{\psi_i\}_{i=1}^\infty$ defined by (1.30). A solution of the regularized problem (9.13) is constructed as the limit of the sequence of finite-dimensional Galerkin's approximations

$$v^{(k)} = \sum_{i=1}^k c_i(t)\psi_i(x) \in \mathscr{P}_k \equiv \text{ span}\{\psi_1, \ldots, \psi_k\}.$$

The coefficients $c_i(t)$ are defined from the system of equations

$$\left(\frac{d}{dt}\Phi_\varepsilon(v^{(k)},m), \psi_j\right)_{2,\Omega} + \left(\mathscr{F}(\nabla v^{(k)}, z), \nabla\psi_j\right)_{2,\Omega} = \left(f_K(v^{(k)},z), \psi_j\right)_{2,\Omega}, \tag{9.16}$$

$j = \overline{1,k}$. The equality

$$\frac{d}{dt}\Phi_\varepsilon(v,m) = (\Phi_\varepsilon)'_v v_t + (\Phi_\varepsilon)'_m m_t = A_\varepsilon(v,m)\,\partial_t v + \partial_t m(z)\,v\,|v|^{m-1}\ln|v| \tag{9.17}$$

allows one to rewrite system (9.16) in the equivalent form

$$\sum_{i=1}^{k} B_{ij}(\mathbf{c}(t), t)c_i'(t) = F_j(\mathbf{c}(t)), \quad c_j(0) = (v_0(x), \psi_j(x))_{2,\Omega}, \quad j = \overline{1,k}, \tag{9.18}$$

with continuous functions

$$\begin{aligned} F_j = & -\left(\mathscr{F}(\nabla v^{(k)}, z), \nabla \psi_j\right)_{2,\Omega} \\ & + \left(f_K(v^{(k)}, z) - \partial_t m(z)\, v^{(k)}\, |v^{(k)}|^{m(z)-1} \ln |v^{(k)}|, \psi_j\right)_{2,\Omega}. \end{aligned} \tag{9.19}$$

The entries of the matrix $B$ have the form

$$B_{ij}(\mathbf{c}(t), t) = \left(A_\varepsilon(v^{(k)}, m)\, \psi_i(x), \; \psi_j(x)\right)_{2,\Omega}, \quad i, j = \overline{1,k}.$$

The determinant of $B$ is the Gram determinant of the linearly independent system $\{A_\varepsilon^{\frac{1}{2}}(v^{(k)}, m)\, \psi_i(x)\}_{i=1}^{k}$. Since $A_\varepsilon \geq \varepsilon > 0$, one can solve system (9.18) with respect to the derivatives $c_j'(t)$ and write the system in the normal form. By Peano's theorem, for every $k \in \mathbb{N}$ there exists at least one solution of system (9.18) on an in interval $(0, T_k)$.

### 9.2.2 A Priori Estimates

Let us introduce the function

$$\Psi_\varepsilon(v, m) = \int_0^v s\, A_\varepsilon(s, m) ds = \frac{\varepsilon}{2} v^2 + \frac{\varepsilon\, l}{l+1} |v|^{l+1} + \frac{m}{m+1} |v|^{m+1} \tag{9.20}$$

and notice that

$$v \frac{d}{dt} \Phi_\varepsilon(v, m) = \frac{d}{dt} \Psi_\varepsilon(v, m) - \frac{\partial_t m}{(m+1)^2} \left[1 - (m+1) \ln |v|\right] |v|^{m+1}. \tag{9.21}$$

**Lemma 9.1** *Let $\Psi_\varepsilon(v_0^{(k)}, m_0) \in L^1(\Omega)$ and $g \in L^{(m(\cdot)+1)'}(Q)$. Assume that $0 \leq \partial_t m \leq \mu$ a.e. in $Q$ with a positive constant $\mu$ Then each of the functions $v^{(k)}$ can be continued from the cylinder $Q_{T_k}$ to $Q$. The continued functions satisfy the estimates*

$$\sup_{(0,T)} \int_\Omega \Psi_\varepsilon(v^{(k)}, m) dx + \sum_{i=1}^{n} \int_Q |D_i v^{(k)}|^{p_i(z)} dz \leq M + \int_\Omega \Psi_\varepsilon(v_0^{(k)}, m_0) dx \tag{9.22}$$

*with an independent of $k$ and $\varepsilon$ constant $M = M(\mu, m^\pm, \sigma^\pm, b^+, K, T)$.*

*Proof* For the sake of simplicity of notation, we denote by $v$ the finite-dimensional Galerkin's approximation $v^{(k)}$ of the solution of problem (9.13) and omit the subindex $\varepsilon$. Multiplying each of the Eq. (9.16) by $c_j(t)$, integrating over the interval $(0,\tau)\subset(0,T_k)$, summing up in $j=\overline{1,k}$, and using the definition of $f_K$, we obtain the inequality

$$\begin{aligned}&\int_\Omega \Psi(v,m)dx\Big|_{t=0}^{t=\tau}+\frac{a^-}{2}\sum_{i=1}^n\int_{Q_\tau}|D_iv|^{p_i(z)}dz\\&\leq j_0(\tau)+\int_{Q_\tau}\frac{\partial_t m}{(m+1)^2}[1-(m+1)\ln|v|]|v|^{m+1}dz+\int_{Q_\tau}gv\,dz\equiv\sum_{i=0}^3 j_i(\tau),\end{aligned}\tag{9.23}$$

where $j_0(\tau)=j_0^{(1)}(\tau)+j_0^{(2)}(\tau)$,

$$\begin{aligned}&j_0^{(1)}(\tau)=\int_{Q_\tau\cap\{\sigma\geq m+1\}}b(z)\left(\min\{|v|,K\}\right)^{\sigma-m-1}|v|^{m+1}dz,\\&j_0^{(2)}(\tau)=\int_{Q_\tau\cap\{\sigma<m+1\}}b(z)|v|^\sigma dz,\qquad j_1(\tau)=\int_{Q_\tau}\frac{\partial_t m}{(m+1)^2}|v|^{m+1}dz,\\&j_2(\tau)=-\int_{Q_\tau}\frac{\partial_t m}{m+1}\ln|v|\,|v|^{m+1}dz,\qquad j_3(\tau)=\int_{Q_\tau}g\,vdz.\end{aligned}$$

Let us introduce the function

$$Y(\tau)=\int_{Q_\tau}|v|^{m+1}dz.$$

By the definition $j_0^{(1)}(\tau)\leq b^+K^{\sigma^+-m^--1}Y(\tau)$, and, by Young's inequality, $j_0^{(2)}(\tau)\leq C\,b^+\,(1+Y(\tau))$. The estimate on $j_1(\tau)$ is immediate: $j_1(\tau)\leq\mu Y(\tau)$. To estimate $j_2(\tau)$ we write it in the form

$$\begin{aligned}j_2(\tau)=&-\int_{Q_\tau\cap\{|v|>1\}}\frac{\partial_t m}{1+m}\left(|v|^{m+1}\ln|v|\right)dz\\&-\int_{Q_\tau\cap\{|v|\leq 1\}}\left(|v|^\beta\ln|v|\right)\frac{\partial_t m}{1+m}|v|^{m+1-\beta}dz\end{aligned}$$

with a constant $\beta\in(0,m^-+1)$. Dropping the first nonpositive term and estimating the second one by Young's inequality, we arrive at the estimate $j_2(\tau)\leq C\,(1+Y(\tau))$ with a constant $C=C(\mu,m^\pm)$. By Young's inequality

$$j_3(\tau)\leq C\Big(\int_{Q_\tau}|g|^{(m+1)'}dz+Y(\tau)\Big).$$

Plugging these estimates to (9.23) and dropping the nonpositive terms on the right-hand side, we obtain Gronwall's inequality for $Y(\tau)$,

$$Y'(\tau) \leq C(1 + Y(\tau)), \quad Y(0) = 0.$$

It follows that

$$Y(t) \leq e^{Ct} - 1$$

with an independent of $v$ constant $C$, which entailes (9.22) with an independent of $k$ constant $M = M(\mu, m^{\pm}, \sigma^{\pm}, b^{+}, K, T)$.

The possibility to continue each of $v \equiv v^{(k)}$ to the same interval $[0, T]$ follows from (9.22) because the function $v(x, T_k)$ possesses the same properties that $v_0^{(k)}$. □

**Corollary 9.1** *The sequence $\{v^{(k)}\}$ contains a subsequence (we assume that it coincides with the whole sequence) such that*

$$\begin{cases} v^{(k)} \rightharpoonup v & \text{in } L^{l+1}(Q) \cap \mathbf{W}(Q), \\ \Psi_\varepsilon(v^{(k)}, m) \to \psi & \star\text{-weakly in } L^{\infty}(0, T; L^1(\Omega)), \\ \mathscr{F}_i(\nabla v^{(k)}, z) \rightharpoonup A_i & \text{in } L^{p_i'(\cdot)}(Q). \end{cases} \tag{9.24}$$

**Lemma 9.2** *The sequence $\left\{\frac{d}{dt}\Phi_\varepsilon(v^{(k)}, m)\right\}_{k=1}^{\infty}$ is uniformly with respect to $k$ and $\varepsilon$ bounded in $L^{1+\frac{1}{l}}(Q) \cap \mathbf{W}'(Q)$.*

*Proof* It suffices to show that

$$\left|\left(\phi, \frac{d}{dt}\Phi_\varepsilon(v^{(k)}, m)\right)_{2,Q}\right| \leq C \qquad \forall \phi \in L^{l+1}(Q) \cap \mathbf{W}(Q)$$

with $\|\phi\|_{l+1,Q} + \|\phi\|_{\mathbf{W}(Q)} \leq 1$ and a constant $C$ independent of $k$, $\varepsilon$ and $\phi$. By (9.16) and (9.22)

$$\left|(\phi, \frac{d}{dt}\Phi_\varepsilon(v^{(k)}, m))_{2,Q}\right| \leq a^{+}\|v\|_{\mathbf{W}(Q)}\|\phi\|_{\mathbf{W}(Q)} + \|\phi\|_{m+1,Q}\|f_K\|_{\frac{m+1}{m},Q},$$

whence the assertion of the lemma. □

**Corollary 9.2** *The sequence $\left\{\frac{d}{dt}\Phi_\varepsilon(v^{(k)}, m)\right\}_{k=1}^{\infty}$ contains a subsequence that converges weakly in $L^{1+\frac{1}{l}}(Q) \cap \mathbf{W}'(Q)$.*

## 9.3 The Energy Inequality

Recall that wherever it doesn't cause a confusion $v \equiv v^{(k)}$ stands for the $k$th Galerkin's approximation of the solution of problem (9.13) with a positive $\varepsilon$ and $K > 1$. Multiplying each of Eq. (9.16) by $c_j'(t)$ and summing up in $j = \overline{1, k}$ we obtain the equality

$$\int_\Omega A_\varepsilon(v,m)v_t^2\,dx+\int_\Omega \mathscr{F}(\nabla v,z)\cdot\nabla v_t dx \\ =\int_\Omega f_K(v,z)\,v_t dx-\int_\Omega m_t v|v|^{m-1}\ln|v|v_t dx. \tag{9.25}$$

Straightforward computations lead to the formulas

$$a_i|D_iv|^{p_i-2}D_iv\cdot D_iv_t=\partial_t\Big(a_i\frac{|D_iv^{(k)}|^{p_i}}{p_i}\Big) \\ +a_i|D_iv|^{p_i}\Big(\frac{1}{p_i^2}-\frac{\ln|D_iv|}{p_i}\Big)p_{it}-a_{it}\frac{|D_iv|^{p_i}}{p_i},$$

$$f_K(v,z)\,v_t=\partial_t(b(z)\,S)+R+g\,v_t$$

with

$$S=\begin{cases}K^{\sigma-m-1}\dfrac{|v|^{m+1}}{m+1} & \text{if } v^2\geq K^2 \text{ and } \sigma\geq m+1,\\ \dfrac{|v|^\sigma}{\sigma} & \text{either if } v^2<K^2 \text{ and } \sigma\geq m+1, \text{ or if } \sigma<m+1,\end{cases}$$

$$R=\begin{cases}K^{\sigma-m-1}\dfrac{-b_t}{m+1}|v|^{m+1}-\sigma_t\dfrac{b|v|^{m+1}}{m+1}K^{\sigma-m-1}\ln\,K\\ \qquad\text{if } v^2\geq K^2 \text{ and } \sigma\geq m+1,\\ -\dfrac{b_t}{\sigma}|v|^\sigma+\dfrac{b}{\sigma^2}(1-\sigma\,\ln\,|v|)|v|^\sigma\sigma_t\\ \qquad\text{either if } v^2<K^2 \text{ and } \sigma\geq m+1, \text{ or if } \sigma<m+1.\end{cases}$$

Let us introduce the energy function

$$\mathscr{E}(t,v)=\int_\Omega\Big(\sum_{i=1}^n a_i\frac{|D_iv|^{p_i}}{p_i}-b\,S\Big)dx. \tag{9.26}$$

Combining the above formulas with (9.25) we obtain the equality

$$\frac{d}{dt}\mathscr{E}(t,v)+\int_\Omega A_\varepsilon(v,m)\,v_t^2dx=\Lambda(t)+\int_\Omega R dx-\int_\Omega P dx+\int_\Omega g\,v_t dx \tag{9.27}$$

in which

$$P(z)=m_t v|v|^{m-1}\ln|v|v_t$$

$$\Lambda(t)=\int_\Omega\sum_{i=1}^n\Big(-a_i|D_iv|^{p_i}\Big(\frac{1}{p_i^2}-\frac{\ln|D_iv|}{p_i}\Big)p_{it}+a_{it}\frac{|D_iv|^{p_i}}{p_i}\Big)dx.$$

**Lemma 9.3** *If condition* (9.9) *is fulfilled, then*

$$-\int_{\Omega} P dx \leq \frac{m_t^+(t)}{2\mu} \int_{\Omega} A_{\varepsilon}(v,m) v_t^2 dx + C\, m_t^+(t)\Big(1+\sum_{i=1}^{n} \int_{\Omega} |D_i v|^{p_i} dx\Big) \quad (9.28)$$

*with an independent of v constant C.*

*Proof* Let us fix $t \in [0, T]$. If $m_t^+(t) = 0$, then $P = 0$ and the assertion is obvious. Assume that $m_t^+(t) > 0$. By the Cauchy inequality

$$\begin{aligned} -\int_{\Omega} P\, dx &\leq \frac{m_t^+(t)}{2\mu} \int_{\Omega} A_{\varepsilon}(v,m) v_t^2 dx + \frac{m_t^+(t)}{2\mu} \int_{\Omega} \frac{|v|^{2m}}{A_{\varepsilon,m}} |\ln |v||^2 dx \\ &\equiv \frac{m_t^+(t)}{2\mu} (I_1 + I_2)\,. \end{aligned}$$

According to (9.9) there is a constant $\alpha \in (0, 1)$ such that $m^+(t) + \alpha + 1 \leq np^-(t)/(n - p^-(t))$. Since $s^{\alpha} \,|\ln s|^2 \to 0$ as $s \to 0+$ and $s^{-\alpha} \,|\ln s|^2 \to 0$ as $s \to \infty$,

$$\begin{aligned} I_2 &\leq \int_{\Omega \cap \{|v|>e\}} |v|^{m+1+\alpha} \Big(|v|^{-\alpha} |\ln |v||^2\Big)\, dx - \int_{\Omega \cap \{|v| \leq e\}} |v|^{m+1-\alpha} \Big(|v|^{\alpha} |\ln |v||^2\Big)\, dx \\ &\leq C\Big(1 + \|v\|_{m^+(t)+\alpha+1,\Omega}^{m^+(t)+\alpha+1}\Big). \end{aligned}$$

Let us make use of the interpolation inequality (1.58)

$$\|v(t)\|_{\lambda(t),\Omega} \leq C\, \|v\|_{m^-(t)+1,\Omega}^{1-s} \|\nabla v\|_{p^-(t),\Omega}^{s}, \quad (9.29)$$

in which the constant $C$ is independent of $v$ and the exponents are chosen in the following way:

$$\begin{aligned} s &= \Big(\frac{1}{m^-(t)+1} - \frac{1}{p^-(t)}\Big)\Big(\frac{1}{m^-(t)+1} - \frac{1}{p_-^*(t)}\Big)^{-1} \in (0,1), \\ \frac{1}{p_-^*(t)} &= \frac{1}{p^-(t)} - \frac{1}{n}, \end{aligned}$$

$$\begin{aligned} &\lambda(t) \in [m^-(t)+1, p_-^*(t)] \quad \text{if } p^-(t) < n, \\ &\lambda(t) \text{ is any number from } [m^-(t)+1, \infty) \quad \text{if } p^-(t) \geq n. \end{aligned}$$

Taking in (9.29) $\lambda(t) = m^+(t) + \alpha + 1$ and applying estimate (9.22), Young's inequality and then (1.9), we obtain:

$$\|v(t)\|_{m^+(t)+1+\alpha,\Omega}^{m^+(t)+1+\alpha} \le C\Big(\int_\Omega |\nabla v|^{p^-}\,dx\Big)^{\frac{s}{p^-(t)}(m^+(t)+1+\alpha)}$$
$$\le C'\Big(1+\int_\Omega |\nabla v|^{p^-}\,dx\Big) \le C''\Big(1+\sum_{i=1}^n\int_\Omega |D_i v|^{p_i}\,dx\Big).$$

□

**Lemma 9.4** *Let the exponents* $\sigma$, $p_i$, $m$ *and the coefficients* $a$, $b$ *satisfy conditions* (9.8) *with nonnegative constants* $C_a$, $C_b$, $C_\sigma$, $C_{pi}$. *If the conditions of* Lemma 9.3 *are fulfilled, then*

$$\mathcal{E}(t,v)+\frac{1}{2}\int_0^t\int_\Omega A_\varepsilon(v,m)\,v_t^2\,dx\,d\tau \le \mathcal{E}(0,v_0^{(k)})+C^*(1+t) \tag{9.30}$$

*with a constant* $C^*$ *depending on* $a^+$, $p_i^\pm$, $\sigma^\pm$, $m^\pm$, $\mu$, *the constants in* (9.6)–(9.8), *the constant* $M$ *in* (9.22) *and* $\|v_0\|_{m_0(\cdot)+1,\Omega}$.

*Proof* Under the assumptions of the Lemma

$$\Lambda(t)\le\sum_{i=1}^n\int_\Omega a_i|D_i v|^{p_i}\Big(\frac{1}{p_i^2}-\frac{\ln|D_i v|}{p_i}\Big)(-p_{it})\,dx.$$

Using (9.28) we rewrite (9.27) in the form

$$\frac{d}{dt}\mathcal{E}(t,v)+\frac{1}{2}\int_\Omega A_\varepsilon(v,m)v_t^2\,dx\le\sum_{i=1}^n J_{p_i}+\int_\Omega R\,dx+\int_\Omega v_t\,g\,dx$$
$$+C\,m_t^+(t)\Big(1+\sum_{i=1}^n\int_\Omega |D_i v|^{p_i}\,dx\Big) \tag{9.31}$$

with

$$J_{p_i}=\int_\Omega\frac{a_i|D_i v|^{p_i}}{p_i^2}\,(1-p_i\ln|D_i v|)\,(-p_{it})\,dx$$

and the constant $C$ from the conditions of Lemma 9.3. Let us introduce the function

$$G(s)=\frac{s^p}{p^2}\,(1-p\ln s)\,,\quad s\in I\equiv(0,\mathrm{e}^{1/p}].$$

It is easy to see that $G(0)=G\left(\mathrm{e}^{1/p}\right)=0$ and $G'(s)=-s^{p-1}\ln s$, whence $\max_I G(s)=G(1)=p^{-2}$. Using this observation we have

$$J_{p_i}=\Big(\int_{\Omega\cap(1\le p_i\ln|D_i v|)}+\int_{\Omega\cap(p_i\ln|D_i v|<1)}\Big)\frac{a_i|D_i v|^{p_i}}{p_i^2}(1-p_i\ln|D_i v|)(-p_{it})\,dx$$

$$\leq \int_{\Omega\cap(p_i \ln|D_i v|<1)} |p_{it}| \frac{a_i |D_i v|^{p_i}}{p_i^2} (1 - p_i \ln |D_i v|)\, dx \leq \frac{a^+ C_p}{(p^-)^2} |\Omega| .$$

Applying (9.8) we estimate $R$ by a constant depending only on $a^{\pm}$, $p_i^{\pm}$, $\sigma^{\pm}$, $m^{\pm}$, $\mu$, $|\Omega|$ and the constants in (9.8). Integrating (9.31) with respect to $t$ and then applying (9.22) and Young's inequality we obtain

$$\begin{aligned}
&\int_0^t \int_\Omega g\, v_t dz = \int_\Omega g\, v\, dx \Big|_{\tau=0}^{\tau=t} - \int_0^t \int_\Omega v\, g_t dx \\
&\leq C\Big(1 + 2 \sup_{(0,T)} \Psi_\varepsilon(v, m) + \sup_{(0,T)} \|g\|_{1+\frac{1}{m},\Omega} + \int_Q |v|^{m+1} dz + \int_Q |g_t|^{1+\frac{1}{m}} dz\Big) \\
&\leq C'.
\end{aligned}$$

Finally,

$$C \int_0^t m_t^+(t)\Big(1 + \sum_{i=1}^n \int_\Omega |D_i v|^{p_i} dx\Big) dt \leq C\,\mu\Big(t + \sum_{i=1}^n \int_Q |D_i v|^{p_i} dz\Big)$$

and the required estimate follows now from (9.22). □

**Corollary 9.3** 1. *If in the conditions of* Lemma 9.4 $m_t = 0$ *and* $g \equiv 0$, *then* (9.30) *takes on the form*

$$\mathscr{E}(t, v^{(k)}) + \int_0^t \int_\Omega A_\varepsilon(v^{(k)}, m) \left(\partial_t v^{(k)}\right)^2 dx d\tau \leq \mathscr{E}(0, v_0^{(k)}) + C^* t. \quad (9.32)$$

2. *If in the conditions of* Lemma 9.4 $g \equiv 0$, *and if the exponents* $p_i$, $\sigma$, $m$ *and the coefficients* $a_i$, $b$ *are independent of* $t$, *then* (9.32) *holds with* $C^* = 0$ *and transforms into equality.*
3. *The constant* $C^*$ *in* (9.30) *and* (9.32) *tends to zero when the constants in* (9.7) *and* (9.8) *tend to zero.*

**Corollary 9.4** *Under the conditions of* Lemma 9.4

$$\sup_{(0,T)} \int_\Omega \sum_{i=1}^n |D_i v^{(k)}|^{p_i(z)} dx + \int_Q A_\varepsilon(v, m)(v_t^{(k)})^2 dz \leq M_1 \quad (9.33)$$

*with an independent of* $k$ *constant* $M_1$. *The constant* $M_1$ *may depend on* $K$ *and the constants in conditions* (9.6)–(9.8).

*Proof* In view of (9.32), it suffices to prove that $\int_\Omega b\, S\, dx$ is uniformly bounded in $(0, T)$. The assertion is obvious if $b(z) \leq 0$. In case that $b^+ \geq 0$, it is a byproduct of the definition of $S$ and inequality (9.22). □

**Lemma 9.5** *Let us assume that condition* (9.10) *are fulfilled. The sequence* $\{v^{(k)}\}$ *contains a subsequence which possesses the properties* (9.24) *and converges a.e. in* $Q$.

*Proof* Let us consider the sequence $\{w_k\}$ with

$$w_k = |v^{(k)}|^{\alpha} \operatorname{sign} v^{(k)}, \qquad \alpha = \frac{1}{2}\left(1 + \max\{m^+, 1\}\right) \geq 1.$$

Let us define $\theta = \max\{m^+, 1\} - m(z)$ and claim that

$$\theta \leq m^- + 1 \quad \text{ if } \quad \sum_{i=1}^{n} \frac{1}{p_i^-} > 1.$$

This condition is automatically fulfilled if $m^+ \leq 1$, otherwise it reads $\frac{m^+-1}{2} \leq m^-$. Applying Hölder's inequality and using (9.22) and Lemma 1.24 we estimate

$$\int_Q |w_{k,t}|\, dz \leq C\Big(\int_Q |v^{(k)}|^{\theta} dx\Big)^{\frac{1}{2}} \Big(\int_Q A_\varepsilon(v^{(k)}, m)(v_t^{(k)})^2 dz\Big)^{\frac{1}{2}} \leq C''.$$

On the other hand, by virtue of (9.33) for any $r \in (1, p^-)$ and every $t \in (0, T)$

$$\begin{aligned} I_k \equiv \int_\Omega |\nabla w_k|^r dx &\leq C\Big(\int_\Omega |\nabla v^{(k)}|^{p^-} dx\Big)^{\frac{r}{p^-}} \Big(\int_\Omega |v^{(k)}|^{\omega} dz\Big)^{1-\frac{r}{p^-}} \\ &\leq C'\Big(\int_\Omega |v^{(k)}|^{\omega} dz\Big)^{1-\frac{r}{p^-}} \quad \text{with } \omega = (\alpha - 1)\frac{rp^-}{p^- - r}. \end{aligned}$$

If $\alpha = 1$ (that is, $m^+ \leq 1$), we simply take $r = p^-$, in this case $I_k$ are uniformly bounded. If $\alpha > 1$ (i.e. $m^+ > 1$), we claim that

$$\frac{m^+ - 1}{2} < \Big(1 - \frac{1}{p^-}\Big)(m^- + 1) \quad \Rightarrow \quad \frac{m^+ - 1}{2(m^- + 1)} + \frac{1}{p^-} < 1 \quad \text{if } \sum_{i=1}^{n} \frac{1}{p_i^-} > 1.$$

This allows us to choose $r$ in such a way that $\omega \leq m^- + 1$ and leads to a uniform estimate on $I_k$. By Lemma 2.6 the sequence $\{w_k\}$ is compact in $L^{\nu}(Q)$ with some $\nu \geq 1$ and, therefore, contains a subsequence that converges almost everywhere. □

**Lemma 9.6** *For every* $\varepsilon$ *the sequence* $\{\Psi_\varepsilon(v^{(k)}, m)\}$ *is precompact in* $C^0([0, T]; L^1(\Omega))$.

*Proof* According to (9.22) $\|\Psi_\varepsilon(v^{(k)}, m)\|_{1,\Omega}(t)$ are uniformly bounded in $(0, T)$. For every $t_1, t_2 \in [0, T]$, $t_1 \leq t_2$,

$$\int_\Omega |\Psi_\varepsilon|_{t=t_2} - \Psi_\varepsilon|_{t=t_1}|dx = \int_\Omega \left| \int_{t_1}^{t_2} \partial_t \Big( \int_0^v s A_\varepsilon(s,m) ds \Big) dt \right| dx$$
$$\le 2\int_{t_1}^{t_2} \int_\Omega \Big( |v|\,|A_\varepsilon(v,m)||v_t| + \Big| \int_0^v s\,(A_\varepsilon)'_m(s,m)\,m_t ds \Big| \Big) dz \equiv 2(I_1 + I_2).$$

By Hölder's inequality and due to (9.22) and (9.33)

$$|I_1| \le C\Big(\int_{t_1}^{t_2} \int_\Omega v^2 A_\varepsilon(v,m) dz\Big)^{\frac{1}{2}} \Big(\int_Q A_\varepsilon(v,m)(v_t)^2 dz\Big)^{\frac{1}{2}} \le C'\sqrt{t_2 - t_1}.$$

To estimate $I_2$ we represent it as the sum of the integrals $I_2^+$, $I_2^-$ over the sets where $|v| \ge 1$ or $|v| \le 1$. Recall that

$$\int_0^v s(A_\varepsilon)'_m(s,m) ds = m\int_0^v s|s|^{m-1} \ln|s| ds = \frac{m}{m+1}\Big(\ln|v| - \frac{1}{m+1}\Big)|v|^{m+1}.$$

Following the proof of Lemma 9.3 and applying Young's inequality, Lemma 1.24 and then (9.33), we find that for small $\alpha = const > 0$

$$|I_2^+| \le C\,\mu\Big(\sup_{s\ge 1} |s|^{-\alpha} \ln s\Big) \int_{t_1}^{t_2} \Big(1 + \int_\Omega |v|^{m^+(t)+\alpha+1} dx\Big) dt \le \mu\, C''\,(t_2 - t_1).$$

On the other hand, by virtue of (9.22)

$$|I_2^-| \le C\,\mu\Big(\sup_{0<s\le 1} s^{\frac{1}{2}} |\ln s|\Big) \int_{t_1}^{t_2} \int_\Omega \Big(1 + |v|^{m+1}\Big)\, dz \le \mu\, C'\,(t_2 - t_1).$$

The assertion follows now from the Arzelá-Ascoli theorem. □

## 9.4 Weak Solution of Regularized Problem

We may choose a subsequence of $\{v^{(k)}\}$ in such a way that (9.24) is fulfilled and, by virtue of Lemmas 9.5 and 9.6,

$$\Phi_\varepsilon(v^{(k)}, m) \to \Phi_\varepsilon(v,m) \text{ a.e. in } Q,$$
$$f_K(v^{(k)}, z) \to f_K(v,z) \text{ a.e. in } Q,$$
$$\Psi_\varepsilon(v^{(k)}, m) \to \Psi_\varepsilon(v,m) \text{ in } C^0([0,T]; L^1(\Omega)).$$

By construction, for every $k \geq s \in \mathbb{N}$ and $\phi^{(s)} \in \operatorname{span}\{\psi_1, \dots, \psi_s\}$

$$\int_Q \Big[\phi^{(s)} \frac{d}{dt}\Phi_\varepsilon(v^{(k)}, m) + \sum_{i=1}^n \mathscr{F}_i(\nabla v^{(k)}, z) \cdot D_i\phi^{(s)} - \phi^{(s)} f_K(v^{(k)}, z)\Big]dz = 0. \tag{9.34}$$

Letting in (9.34) $k \to \infty$ and then $s \to \infty$, we find that $\forall \phi \in L^{l+1}(Q) \cap \mathbf{W}(Q)$

$$\int_Q \Big[\phi \frac{d}{dt}\Phi_\varepsilon(v, m) + \sum_{i=1}^n A_i \cdot D_i\phi - \phi\, f_K(v, z)\Big]dz = 0. \tag{9.35}$$

Let $\phi(x) \in C_0^\infty(\Omega)$. Fix some $t_1, t_2 \in (0, T)$ $(t_1 < t_2)$, $s \in \mathbb{N}$, and test (9.35) with the function

$$\zeta(x,t) = \phi(x)\xi_s(t), \quad \xi_s(t) = \begin{cases} 0 & \text{for } t \geq t_2 \\ s(t_2 - t) & \text{for } t \in (t_2 - 1/s, t_2), \\ 1 & \text{for } t \in (t_1 + 1/s, t_2 - 1/s), \\ s(t - t_1) & t \in (t_1, t_1 + 1/s), \\ 0 & \text{for } t \leq t_1, \end{cases}$$

which gives

$$\begin{aligned} \int_0^T \Big(\int_\Omega \zeta \frac{d}{dt}\Phi_\varepsilon(v, m)dx\Big)dt &= \int_0^T \frac{d}{dt}\Big(\int_\Omega \zeta\Phi_\varepsilon(v, m)dx\Big)dt - \int_Q \zeta_t\Phi_\varepsilon(v, m)dz \\ &= s\int_{t_2-\frac{1}{s}}^{t_2}\int_\Omega \Phi_\varepsilon(v, m)\phi(x)dxdt - s\int_{t_1}^{t_1+\frac{1}{s}}\int_\Omega \Phi_\varepsilon(v, m)\phi(x)dxdt \\ &= -\int_{t_1}^{t_2}\int_\Omega \Big[\sum_{i=1}^n A_i \cdot D_i\phi(x) - \phi(x)\, f_K(v, z)\Big]dz. \end{aligned}$$

The inclusion $\Phi_\varepsilon(v, m) \in L^{1+\frac{1}{l}}(\Omega)$ for a.e. $t \in (0, T)$ allows one to pass to the limit as $s \to \infty$ and apply the Lebesgue differentiation theorem: for a.e. $\tau \in (0, T)$

$$\lim_{s\to\infty} \frac{1}{1/s}\int_\tau^{\tau+1/s}\Big(\int_\Omega \Phi_\varepsilon(v, m)\phi(x)dx\Big) = \int_\Omega \Phi_\varepsilon(v, m)\phi(x)dx\Big|_{t=\tau},$$

It follows that for a.e. $t_1, t_2 \in [0, T]$

$$\int_\Omega \Phi_\varepsilon(v, m)\phi(x)dx\Big|_{t=t_1}^{t=t_2} = -\int_{t_1}^{t_2}\int_\Omega \Big[\sum_{i=1}^n A_i \cdot D_i\phi - \phi\, f_K(v, z)\Big]dz.$$

By the property of absolute continuity of the integral the right-hand side is continuous, which gives the inclusion $\Phi_\varepsilon(v,m) \in C^0_w(0,T;L^{1+\frac{1}{\gamma}}(\Omega))$. To check that $A_i = \mathscr{F}_i(\nabla v, z)$. we rely on the monotonicity of $\mathscr{F}(\mathbf{s}, z)$ with respect to $\mathbf{s}$. Choosing the limit function $v$ for the test-function in (9.35), we obtain the equality

$$\int_Q \Big[ v\frac{d}{dt}\Phi_\varepsilon(v,m) + \sum_{i=1}^n A_i \cdot D_i v - v\, f_K(v,z)\Big] dz = 0. \tag{9.36}$$

Due to monotonicity [see inequality (4.35)], for every $\zeta \in \mathbf{W}(Q)$ and $k \in \mathbb{N}$

$$X^{(k)} := \int_Q \Big(\mathscr{F}(\nabla v^{(k)}, z) - \mathscr{F}(\nabla\zeta, z)\Big)\cdot \nabla(v^{(k)} - \zeta) dz \geq 0.$$

Following the derivation of estimate (9.22), for every $k = 1,2,\dots$ we obtain the equality

$$\begin{aligned}&\int_\Omega \Psi_\varepsilon(v^{(k)}, m) dx\Big|_{t=0}^{t=T} - \int_Q \frac{\partial_t m}{(m+1)^2}\Big[1-(m+1)\ln|v^{(k)}|\Big]|v^{(k)}|^{m+1} dz\\ &+ \int_Q \Big[\mathscr{F}(\nabla v^{(k)}, z)\cdot\nabla v^{(k)} + v^{(k)} f_K(v^{(k)}, z)\Big] dz = 0.\end{aligned} \tag{9.37}$$

On the other hand,

$$\begin{aligned}\mathscr{F}(\nabla v^{(k)}, z)\cdot\nabla v^{(k)} &= \mathscr{F}(\nabla v^{(k)}, z)\cdot\nabla(v^{(k)}-\zeta) + \mathscr{F}(\nabla v^{(k)}, z)\cdot\nabla\zeta\\ &= \Big(\mathscr{F}(\nabla v^{(k)}, z) - \mathscr{F}(\nabla\zeta, z)\Big)\cdot\nabla(v^{(k)}-\zeta)\\ &\quad + \mathscr{F}(\nabla\zeta, z)\cdot\nabla(v^{(k)}-\zeta) + \mathscr{F}(\nabla v^{(k)}, z)\cdot\nabla\zeta.\end{aligned}$$

Gathering these two equalities we obtain:

$$\begin{aligned}0 \leq X_i^{(k)} &= -\int_\Omega \Psi_\varepsilon(v^{(k)}, m)\, dx\Big|_{t=0}^{t=T}\\ &- \int_Q v^{(k)} f_K(v^{(k)}, z) dz + \int_{Q_\tau} \frac{\partial_t m}{(m+1)^2}\Big[1-(m+1)\ln|v^{(k)}|\Big]|v^{(k)}|^{m+1} dz\\ &- \int_Q (\mathscr{F}(\nabla v^{(k)}, z) - \mathscr{F}(\nabla\zeta, z))\cdot\nabla\zeta dz - \int_Q \mathscr{F}(\nabla\zeta, z)\cdot\nabla v^{(k)} dz.\end{aligned} \tag{9.38}$$

By the property of lower semicontinuity of the norm

$$-\liminf_{k\to\infty}\int_\Omega \Psi_\varepsilon(v^{(k)}(x,T), m(x,T)) dx \leq -\int_\Omega \Psi_\varepsilon(v(x,T), m(x,T)) dx,$$

while

$$\liminf_{k\to\infty}\int_{\Omega}\Psi_{\varepsilon}(v_0^{(k)},m)dx=\int_{\Omega}\Psi_{\varepsilon}(v_0,m(x,0))dx$$

because $v_0\in L^{l+1}(\Omega)$. For every fixed $\varepsilon>0$

$$v\in L^{l+1}(Q)\cap\mathbf{W}(Q),\qquad \frac{d}{dt}\Phi_{\varepsilon}(v,m)\in L^{1+\frac{1}{l}}(Q)\cap\mathbf{W}'(Q),$$

and for a.e. $t_1,t_2\in[0,T]$ the integration-by-parts formula holds:

$$\begin{aligned}&\int_{t_1}^{t_2}\int_{\Omega}v\frac{d}{dt}\Phi_{\varepsilon}(v,m)dz\\&=\int_{\Omega}\Psi_{\varepsilon}(v,m)\,dx\Big|_{t=t_1}^{t=t_2}-\int_{t_1}^{t_2}\int_{\Omega}\frac{\partial_t m}{(m+1)^2}\,[1-(m+1)\ln|v|]\;|v|^{m+1}dz.\end{aligned}\tag{9.39}$$

According to (9.21) and (9.36) $\Psi_{\varepsilon}(v,m)\in C^0([0,T];L^1(\Omega))$, and (9.39) is true for every $t\in[0,T]$. Plugging (9.39) into (9.36) we arrive at the inequality

$$\begin{aligned}0\le&-\int_{\Omega}\Psi_{\varepsilon}(v(x,T),m(x,T))dx+\int_{\Omega}\Psi_{\varepsilon}(v_0,m_0)dx\\&-\int_{Q}\Big[\sum_i A_i\cdot D_iv+v\,f_K(v,z)\Big]dz\\&+\int_{Q}\frac{\partial_t m}{(m+1)^2}\,[1-(m+1)\ln|v|]\;|v|^{m+1}dz+\sum_i\int_{Q}A_i\cdot D_i(v-\zeta)dz\\&-\int_{Q}\mathscr{F}(\nabla\zeta,z)\cdot\nabla(v-\zeta)dz=\sum_i\int_{Q}(A_i-\mathscr{F}_i(\nabla\zeta,z))\,D_i(v-\zeta)dz.\end{aligned}$$

We may now take $\zeta=v\pm\lambda\,\eta$ with arbitrary $\lambda>0$ and $\eta\in L^{l+1}(Q)\cap\mathbf{W}(Q)$. Simplifying and letting $\lambda\to 0$ we obtain the inequalities

$$\pm\sum_i\int_{Q}[A_i-\mathscr{F}_i(\nabla v,z)]\cdot D_i\eta dz\ge 0,$$

which are impossible unless $A_i(z)=\mathscr{F}_i(\nabla v,z)$ a.e. in $Q$. The proof of Theorem 9.2 is completed.

## 9.5 Bounded Solutions of Regularized Problem

Let us show that the solutions of problem (9.13) are uniformly bounded on a time interval whose choice depends on $\|v_0\|_{\infty,\Omega}$. This estimate would mean that for big $K$ the solutions of problem (9.13) are, in fact, independent of $K$ and solve the same problem but with $f_K$ substituted by $f$.

**Theorem 9.3** *Let in the conditions of* Theorem 9.2 $v_0 \in L^\infty(\Omega)$. *There exist* $T_0$ *and* $A = const$, *which depend on* $\|v_0\|_{\infty,\Omega}$, $\|g\|_{\infty,Q}$, $m^\pm$, $\sigma^\pm$, $\mu$, *but are independent of* $\varepsilon$, *such that the solutions of problem* (9.13) *satisfy the estimate*

$$\|v(t)\|_{\infty,\Omega} \le \left(1 + \|v_0\|_{\infty,\Omega}\right) \mathrm{e}^{At} \quad \text{for } t \in [0, T_0]. \tag{9.40}$$

*If* $K$ *is chosen according to the condition*

$$(1 + \|v_0\|_{\infty,\Omega})\mathrm{e}^{AT_0} \le K,$$

*then the solutions* $v$ *solve problem* (9.13) *in* $Q_{T_K}$ *with* $f_K \equiv f$.

The proof relies on the following technical assertion.

**Lemma 9.7** *Let* $K > 1$, $m > 1$, $l > 1$ *and* $s \ge 1$ *be given constants. For every constant* $C_s > 2s + max\{2; m+1; l+1\}$

$$\begin{aligned} G(v) \equiv C_s \int_0^{\max\{0,v\}} & \rho\, A_\varepsilon(\rho, m) \min\{K^{2(s-1)}, \rho^{2(s-1)}\}\, d\rho \\ & - v \max\{0, v\}\, A_\varepsilon(v, m) \min\{K^{2(s-1)}, v^{2(s-1)}\} \ge 0. \end{aligned}$$

*Proof* Let us write $G$ in the form

$$G(v) = \begin{cases} C_s \displaystyle\int_0^v \rho^{2s-1} A_\varepsilon(\rho, m)\, d\rho - A_\varepsilon(v, m)\, v^{2s} & \text{for } v \in (0, K), \\ C_s \left( \displaystyle\int_0^K \rho^{2s-1} A_\varepsilon(\rho, m)\, d\rho + \int_K^v \rho^{2s-1} A_\varepsilon(s, m) ds \right) & \\ \qquad\qquad - v^2 A_\varepsilon(v, m) K^{2(s-1)} & \text{if } v \ge K, \\ 0 & \text{if } v \le 0. \end{cases}$$

A straightforward calculation gives

$$G'(v) = v^{2s-1}\left(\varepsilon\, l\, (C_s - 2s(l-1))\, v^{l-1} + m\, v^{m-1}(C_s - m) + \varepsilon\, C_s\right) \ge 0$$

if $v \in (0, K]$ and

$$G'(v) \ge \varepsilon\, v^{2s-1}(C_s - 2) + \varepsilon\, l\, v^{2s+l-2}(C_s - (l+1)) + m v^{m+2s-2}(C_s - (m+1)) > 0$$

if $v > K$. The assertion follows because $G(v)$ is nondecreasing and $G(0) = 0$. □

*Proof* (*Proof of* Theorem 9.3) It is sufficient to show that

$$\|\max\{0, v\}\|_{\infty,\Omega} \leq \left(1 + \|\max\{0, v_0\}\|_{\infty,\Omega}\right) e^{At}$$

with a suitable constant $A$. Let us fix some $K > 1$ and test (9.35) with the function

$$\phi_s(v) = \max\{0, v\} \min\{K^{2(s-1)}, v^{2(s-1)}\}.$$

It is easy to see that

$$\begin{aligned}\phi_s(v) \frac{d\Phi_\varepsilon(v, m)}{dt} &= \phi_s(v)\, A_\varepsilon(v, m) v_t + \phi_s(v) m_t |v|^m \ln |v| \\ &= \frac{d}{dt}\Big(\int_0^v \phi_s(\rho) A_\varepsilon(\rho, m) d\rho\Big) + m_t \Big(\phi_s(v) \int_0^{\max\{0;v\}} (A_\varepsilon)'_m(\rho, m) d\rho \\ &\quad - \int_0^{\max\{0;v\}} \phi_s(\rho)(A_\varepsilon)'_m(\rho, m)\, d\rho\Big).\end{aligned}$$

Let us consider the function

$$H(v) = \phi_s(v) \int_0^v (A_\varepsilon)'_m(\rho, m)\, d\rho - \int_0^v \phi_s(\rho)(A_\varepsilon)'_m(\rho, m)\, d\rho.$$

It is continuous for $v \geq 0$, continuously differentiable on $[0, K) \cup (K, \infty)$, $H(0) = 0$ and

$$H'(v) = \phi'_s(v) \int_0^v (A_\varepsilon)'_m(\rho, m) d\rho = v^m \ln v \times \begin{cases} (2s-1) v^{2(s-1)} & \text{if } v \in [0, K), \\ K^{2(s-1)} & \text{if } v > K. \end{cases}$$

It follows that $H(v)$ is monotone decreasing on the interval $(0, 1)$ and monotone increasing for $v > 1$. $H(v)$ attains its minimal value at the point $v = 1$: $H(1) = -\frac{2s-1}{(2s+m-1)^2}$. Since $0 \leq m_t \leq \mu$ a.e. in $Q$, we estimate

$$\phi_s(v) \frac{d\Phi_\varepsilon(v, m)}{dt} \geq \frac{d}{dt}\Big(\int_0^v \phi_s(\rho) A_\varepsilon(\rho, m)\, d\rho\Big) - \mu \frac{2s-1}{(2s+m-1)^2}.$$

Further, $\mathscr{F}_i(\nabla v)\, D_i \phi_s(v) \geq 0$ a.e. in $Q$. Substituting these inequalities in (9.35) we arrive at the following estimate: for every $t, t+h \in (0, T)$

$$\begin{aligned}
&\frac{1}{h}\int_t^{t+h}\frac{d}{dt}\left(\int_\Omega\int_0^{\max\{0;v\}}\phi_s(\rho)\,A_\varepsilon(\rho,m)\,d\rho\,dx\right)dt-\mu\frac{(2s-1)|\Omega|}{(2s+m^--1)^2}\\
&\qquad\le\frac{b^+}{h}\int_t^{t+h}\int\limits_{\Omega\cap\{\sigma\ge m+1\}}\phi_s(v)\,v^m\,(\min\{v,\ K\})^{\sigma-m-1}\,dz\\
&\qquad\quad+\frac{b^+}{h}\int_t^{t+h}\int\limits_{\Omega\cap\{\sigma<m+1\}}\phi_s(v)\,v^{\sigma-1}\,dz+\frac{1}{h}\int_t^{t+h}\int\limits_\Omega\phi_s(v)\,g(z)\,dz.
\end{aligned}\tag{9.41}$$

Let us denote $v_+ = \max\{0, v\}$. Applying Young's inequality and the fact that $m^-v_+^m \le v_+A_\varepsilon(v_+, m)$ we estimate the integrands of the second and third terms on the right-hand side of (9.41) by means of the inequalities

$$\begin{aligned}
\phi_s(v)v^{\sigma-1}&=\begin{cases}v_+^\sigma v^{2(s-1)} & \text{if } v\le K\\ v_+^\sigma K^{2(s-1)} & \text{if } v>K>1\end{cases}\\
&\le\begin{cases}\frac{\sigma+1+2(s-1)}{m+1+2(s-1)}v_+^{m+1}v^{2(s-1)}+\frac{m-(\sigma+1)}{m+1+2(s-1)} & \text{if } v<K\\ v_+^{m+1}K^{2(s-1)} & \text{if } v>K>1\end{cases}\\
&\le\begin{cases}v^m\phi_s(v)+\frac{m-(\sigma+1)}{m+1+2(s-1)} & \text{if } v\le K,\\ v^m\phi_s(v) & \text{if } v>K>1,\end{cases}
\end{aligned}$$

$$\begin{aligned}
\phi_s(v)&=\begin{cases}v^{-m}\,(v^m\phi_s(v)) & \text{if } v\ge K>1\\ v^{2s-1} & \text{if } v<K\end{cases}\\
&\le\begin{cases}\frac{1}{m^-}v\,\phi_s(v)A_s(v) & \text{if } v\ge K>1\\ \frac{2s-1}{2(s-1)+m+1}v^{2(s-1)+m+1}+\frac{m}{2(s-1)+m+1} & \text{if } v<K\end{cases}\\
&\le\begin{cases}\frac{1}{m^-}v\,\phi_s(v)A_s(v) & \text{if } v\ge K>1\\ \frac{1}{m^+}\frac{2s-1}{2(s-1)+m+1}v\,\phi_s(v)A_\varepsilon(v)+\frac{m}{2(s-1)+m+1} & \text{if } v<K.\end{cases}
\end{aligned}$$

Let us introduce the function

$$I(t)=\int_\Omega\int_0^{v_+}\rho\,A_\varepsilon(\rho,x)\min\{K^{2(s-1)},\ \rho^{2(s-1)}\}\,d\rho dx.$$

By virtue of Lemma 9.7 with $C_s = 2s+(l+1)$ inequality (9.41) entails the inequality

$$
\begin{aligned}
\frac{1}{h}\int_t^{t+h}\frac{dI(\tau)}{d\tau}\,d\tau \leq & \frac{C_s}{m^-}\Big(\frac{1}{h}\int_t^{t+h}\Big(\|g(\cdot,\tau)\|_{\infty,\Omega}+b^+\Big(1+K^\gamma\Big)\Big)I(\tau)d\tau\Big)\\
& + \frac{m^+|\Omega|}{2(s-1)+m^-+1}\Big(\frac{1}{h}\int_t^{t+h}\|g(\cdot,\tau)\|_{\infty,\Omega}\,d\tau\Big)\\
& + \mu\frac{(2s-1)|\Omega|}{(2s+m^--1)^2}+b^+|\Omega|\frac{m^+-(\sigma^-+1)}{m^-+2(s-1)},
\end{aligned}
$$

$\gamma = \max\{0, \sigma - m - 1\}$. Letting $h \to 0$, we arrive at the differential inequality for $I(t)$: for a.e. $t \in (0, T)$

$$
I'(t) \leq C_s M_1\, I(t) + M_2, \qquad I(0) = I_0,
$$

with the constants

$$
\begin{aligned}
M_1 &= \frac{1}{m^-}\left(G + b^+\left(1+K^\gamma\right)\right), \qquad G = \operatorname{ess}\sup_{(0,T)}\|g\|_{\infty,\Omega},\\
M_2 &= \mu\frac{(2s-1)|\Omega|}{(2s+m^--1)^2}+b^+\,|\Omega|\,\frac{m^+-(\sigma^-+1)}{m^-+2(s-1)}+\frac{m^+|\Omega|\,G}{2(s-1)+m^-+1}\sim\frac{1}{s}
\end{aligned}
$$

as $s \to \infty$. The direct integration of this inequality gives

$$
I^{\frac{1}{2C_s}}(t) \leq \mathrm{e}^{M_1 t}\left(I^{\frac{1}{2C_s}}(0) + (M_2\,t)^{\frac{1}{2C_s}}\right). \tag{9.42}
$$

Let us now pass to the limit as $s \to \infty$. Fix some $t \in (0, T]$, denote

$$
\Omega_+ = \{x \in \Omega : \ v(x,t) > K\} \qquad \Omega_- = \{x \in \Omega : \ v(x,t) \leq K\}
$$

and represent $I(t)$ as the sum of the integrals $I_\pm(t)$ taken over the sets $\Omega_\pm$. A straightforward calculation shows that

$$
\begin{aligned}
I_+(t) &\geq K^{2(s-1)}\left(K^{m^-}+\frac{\varepsilon}{2}K^2+\varepsilon\,K^l\right)|\Omega_+|,\\
I_-(t) &\geq \frac{m^-}{2(s-1)+m^++1}\int_{\Omega_-}|v|^{2(s-1)+m+1}\,dx.
\end{aligned}
$$

Since $\|v_0\|_{\infty,\Omega} < K$,

$$
I(0) = I_-(0) \quad \text{and} \quad \lim_{s\to\infty} I^{\frac{1}{2C_s}}(0) = \|v_0\|_{\infty,\Omega}.
$$

Letting $s \to \infty$ in (9.42) we also obtain the inequalities

$$K \leq \lim_{s\to\infty} I_+^{\frac{1}{2C_s}}(t) \leq e^{M_1 t}\left(1 + \|v_0\|_{\infty,\Omega}\right) \quad \text{if } |\Omega_+| \neq 0,$$
$$\|v(t)\|_{\infty,\Omega_-} \leq e^{M_1 t}\left(1 + \|v_0\|_{\infty,\Omega}\right),$$

whence

$$K + \|v(t)\|_{\infty,\Omega_-} \leq 2\, e^{M_1 t}\left(1 + \|v_0\|_{\infty,\Omega}\right). \tag{9.43}$$

Given $K > 1 + 2\left(1 + \|v_0\|_{\infty,\Omega}\right)$, let us set

$$T_K = \frac{1}{M_1} \ln\left(1 + \frac{1}{1 + 2(1 + \|v_0\|_{\infty,\Omega})}\right). \tag{9.44}$$

It is easy to see that for $t \in [0, T_K]$ inequality (9.43) is impossible unless $|\Omega_+| = 0$, which means that $\Omega = \Omega_-$ for all $t \in [0, T_K]$. Estimate (9.40) follows now from (9.42) with $I(t) \equiv I_-(t)$ as $s \to \infty$. □

**Corollary 9.5** *In the special cases when*

$$\textit{either} \quad \sup_Q(\sigma(z) - m(z) - 1) \leq 0, \quad \textit{or} \quad b^+ = 0,$$

*the constant $M_1$ and, respectively, $T_K$ are independent of $K$. This means that estimate* (9.40) *remains true for every finite $T$.*

Now we are in position to show that for small $T_0$ the parameter $K$ is a dummy and can be omitted. Let us consider the problem with the unique regularization parameter $\varepsilon$:

$$\begin{cases} \partial_t \Phi_\varepsilon(v, x) = \operatorname{div} \mathscr{F}(\nabla v, z) + f(v, z) & \text{in } Q, \\ v(x, 0) = v_0(x) \text{ in } \Omega, \quad v = 0 \text{ on } \Gamma. \end{cases} \tag{9.45}$$

**Theorem 9.4** *Let us assume that the conditions of* Theorem 9.2 *are fulfilled. If, in addition, $v_0 \in L^\infty(\Omega)$, then there exists $T_0 \in (0, T]$ such that in the cylinder $Q_{T_0}$ the solution of problem* (9.13) *solves problem* (9.45) *and satisfies the estimate*

$$\|v(t)\|_{\infty,\Omega} \leq \left(1 + \|v_0\|_{\infty,\Omega}\right) e^{At} \quad \textit{for } t \in [0, T_0] \tag{9.46}$$

*with an independent of $\varepsilon$ constant $A$.*

*Proof* It follows from Theorem 9.3 that for every $K > 2(1 + \|v_0\|_{\infty,\Omega})$ the solutions of problem (9.13) are bounded in $O_{T_K}$ independently of $\varepsilon$, which means that there is an interval $[0, T_0]$ where $\|v(t)\|_{\infty,\Omega} < K$. Thus, $f_K(v, z) \equiv f(v, z)$ in $Q_{T_0}$ and $v$ is a bounded solution of problem (9.45). □

The constructed solution is defined on an interval $(0, T_0)$ with $T_0$ from Theorem 9.3. Since the function $v(x, T_0)$ possesses the same properties that the initial function $v_0$, one may take it for the initial datum and repeat all the above arguments to show that the solution of problem (9.45) can be continued to a time interval $(T_0, T_1)$ with $T_1$ depending on $\|v(\cdot, T_0)\|_{\infty,\Omega}$. Continuing this process we obtain the sequence $\{T_k\}$ and a solution of problem (9.45) in the cylinders $Q_{T_k}$. If $T_k \to \infty$, the solution exists globally in time, otherwise $\lim T_k = T^* < \infty$ and problem (9.45) admits a local in time solution.

Let us specially notice that in the cases mentioned in Corollary 9.5 the solutions of problem (9.45) exist globally in time and are uniformly with respect to $\varepsilon$ bounded in every cylinder $Q$.

## 9.6 Proof of the Existence Theorem

Let us denote by $\{v_\varepsilon\}$ the sequence of strong bounded solutions of problem (9.45).

**Lemma 9.8** *Let the conditions of* Theorem 9.3 *be fulfilled. If* $v_0 \in W_0^{1,q}(\Omega) \cap L^\infty(\Omega)$ *and conditions* (9.8) *hold, then there exists* $T_0$ *such that in the cylinder* $Q_{T_0}$ *problem* (9.45) *has a solution* $v_\varepsilon$ *which satisfies the inequality*

$$\sup_{(0,T_0)} \int_\Omega \sum_{i=1}^n |D_i v_\varepsilon|^{p_i(z)}\, dx + \frac{1}{2}\int_0^{T_0}\int_\Omega A_\varepsilon(v_\varepsilon, m)\,(\partial_t v_\varepsilon)^2\, dx d\tau \le C \qquad (9.47)$$

*with an independent of* $\varepsilon$ *constant* $C$.

*Proof* It is sufficient to fix $K$ and $T_K$ according to (9.43) and (9.44) and then let $k \to \infty$ in estimate (9.47). □

**Lemma 9.9** *Under the foregoing conditions the sequence* $\{v_\varepsilon\}$ *contains a subsequence that converges pointwise as* $\varepsilon \to 0$.

*Proof* We imitate the proof of Lemma 9.5 taking into account uniform boundedness of $v_\varepsilon$. For the functions $w_\varepsilon = |v_\varepsilon|^{\alpha-1}v_\varepsilon$ with $\alpha = (m^+ + 1)/2$

$$\begin{aligned}
|\partial_t w_\varepsilon| &= \alpha|v_\varepsilon|^{\alpha-1}|\partial_t v_\varepsilon| \le C\,|v_\varepsilon|^{\alpha-1-(m-1)/2}\left|\sqrt{A_\varepsilon(v_\varepsilon, m)}\,\partial_t v_\varepsilon\right| \\
&\le C'\left|\sqrt{A_\varepsilon(v_\varepsilon, m)}\,\partial_t v_\varepsilon\right|, \\
|\nabla w_\varepsilon|^{p^-} &\le C\,|v_\varepsilon|^{(\alpha-1)p^-}|\nabla v_\varepsilon|^{p^-} \le C'\,|\nabla v_\varepsilon|^{p^-}.
\end{aligned}$$

By Lemma 2.6 there is a subsequence of $\{v_\varepsilon\}$ that converges in $L^\nu(Q_{T_0})$ with some $\nu > 1$ and, respectively, pointwise in $Q_{T_0}$. □

The main a priori estimates derived for the solutions of problems (9.13) are independent of $\varepsilon$, do not depend on $K$, and remain true for the solutions $v_\varepsilon$ of problem (9.45). Gathering the uniform estimates on $v_\varepsilon$ we extract a subsequence (for which we keep the same notation) such that

$$\begin{cases} v_\varepsilon \to v & \text{weakly in } \mathbf{W}(Q_{T_0}) \text{ and a.e. in } Q_{T_0}, \\ \Phi_\varepsilon(v_\varepsilon, m) \to |v|^{m(z)-1}v & \text{a.e. in } Q_{T_0}, \\ \dfrac{d}{dt}\Phi_\varepsilon(v_\varepsilon, m) \to \dfrac{d}{dt}\left(|v|^{m(z)-1}v\right) & \text{weakly in } \mathbf{W}'(Q_{T_0}), \\ \mathscr{F}_i(\nabla v_\varepsilon, z) \to V_i & \text{weakly in } L^{p_i'(\cdot)}(Q_{T_0}), \\ f(v_\varepsilon, z) \to f(v, z) & \star\text{-weakly in } L^\infty(Q_{T_0}). \end{cases}$$

Letting $\varepsilon \to 0$ in (9.15) we conclude that for every $\phi \in L^{l+1}(Q) \cap \mathbf{W}(Q_{T_0})$

$$\int_{Q_{T_0}} \left[\phi \frac{d}{dt}\Phi(v, m) + \sum_{i=1}^n V_i \cdot D_i\phi + f(v, z)\,\phi\right] dx = 0. \tag{9.48}$$

Repeating the proof of Lemma 9.6 we show that the family $\{\Psi_\varepsilon(v_\varepsilon, m)\}$ is compact in $C^0([0, T]; L^1(\Omega))$ and

$$\Psi_\varepsilon(v_\varepsilon, m) \to \frac{m}{m+1}|v|^{m+1} \in C^0([0, T]; L^1(\Omega)).$$

Moreover, $v$ satisfies estimate (9.40) and the formula of integration by parts holds:

$$\begin{aligned} \int_{t_1}^{t_2}\int_\Omega v\frac{d}{dt}\left(|v|^{m-1}v\right)\,dz = \int_\Omega \frac{m}{m+1}|v|^{m+1}\,dx\Big|_{t=t_1}^{t=t_2} \\ - \int_{t_1}^{t_2}\int_\Omega \frac{\partial_t m}{(m+1)^2}\,[1-(m+1)\ln|v|]\,|v|^{m+1}\,dz. \end{aligned}$$

The monotonicity arguments show that $V_i = a_i|D_i v|^{p_i(z)-2}D_i v$. Condition (9.5) is proved in exactly the same way as in the proof of Theorem 9.2.

Given $\phi \in \mathbf{W}(Q_{T_0})$, there is a sequence $\phi_\delta \in L^{l+1}(Q) \cap \mathbf{W}(Q_{T_0})$ that converges to $\phi$ in $\mathbf{W}(Q_{T_0})$. Testing (9.48) with $\phi_\delta$ we then pass to the limit as $\delta \to 0$, which gives the integral identity (9.4).

Let us assume now that $m^- \geq 1$. By virtue of (9.17), (9.46) and (9.47)

$$\left|\frac{d}{dt}\Phi_\varepsilon(v_\varepsilon, m)\right|^2 \leq 2\,A_\varepsilon(v_\varepsilon, m)\left(A_\varepsilon(v_\varepsilon, m)\,|\partial_t v_\varepsilon|^2\right) + 2\,|v_\varepsilon||\ln|v_\varepsilon||\,m_t \leq C$$

with an independent of $\varepsilon$ constant $C$. It follows that the family $\left\{\frac{d}{dt}\Phi_\varepsilon(v_\varepsilon, m)\right\}$ contains a subsequence which converges weakly in $L^2(Q_{T_0})$. Finally, the possibility of continuation of the constructed solution to the arbitrary time interval follows from the maximum principle for the solutions of the regularized problems given in Theorem 9.3.

The proof of Theorem 9.1 is completed.

## 9.7 The Energy Relations for Strong Solutions

In this section we prove several energy estimates for the constructed strong solution of problem (9.1).

**Theorem 9.5** *Under the conditions of* Theorem 9.1 *the solution of problem* (9.3) *satisfies the energy relations:*

1. *for every* $t \in (0, T_0)$

$$\int_{\Omega} \frac{m}{m+1}|v(x,t)|^{m+1}dx + \sum_{i=1}^{n}\int_{Q} a_i|D_i v|^{p_i}dz = \int_{Q} v f(v,z)dz \\ + \int_{\Omega} \frac{m_0}{m_0+1}|v_0|^{m_0+1}dx + \int_{Q} \frac{\partial_t m}{(m+1)^2}\left[1-(m+1)\ln|v|\right]|v|^{m+1}dz, \tag{9.49}$$

2. *for a.e.* $t \in (0, T_0)$

$$\frac{d}{dt}\left(\int_{\Omega} \frac{m}{m+1}|v|^{m+1}dx\right) + \sum_{i=1}^{n}\int_{\Omega} a_i|D_i v|^{p_i}dz = \int_{\Omega} v f(v,z)dz \\ + \int_{\Omega} \frac{\partial_t m}{(m+1)^2}\left[1-(m+1)\ln|v|\right]|v|^{m+1}dz. \tag{9.50}$$

*Proof* To obtain the first energy equality if suffices to take the solution $v$ for the test function in identity (9.4), the second equality follows from first one by the Lebesgue differentiation theorem. □

Let us introduce the function

$$E(t,u) = \int_{\Omega} \sum_{i=1}^{n} a_i \frac{|D_i u|^{p_i}}{p_i}dx - \int_{\Omega} b\frac{|u|^{\sigma}}{\sigma}dx.$$

**Theorem 9.6** *Let in the conditions of* Theorem 9.1 $b(z) \geq 0$ *in* $Q$. *For a.e.* $t \in (0,t)$

$$E(t,v) + \int_{Q} m(z)|v|^{m(z)-1}v_t^2\,dxd\tau \leq E(0,v_0) + C^*(1+t) \tag{9.51}$$

*with the constant* $C^*$ *from* (9.30). $C^* = 0$ *in the special case when* $p_i$, $\sigma$, $m$ *are independent of* $t$ *and the coefficients* $a_i$, $b$ *satisfy the inequalities* $a_{it}(z) \leq 0$, $b_t(z) \geq 0$ *in* $Q$.

*Proof* To derive (9.51) we revert to the sequence of finite-dimensional approximations for the solution of the regularized problem (9.13).

**Lemma 9.10** *Let in the conditions of* Theorem 9.2, $b(z) \geq 0$. *For every* $t \in [0, T]$ *and* $k \in \mathbb{N}$

$$\int_\Omega b\, S^{(k)}\, dx \leq \int_\Omega \frac{b}{\sigma} \left(\min\{|v^{(k)}|, K\}\right)^\sigma dx + C \left( meas\,\{x \in \Omega : |v^{(k)}| > K\}\right)^{\frac{\alpha}{m^+(t)+\alpha+1}}.$$

*Proof* Fix some $t > 0$ and denote by $\chi_{\{|v^{(k)}|>K\}}$ the characteristic function of the set $\{x \in \Omega : |v^{(k)}| > K\}$. Since $K > 1$, by the definition of $S^{(k)}$

$$S^{(k)}(z) \leq \frac{1}{\sigma(z)} \left(\min\{|v^{(k)}|, K\}\right)^{\sigma(z)} + \frac{K^{\sigma^+(t)-m^-(t)-1}}{m^-(t)+1} |v^{(k)}|^{m^+(t)+1} \chi_{\{|v^{(k)}|>K\}} + \frac{1}{\sigma^-(t)} |v^{(k)}|^{m^+(t)+1} \chi_{\{|v^{(k)}|>K\}}.$$

It follows that for $b(z) \geq 0$

$$\int_\Omega bS^{(k)} dx \leq C \int_\Omega |v^{(k)}|^{m^+(t)+1} \chi_{\{|v^{(k)}|>K\}} dx + \int_\Omega \frac{b}{\sigma} \left(\min\{|v^{(k)}|, K\}\right)^{\sigma(z)} dx \equiv I_1^{(k)} + I_2^{(k)}.$$

Using Hölder's inequality and Lemma 1.24

$$\begin{aligned} I_1^{(k)} &\leq C\, \|v^{(k)}\|_{m^+(t)+\alpha+1,\Omega} \left( \text{meas}\,\{x \in \Omega : |v^{(k)}| > K\}\right)^{\frac{\alpha}{m^+(t)+\alpha+1}} \\ &\leq C\Big(\int_\Omega \sum_{i=1}^n |D_i v^{(k)}|^{p_i^-(t)}\Big)^{\frac{1}{p^-(t)}} \left( \text{meas}\,\{x \in \Omega : |v^{(k)}| > K\}\right)^{\frac{\alpha}{m^+(t)+\alpha+1}} \end{aligned}$$

and the assertion follows now from the uniform in $k$ estimate (9.22). □

**Lemma 9.11** *Let* $b(z) \geq 0$, $v_\varepsilon = \lim v^{(k)}$ *be the solution of problem* (9.45), *constructed in* Theorem 9.4, *and* $\mathscr{E}(t, v_\varepsilon)$ *be the corresponding energy function defined in* (9.26). *For every* $t \in [0, T_0)$

$$\mathscr{E}(t, v_\varepsilon) + \int_Q A_\varepsilon(v_\varepsilon, m) v_{\varepsilon t}^2\, dz \leq \mathscr{E}(0, v_0) + C^*(1+t) \tag{9.52}$$

*with the constant* $C^*$ *from* (9.30).

*Proof* The assertion follows from (9.30) as $k \to \infty$. By the choice of the basis $\mathscr{E}(0, v_0^{(k)}) \to \mathscr{E}(0, v_0)$. By virtue of the Fatou lemma we may let $k \to \infty$ in the nonnegative terms on the left-hand side of (9.30). By the dominated convergence theorem, for every $K > 1$

$$
\begin{aligned}
&\lim_{k\to\infty} \int_{\Omega} b\, S^{(k)}\, dx \\
&\leq \int_{\Omega} \frac{b}{\sigma} \left(\min\{|v_\varepsilon|, K\}\right)^{\sigma}\, dx + C\, \left( \operatorname{meas}\{x \in \Omega : \, |v_\varepsilon| > K\}\right)^{\frac{\alpha}{m^+(t)+\alpha+1}} .
\end{aligned}
$$

For $K > 2(1 + \|v_0\|_{\infty,\Omega})$ we have $\|v_\varepsilon(t)\|_{\infty,\Omega} < K$ on $[0, T_0)$ (see the proof of Theorem 9.3), whence the required estimate. □

To complete the proof of Theorem 9.5 we repeat these arguments for the sequence $\{v_\varepsilon\}$ of solutions of problems (9.45), which are uniformly bounded and satisfy estimate (9.52). □

# Chapter 10
# Anisotropic Equations with Double Nonlinearity: Blow-up and Vanishing

This chapter is devoted to the study of qualitative properties of strong solutions of the Dirichlet problem for the doubly nonlinear anisotropic parabolic equation with variable nonlinearity

$$\begin{cases} \dfrac{d}{dt}\left(|v|^{m(z)} \text{ sign } v\right) \\ \qquad = \displaystyle\sum_{i=1}^{n} D_i\left(a_i(z)|D_i v|^{p_i(z)-2} D_i v\right) + b(z)|v|^{\sigma(z)-2} v + g \quad \text{in } Q_T, \\ v = 0 \text{ on } \Gamma_T, \qquad v(x,0) = v_0(x) \text{ in } \Omega \end{cases} \tag{10.1}$$

under the assumptions on the data made in Chap. 9. Rewriting Eq. (10.1) in the formally equivalent form

$$m|v|^{m-1} v_t = \sum_{i=1}^{n} D_i\left(a_i|D_i v|^{p_i-2} D_i v\right) + b|v|^{\sigma-2} v - |v|^{m-1} v \ln|v|\, m_t + g$$

makes it clear that in dependence on the sign of $m_t$ and the values of the solution $v$ the logarithmic nonlinear term on the right-hand side acquires a special importance because it may prevent or cause the effects of blow-up and extinction.

We will derive sufficient conditions on the data that ensure the presence of the following effects.

- *Finite time blow-up*: there exists a finite moment $t^*$ such that

$$\Sigma(t) \equiv \int_{\Omega} |v|^{m(x,t)+1}\, dx \to \infty \quad \text{as } t \to t^*,$$

(which implies that $\|v(\cdot, t)\|_{\infty,\Omega}$ becomes infinite at the same moment).



S. Antontsev and S. Shmarev, *Evolution PDEs with Nonstandard Growth Conditions*, Atlantis Studies in Differential Equations 4,
DOI 10.2991/978-94-6239-112-3_10

- *Extinction in a finite time*: there is a finite $t_* > 0$ such that $\Sigma(t) \equiv 0$ for all $t \geq t_*$. In particular, this effect may take place even in the case when the equation eventually becomes linear. This happens if $m \to 1, \sigma \to 2$ and $p_i \to 2$ as $t \to \infty$ and the rates of convergence satisfy some additional conditions.
- In case that the solution exists for all times but doesn't vanish in a finite time, we estimate the rate of growth or decay of $\Sigma(t)$ as $t \to \infty$.

Examples of conditions on the structure of the equation that provide these effects can be furnished by the model equation

$$\frac{d}{dt}\left(|v|^{m(t)-1}v\right) = \left(|v_x|^{p(t)-2}v_x\right)_x + b\,|v|^{\sigma(t)-2}v, \quad (x,t) \in (0,1)\times(0,T], \quad (10.2)$$

with the constant coefficient $b$ and the independent of $x$ nonlinearity exponents $m(t) > 0$, $p(t) > 1$ and $\sigma(t) > 1$. In this simple case

- if $b > 0$, $\sigma(t) > \max\{m(t)+1,\ p(t)\}$, $m'(t) \in [-\mu, 0]$ with some constant $\mu > 0$, and if
$$\int_0^1 \left(\frac{1}{p(0)}|v_0'|^{p(0)} - \frac{1}{\sigma(0)}|v|^{\sigma(0)}\right) dx < 0,$$
then every the solution blows-up for the sufficiently small $\mu$;
- if $b = 0$, $m'(t) \in [0, \mu]$ and $\dfrac{m(t)p(t)}{m(t)+1} < 1$, then the solution vanishes in a finite time, provided that $\mu$ is sufficiently small;
- if $b < 0$, $m'(t) \in [0, \mu]$ and $\dfrac{1}{\sigma(t)} + \dfrac{m(t)}{p(t)} > 1$, then the solution vanishes in a finite time if $\mu$ is sufficiently small.

The specific form of the main energy identity proved in Theorem 9.5 allows for an easy extension of the results concerning the finite-time extinction and the large-time behavior to the equations with the lower-order terms of critical growth:

$$\frac{d}{dt}\left(|v|^{m(z)-1}v\right) = \sum_{i=1}^{n} D_i\left(a_i|D_i v|^{p_i(z)-2}D_i v\right) + c(z)|v|^{m(z)-1}v + b(z)|v|^{\sigma(z)-2}v + g,$$

where $b(z) \leq 0$ and $c(z) \leq \gamma(t)$ with a bounded function $\gamma(t)$.

The presentation follows [57, 60]. The results are obtained by means of analysis of the differential inequalities for suitable energy functions associated with the solution. In the case that the exponents of nonlinearity are constant or time-independent we recover some of the results available in the literature for doubly nonlinear equations (see, e.g., [96]). The results of this chapter remain true in the special case $m = 1$ and extend the results of Chap. 7 to Eq. 7.41) with $K = 1$ and time-dependent exponents $p_i$ and $\sigma$.

Throughout this chapter we denote by $\mathbf{V}$, $\mathbf{W}$ and $\mathbf{W}'$ for the function spaces introduced in Sect. 9.1 of Chap. 9.

## 10.1 Strong Solutions

The solutions of problem (10.1) are strong solutions in the sense of Definition 9.1.

It is assumed that the coefficients of Eq. (10.1) and the exponents of nonlinearity are subject to conditions (9.6) and (9.8). As for condition (9.7), in this chapter we substitute it by the weaker restriction

$$m_t \in L^{\infty}(Q_T), \quad \exists\, \mu = const > 0: \quad |m_t| \le \mu \ \text{ a.e. in } Q_T. \tag{10.3}$$

It is shown in Theorem 9.1 that if conditions (9.6) and (9.8) are fulfilled, $m(z)$ satisfies (10.3) and is nondecreasing in $t$ [cf. with condition (9.7)], then problem (10.1) has a strong solution. Moreover, by Theorems 9.5 and 9.6 for the energy function

$$E(t,v) = \int_{\Omega} \left( \sum_{i=1}^{n} a_i \frac{|D_i v|^{p_i}}{p_i} - b \frac{|v|^{\sigma}}{\sigma} \right) dx \tag{10.4}$$

the following relations hold: $\forall$ a.e. $t \in (0, T^*)$

$$\begin{aligned} &\frac{d}{dt}\left( \int_{\Omega} \frac{m}{m+1} |v|^{m+1}\, dx \right) + \sum_{i=1}^{n} \int_{\Omega} a_i |D_i v|^{p_i}\, dz + \int_{\Omega} b\, |v|^{\sigma}\, dx \\ &\quad = \int_{\Omega} g\, v\, dx + \int_{\Omega} \frac{\partial_t m}{(m+1)^2} \left[1 - (m+1) \ln |v|\right] |v|^{m+1}\, dx; \end{aligned} \tag{10.5}$$

if $b(z) \ge b^- > 0$, $g(z) \equiv 0$ and $0 \le \sigma_t(z) \le C_{\sigma}$ in the cylinder $Q_{T^*}$, then

$$E(t,v) \le E(0,v_0) + C^* t + \mathscr{M}(\mu) \quad \text{for a.e. } t \in (0, T^*), \tag{10.6}$$

where $\mathscr{M}(s) \to 0$ as $s \to 0$ and

$$C^* = |\Omega| \left( n \frac{a^+ C_p}{(p^-)^2} + \left(1 + \frac{1}{\mathrm{e}}\right) \frac{b^+ C_{\sigma}}{(\sigma^-)^2} \right).$$

The constants $a^{\pm}$, $b^{\pm}$, $p^{\pm}$, $\sigma^{\pm}$ are introduced in (1.1), the constants $C_p$ and $C_{\sigma}$ are defined in (9.8).

The derivation of the energy estimates (10.5) and (10.6) was tightly related to the method of regularization of the degenerate Eq. (10.1). The sequence of solutions of the regularized problems converges to a strong solution of problem (10.1), and the energy relations (10.5) and (10.6) are obtained as the limits of the corresponding relations for the solutions of the regularized problems. It is not clear, however, whether the conditions on the data imposed in Theorem 9.1 are indeed essential for the validity of the energy estimates. Since the study of the localization and blow-up properties of solutions to problem (10.1) relies solely on the analysis of the energy relations (10.5) and (10.6) and is practically independent of the existence theorem, it is convenient to introduce the *energy strong solution* of problem (10.1): a function

$$v \in C^0(0,T; L^{m(\cdot,\cdot)+1}(\Omega)) \cap \mathbf{W}(Q_T)$$

is called *energy solution* of problem (10.1) if it satisfies (9.4) and (9.5) and the energy relations (10.5) and (10.6).

All further considerations are confined to the class of energy strong solutions of problem (10.1), which is surely nonempty by virtue of Theorem 9.1.

## 10.2 Sufficient Conditions of the Finite Time Blow-up

Given an energy solution $v(z)$ of problem (10.1), we define the blow-up moment $t^*$ as follows:

$$\Sigma(t) \equiv \int_\Omega \frac{m}{m+1}|v|^{m+1}\,dx \to \infty \quad \text{as } t \to t^*, \quad m = m(x,t). \tag{10.7}$$

We begin with proving several technical lemmas.

**Lemma 10.1** *Let $\mu^+$ and $\mu^-$ be positive constants, $1 < \mu^- \le \mu^+ < \infty$. If a nonnegative function $f(t)$ satisfies the inequality*

$$\begin{cases} \lambda f'(t) \ge C\min\left\{f^{\mu^+}(t),\ f^{\mu^-}(t)\right\} & \text{for a.e. } t \ge 0, \\ f(0) = f_0 > 0, \quad \lambda = const > 0, \end{cases} \tag{10.8}$$

*then there exists a finite moment $t^*$ such that $f(t) \to \infty$ as $t \to t^*$.*

*Proof* Let us denote $t^* = \sup\{t > 0 : \ f(t) < \infty\}$. Introduce the function $w(t) := \dfrac{f(t)}{f_0}$, which satisfies the conditions

$$\begin{cases} \lambda w'(t) \ge D\min\left\{w^{\mu^+}(t),\ w^{\mu^-}(t)\right\} & \text{for a.e. } t \in (0,t^*), \\ w(0) = 1, \quad D = C\min\left\{f_0^{\mu^+-1},\ f_0^{\mu^--1}\right\}. \end{cases} \tag{10.9}$$

By virtue of (10.9) $w(t)$ is monotone increasing, whence $w(t) > 1$ for all $t \in (0,t^*)$. Then (10.9) reads

$$D\, w^{\mu^-}(t) \leq \lambda\, w'(t) \quad \text{for a.e. } t \in (0, t^*), \qquad w(0) = 1,$$

and can be directly integrated:

$$w(t) \geq \frac{\lambda^{\frac{1}{\mu^- - 1}}}{\left(\lambda - D(\mu^- - 1)\, t\right)^{\frac{1}{\mu^- - 1}}}.$$

It follows that

$$\begin{aligned} w(t) \to \infty \text{ as } t \to \frac{\lambda}{D(\mu^- - 1)} &= \frac{\lambda\, f_0}{C(\mu^- - 1) \min\left\{f_0^{\mu^+},\, f_0^{\mu^-}\right\}} \\ &\equiv \theta^*(f_0) < \infty, \end{aligned} \tag{10.10}$$

which means that $t^* \leq \theta^*$. □

**Lemma 10.2** *Let $\mu^\pm$ be positive constants, $1 < \mu^- \leq \mu^+ < \infty$. If a continuous function $f(t)$ satisfy the conditions*

$$\begin{cases} (C_1 t - C_2) + \lambda\, f'(t) \geq C \min\left\{f^{\mu^+}(t),\, f^{\mu^-}(t)\right\} & \textit{for a.e. } t \geq 0, \\ f(0) = f_0 > 0, \end{cases} \tag{10.11}$$

*with positive constants $C_1$ and $C_2$, then there exists $f_* > 0$ such that for every initial value $f_0 > f_*$ there is a finite $t^*$ such that $f(t) \to \infty$ as $t \to t^* \leq C_1/C_2$.*

*Proof* Let $\theta^*(s)$ be the monotone decreasing function defined in (10.10) and let $f_*$ be the solution of the equation $C_1(1 + \theta^*(f_*)) = C_2$:

$$f_* = \begin{cases} \left(\dfrac{\lambda\, C_1}{C C_2(\mu^- - 1)}\right)^{\frac{1}{\mu^+ - 1}} & \text{if } \dfrac{\lambda\, C_1}{C C_2(\mu^- - 1)} < 1, \\ \left(\dfrac{\lambda\, C_1}{C C_2(\mu^- - 1)}\right)^{\frac{1}{\mu^- - 1}} & \text{otherwise.} \end{cases}$$

Under this choice of $f_*$ we have that $C_1 t - C_2 < 0$ on the interval $(0, \theta^*(f_0))$ for every $f_0 > f_*$. It follows that instead of (10.11) $f(t)$ satisfies inequality (10.8) and, by virtue of Lemma 10.1,

$$f(t) \to \infty \quad \text{as } t \to \frac{\lambda\, f_0}{C(\mu^- - 1) \min\left\{f_0^{\mu^+},\, f_0^{\mu^-}\right\}} < \theta^*(f_*, \mu^-).$$

□

**Theorem 10.1** *Let $v(z)$ be a strong energy solution of problem* (10.1) *in a cylinder $Q$, $T < \infty$. Assume that $g \equiv 0$ and there exist constants $b^- > 0$ and $\delta > 1$ such* *that* $b(z) \geq b^-$ *in* $Q$

$$\sigma^-(t) > \delta \max\left\{m^+(t)+1,\ \max_{i=1,\ldots,n} p_i^+(t)\right\} \quad \forall\,(x,t)\in Q_T. \tag{10.12}$$

*Let* $C^*$ *and* $E(t,v)$ *denote the constant and the energy functional from* Theorem 9.5.

(i) *If* $C^* = 0$, $m_t = 0$ *and* $E(0, v_0) < 0$, *the solution blows-up in a finite time* $t^*$ *with*

$$t^* \leq \frac{1}{b^-(\sigma^- - p^+)\left(1 - \frac{m^+ + 1}{\sigma^-}\right) \min\left\{\Sigma^{\mu^+}(0),\ \Sigma^{\mu^-}(0)\right\}},$$

$$\mu^{\pm} = \left(\frac{\sigma(z)}{m(z)+1}\right)^{\pm}.$$

(ii) *If* $C^* > 0$, $m_t = 0$ *and* $E(0, v_0) < 0$, *there exists* $\Sigma_*(0)$ *such that for every* $\Sigma(0) > \Sigma_*(0)$ *the solution blows-up in finite time.*

(iii) *If* $-\mu \leq m_t \leq 0$, $C^* > 0$ *and* $E(0, v_0) < 0$,

$$E(0, v_0) + \mu\,|\Omega| \left(1 + \frac{m^+ + 1}{m^- + 1}\right) \mathrm{e}^{\frac{m^+ + 1}{m^- + 1}} < 0,$$

*there exist* $\mu^* > 0$ *and* $\Sigma_*(0)$ *such that for* $0 \leq \mu < \mu^*$ *and* $\Sigma(0) > \Sigma_*(0)$ *the solution blows-up in a finite time.*

*Proof* (i)-(ii) The assumption $m \equiv m(x)$ allows us to drop in (10.5) and (10.6) the terms depending on $m_t$ and to write these relations in the form

$$\forall \text{ a.e. } t \in (0,T) \qquad \Sigma'(t) = \int_\Omega \left(-\sum_{i=1}^n a_i |D_i v|^{p_i} + b\,|v|^\sigma\right) dx, \tag{10.13}$$

$$E(t, v) \leq E(0, v_0) + C^* t. \tag{10.14}$$

Let us multiply (10.13) by a constant $\lambda > 0$ and add to (10.14). Dropping the nonnegative term on the left-hand side, we obtain the inequality

$$E(t,v) + \lambda \int_\Omega \left(-\sum_{i=1}^n a_i |D_i v|^{p_i} + b\,|v|^\sigma\right) dx \leq \lambda \Sigma'(t) + (E(0, v_0) + C^* t), \tag{10.15}$$

whence

$$\int_{\Omega}\left(\sum_{i=1}^{n}\left(\frac{1}{p_i}-\lambda\right)a_i|D_iv|^{p_i}+b\left(\lambda-\frac{1}{\sigma}\right)|v|^{\sigma}\right)dx \\ \leq \lambda\Sigma'(t)+(E(0,v_0)+C^*t) \tag{10.16}$$

Assumption (10.12) allows one to choose $\lambda$ in such a way that

$$\frac{1}{\sigma(z)}\leq\frac{1}{\sigma^-(t)}<\lambda\leq\frac{1}{p^+(t)}\leq\frac{1}{p(z)}.$$

Plugging this inequality to (10.16) we obtain the inequality

$$\int_{\Omega}\left(a^-\left(\frac{1}{p^+}-\lambda\right)\sum_{i=1}^{n}|D_iv|^{p_i}+b^-\left(\lambda-\frac{1}{\sigma^-}\right)|v|^{\sigma}\right)dx \\ \leq\lambda\Sigma'(t)+(E(0,v_0)+C^*t),$$

in which the first term on the left-hand side is nonnegative and can be omitted:

$$b^-\left(\frac{1}{p^+}-\frac{1}{\sigma-}\right)\int_{\Omega}|v|^{\sigma}\,dx\leq\lambda\Sigma'(t)+(E(0,v_0)+C^*t). \tag{10.17}$$

According to (10.12) $1+m^+<\sigma^-$ in $Q_T$. Using the Hölder inequality (1.9) and (1.17), we have that for a.e. $t\in(0,T)$

$$\Sigma(t)\leq\int_{\Omega}|v|^{m+1}\,dx\leq\max\left\{\|v\|_{m(\cdot,t)+1}^{m^+(t)+1},\|v\|_{m(\cdot,t)+1}^{m^-(t)+1}\right\} \\ \leq\max\left\{\left(\int_{\Omega}v^{\sigma(z)}\,dx\right)^{\frac{m^+(t)+1}{\sigma^-(t)}},\left(\int_{\Omega}v^{\sigma(z)}\,dx\right)^{\frac{m^-(t)+1}{\sigma^+(t)}}\right\},$$

whence

$$\min\left\{\Sigma^{\frac{\sigma^-(t)}{m^+(t)+1}}(t),\ \Sigma^{\frac{\sigma^+}{m^-(t)+1}}(t)\right\}\leq\int_{\Omega}|v|^{\sigma(z)}\,dx.$$

Set

$$\mu^+=\frac{\sigma^+}{m^-+1},\quad\mu^-=\frac{\sigma^-}{m^++1},\quad C=b^-\left(\frac{1}{p^+}-\frac{1}{\sigma^-}\right). \tag{10.18}$$

Combining the last inequality with (10.17), we arrive at the differential inequality for $\Sigma(t)$:

$$\begin{cases} C \min\left\{\Sigma^{\mu^+}(t),\ \Sigma^{\mu^-}(t)\right\} \leq \lambda\, \Sigma'(t) + (E(0, v_0) + C^* t) \\ \qquad\qquad \text{for a.e. } t \in (0, T), \\ \Sigma(0) = \displaystyle\int_\Omega \frac{m_0}{m_0+1} |v_0|^{m_0+1}\, dx. \end{cases} \tag{10.19}$$

The assertions (i) and (ii) follow now from Lemmas 10.1 and 10.2.

(iii) It is assumed that $-\mu \leq m_t \leq 0$ a.e. in $Q_T$. Let us denote

$$J(t) = \int_\Omega \frac{\partial_t m}{(m+1)^2} \left[1 - (m+1) \ln |v|\right] |v|^{m+1}\, dx.$$

Noting that the integrand of $J$ is nonnegative on the set where $\ln |v| > \dfrac{1}{m+1}$, we estimate it as follows: $\forall\, t \in (0, T)$

$$\begin{aligned} J(t) &\geq \int_{\Omega \cap \left\{\ln |v| < \frac{1}{m+1}\right\}} \frac{\partial_t m}{(m+1)^2} \left[1 - (m+1) \ln |v|\right] |v|^{m+1}\, dx \\ &\geq \int_{\Omega \cap \left\{\ln |v| < \frac{1}{m+1}\right\}} \frac{\partial_t m}{(m+1)^2} \left(1 + \frac{m^+ + 1}{m^- + 1}\right) e^{\frac{m^+ + 1}{m^- + 1}}\, dx \\ &\geq \left(1 + \frac{m^+ + 1}{m^- + 1}\right) e^{\frac{m^+ + 1}{m^- + 1}} \int_\Omega \frac{\partial_t m}{(m+1)^2}\, dx \geq -\mu\, L \end{aligned}$$

with the constant $L = |\Omega| \left(1 + \dfrac{m^+ + 1}{m^- + 1}\right) e^{\frac{m^+ + 1}{m^- + 1}} > 1$. The energy relations (10.5), (10.6) take the form

$$\begin{aligned} \forall \text{ a.e. } t \in (0, T) \qquad & \int_\Omega \left(-\sum_{i=1}^n a_i |D_i v|^{p_i} + b\, |v|^\sigma\right) dx \leq \Sigma'(t) + \mu\, L, \\ & E(t, v) \leq E(0, v_0) + C^* t + \mathscr{M}(\mu), \quad \mathscr{M}(\mu) \to 0 \text{ as } \mu \to 0. \end{aligned}$$

Gathering them and following the derivation of (10.19) we arrive at the inequality

$$\begin{cases} C \min\left\{\Sigma^{\mu^+}(t),\ \Sigma^{\mu^-}(t)\right\} \\ \qquad \leq \lambda\, \Sigma'(t) + (\lambda\, \mu\, L + \mathscr{M}(\mu) + E(0, v_0)) + C^* t \quad \text{in } (0, T), \\ \Sigma(0) = \displaystyle\int_\Omega \frac{m_0}{m_0+1} |v_0|^{m_0+1}\, dx > 0. \end{cases}$$

Since $E(0, v_0) < 0$ and $\lambda \mu L + \mathscr{M}(\mu) \to 0$ as $\mu \to 0$, assertion (iii) follows from Lemma 10.2. □

**Corollary 10.1** Theorem 10.1 *states that given any negative initial energy* $E(0, v_0)$ *every energy solution blows-up in a finite time, provided that* $m_t$ *is nonpositive and appropriately small, and that* $\Sigma(0)$ *is sufficiently big. On the other hand, it is easy to infer from the proof of* Lemmas 10.1 *and* 10.2 *that for every fixed* $\mu$ *and* $\Sigma(0)$ *one may indicate the threshold value* $E_* < 0$ *such that the every energy strong solution blows-up if* $E(0, v_0) < E_*$.

**Corollary 10.2** *Let in the conditions of* Theorem 10.1 $m \equiv 1$. *Then item* (ii) *gives sufficient conditions for the finite time blow-up of solutions to the equation*

$$u_t = \sum_{i=1}^{n} D_i \left(a_i |D_i u|^{p_i(z)-2} D_i u\right) - c|u|^{\sigma(z)-2} u.$$

## 10.3 Embedding Inequalities in Anisotropic Spaces

Let us introduce the functions

$$\begin{aligned}
\Sigma(t) &= \int_\Omega \frac{m(z)}{1+m(z)} |v(z)|^{m(z)+1}\, dx,\\
\Lambda(t) &= \sum_{i=1}^{n} \int_\Omega |D_i v(z)|^{p_i(z)}\, dx,\\
\Theta(t) &= \int_\Omega \left( \sum_{i=1}^{n} |D_i v(z)|^{p_i(z)} + |v(z)|^{\sigma(z)} \right) dx
\end{aligned}$$

and denote

$$\frac{1}{p^{\pm}(t)} = \frac{1}{n} \sum_{i=1}^{n} \frac{1}{p_i^{\pm}(t)}. \tag{10.20}$$

For any given positive constant $L$ we denote

$$\mathscr{B}_L = \left\{ v \in \mathbf{W}(Q) : \sup_{(0,T)} \Sigma(t) + \int_0^T \Lambda(t)\, dt \le L \right\}.$$

**Lemma 10.3** *Let the exponents $m(z)$, $p_i(z)$ satisfy conditions* (9.6). *If*

$$m^+(t)+1 \le \begin{cases} \dfrac{n\,p^-(t)}{n-p^-(t)} & \text{if } p^-(t)<n, \\ \text{any number from } [1,\infty) & \text{if } p^-(t)\ge n, \end{cases}$$

*then for every $v\in\mathscr{B}_L$ the energy functions $\Sigma(t)$, $\Lambda(t)$ satisfy the inequality*

$$\forall \text{ a.e. } t\in(0,T)\quad \Sigma^{\nu^+(t)}(t)\le C\,\Lambda(t)\quad \text{with } \nu^+(t)=\frac{p^+(t)}{m^-(t)+1} \tag{10.21}$$

*with $p^+(t)$ defined in* (10.20) *and a constant $C$ depending on the constants in* (9.6), *$n$, $|\Omega|$ and $L$, but independent of $v$.*

*Proof* Given $u\in\mathscr{B}_L$, let us introduce the function $v=\gamma\,u$ with a constant $\gamma\in(0,1]$. Denote by $\Sigma_v(t)$ and $\Lambda_v(t)$ the energy functions associated with $v$. It is easy to calculate that

$$\Sigma_v(t)+\int_0^T \Lambda_v(t)\,dt \le \left(\gamma^{m^-+1}+\max_i \gamma^{p_i^-}\right)\left(\Sigma(t)+\int_0^T \Lambda(t)\,dt\right)\le 2\,\gamma\,L,$$

whence $v\in\mathscr{B}_{2\gamma L}$. If we choose $\gamma$ from the equality $2\,\gamma\,L\,(1+m^+)=m^-$, then

$$v\in\mathscr{B}_1,\qquad \sup_{t\in(0,T)}\|v\|_{m(\cdot,t)+1,\Omega}\le 1. \tag{10.22}$$

Let us consider first the case when $v$ satisfies conditions (10.22). According to (1.6)

$$\Sigma(t)\le \frac{m^+}{m^-+1}\max\left\{\|v(\cdot,t)\|^{m^++1}_{m+1,\Omega},\,\|v(\cdot,t)\|^{m^-+1}_{m+1,\Omega}\right\}\le \frac{m^+}{m^-+1}\|v(\cdot,t)\|^{m^-+1}_{m+1,\Omega}. \tag{10.23}$$

By Lemma 1.24 for every $t\in[0,T]$

$$\begin{aligned}\|v(\cdot,t)\|_{m+1,\Omega} &\le C\,\|v(\cdot,t)\|_{m^+(t)+1,\Omega} \\ &\le C\left(\prod_{i=1}^n \|D_i v(\cdot,t)\|_{p_i^-(t),\Omega}\right)^{\frac{1}{n}} \le C\left(\prod_{i=1}^n \|D_i v(\cdot,t)\|_{p_i(\cdot,t),\Omega}\right)^{\frac{1}{n}}.\end{aligned}$$

The right-hand side of this inequality can be estimated by means of (1.6): for every fixed $t \in [0, T]$

$$\begin{aligned}\|D_i v\|_{p_i,\Omega} &\le \max\left\{\left(\int_\Omega |D_i v|^{p_i}\,dx\right)^{\frac{1}{p_i^-(t)}}, \left(\int_\Omega |D_i v|^{p_i}\,dx\right)^{\frac{1}{p_i^+(t)}}\right\}\\ &\le \max\left\{\Lambda^{\frac{1}{p_i^-(t)}}(t), \Lambda^{\frac{1}{p_i^+(t)}}(t)\right\},\end{aligned}$$

whence

$$\begin{aligned}\|v(\cdot,t)\|_{m+1,\Omega} &\le C\left(\prod_{i=1}^n \max\left\{\Lambda^{\frac{1}{p_i^-(t)}}(t), \Lambda^{\frac{1}{p_i^-(t)}}(t)\right\}\right)^{\frac{1}{n}}\\ &\le C\begin{cases}\left(\prod_{i=1}^n \Lambda^{\frac{1}{p_i^+(t)}}(t)\right)^{\frac{1}{n}} & \text{if } \Lambda(t)\le 1,\\ \left(\prod_{i=1}^n \Lambda^{\frac{1}{p_i^-(t)}}(t)\right)^{\frac{1}{n}} & \text{if } \Lambda(t)>1,\end{cases} \equiv C\begin{cases}\Lambda^{\frac{1}{p^+(t)}}(t) & \text{if } \Lambda(t)\le 1,\\ \Lambda^{\frac{1}{p^-(t)}}(t) & \text{if } \Lambda(t)>1.\end{cases}\end{aligned}$$

Using (10.23) and (1.6) we obtain the inequality

$$\begin{aligned}\left(\frac{1+m^-}{m^+}\Sigma(t)\right)^{\frac{p^+(t)}{1+m^-(t)}} &\le \|v(\cdot,t)\|_{m+1,\Omega}^{p^+(t)}\\ &= \min\left\{\|v(\cdot,t)\|_{m+1,\Omega}^{p^-(t)}, \|v\|_{m+1,\Omega}^{p^+(t)}\right\} \le C\,\Lambda(t),\end{aligned}$$

which yields (10.21), provided that $v(z)$ satisfies (10.22). In the general case (10.21) holds for the function $w = \gamma\, v$ with $\gamma = m^-/(2L(1+m^+))$, which gives estimate (10.21) for $v$. □

Let us accept the notation

$$\mathcal{G}_L = \left\{v \in \mathbf{W}(Q)\cap L^\sigma(Q): \sup_{(0,T)} \Sigma(t) + \int_0^T \Theta(t)\,dt \le L\right\}.$$

**Lemma 10.4** *Let the exponents $m$, $p_i$, $\sigma$ satisfy conditions* (9.6). *If*

$$\begin{cases}\sigma^- + m^- > 2,\\ \dfrac{m^+(t)\,\sigma^+(t)}{\sigma^-(t)-1} \le \dfrac{n\,p^-(t)}{n-p^-(t)} & \text{if } p^-(t)<n,\end{cases} \tag{10.24}$$

*then for every $u \in \mathcal{G}_L$*

$$\forall \text{a.e. } t\in(0,T) \qquad \Sigma(t) \le C\,\Theta^{\frac{1}{\nu^+(t)}}(t), \qquad \frac{1}{\nu^+(t)} = \frac{1}{\sigma^+(t)} + \frac{m^-(t)}{p^+(t)} \tag{10.25}$$

*with $p^+(t)$ defined in* (10.20) *and a constant $C$ depending on $n$, $|\Omega|$, the constants in* (9.6) *and $L$, but independent of $u$.*

*Proof* The proof imitates the proof of the previous lemma. It is sufficient to prove the assertion for the functions $u \in \mathbf{W}(Q) \cap L^{\sigma}(Q)$ satisfying the inequality

$$\sup_{t\in(0,T)} \|u\|_{m(\cdot,t)+1,\Omega} \leq 1. \tag{10.26}$$

Observe first that by virtue of Lebesgue's differentiation theorem $\Theta(t)$ is well-defined for a.a. $t \in (0, T)$ and every $u \in \mathcal{G}_L$ . Following the proof of Lemma 10.3 we find that under assumption (10.24) for a.e. $t \in (0, T)$

$$\begin{aligned}
\||u|^{m(\cdot,t)}\|_{\frac{\sigma(\cdot,t)}{\sigma(\cdot,t)-1},\,\Omega} &\leq C\||u|^{m(\cdot,t)}\|_{\frac{\sigma^-(t)}{\sigma^-(t)-1},\,\Omega} \\
&= C\left(\int_{\Omega}|u|^{m(\cdot,t)\frac{\sigma^-(t)}{\sigma^-(t)-1}}\,dx\right)^{\frac{\sigma^-(t)-1}{\sigma^-(t)}} \\
&\leq C\max\left\{\|u\|^{m^-(t)}_{\frac{m(\cdot,t)\sigma^-(t)}{\sigma^-(t)-1},\,\Omega},\ \|u\|^{m^+(t)}_{\frac{m(\cdot,t)\sigma^-(t)}{\sigma^-(t)-1},\,\Omega}\right\} \\
&\leq C\max\left\{\|u\|^{m^-(t)}_{\frac{m^+(t)\sigma^-(t)}{\sigma^-(t)-1},\,\Omega},\ \|u\|^{m^+(t)}_{\frac{m^+(t)\sigma^-(t)}{\sigma^-(t)-1},\,\Omega}\right\} \\
&\leq C\max\left\{\max\left\{\Lambda^{\frac{m^-(t)}{p^+(t)}},\Lambda^{\frac{m^-(t)}{p^-(t)}}\right\},\max\left\{\Lambda^{\frac{m^+(t)}{p^+(t)}},\Lambda^{\frac{m^+(t)}{p^-(t)}}\right\}\right\} \\
&\leq C\begin{cases}\Lambda^{\frac{m^-(t)}{p^+(t)}} & \text{if } \Lambda \leq 1,\\ \Lambda^{\frac{m^+(t)}{p^-(t)}} & \text{if } \Lambda \geq 1.\end{cases}
\end{aligned}$$

Combining this inequality with Hölder's inequality (1.9) we obtain the estimate

$$\begin{aligned}
\int_{\Omega}|u(x,t)|^{m+1}\,dx &\leq \||u(\cdot,t)|^{m(\cdot,t)}\|_{\frac{\sigma(\cdot,t)}{\sigma(\cdot,t)-1},\Omega}\|u(\cdot,t)\|_{\sigma(\cdot,t),\Omega} \\
&\leq C\max\left\{\left(\int_{\Omega}|u|^{\sigma(z)}\,dx\right)^{\frac{1}{\sigma^+(t)}},\left(\int_{\Omega}|u|^{\sigma(z)}\,dx\right)^{\frac{1}{\sigma^-(t)}}\right\} \\
&\qquad\times\max\left\{\Lambda^{\frac{m^-(t)}{p^+(t)}}(t),\Lambda^{\frac{m^-(t)}{p^-(t)}}(t)\right\} \\
&\leq C\begin{cases}\Theta^{\frac{1}{\nu^+(t)}}(t) & \text{if } \Theta(t) < 1,\\ \Theta^{\frac{1}{\nu^-(t)}}(t) & \text{if } \Theta(t) \geq 1.\end{cases}
\end{aligned}$$

Using assumption (10.26) we arrive at the inequality

$$\begin{aligned}\left(\frac{1+m^-}{m^+}\Sigma(t)\right)^{\nu^+(t)} &\leq \left(\int_\Omega |u(z)|^{m+1}\,dx\right)^{\nu^+(t)}\\ &\leq \min\left\{\left(\int_\Omega |u(z)|^{m+1}\,dx\right)^{\nu^+(t)}, \left(\int_\Omega |u(z)|^{m+1}\,dx\right)^{\nu^-(t)}\right\}\\ &\leq C\,\Theta(t).\end{aligned}$$

The general case reduces to the considered one by means of scaling $v = \gamma\, u$ with a constant $\gamma \in (0, 1]$ such that

$$\Sigma_v(t) + \int_0^T \Theta_v(t)\,dt \leq \left(\gamma^{m^-+1} + \max_i \gamma^{p_i^-} + \gamma^{\sigma^-}\right)\left(\Sigma(t) + \int_0^T \Theta(t)\,dt\right)$$
$$\leq 3\gamma\, L, \int_\Omega |v|^{m+1}\,dx \leq \frac{1+m^+}{m^-}\Sigma_v(t) \leq 3\frac{1+m^+}{m^-}\gamma\, L \leq 1.$$

□

## 10.4 Ordinary Differential Inequalities for the Energy Function

Let $u \in \mathbf{W}(Q)$ be an energy solution of problem (10.1). We assume that

$$g(z) \in L^\infty(0, T; L^1(\Omega)), \qquad b(z) \leq 0 \text{ and } 0 \leq m_t \leq \mu \text{ a.e. in } Q_T. \tag{10.27}$$

Under these assumptions (10.5) takes on the form: $\forall$ a.e. $t \in (0, T)$

$$\begin{aligned}\Sigma'(t) &+ \int_\Omega \left(a^-\sum_{i=1}^n |D_i u(z)|^{p_i(z)} - b(z)|u(z)|^{\sigma(z)}\right)dx\\ &\leq \int_\Omega u\, g(z)\,dx + \int_\Omega \frac{m_t}{(m+1)^2}(1-(m+1)\ln|u|)|u|^{m+1}\,dx.\end{aligned} \tag{10.28}$$

Moreover, we will assume that the solution is bounded for all $t \in (0, T)$ (see Theorem 9.5): there exists a constant $l < \infty$ such that

$$\operatorname*{ess\,sup}_{(0,T)} \|u(\cdot, t)\|_{\infty,\Omega} \leq l. \tag{10.29}$$

This assumption yields boundedness of the energy function $\Sigma(t)$:

$$\Sigma(t) = \int_{\Omega} |u|^{m+1} dx \le |\Omega| \max\left\{l^{m^{+}+1}, l^{m^{-}+1}\right\} \equiv q. \tag{10.30}$$

Let us observe that

$$\begin{aligned}
&\int_{\Omega} \frac{m_t}{(m+1)^2}(1-(m+1)\ln |u|)|u|^{m+1} \\
&\quad = \int_{\Omega} \frac{m_t}{(m+1)^2}|u|^{m+1} dx - \int_{\Omega} \frac{m_t}{m+1} \ln |u||u|^{m+1} dx \\
&\quad \le \operatorname*{ess\,sup}_{\Omega} \left(\frac{m_t}{(m+1)^2}\right) \Sigma(t) - \operatorname*{ess\,sup}_{\Omega} \left(\frac{m_t}{m+1}\right) \int_{\Omega\cap\{|u|<1\}} |u|^{m+1} \ln |u| \, dx \\
&\quad \equiv K_1(t)\Sigma(t) + K_2(t)\, I(t).
\end{aligned}$$

Fix $t \in (0, T)$, some $\alpha \in (0, m^{-}(t)+1)$ and represent the integral $I(t)$ in the form

$$I(t) = \int_{\Omega\cap\{|u|<1\}} |u|^{m+1-\alpha} \left(|u|^{\alpha} |\ln |u||\right) dx.$$

It is easy to see that the function $s^{\alpha}|\ln s|$ with $\alpha \in (0, m^{-}(t)+1)$, defined on the interval $(0, 1)$, vanishes as $s \to 0$, $s \to 1$ and attains the maximum value at the point $s = \mathrm{e}^{-\frac{1}{\alpha}}$:

$$\max_{[0,1]} \left(s^{\alpha} |\ln s|\right) = \frac{1}{\alpha\,\mathrm{e}}.$$

Using this fact, applying Hölder's inequality (1.9) and then (1.6), we estimate

$$\begin{aligned}
I(t) &\le \frac{1}{\alpha\,\mathrm{e}} \int_{\Omega\cap\{|u|<1\}} |u|^{m+1-\alpha} \, dx \le \frac{1}{\alpha\,\mathrm{e}} \int_{\Omega} |u|^{m+1-\alpha} \, dx \\
&\le \frac{2}{\alpha\,\mathrm{e}} \|1\|_{\frac{m(\cdot,t)+1}{\alpha},\Omega} \left\||u|^{m+1-\alpha}\right\|_{\frac{m(\cdot,t)+1}{m(\cdot,t)+1-\alpha},\Omega} \\
&\le \frac{2}{\alpha\,\mathrm{e}} \|1\|_{\frac{m(\cdot,t)+1}{\alpha},\Omega} \max\left\{\Sigma^{1-\frac{\alpha}{m^{+}(t)+1}}(t), \Sigma^{1-\frac{\alpha}{m^{-}(t)+1}}(t)\right\}.
\end{aligned}$$

Recall that according to the definition of the Luxemburg norm

$$\frac{2}{\alpha\,\mathrm{e}} \|1\|_{\frac{m(\cdot,t)+1}{\alpha},\Omega} \le \frac{2}{\alpha\,\mathrm{e}} \max\left\{|\Omega|^{\frac{\alpha}{m^{+}(t)+1}}, |\Omega|^{\frac{\alpha}{m^{-}(t)+1}}\right\} \equiv \lambda.$$

Applying Lemmas 10.3 and 10.4 to the left-hand side of (10.28), we continue it in the following way: $\forall$ a.e. $t \in (0, T)$

$$\Sigma'(t) + C\Sigma^{\nu^+(t)}(t) \le l \int_\Omega |g(z)|\, dx + K_1(t)\Sigma(t) + \lambda K_2(t) \max\left\{\Sigma^{1-\frac{\alpha}{m^+(t)+1}}(t),\ \Sigma^{1-\frac{\alpha}{m^-(t)+1}}(t)\right\}, \tag{10.31}$$

where $\nu^+(t)$ is the variable exponent defined by formula (10.21) if $b^+ = 0$, and by formula (10.25) if $b^+ < 0$. Let

$$M = \max\{1, q\}, \qquad Z(t) = \frac{\Sigma(t)}{M} \le 1. \tag{10.32}$$

Dividing (10.31) by $M$ we arrive at the inequality for $Z(t)$

$$\begin{cases} Z'(t) + CM^{\nu^+(t)-1}Z^{\nu^+(t)}(t) \le \dfrac{l}{M}\displaystyle\int_\Omega g(x,t)\, dx \\ \quad +K_1(t)\, Z(t) + \lambda K_2(t)\, M^{-\frac{\alpha}{m^+(t)+1}} Z^{1-\frac{\alpha}{m^-(t)+1}}(t) \\ C = const \text{ is defined in (10.21) if } b^+ = 0, \text{ or in (10.23) if } b^+ < 0. \end{cases} \tag{10.33}$$

Finally, multiplying (10.33) by the function $e^{-K(t)}$ with

$$K(t) \equiv \int_0^t K_1(\tau)\, d\tau = \int_0^t \operatorname*{ess\,sup}_\Omega \left(\frac{m_t}{(m+1)^2}\right) d\tau, \tag{10.34}$$

we obtain the differential inequality for $S(t) \equiv Z(t)\, e^{-K(t)}$: $\forall$ a.e. $t \in (0, T)$

$$S'(t) + D(t)\, S^{\nu^+(t)}(t) \le \phi(t) + \rho(t)\, S^{1-\frac{\alpha}{m^-(t)+1}}(t) \tag{10.35}$$

with the right-hand side

$$\phi(t) = \frac{l\, e^{-K(t)}}{M} \int_\Omega g(z)\, dx \tag{10.36}$$

and the coefficients

$$D(t) = C\, (e\, M)^{-K(t)\,(1-\nu^+(t))},$$
$$\rho(t) = \lambda K_2(t)\, M^{-\frac{\alpha}{m^+(t)+1}} \exp\left(K(t)\left((1 - \frac{\alpha}{m^-(t)+1}\right)\right). \tag{10.37}$$

The derivation of inequality (10.35) remains true if the exponent $\alpha$ depends on $t$. Let us choose $\alpha \equiv \alpha(t) \in (0, m^-(t) + 1)$ in the special way:

$$\nu^+(t) = 1 - \frac{\alpha(t)}{m^-(t) + 1} \qquad \Leftrightarrow \qquad \alpha(t) = (m^-(t) + 1)(1 - \nu^+(t)).$$

Then the second term on the right-hand side of (10.35) transforms into $\rho(t)\, S^{\nu^+(t)}(t)$ and (10.35) leads to the inequality

$$\forall\, \text{a.e. } t \in (0, T) \qquad S'(t) + (D(t) - \rho(t))\, S^{\nu^+(t)}(t) \le \phi(t). \tag{10.38}$$

The above arguments are summarized in the following lemma.

**Lemma 10.5** *Let* $u \in \mathbf{W}(Q_T)$ *be a bounded strong energy solution of problem* (10.1). *Assume that the conditions of* Lemma 10.3 or 10.4, *and conditions* (10.27)-(10.29) *are fulfilled. Then the function*

$$S(t) = \frac{e^{-K(t)}}{M} \int_\Omega |u|^{m(x,t)+1}\, dx$$

*satisfies the differential inequality* (10.38) *with* $K(t)$, $D(t)$, $\rho(t)$ *and* $M$ *defined in* (10.34), (10.37) *and* (10.32).

**Lemma 10.6** *Let the conditions of* Lemma 10.5 *be fulfilled. Assume that in condition* (10.30) $q \le 1$. *If the functions* $m(z)$, $m_t(z)$ *satisfy the inequality*

$$4\, \frac{m_t^+(t)}{\alpha(t)\, e}\, e^{K(t)} \max\{1, |\Omega|\} \le C \qquad \forall t \in (0, T) \tag{10.39}$$

*with the constant* $C$ *from* (10.33), $m_t^+(t) = \operatorname{ess\,sup}_\Omega m_t(x, t)$, *then the strong energy function* $S(t)$ *for a.e.* $t \in (0, T)$ *satisfies the differential inequality*

$$S'(t) + \frac{1}{2} D(t)\, S^{\nu^+(t)}(t) \le \phi(t), \quad D(t) = C\, e^{-K(t)(1-\nu^+(t))}, \tag{10.40}$$

*with the right-hand side* $\phi(t)$ *defined in* (10.36).

*Proof* It suffices to check that

$$\rho(t) \le \frac{1}{2} D(t) \quad \Longleftrightarrow \quad 2\lambda\, K_2(t) \le C\, e^{-K(t)} M^{-K(t)\frac{\alpha(t)}{m^-(t)+1} + \frac{\alpha(t)}{m^+(t)+1}}.$$

In the conditions of the lemma $q \le 1$ and $M = 1$, so that the required inequality follows from the inequality

$$2\lambda\, K_2(t) \le C\, e^{-K(t)}.$$

Since

$$\frac{\alpha(t)}{m^{\pm}(t)+1} = \frac{m^{-}(t)+1}{m^{\pm}(t)+1}(1-\nu^{+}(t)) \quad \Longrightarrow \quad |\Omega|^{\frac{\alpha(t)}{m^{\pm}(t)+1}} \leq \max\{1, |\Omega|\},$$

the last inequality if fulfilled by virtue of (10.39). □

*Remark 10.1* Assumption (10.30) is merely technical and can be relaxed. Nonetheless, we prefer to keep it for the sake of simplicity of presentation.

*Remark 10.2* 1. Condition (10.39) is easy to satisfy if $m$ is independent of $x$. In this case

$$K(t) = \int_0^t \frac{m_t}{(m+1)^2}\,d\tau = \frac{1}{m(0)-1} - \frac{1}{m(t)+1} < \frac{1}{m(0)+1} < 1$$

and (10.39) becomes

$$\frac{4m'(t)}{(m^{-}(t)+1)(1-\nu^{+}(t))}\max\{1, |\Omega|\} \leq C.$$

If $b^{+}=0$, the last inequality is equivalent to the inequality

$$\frac{4m'(t)}{m^{-}(t)+1-p^{+}(t)}\max\{1, |\Omega|\} \leq C, \tag{10.41}$$

which relates the admissible growth of $m(t)$ to the behavior of $p(x,t)$.

2. It is to be specially noted that inequality (10.40) remains true even when $\nu^{+}(t) \to 1$ as $t \to \infty$, i.e., when $\alpha(t) \to 0$. In the case $m \equiv m(t)$, $b^{+}=0$ the admissible behavior of the exponents $m(t)$ and $p(x,t)$ is defined by inequality (10.41).

## 10.5 Extinction in a Finite Time

### *10.5.1 The Regular Case*

**Theorem 10.2** (Vanishing in a finite time) *Let $v(z)$ be a bounded energy strong solution of problem* (10.1). *Assume that $v$ satisfies the energy estimate* (10.5) *and that either the conditions of* Lemma 10.3, *or the conditions of* Lemma 10.4 *are fulfilled. Let $\nu^{+}(t)$ be the exponent defined in* (10.21) *if $b^{+}=0$, or in* (10.25) *if $b^{+}<0$. If*

(a) *the conditions of* Lemma 10.6 *are fulfilled,*
(b) $1 > \nu_0 \geq \nu^{+}(t)$ *and* $\dfrac{1}{n} + \dfrac{1}{m^{+}(t)+1} \geq \dfrac{1}{p^{-}(t)}$ *if $n > p^{-}(t)$,*

(c) $g \equiv 0$ *a.e. in* $Q_T$,
(d) *in the case* $K(t) \not\equiv 0$ *the initial value* $\Sigma(0)$ *satisfies the inequality*

$$\Sigma^{1-\nu_0}(0) < \frac{1}{2}C(1-\nu_0)\int_0^\infty e^{-K(\tau)(1-\nu^+(\tau))}\,d\tau \tag{10.42}$$

*with* $K(t)$ *from* (10.34) *and the constant* $C$ *from* (10.33),

*then there exists a finite* $t_*$ *such that* $\|v\|_{m+1,\Omega}(t) \equiv 0$ *for all* $t \geq t_*$.

*Proof* The assertion of the theorem will follow if we prove that $S(t) \equiv 0$ for all $t$ from some $t_*$ on. Under the conditions of the theorem the energy function

$$S(t) = e^{-K(t)}\int_\Omega \frac{m}{m+1}|u|^{m+1}\,dx$$

satisfies the differential inequality (10.40) with $\phi \equiv 0$. Let us define

$$t_* = \inf\{t > 0 : \ \forall\,\theta \in [0,t) \ \ S(\theta) > 0\}.$$

A straightforward integration of (10.40) over the interval $(0,t)$ with $t \leq t_*$ yields the relation (in the case $K \neq 0$)

$$0 \leq S^{1-\nu_0}(t) \leq \Sigma^{1-\nu_0}(0) - \frac{1}{2}C\,(1-\nu_0)\int_0^t e^{-K(\tau)(1-\nu^+(\tau))}\,d\tau \equiv \mathscr{R}(t).$$

$\mathscr{R}(t)$ is a continuous function of $t$, $\mathscr{R}(0) > 0$ by the definition and, due to (10.42), $\mathscr{R}(t) < 0$ for the sufficiently big $t$. It follows from that there is a finite $t_*$ such that $\mathscr{R}(t_*) = 0$ and, respectively, $\Sigma(t_*) = 0$. Since $\Sigma(t)$ is nonnegative and decreasing, it is necessary that $\Sigma(t) \equiv 0$ for all $t \geq t_*$. In the case $K = 0$ integration of (10.40) leads to the inequality

$$0 \leq \Sigma^{1-\nu_0}(t) \leq \Sigma^{1-\nu_0}(0) - C(1-\nu_0)t,$$

and the conclusion follows in the same way with $t_* = \Sigma^{1-\nu_0}(0)/(C(1-\nu_0))$. □

**Theorem 10.3** (Vanishing at a prescribed moment) *Let the conditions of* Theorem 10.2 *be fulfilled and* $g(z) \equiv 0$ *for a.e.* $t \geq t_g$. *Set*

$$R^{1-\nu_0}(t) = \frac{1}{2}C(1-\nu_0)\int_t^{t_g} e^{1-\nu^+(\tau)}\,d\tau.$$

*If the integral*

$$\mathscr{N}(t_g) = \frac{1}{M}\int_0^{t_g}\frac{\phi(\tau)}{R(\tau)}\,d\tau$$

*is convergent and*

$$\Sigma(0) - R(0)(M - \mathcal{N}(t_g)) \leq 0,$$

*then*

$$\|v(\cdot, t)\|_{m+1,\Omega} \equiv 0 \textit{ in } \Omega \textit{ for all } t \geq t_g.$$

*Proof* The energy function $S(t) = \Sigma(t)\mathrm{e}^{-\bar{K}(t)}$ satisfies the nonlinear differential inequality (10.40) with the nonzero right-hand side $\phi(t)$. The proof is based on comparison of $S(t)$ with the solution of the homogeneous differential equation

$$R'(t) + \frac{1}{2}D(t)R^{\nu_0}(t) = 0, \qquad R(0) = R_0$$

with the initial value chosen according to the condition

$$R_0^{1-\nu_0} = \frac{1}{2}C(1-\nu_0)\int_t^{t_g} \mathrm{e}^{1-\nu^+(\tau)}\,d\tau.$$

We refer to the proof of Theorem 6.3 for the details of this argument. □

## 10.5.2 The Limit Case

The differential inequality (10.40) remains true if $\nu^+(t) \to 1$ as $t \to \infty$, but the proof of Theorem 10.2 is no longer valid because $\nu^+(t)$ is not separated away from the critical value. Nonetheless, under certain assumptions on $\nu^+(t)$ the solutions of problem (10.1) still may display the property of vanishing in a finite time. We will need the following technical lemma.

**Lemma 10.7** *Let a nonnegative function $S(t)$ satisfy the conditions*

$$\begin{cases} S'(t) + \frac{1}{2}C\,\mathrm{e}^{-K(t)(1-\nu^+(t))}S^{\nu^+(t)}(t) \leq 0 \quad \textit{for a.e. } t \geq 0 \\ \qquad\qquad \textit{with } \nu^+(t) \in (0,1) \textit{ and } C = const > 0, \\ 0 < S(t) \leq S(0) < \infty, \end{cases} \tag{10.43}$$

*$K(t)$ is a nonnegative function, $K(0) = 0$. If the exponent $\nu^+(t)$ is monotone increasing, then $S(t) \equiv 0$ for all $t \geq t_*$ with $t_*$ defined from the equality*

$$C\int_0^{t_*} \frac{S^{\nu^+(s)-1}(0)}{\mathrm{e}^{K(s)(1-\nu^+(s))}}\,ds = 2\int_0^{\infty} \frac{dz}{\mathrm{e}^{z(1-\nu^+(z))}}.$$

*Proof* We adapt the proof of Lemma 6.7. Let us consider the function $J(t) = S(t)/S(0)$ which satisfies the conditions

$$\begin{aligned} \forall \text{a.e.}\, t > 0 \quad & J'(t) + C\,\mathrm{e}^{-K(t)\,(1-\nu^+(t))} S^{\nu^+(t)-1}(0) J^{\nu^+(t)} \le 0, \\ & J(0) = 1, \quad J'(t) \le 0. \end{aligned}$$

Introducing the new independent variable

$$\tau = C \int_0^t \left( S(0)\,\mathrm{e}^{K(\theta)} \right)^{\nu^+(\theta)-1} d\theta,$$

the new exponent $a(\tau) \equiv \nu^+(t)$ and the new sought function $I(\tau) \equiv J(t)$, we find that

$$\begin{aligned} \forall \text{a.e. } \tau > 0 \qquad & I'(\tau) + I^{a(\tau)} \le 0, \\ & I(0) = 1, \quad I(\tau) \ge 0, \quad I'(\tau) \le 0. \end{aligned} \tag{10.44}$$

By monotonicity of $I(\tau)$ there is an interval $[0, \varepsilon)$ where $I(\tau) > 0$, otherwise $I(\tau) \equiv 0$ for all $\tau \ge 0$. Since $I(\tau) \le 1$ and $a(\tau) \in (0, 1)$, then $I(\tau) \le I^{a(\tau)}(\tau)$, which leads to the inequality

$$I'(\tau) + I(\tau) \le I'(\tau) + I^{a(\tau)}(\tau) \le 0. \tag{10.45}$$

This inequality yields $I(\tau) \le e^{-\tau}$, whence

$$\tau \le -\ln I(\tau) \quad \text{and} \quad a(\tau) \le a\left(-\ln I(\tau)\right). \tag{10.46}$$

Combining this inequality with (10.44) we have that

$$I'(\tau) + I^{a(-\ln I(\tau))}(\tau) \le 0. \tag{10.47}$$

The straightforward integration of this inequality over the interval $(0, \tau)$ gives:

$$\int_1^{I(\tau)} \frac{dr}{r^{a(-\ln r)}} \le -\tau.$$

Introducing the new variable $z = -\ln r$, we rewrite it in the form

$$\int_0^{-\ln I(\tau)} \frac{dz}{\mathrm{e}^{z(1-a(z))}} \ge \tau \qquad \text{for } \tau \ge 0.$$

According to the choice of $t_*$

$$\int_0^{-\ln I(\tau_*)} \frac{dz}{\mathrm{e}^{z(1-a(z))}} \ge \tau_* = \int_0^{\infty} \frac{dz}{\mathrm{e}^{z(1-a(z))}},$$

which is impossible unless $I(\tau_*) = S(t_*)/S(0) = 0$. □

**Theorem 10.4** (Extinction of solutions of eventually linear equations) *Let $v(z)$ be a strong energy solution of problem* (10.1) *with $\Sigma_0 = \int_\Omega |v_0|^{m(x,0)+1} dx < \infty$ and $g \equiv 0$. Assume that $|v| \le l$ a.e. in $Q_T$ for some $l = const. \ge 1$, $m_t \ge 0$ a.e. in $Q_T$, and that one of the following conditions holds:*

1. $b(z) \le 0, \quad \nu^+(t) = \dfrac{m^+(t)\, p^+(t)}{1 + m^-(t)} \to 1$ *as* $t \to \infty$, $\nu^+(t)$ *is monotone increasing,*
2. $b(z) < b^+ < 0, \quad \nu^+(t) = \dfrac{(\sigma^+(t) + m^+(t) - 1)\, p^+(t)}{\sigma^+(t) + m^+(t)} \to 1$ *as* $t \to \infty$, $\nu^+(t)$ *is monotone increasing.*

*If*

$$\int_0^\infty \left(\Sigma_0 e^{K(s)}\right)^{\nu^+(s)-1} ds = \infty, \qquad \int_0^\infty \frac{ds}{e^{s(1-\nu^+(s))}} < \infty$$

*with $K(t)$ defined in* (10.34), *then there exists $t_* \in (0, \infty)$ such that $v(x,t) \equiv 0$ in $\Omega$ for all $t \ge t_*$, provided that $\mu$ and $\Sigma_0$ are sufficiently small.*

*Proof* Since the energy function satisfies the differential inequality (10.40), the assertion immediately follows from Lemma 10.7. □

### *10.5.3 Equations with Critical Growth*

Let $v(z) \in C^0(0, T; L^{m(x,t)+1}(\Omega)) \cap \mathbf{W}(Q_T)$ be a strong energy solution of the problem

$$\begin{cases} \dfrac{d}{dt}\left(|v|^{m(z)-1} v\right) = \displaystyle\sum_{i=1}^n D_i \left(a_i(z) |D_i v|^{p_i(z)-2} D_i v\right) \\ \qquad\qquad + c(z)|v|^{m(z)-1} v + b(z)|v|^{\sigma(z)-2} v + g \quad \text{in } Q, \\ v = 0 \text{ on } \Gamma, \\ v(x,0) = v_0(x) \text{ in } \Omega. \end{cases} \tag{10.48}$$

Assume that the data of the problem satisfy the foregoing assumptions and that $c(z) \le \gamma(t)$ a.e. in $Q_T$ with a bounded function $\gamma(t)$. Taking $v$ for the test-function, we arrive at inequality (10.31) with the coefficient $K_1(t)$ substituted by

$$\widetilde{K}_1(t) = \gamma(t) + \operatorname*{ess\,sup}_\Omega \left(\frac{m_t}{(m+1)^2}\right)$$

and the coefficient $K(t)$ given by the formula

$$\widetilde{K}(t) = \int_0^t \left(\gamma(\tau) + \operatorname*{ess\,sup}_{\Omega} \left(\frac{m_t}{(m+1)^2}\right)\right) d\tau.$$

The differential inequality for the energy function $\widetilde{S}(t) = \dfrac{\mathrm{e}^{-\widetilde{K}(t)}}{M} \displaystyle\int_\Omega |v|^{m(x,t)+1}\, dx$ takes the form

$$\forall \text{ a.e. } t \in (0,T) \quad \widetilde{S}'(t) + \frac{1}{2}\, \widetilde{D}(t)\, \widetilde{S}^{\nu^+(t)}(t) \le \widetilde{\phi}(t), \qquad \widetilde{S}(0) = \frac{1}{M}\int_\Omega |v_0|^{m_0+1}\, dx,$$

with the coefficient and the right-hand side

$$\widetilde{D}(t) = C\, \mathrm{e}^{-\widetilde{K}(t)(1-\nu^+(t))}, \qquad \widetilde{\phi}(t) = \frac{l\, \mathrm{e}^{-\widetilde{K}(t)}}{M} \int_\Omega g(z)\, dx.$$

The conclusions about the possibility of the finite-time extinction of the solution follow from the proofs of Theorems 10.2 and 10.3.

## 10.6 Large Time Behavior

The results of Sect. 10.5 refer to the cases when either $\nu^+(t) \in (0, \nu_0)$ with $\nu_0 < 1$, or $\nu^+(t)$ is monotone increasing and tends to the critical value $\nu_0 = 1$ as $t \to \infty$. If neither of these conditions is fulfilled, the analysis of the differential inequality for the energy function leads to the following results.

**Theorem 10.5** *Let the conditions of* Theorem 10.2 *be fulfilled with* $g \equiv 0$ *in* $Q_T$ *and* $v$ *be a bounded energy solution of problem* (10.1). *Let* $\nu^+(t) : [0,\infty) \mapsto (0, \nu^+]$ *be the function defined in* (10.21) *if* $b^+ < 0$, *and in* (10.25) *if* $b^+ = 0$.

1. *If* $\nu^+ = 1$, *then*

$$\Sigma(t) \equiv \int_\Omega |v|^{m(z)+1}\, dx \le \Sigma(0) \exp\left(K(t) - \frac{1}{2} C \int_0^t \left(\Sigma(0)\, \mathrm{e}^{K(s)}\right)^{\nu^+(s)-1} ds\right)$$

   *with the function* $K(t)$ *from* (10.34) *and the constant* $C$ *from* (10.33).
2. *If* $\nu^+ > 1$, *then*

$$\Sigma(t) \le \frac{\Sigma(0)\, \mathrm{e}^{K(t)}}{\left(1 + \frac{1}{2} C(\nu^+ - 1) \int_0^t \left(\Sigma(0)\, \mathrm{e}^{K(s)}\right)^{\nu^+(s)-1} ds\right)^{\frac{1}{\nu^+-1}}}.$$

3. *If* $\nu^+ < 1$, *then*

$$\Sigma(t) \le \Sigma(0)\ \max\left\{0,\ \left(1 - \frac{1}{2}C(1-\nu^+)\int_0^t \left(\Sigma(0)\,\mathrm{e}^{K(s)}\right)^{\nu^+(s)-1} ds\right)^{\frac{1}{1-\nu^+}}\right\}.$$

*Proof* By Lemma 10.6 the energy function $S(t) = \dfrac{\Sigma(t)}{M}\mathrm{e}^{-K(t)}$ satisfies the homogeneous differential inequality (10.40). Since $S \in [0, 1]$, we may continue (10.40) as follows:

$$\forall \text{a.e. } t \in (0, T) \qquad S' + \frac{1}{2}C\,\mathrm{e}^{-K(t)(1-\nu^+(t))}\, S^{\nu^+} \le 0.$$

The assertions follow now by the direct integration of the last inequality. □

**Corollary 10.3** *The large time behavior of energy solutions of problem* (10.48) *is also described by the estimates of* Theorem 10.5, *but with the coefficient* $K(t)$ *substituted by* $\widetilde{K}(t)$*—see* Sect. 10.5.3.

# Chapter 11
# Wave Equation with $p(x, t)$-Laplacian

Let $\Omega \subset \mathbb{R}^n$ be a bounded domain with the Lipschitz-continuous boundary $\Gamma$. We consider the initial and boundary value problem

$$
\begin{aligned}
u_{tt} &= \mathscr{L}u + f(x,t) \quad \text{in } Q_T = \Omega \times (0,T), &\quad (11.1)\\
u(x,0) &= u_0(x), \quad u_t(x,0) = u_1(x) \quad \text{in } \Omega, &\quad (11.2)\\
u &= 0 \quad \text{on } \Gamma_T = \partial\Omega \times (0,T), &\quad (11.3)
\end{aligned}
$$

where $\mathscr{L}$ denotes the differential operator

$$
\mathscr{L}u = \operatorname{div}\left(a(x,t)\,|\nabla u|^{p(x,t)-2}\,\nabla u - \varepsilon\nabla u_t\right) + b(x,t)\,|u|^{\sigma(x,t)-2}\,u,
$$

depending on a parameter $\varepsilon > 0$. The coefficients $a(x,t)$, $b(x,t)$, the exponents $p(x,t)$, $\sigma(x,t)$ and the source term $f(x,t)$ are given functions.

We are interested in the local and global in time existence of solutions of problem (11.1)–(11.3), and in the effect of finite time blow-up. Problem (11.1)–(11.3) appears in the mathematical modelling of nonlinear viscoelastic processes [72]. For hyperbolic equations of this type with constant $p$ and $\sigma$, the issues of local and global existence and the possibility of blow-up have been studied in numerous works—see, e.g., [118, 136, 141, 145, 178, 190, 204, 205, 261, 227, 265, 273, 274] and the further references therein.

It is noteworthy that the viscous term $\varepsilon\Delta u_t$ plays a key role in the proof of global existence in all works dealing with energy solutions in the case $p = const \neq 2$. However, the role of this term is two fold. On one hand, it is of crucial importance in deriving the a priori estimates used in the proof of existence, on the other hand, its presence makes much more complicated the study of the blow-up effect. Equations with constant exponents $p$, $\sigma$ and without viscous terms, i.e., with $\varepsilon = 0$, were studied in [94, 201, 229].



S. Antontsev and S. Shmarev, *Evolution PDEs with Nonstandard Growth Conditions*, Atlantis Studies in Differential Equations 4,
DOI 10.2991/978-94-6239-112-3_11

Equations of the type (11.1) with variable nonlinearity were studied in [25, 26, 27, 154, 218, 249]. Paper [218] addresses the question of blow-up of solutions of the Dirichlet problem for the semilinear equation with local or nonlocal low-order terms

$$u_{tt} = \Delta u + f(u),$$

where

$$f(u) = b(x)u^{p(x)} \quad \text{or} \quad f(u) = b(x)\int_{\Omega} u^{q(y)}(y,t)\,dy.$$

In [154], the existence theorem for problem (11.1)–(11.3) was proved in the case of independent of $t$ exponent $p$:

$$u_{tt} = \operatorname{div}\left(|\nabla u|^{p(x)-2}\nabla u\right) + \varepsilon\Delta u_t + f(x,u), \qquad \varepsilon > 0.$$

Papers [65, 66] discuss the questions of asymptotic stability as $t \to \infty$ and global nonexistence of solutions to the Dirichlet problem for the system of nonlocal equations with nonstandard growth:

$$u_{tt} - M\left(\int_{\Omega}\frac{|\nabla u(x,t)|^{p(x)}}{p(x)}\,dx\right)\Delta_{p(x)}u - g(t)\Delta_{p(x)}u_t + Q(t,x,u,u_t) + f(t,x,u) = 0.$$

Here the unknown is the vector-valued function $u = (u_1,\dots,u_N)$ (the displacement), $N \geq 1$, the problem domain is the cylinder $\Omega \times (0,T]$ with bounded $\Omega \subset \mathbb{R}^n$. It is assumed that the function $M$ satisfies the growth conditions

$$M(\tau) = a + b\gamma\tau^{\gamma-1}, \quad \tau \geq 0, \quad a, b \geq 0, \quad a + b > 0, \quad \gamma > 1$$

and $g \in L^1_{loc}[0,\infty) \to [0,\infty)$.

Solvability of the problem

$$\begin{aligned} &u_{tt} - \Delta_{p(x)}u - \Delta u_t + g(u) = f(x,t) \;\text{ in } Q_T = \Omega \times (0,T] \\ &u(x,0) = u_0(x),\, u_t(x,0) = u_1(x) \text{ in } \Omega, \\ &u = 0 \text{ on } \partial\Omega \times (0,T] \end{aligned}$$

is studied in [249]. It is proved that for $p(x) \in C_{\log}(\Omega)$, $p(x) \in (n/2, n)$, the problem has a solution in the class of functions

$$u \in L^{\infty}\left(0,T; W_0^{1,p(x)}(\Omega)\right), \quad u_t \in L^{\infty}\left(0,T; L^2(\Omega)\right) \cap L^{\infty}\left(0,T; H_0^1(\Omega)\right).$$

The bulk of material of this chapter is based on the papers [17, 25, 26, 27, 38] and covers the questions of existence and blow-up of solutions for problem (11.1)–(11.3) with variable coefficients $a$, $b$ and exponents $p$, $\sigma$.

In Sect. 11.1 we construct a weak energy solution of problem (11.1)–(11.3) with $\varepsilon > 0$. The solution is obtained as the limit of a sequence of Galerkin's approximations. This approach requires derivation of local and global in time uniform a priori estimates for the members of the approximating sequence. For further convenience, we also derive estimates on the energy functional and establish sufficient conditions for its non-positivity, an important property used later on in the proof of the finite time blow-up. The limit passage in the sequence of approximations and the proof of existence of a solution are based on the monotonicity arguments. In Sect. 11.2 we derive conditions that guarantee non-existence of global in time energy solutions of the problems with $\varepsilon > 0$ and $\varepsilon = 0$. The latter result is conditional: it is established under the assumption that the problem admits energy solution.

Equation (11.1) without damping is studied in Sect. 11.3. Such equations were studied in papers [94, 201, 229], where the existence theorem for measure solutions was proved for the equation in divergence form

$$u_{tt}(x,t) = \operatorname{div}_x S(\nabla_x u(x,t)).$$

Equation (11.1) with variable $p(x,t)$, $\sigma(x,t)$ and $\varepsilon = 0$ was studied in [17]. Following this work, we construct Young measure solutions of problem (11.1)–(11.3) with $\varepsilon = 0$ as a limit as $\varepsilon \to 0$ of the sequence of solutions to the same problem with $\varepsilon > 0$. We derive then the conditions on the exponents of nonlinearity which provide a finite time blow-up of the Young measure solutions corresponding the initial data with negative energy.

We leave open the challenging question of existence of energy solutions in the case $\varepsilon = 0$.

## 11.1 Local and Global Existence of Weak Solutions

The natural function space for the study of problem (11.1)–(11.3) $\mathbf{U}(Q_T)$ was introduced in Sect. 1.4.2.

**Definition 11.1** A function $u : Q_T \to \mathbb{R}$ is called weak energy solution of problem (11.1)–(11.3) if

(i) $u \in \mathbf{U}(Q_T) \cap C([0,T]; L^2(\Omega))$,

(ii) $\forall\, \eta(x) \in C_0^\infty(\Omega)$ $(\eta(x), u(x,t) - u_0)_{H_0^1(\Omega)} \to 0$ as $t \to 0$,

(iii) $\forall\, \eta(x) \in C_0^\infty(\Omega)$ $\exists\, \{t_k\} \to 0 :$ $(u_t(x,t_k) - u_1(x), \eta(x))_{2,\Omega} \to 0$ as $k \to \infty$,

(iv) $\forall \phi \in \mathbf{U}(Q_T) \cap C([0,T]; L^2(\Omega))$ with $\phi(x,T) = 0$

$$\int_{Q_T}\left(-u_t\phi_t+\left(a\,|\nabla u|^{p-2}\,\nabla u+\varepsilon\nabla u_t\right)\cdot\nabla\phi-b\,|u|^{\sigma-2}\,u\phi\right)dz$$
$$=\int_{\Omega}u_1\phi(\cdot,0)dx+\int_{Q_T}f\phi\,dz. \tag{11.4}$$

Assume that the coefficient $a$ and the exponents $p$, $\sigma$ satisfy the conditions

$$\begin{array}{rll} & 0<a_-\leq a(x,t)\leq a_+<\infty, & |a_t|\leq C_a,\\ p\in C_{\log}(Q_T), & 1<p^-\leq p(x,t)\leq p^+<\infty, & |p_t|=-p_t\leq C_p,\\ & 1<\sigma^-\leq\sigma(x,t)\leq\sigma^+<\infty, & 0\leq\sigma_t\leq C_\sigma.\end{array} \tag{11.5}$$

Assume also that one of the following conditions holds:

$$\begin{array}{lll}\text{(a)} & 0<b_-\leq -b(x,t)\leq b_+<\infty, & 0\leq b_t,\\ \text{(b)} & 0<b_-\leq b(x,t)\leq b_+<\infty, & 0\leq b_t, \quad \begin{cases}\text{either} & \sigma^+\leq 2\\ \text{or} & \sigma^+<p^-.\end{cases}\end{array} \tag{11.6}$$

**Theorem 11.1** (Global in time existence) *Let $a$, $p$, $\sigma$ satisfy conditions* (11.5) *and $b$ satisfy* (11.6). *If*

$$\begin{aligned}&u_0\in L^2(\Omega)\cap L^{\sigma(x,0)}\cap H_0^1(\Omega)\cap W^{1,p(x,0)}(\Omega),\\&u_1\in L^2(\Omega),\ f\in L^2(Q_T),\end{aligned} \tag{11.7}$$

*then problem* (11.1)–(11.3) *has at least one global in time energy weak solution.*

Local in time solvability is established under different conditions on the data.

**Theorem 11.2** (Local in time existence) *Let $a$, $p$, $\sigma$ satisfy conditions* (11.5) *and $b$ satisfies the conditions*

$$0<b_-\leq b(x,t)=b(x,t)\leq b_+<\infty,\quad |b_t|\leq C_b. \tag{11.8}$$

*If*

$$2<\sigma^-\leq\sigma^+<\frac{n+2}{n}p^-, \tag{11.9}$$

*then for every $u_0$, $u_1$, $f$ satisfying* (11.7) *there is $T_{\max}<T$ such that problem* (11.1)–(11.3) *has at least one energy solution in $Q_{T_{\max}}$.*

*Remark 11.1* It turns out that the solutions possess better regularity than stated in Theorems 11.1 and 11.2. In fact, it will be proved that

$$u_t,\ |\nabla u_t|\in L^2(Q_T),\qquad |\nabla u|^{p(x,t)}\in L^\infty(0,T;L^1(\Omega)).$$

### 11.1.1 Galerkin's Approximations

The sequence of finite-dimensional Galerkin's approximations has the form

$$u^{(m)} \equiv \sum_{k=1}^{m} u_k(t)\psi_k(x), \qquad u_k(t) = (u(x,t),\ \psi_k(x))_{2,\Omega},$$

where $\{\psi_m\}$ is the orthogonal basis composed of the eigenfunctions of problem (1.30). By the choice of the basis $\{\psi_m\}$

$$\begin{aligned} u_1^{(m)} &= \sum_{i=1}^{m}(u_1, \psi_i)_{2,\Omega}\psi_i \to u_1 \text{ in } L^2(\Omega), \\ u_0^{(m)} &= \sum_{i=1}^{m}(u_0, \psi_i)_{2,\Omega}\psi_i \to u_0 \text{ in } L^{\sigma(x,0)}(\Omega) \cap H_0^1(\Omega) \cap W_0^{1,p(x,0)}(\Omega). \end{aligned} \tag{11.10}$$

The coefficients $u_k(t)$ are defined from the system of equations

$$\int_\Omega (u_{tt}^{(m)} - \mathscr{L}u^{(m)} - f)\psi_k\,dx = 0, \qquad k = 1, \dots, m. \tag{11.11}$$

Equalities (11.11) and the initial conditions (11.10) lead to the Cauchy problem for the system of 2nd-order ODEs for the coefficients $u_k(t)$

$$\begin{cases} u_k'' = \mathscr{F}_k(t, u_1, \dots, u_m, u_1', \dots, u_m'), & k = 1, \dots, m, \\ u_k(0) = (u_0, \psi_k)_{2,\Omega}, \qquad u_k'(0) = (u_1, \psi_k)_{2,\Omega}, \end{cases} \tag{11.12}$$

where

$$\begin{aligned} \mathscr{F}_k = &\int_\Omega \left[-\left(a|\nabla u^{(m)}|^{p-2}\nabla u^{(m)} + \varepsilon\nabla u_t^{(m)}\right)\cdot\nabla\psi_k\right] dx \\ &+ \int_\Omega \left[b|u^{(m)}|^{\sigma-2}u^{(m)}\psi_k + f\psi_k\right] dx. \end{aligned}$$

For every finite $m$ problem (11.12) has a solution on an interval $(0, T_m)$.

Let us show that each of solutions $u_m$ can be continued to the maximal existence interval $[0, T]$. Multiplication of the $k$th equation in (11.12) by $u_k'(t)$ and summation in $k$ lead to the equality

$$\frac{d}{dt}\int_{\Omega}\left[\frac{1}{2}(u_t^{(m)})^2+\frac{a}{p}|\nabla u^{(m)}|^p-\frac{b}{\sigma}|u^{(m)}|^{\sigma}\right]dx+\varepsilon\int_{\Omega}\left|\nabla u_t^{(m)}\right|^2dx$$
$$=\int_{\Omega}\left[a_t\frac{|\nabla u^{(m)}|^p}{p}+\frac{a|\nabla u^{(m)}|^p}{p^2}\left(1-p\ln|\nabla u^{(m)}|\right)|p_t|\right]dx$$
$$+\int_{\Omega}\left(-\frac{b_t|u^{(m)}|^{\sigma}}{\sigma}+\frac{b|u^{(m)}|^{\sigma}}{\sigma^2}\left(1-\sigma\ln|u^{(m)}|\right)\sigma_t\right)dx+\int_{\Omega}fu_t^{(m)}dx. \tag{11.13}$$

For the sake of simplicity of notations, throughout the rest of this section we omit the super-index $m$. Let us introduce the energy functional

$$E(t)=\int_{\Omega}\left[\frac{1}{2}u_t^2+\frac{a}{p}|\nabla u|^p-\frac{b}{\sigma}|u|^{\sigma}\right]dx \tag{11.14}$$

and rewrite (11.13) in the form

$$E'(t)+\varepsilon\int_{\Omega}|\nabla u_t(\cdot,t)|^2dx=\Lambda(t), \tag{11.15}$$

where $\Lambda\equiv\Lambda_1+\Lambda_2+\Lambda_3$ and

$$\begin{aligned}\Lambda_1&=\int_{\Omega}\left[a_t\frac{|\nabla u|^p}{p}+\frac{a|\nabla u|^p}{p^2}(1-p\ln|\nabla u|)|p_t|\right]dx,\\ \Lambda_2&=\int_{\Omega}\left(-\frac{b_t|u|^{\sigma}}{\sigma}+\frac{b|u|^{\sigma}}{\sigma^2}(1-\sigma\ln|u|)\sigma_t\right)dx,\\ \Lambda_3&=\int_{\Omega}fu_tdx.\end{aligned} \tag{11.16}$$

**Lemma 11.1** *Let the functions $a$, $b$, $\sigma$ satisfy* (11.5) *and* (11.8). *If*

$$a_t\le 0,\quad b_t\ge 0,\quad p_t=\sigma_t=f=0,$$

*then*

$$E(t)+\varepsilon\int_0^t\int_{\Omega}|\nabla u_t(x,s)|^2dxds\le E(0),\quad\forall t\ge 0. \tag{11.17}$$

*If $a_t=b_t=0$, then inequality* (11.17) *transforms into the equality.*

*Proof* The assertion immediately follows from (11.15)–(11.16). □

**Lemma 11.2** *Let the functions $a$, $b$, $p$, $\sigma$ satisfy assumptions* (11.5) *and* (11.8). *If $a_t\le 0$, $b_t\ge 0$ and $f=0$, then*

$$E(t)+\int_0^t\int_{\Omega}\varepsilon|\nabla u_t(x,s)|^2dxds\le E(0)+Ct \tag{11.18}$$

*with the constant*

$$C = \left( \frac{a_+}{(p^-)^2} C_p + \frac{b_+}{(\sigma^-)^2} C_\sigma \right) |\Omega|.$$

*Proof* It is easy to see that the function

$$F(\eta) = \frac{\eta^p}{p^2} (1 - p \ln \eta), \qquad 0 \le \eta \le \mathrm{e}^{\frac{1}{p}},$$

possesses the properties

$$F(0) = F\left(\mathrm{e}^{\frac{1}{p}}\right) = 0, \quad F'(\eta) = -\eta^{p-1} \ln \eta, \qquad \max_{0 \le \eta \le \mathrm{e}^{\frac{1}{p}}} F(\eta) = F(1) = \frac{1}{p^2}.$$

Using these properties we evaluate $\Lambda_1$, $\Lambda_2$ in the following way:

$$\begin{aligned} \Lambda_1 &= \int_\Omega \left[ a_t \frac{|\nabla u|^p}{p} + \frac{a |\nabla u|^p}{p^2} (1 - p \ln |\nabla u|) |p_t| \right] dx \\ &\le \int_{\Omega \cap (p \ln |\nabla u| \le 1)} \frac{a |\nabla u|^p}{p^2} (1 - p \ln |\nabla u|) |p_t| dx \\ &\le \frac{a_+}{p^-} C_p |\Omega| = C_1, \end{aligned}$$

$$\begin{aligned} \Lambda_2 &= \int_\Omega \left[ -\frac{b_t |u|^\sigma}{\sigma} + \frac{b |u|^\sigma}{\sigma^2} (1 - \sigma \ln |u|) \sigma_t \right] dx \\ &\le \int_{\Omega \cap (\sigma \ln |u| \le 1)} \frac{b |u|^\sigma}{\sigma^2} (1 - \sigma \ln |u|) \sigma_t dx \\ &\le \frac{b_+}{\sigma^-} C_\sigma |\Omega| = C_2. \end{aligned}$$

Integration of the energy relation (11.15) in $t$ gives then the inequality

$$E(t) + \int_0^t \int_\Omega \varepsilon |\nabla u_t|^2 \, dx ds \le E(0) + C t, \qquad C = C_1 + C_2.$$

□

**Lemma 11.3** (Global estimates in the case $b \le 0$) *Let conditions* (11.5)–(11.7) *be fulfilled. Then for every finite T*

$$\Psi(T) = \sup_{0 \le t \le T} \int_\Omega \left[ u_t^2 + |\nabla u|^p + |u|^\sigma \right] dx + \varepsilon \int_0^T \int_\Omega |\nabla u_t|^2 \, dx ds \le C \quad (11.19)$$

*with a constant $C$ depending on $\|f\|^2_{2,Q_T}$, $\|u_1\|_{2,\Omega}$, $\|u_0\|_{\sigma(\cdot,0),\Omega}$, $\|u_0\|_{W_0^{1,p(x,0)}(\Omega)}$, $\|u_0\|_{H_0^1(\Omega)}$, $|\Omega|$, $T$, but independent of $m$.*

*Proof* Since in the present case all terms of the energy functional $E$ are nonnegative, we get

$$\begin{aligned}\Lambda_1 &= \int_\Omega \left[a_t \frac{|\nabla u|^p}{p} + \frac{a|\nabla u|^p}{p^2}(1 - p\ln|\nabla u|)\,|p_t|\right] dx \\ &\leq \frac{C_a}{a_-}E(t) + \int_{\Omega\cap(p\ln|\nabla u|\leq 1)} \frac{a|\nabla u|^p}{p^2}(1 - p\ln|\nabla u|)\,|p_t| dx \\ &\leq \frac{C_a}{a_-}E(t) + \frac{a_+}{p^-}e|\,\Omega| \leq C(E(t)+1),\end{aligned}$$

$$\begin{aligned}\Lambda_2 &= \int_\Omega \left(-\frac{b_t|u|^\sigma}{\sigma} + \frac{b|u|^\sigma}{\sigma^2}(1 - \sigma\ln|u|)\,\sigma_t\right) dx \\ &\leq \frac{C_b}{\sigma^-}E(t) + \int_{\Omega\cap(\sigma ln|u|\leq 1)} \frac{b|u|^\sigma}{\sigma^2}(1 - \sigma\ln|u|)\,\sigma_t dx \\ &\leq \frac{C_b}{\sigma^-}E(t) + \frac{b_+}{\sigma^-}e|\Omega| \leq C(E(t)+1),\end{aligned}$$

$$|\Lambda_3| \leq E(t) + \|f\|^2_{2,\Omega}.$$

These estimate yield the inequality

$$\Lambda \leq C(E(t) + \|f\|^2_{2,\Omega} + 1),$$

whence

$$E'(t) + \varepsilon \int_\Omega |\nabla u_t(\cdot,t)|^2\, dx \leq C(E(t) + \|f\|^2_{2,\Omega} + 1).$$

The second term on the left-hand side can be dropped, the estimate on $E(t)$ follows then from the Gronwall lemma. □

**Lemma 11.4** (Global estimates: $0 \leq b$ and either $\sigma^+ \leq 2$, or $2 < \sigma^+ < p^-$) *Estimate* (11.19) *remains true if the condition $b(x,t) \leq 0$ is replaced by the condition*

$$0 \leq b(x,t) \textit{ and either } \sigma^+ \leq 2, \textit{ or } 2 < \sigma^+ < p^-.$$

*Proof* Denote

$$\widetilde{E}(t) = \int_\Omega \left[\frac{1}{2}u_t^2 + \frac{a}{p}|\nabla u|^p\right] dx$$

and rewrite the energy relation (11.15) in the form

$$\widetilde{E}'(t) + \varepsilon \int_{\Omega} |\nabla u_t|^2 \, dx = \left( \int_{\Omega} \frac{b}{\sigma} |u|^{\sigma} \, dx \right)' + \Lambda.$$

Integration of this equality in $t$ gives

$$\widetilde{E}(t) + \varepsilon \int_0^t \int_{\Omega} |\nabla u_t|^2 \, dx ds = \int_{\Omega} b \frac{|u|^{\sigma}}{\sigma} dx \bigg|_0^t + \int_0^t \Lambda ds + \widetilde{E}(0) \tag{11.20}$$

with

$$\left| \int_{\Omega} b \frac{|u|^{\sigma}}{\sigma} dx \bigg|_0^t \right| \le C \left( \int_{\Omega} |u|^{\sigma} \, dx + 1 \right), \qquad |\Lambda| \le C\widetilde{E}(t) + C \int_{\Omega} |u|^{\sigma} \, dx.$$

To complete the estimate we have to evaluate the last term on the right-hand side. In the case $\sigma^+ \le 2$ we use the inequalities

$$\begin{aligned} \left| \int_{\Omega} \frac{b}{\sigma} |u|^{\sigma} \, dx \right| &\le C \int_{\Omega} |u|^{\sigma} \, dx \le C \left( 1 + \int_{\Omega} |u|^2 \, dx \right) \\ &\le 2C \left( 1 + t \int_0^t \int_{\Omega} u_t^2 \, dx ds + \int_{\Omega} |u_0|^2 \, dx \right) \\ &\le \widetilde{C} \left( 1 + t \int_0^t \widetilde{E}(s) ds \right). \end{aligned}$$

Otherwise $2 < \sigma^+ < p^-$, and we use the embedding $L^{\sigma^+}(\Omega) \subset W_0^{1,p^-}(\Omega)$ and Young's inequality:

$$\begin{aligned} \int_{\Omega} |u|^{\sigma^+} dx &\le C \left( \int_{\Omega} |\nabla u|^{p^-} \, dx \right)^{\frac{\sigma^+}{p^-}} \\ &\le C \left( \int_{\Omega} \nabla u|^{p} dx \right)^{\frac{\sigma^+}{p^-}} + C \\ &\le \delta \widetilde{E}(t) + C(\delta) \end{aligned}$$

with $\delta \in (0, 1)$, $\sigma^+ < p^- < \frac{np^-}{n-p^-}$. Since

$$\widetilde{E}(t) + \varepsilon \int_0^t \int_{\Omega} |\nabla u_t(\cdot, s)|^2 \, dx ds \le C \left( \int_0^t \widetilde{E}(s) ds + 1 \right)$$

for the sufficiently small $\delta > 0$, the result follows from Gronwall's inequality and (11.20). □

**Lemma 11.5** (Local estimates: $2 < \sigma^- \le \sigma^+ < \frac{n+2}{n}p^-$) *Let $a$, $b$, $p$, $\sigma$ satisfy the conditions of* Theorem 11.2. *Then there exists $T_{\max} > 0$ such that for every $t \in (0, T_{\max})$*

$$\Psi(t) = \sup_{0\le s\le t} \int_\Omega \left[u_t^2 + |\nabla u|^p + |u|^\sigma\right] dx + \varepsilon \int_0^t \int_\Omega |\nabla u_t|^2\, dx ds \le C \quad (11.21)$$

*with a constant $C$ depending on $T_{\max}$, $\|f\|^2_{2,Q_T}$, $\|u_1\|_{2,\Omega}$, $\|u_0\|_{\sigma(\cdot,0),\Omega}$, $\|u_0\|_{W^{1,p(\cdot,0)}(\Omega)}$, $\|u_0\|_{H_0^1(\Omega)}$, but independent of $m$.*

*Proof* We will use the energy relation (11.20) and the inequalities

$$\begin{aligned} \left|\int_0^t \Lambda\, dt\right| &\le C \int_0^t \left(\widetilde{E}(s) + \int_\Omega |u|^{\sigma(x,t)}\, dx\right) ds, \\ \int_\Omega |u|^{\sigma(x,t)}\, dx &\le \int_\Omega |u|^{\sigma^+}\, dx + |\Omega|, \\ \int_\Omega |\nabla u|^{p^-}\, dx &\le \int_\Omega |\nabla u|^{p(x,t)}\, dx + |\Omega|. \end{aligned} \quad (11.22)$$

Combining the interpolation inequality (1.58) and Young's inequality we find that

$$\begin{aligned} \int_\Omega |u|^{\sigma^+}\, dx &\le C\left(\int_\Omega |\nabla u|^{p^-}\, dx\right)^{\frac{\sigma^+}{p^-}\theta} \left(\int_\Omega |u|^2\, dx\right)^{\frac{\sigma^+}{2}(1-\theta)} \\ &\le \delta \int_\Omega |\nabla u|^{p(x,t)}\, dx + C_\delta \left(\int_\Omega |u|^2\, dx\right)^\gamma + C \end{aligned} \quad (11.23)$$

with $\delta \in (0,1)$ and the parameters

$$\frac{\sigma^+}{p^-}\theta = \frac{(\sigma^+ - 2)\, n}{np^- - 2(n-p^-)} < 1 \text{ if} \sigma^+ < \frac{n+2}{n}p^-, \quad \gamma = \frac{\sigma^+ p^-}{2(p^- - \theta\sigma^+)} > 1.$$

On the other hand,

$$\int_\Omega u^2 dx \le C\left(\int_0^t \int_\Omega |u_t|^2\, dxds\right) + \int_\Omega |u_0|^2\, dx \le C\left(\int_0^t \widetilde{E}(s)ds + 1\right).$$

For a suitably small $\delta$, a combination of (11.20), (11.22)–(11.23) leads to the inequality

$$\widetilde{E}(t) \le C\left[\left(\int_0^t \widetilde{E}(s)ds\right)^\gamma + 1\right],$$

whence

$$\widetilde{E}(t) + 1 \le \frac{\widetilde{E}(0) + 1}{\left(1 - tC(\widetilde{E}(0)+1)^{\gamma-1}(\gamma-1)\right)^{\frac{1}{\gamma-1}}} < \infty.$$

It is then necessary that

$$t < T_{\max} = \frac{1}{C(\widetilde{E}(0)+1)^{\gamma-1}(\gamma-1)}.$$

□

*Remark 11.2* The constants $C$ in the estimates of Lemmas 11.3 and 11.4 and in (11.21) are independent of $\varepsilon$.

*Remark 11.3* The estimates of Lemmas 11.3 and 11.4 imply the inequality

$$\Lambda(T) = \sup_{t\in[0,T]} \int_\Omega \left(u^2 + |u|^{\frac{2n}{n-2}} + |u|^q\right) dx \le C \tag{11.24}$$

with $q \le \dfrac{np^-}{n-p^-}$ if $p^- < n$ and $q < \infty$ if $p^- \ge n$.

Applying the easily verified inequality

$$\int_\Omega |\nabla u(x,t)|^2\, dx \le 2\left(\int_\Omega |\nabla u_0(x)\ ^2\, dx + t\int_0^t \int_\Omega |\nabla u_t|^2\, dxds\right)$$

and the embedding inequality

$$\int_0^T \left(\int_\Omega |u_t|^q\, dx\right)^{\frac{2}{q}} dt \le C\int_{Q_T} |\nabla u_t|^2\, dxdt$$

with

$$1 \le q \le \frac{2n}{n-2} \text{ if } n > 2, \quad 1 \le q < \infty \text{ if } n = 2,$$

we arrive at the estimate

$$\begin{aligned} \Lambda(T) + \sup_{0\le s\le T} \int_\Omega \left[u_t^2(x,s) + |\nabla u|^2 + |\nabla u|^p + |u|^\sigma\right] dx \\ + \varepsilon \int_0^T \int_\Omega |\nabla u_t|^2\, dxdt + \int_0^T \left(\int_\Omega |u_t|^q\, dx\right)^{\frac{2}{q}} dt \le K \end{aligned} \tag{11.25}$$

with an independent of $m$ constant $K$. Under the conditions of Lemmas 11.3 and 11.4 this estimate holds on any finite interval $[0, T]$. If the conditions of Lemma 11.5 are fulfilled, then the estimate is local in time and is true on an interval $[0, T_0]$.

### 11.1.2 Compactness and Passage to the Limit

**Lemma 11.6** *Let* (11.25) *be true. Then*

$$\left|\int_{Q_T} u_{tt}^{(m)}\phi dxdt\right| \leq C\|\phi\|_{L^{p^+}(0,T;H_0^s(\Omega))}, \qquad T < T_0, \tag{11.26}$$

*with an independent of m constant C.*

*Proof* By the definition of $u^{(m)}$, for every $\phi \in L^{p^+}(0,T;H_0^s(\Omega))$ $(u_{tt}^{(m)},\phi)_{2,Q_T} = (u_{tt}^{(m)},\phi^{(m)})_{2,Q_T}$, where

$$\phi^{(m)} = \sum_{i=1}^{m} \phi_i(t)\psi_i(x) \in \mathscr{P}_m \equiv \text{span}\,\{\psi_1,\ldots,\psi_m\}$$

and $\phi_i(t) = (\phi,\psi_i)_{2,\Omega}$. Then

$$\begin{aligned}\int_{Q_T} u_{tt}^{(m)}\phi\, dz &= \int_{Q_T} u_{tt}^{(m)}\phi^{(m)}\, dz\\ &= -\int_{Q_T}\left((a|\nabla u^{(m)}|^{p-2}\nabla u^{(m)} + \varepsilon\nabla u_t^{(m)}\right)\nabla\phi^{(m)}\, dz\\ &\quad -\int_{Q_T}\left(b|u^{(m)}|^{\sigma-2}u^{(m)} + f\right)\phi^{(m)} dz := J. \end{aligned}\tag{11.27}$$

Following the proof of Lemma 4.4 we estimate

$$\begin{aligned}|J| &\leq a_+\||\nabla u^{(m)}|^{p(\cdot)-1}\|_{\frac{p(\cdot)}{p(\cdot)-1},Q_T}\|\nabla\phi^{(m)}\|_{p(\cdot),Q_T} + \varepsilon\|\nabla u_t^{(m)}\|_{2,Q_T}\|\nabla\phi^{(m)}\|_{2,Q_T}\\ &\quad + b_+\||u^{(m)}|^{\sigma-1}\|_{\frac{\sigma(\cdot)}{\sigma(\cdot)-1},Q_T}\|\phi^{(m)}\|_{\sigma(\cdot),Q_T} + \|f\|_{2,Q_T}\|\phi^{(m)}\|_{2,Q_T}\\ &\leq C(K)\|\phi^{(m)}\|_{L^{p^+}(0,T;H_0^s(\Omega))}\\ &\leq C(K)\|\phi\|_{L^{p^+}(0,T;H_0^s(\Omega))}\end{aligned}$$

with the constant $K$ from (11.25). □

**Corollary 11.1** $\|u_{tt}^{(m)}\|_{L^{(p^+)'}(0,T;H^{-s}(\Omega))} \leq C$ *uniformly with respect to m.*

**Lemma 11.7** *The sequence* $\{u_t^{(m)}\}$ *is precompact in* $L^{\mu}(Q_T)$ *with some* $\mu > 1$, *the sequence* $\{u^{(m)}\}$ *is precompact in* $C(0,T;L^2(\Omega))$.

*Proof* Without loss of generality we may assume that $p^+ > 2$. The sequence $\{u_t^{(m)}\}$ is uniformly bounded in $L^2(0,T;H_0^1(\Omega)) \subset L^{(p^+)'}(0,T;H_0^1(\Omega))$, $\{(u_t^{(m)})_t\}$ is uniformly bounded in $L^{(p^+)'}(0,T;H^{-s}(\Omega))$. Since the embedding $H_0^1(\Omega) \subset L^2(\Omega)$

is compact and $L^2(\Omega) \subset H^{-s}(\Omega)$, it follows from Lemma 2.6 that there is $\mu > 1$ such that the sequence $\{u_t^{(m)}\}$ is relatively compact in $L^{\mu}(Q_T)$.

The second claim also follows from Lemma 2.6 because (11.25) yields uniform boundedness of $\|u^{(m)}\|_{L^\infty(0,T;H_0^1(\Omega))}$ and $\|u_t^{(m)}\|_{L^\infty(0,T;L^2(\Omega))}$, whereas the embedding $H_0^1(\Omega) \subset L^2(\Omega)$ is compact. □

It follows from Corollary 11.1, Lemma 11.7 and (11.25) that there are functions $u$, $U$, $\eta$ and a subsequence of $\{u^{(m)}\}$ (for which we keep the same notation) such that

$$
\begin{aligned}
&u^{(m)} \to u \text{ in } C([0,T]; L^2(\Omega)) \text{ and a.e. in } Q_T,\\
&u_t^{(m)} \to u_t \text{ in } L^{\mu}(Q_T) \text{ and } * \text{-weak in } L^\infty(0,T; L^2(\Omega)),\\
&u_t^{(m)}(x,T) \to U \text{ in } L^2(\Omega),\\
&\nabla u^{(m)} \rightharpoonup \nabla u \text{ in } \left(L^{p(\cdot)}(Q_T)\right)^n \text{ and } * \text{-weak in } L^\infty(0,T; L^2(\Omega)),\\
&\nabla u_t^{(m)} \rightharpoonup \nabla u_t \text{ in } \left(L^2(Q_T)\right)^n,\\
&u_{tt}^{(m)} \rightharpoonup u_{tt} \text{ in } L^{(p^+)'}(0,T; H^{-s}(\Omega)),\\
&\mathscr{A}(z, \nabla u^{(m)}) \equiv a|\nabla u^{(m)}|^{p(x,t)-2}\nabla u^{(m)} \rightharpoonup \eta \text{ in } \left(L^{p'(\cdot)}(Q_T)\right)^n,\\
&b|u^{(m)}|^{\sigma(x,t)-2}u^{(m)} \rightharpoonup b|u|^{\sigma(x,t)-2}u \text{ in } L^{\sigma'(\cdot)}(Q_T) \text{ and a.e. in } Q_T.
\end{aligned}
\tag{11.28}
$$

We will rely on the following known result.

**Lemma 11.8** ([104]). *Let $\Omega$ be any bounded domain in $\mathbb{R}^n$, $\{\psi_k\}$ be an orthogonal basis in $L^2(\Omega)$. Then for every $\delta > 0$ there exists a positive number $N_\delta$ such that*

$$
\|u\|_{2,\Omega} \le \left(\sum_{k=1}^{N_\delta} (u, \psi_k)_\Omega^2\right)^{\frac{1}{2}} + \delta \|u\|_{W_0^{1,q}(\Omega)}
$$

*for all $u \in W_0^{1,q}(\Omega)$ with $2 \le q < \infty$.*

Applying Lemma 11.8 to the functions $u_t^{(m)} - u_t$ and using the uniform estimates (11.25) and relations (11.28) we find that for every $\delta > 0$ there is $N_\delta \ge m$ such that

$$
\|u_t^{(m)} - u_t\|_{2,Q_T}^2 \le \sum_{k=1}^{N_\delta} \int_0^T \left(u_t^{(m)} - u_t, \psi_k\right)_{2,\Omega}^2 dt + \delta \|u_t^{(m)} - u_t\|_{L^2(0,T;H_0^1(\Omega))}^2 \le C\delta.
$$

It follows that

$$
u_t^{(m)} \to u_t \text{ in } L^2(Q_T). \tag{11.29}
$$

Let us take in (11.27) $\phi_j \in \mathscr{P}_j$ with $j \le m$. Letting $m \to \infty$ and using (11.28) we arrive at the equality

$$\int_{Q_T} u_{tt}\phi \, dz = -\int_{Q_T} (\eta + \varepsilon \nabla u_t) \nabla \phi \, dz - \int_{Q_T} \left(b|u|^{\sigma-2}u + f\right) \phi dz. \quad (11.30)$$

Since $j \in \mathbb{N}$ is arbitrary, (11.30) is fulfilled for every $\phi \in \mathbf{U}(Q_T)$, which means that $u_{tt} \in \mathbf{U}'(Q_T)$.

**Lemma 11.9** *If $u \in \mathbf{U}(Q_T)$ and $u_{tt} \in \mathbf{U}'(Q_T)$, then $u_t u \in C([0, T]; L^1(\Omega))$ after possible redefining of $u_t$ on a set of zero measure in $[0, T]$.*

*Proof* Let $\{u_h\}$ be a sequence of smooth functions approximating $u$ in the norm of $\mathbf{U}(Q_T)$ and $(u_h)_{tt} \to u_{tt}$ in $\mathbf{U}'(Q_T)$. We have:

$$\begin{aligned} &\int_\Omega \left(u_{h_1}(u_{h_1})_t - u_{h_2}(u_{h_2})_t\right) dx \Big|_{\tau=t_1}^{\tau=t_2} \\ &= \int_{Q_{t_2}\setminus Q_{t_1}} \left(u_{h_1}(u_{h_1})_{tt} - u_{h_2}(u_{h_2})_{tt}\right) dz + \int_{Q_{t_2}\setminus Q_{t_1}} \left((u_{h_1})_t^2 - (u_{h_2})_t^2\right) dz := J_1 + J_2. \end{aligned}$$

Since the sequences $\{u_{t,h}\}$ and $\{u_h\}$ are fundamental in $L^2(Q_T)$ and $\mathbf{U}(Q_T)$, we may estimate

$$|J_1| \le \|u_{h_1} - u_{h_2}\|_{\mathbf{U}} \|(u_{tt})_h\|_{\mathbf{U}'} + \|u_{h_2}\|_{\mathbf{U}} \|(u_{tt})_{h_1} - (u_{tt})_{h_2}\|_{\mathbf{U}'} \to 0,$$

$$|J_2| \le \|u_{h_1,t} - u_{h_2,t}\|_{2,Q_T} \left(\|u_{h_1,t}\|_{2,Q_T} + \|u_{h_2,t}\|_{2,Q_T}\right) \to 0 \quad \text{as } h_1, h_2 \to 0.$$

It follows that $\{u_{ht}u_h\}$ is a Cauchy sequence in $C([0, T]; L^1(\Omega))$ and there is a function $\zeta \in C([0, T]; L^1(\Omega))$ such that $\|u_{ht}u_h - \zeta\|_{C([0,T];L^1(\Omega))} \to 0$ as $h \to 0$. On the other hand, by (11.28) $\|u_{ht}u_h - u_t u\|_{L^1(\Omega)}(t) \to 0$ as $h \to 0$ for a.e. $t \in (0, T)$. Since $u \in C([0, T]; L^2(\Omega))$, this means that $u_t u \in C([0, T]; L^1(\Omega))$ after possible redefining of $u_t$ on a set of zero measure in $[0, T]$. □

It follows that $u_t^{(m)}(x, T)u^{(m)}(x, T) \to Uu(x, T) \equiv u_t(x, T)u(x, T) \in C([0, T]; L^1(\Omega))$. Let us take a test-function $\phi \in \mathscr{P}_j$, $j \le m$, $\phi(x, T) = 0$, and write (11.27) in the form

$$\begin{aligned} &\int_{Q_T} \left(-u_t^{(m)}\phi_t + \left(\mathscr{A}(z, \nabla u^{(m)}) + \varepsilon \nabla u_t^{(m)}\right) \cdot \nabla \phi\right) dz \\ &\quad - \int_{Q_T} \left(b|u^{(m)}|^{\sigma-2}u^{(m)} + f\right) \phi dz + \int_\Omega u_t^{(m)} \phi \, dx \Big|_{t=0} = 0. \end{aligned} \quad (11.31)$$

By virtue of (11.28) and (11.29) for every $\phi \in \mathscr{P}_j$ we may pass in (11.31) to the limit as $m \to \infty$:

$$\int_{Q_T} \left( -u_t \phi_t + (\eta + \varepsilon \nabla u_t) \nabla \phi - b\,|u|^{\sigma-2} u\phi + f\phi \right) dz \\ - \int_{\Omega} u_1 \phi(x, 0)\, dx = 0. \tag{11.32}$$

Let us define the function

$$\zeta_{k,\tau}(t) = \begin{cases} 1 & \text{if } t \in [0, \tau - \frac{1}{k}], \\ k(\tau - t) & \text{if } t \in [\tau - \frac{1}{k}, \tau], \\ 0 & \text{if } t \geq \tau, \end{cases} \quad k \in \mathbb{N}, \quad \tau \in (0, T]. \tag{11.33}$$

For $\phi = u^{(m)} \zeta_{k,T}(t)$ equality (11.31) transforms into

$$k \int_{T-\frac{1}{k}}^{T} \int_{\Omega} u^{(m)} u_t^{(m)}\, dz \\ + \int_{Q_T} \zeta_{k,T}(t) \left( -(u_t^{(m)})^2 + \mathscr{A}(z, \nabla u^{(m)}) \cdot \nabla u^{(m)} + \varepsilon \nabla u_t^{(m)} \cdot \nabla u^{(m)} \right) dz \\ - \int_{Q_T} \zeta_{k,T}(t) \left( b|u^{(m)}|^{\sigma} + f u^{(m)} \right) dz - \int_{\Omega} u_t^{(m)} u^{(m)}\, dx \Big|_{t=0} = 0.$$

Since $u^{(m)} u_t^{(m)} \in C^0([0, T]; L^1(\Omega))$ by Lemma 11.9, it follows from the Lebesgue differentiation theorem that in the last equality one may pass to the limit as $k \to \infty$:

$$\int_{Q_T} \left( -(u_t^{(m)})^2 + \mathscr{A}(z, \nabla u^{(m)}) \cdot \nabla u^{(m)} + \varepsilon \nabla u_t^{(m)} \cdot \nabla u^{(m)} \right) dz \\ - \int_{Q_T} \left( b|u^{(m)}|^{\sigma} + f u^{(m)} \right) dz + \int_{\Omega} u_t^{(m)} u^{(m)}\, dx \Big|_{t=0}^{t=T} = 0. \tag{11.34}$$

Choosing a sequence $\phi_j \to u$ in $\mathbf{U}(Q_T)$, letting $j \to \infty$ and using (11.28) and Lemma 11.9, from (11.32) we obtain the equality

$$\int_{Q_T} \left( -u_t^2 + \eta \cdot \nabla u + \varepsilon \nabla u_t \cdot \nabla u - b|u|^{\sigma} - f u \right) dz \\ + \int_{\Omega} u_t(x, T) u(x, T)\, dx - \int_{\Omega} u_1 u_0\, dx = 0. \tag{11.35}$$

To prove that $\eta = \mathscr{A}(z, \nabla u)$ a.e. in $Q_T$ we use monotonicity of the operator $\mathscr{A}$. Let $\phi \in \mathscr{P}_m$. Following the proof of Lemma 4.15 we find that

$$\mathscr{A}(z, \nabla u^{(m)}) \cdot \nabla u^{(m)} \geq \mathscr{A}(z, \nabla u^{(m)}) \cdot \nabla \phi + \mathscr{A}(z, \nabla \phi) \cdot \nabla (u^{(m)} - \phi).$$

Substituting this inequality into (11.34), passing to the limit as $m \to \infty$ and then plugging (11.35) we obtain the inequality

$$\begin{aligned} 0 &\geq \int_{Q_T} \left( -u_t^2 - b|u|^\sigma - fu + \varepsilon(\nabla u, \nabla u_t) + \eta \cdot \nabla \phi + \mathscr{A}(z, \nabla \phi) \cdot \nabla (u - \phi) \right) dz \\ &\quad + \int_\Omega u_t(x,T) u(x,T)\, dx - \int_\Omega u_1 u_0 \, dx \\ &= \int_{Q_T} (\mathscr{A}(z, \nabla \phi) - \eta) \cdot \nabla (u - \phi)\, dz. \end{aligned}$$

The conclusion immediately follows as in the proof of Lemma 4.15.

To complete the proofs of Theorems 11.1 and 11.2 we take an arbitrary test-function $\phi \in \mathbf{U}(Q_T) \cap C([0,T]; L^2(\Omega))$ such that $\phi(x,T) = 0$ and a sequence $\mathscr{P}_j \ni \phi_j \to \phi$ in $\mathbf{U}(Q_T)$. Equality (11.32) with $\phi = \phi_j$ gives (11.4) as $j \to \infty$. Item (i) Definition 11.1 is fulfilled by construction [see (11.28)]. The first initial condition (ii) follows from the inclusion $u \in C([0,T]; L^2(\Omega))$ and the estimate $\varepsilon \|\nabla u_t\|^2_{2,Q_T} \leq C$ with $\varepsilon > 0$.

**Lemma 11.10** *For every $\eta(x) \in C_0^\infty(\Omega)$ there exists a sequence $t_k \to 0$ such that*

$$(\eta, u_t(x, t_k) - u_1(x))_{2,\Omega} \to 0 \, as \, k \to \infty.$$

*Proof* Let us take for the test-function in (11.32) $\phi = \eta(x)\zeta_{k,\tau}(t)$ with $\eta \in C_0^\infty(\Omega)$. We have

$$\begin{aligned} k \int_{\tau - \frac{1}{k}}^{\tau} \int_\Omega u_t \eta(x)\, dx + \int_{Q_T} \zeta_{k,\tau} \left( (\mathscr{A}(z, \nabla u) + \varepsilon \nabla u_t) \nabla \eta - b\, |u|^{\sigma - 2} u \eta + f \eta \right) dz \\ - \int_\Omega u_1 \eta \, dx = 0. \end{aligned}$$

Since $u_t \in L^2(Q_T)$, by the Lebesgue differentiation theorem for a.e. $\tau \in (0,T)$ there exists

$$\lim_{k \to \infty} k \int_{\tau - \frac{1}{k}}^{\tau} \int_\Omega u_t \eta(x)\, dx = \int_\Omega u_t(x, \tau) \eta(x)\, dx.$$

Passing to the limit as $k \to \infty$ we obtain the inequality: for a.e. $\tau \in (0, T)$

$$\begin{aligned}
&\left| \int_{\Omega} (u_t(x,\tau) - u_1(x))\eta(x)\,dx \right| \\
&\quad \leq \left| \int_{Q_\tau} \left( (\mathscr{A}(z, \nabla u) + \varepsilon \nabla u_t) \nabla \eta - b\,|u|^{\sigma-2} u\eta + f\eta \right) dz \right| \\
&\quad \leq C\|\eta\|_{\mathbf{W}(Q_\tau)} \left( \|u_t\|_{L^2(0,\tau;H_0^1(\Omega))} + \|f\|_{2,Q_\tau} \right) \to 0 \quad \text{as}\, \tau \to 0. \qquad (11.36)
\end{aligned}$$

□

*Remark 11.4* Notice that by virtue of (11.25)

$$\varepsilon \int_{Q_\tau} \nabla u \cdot \nabla \eta\,dz \leq \sqrt{\varepsilon} \left( \varepsilon \|\nabla u\|_{2,Q_\tau}^2 \right)^{\frac{1}{2}} \|\nabla \eta\|_{2,Q_\tau} \leq \sqrt{\varepsilon K}\|\nabla \eta\|_{2,Q_\tau}$$

which means that the right-hand side of (11.36) tends to zero as $\tau \to 0$ uniformly with respect to $\varepsilon$.

**Lemma 11.11** (The energy identity) *The weak solutions constructed in* Theorems 11.1 and 11.2 *satisfy the energy identity: for every* $t \in (0, T_{\max})$

$$\int_{\Omega} u_t u\,dx \Big|_{\tau=0}^{\tau=t} + \int_{Q_t} \left( a|\nabla u|^p + \varepsilon \nabla u \cdot \nabla u_t + b|u|^\sigma - fu \right) dz = 0. \qquad (11.37)$$

*Proof* Equality (11.37) follows from the proof of Lemma 11.10 with the test-function $\phi = u(x,t)\zeta_{k,\tau}(t)$. □

## 11.2 Nonexistence of Global Solutions: Finite Time Blow up

We consider first problem (11.1)–(11.3) with $\varepsilon > 0$.

**Theorem 11.3** *Let u be a weak energy solution of problem* (11.1)–(11.3). *Assume that the conditions of* Lemma 11.1 *are fulfilled and that the data satisfy the conditions*

$$E(0) \leq 0, \quad 0 < (u_0, u_1)_{2,\Omega}, \quad 2 \leq p^- \leq p^+ < \lambda < \sigma^-. \qquad (11.38)$$

*Then there exists* $T_{\max} < \infty$ *such that*

$$\Phi(t) = \|u(t)\|_{2,\Omega}^2 + \varepsilon \int_0^t \int_{\Omega} |\nabla u|^2\,dxds \to \infty \quad \text{as}\, t \to T_{\max}.$$

*Proof* It is straightforward to check that

$$\Phi' = 2\,(u, u_t)_{2,\Omega} + \varepsilon \int_{\Omega} |\nabla u|^2\,dx,$$
$$\Phi'' = 2\,\|u_t\|_{2,\Omega}^2 + 2\int_{\Omega} \left(-a\,|\nabla u|^p + b\,|u|^{\sigma}\right) dx.$$

Using inequality (11.17) of Lemma 11.1, we calculate:

$$\begin{aligned}\Phi'' &\geq 2\,\|u_t(t)\|_{2,\Omega}^2 + 2\int_{\Omega} \left(-a\,|\nabla u|^p + b\,|u|^{\sigma}\right) dx \\ &\quad + 2\lambda\left(E(t) + \varepsilon \int_0^t \int_{\Omega} |\nabla u_t|^2\,dxds\right) \\ &= (2+\lambda)\,\|u_t(t)\|^2 + 2\int_{\Omega}\left(\left(\frac{\lambda}{p} - 1\right) a\,|\nabla u|^p + b\left(1 - \frac{\lambda}{\sigma}\right)|u|^{\sigma}\right) dx \\ &\quad + 2\lambda\varepsilon \int_0^t \int_{\Omega} |\nabla u_t|^2\,dxds > 0\end{aligned} \tag{11.39}$$

for some $\lambda > 2$ and $p^+ < \lambda < \sigma^-$. It follows that

$$\Phi'(t) > 0 \quad \text{if}\, \Phi'(0) \geq 2(u_0, u_1)_{2,\Omega} > 0.$$

Since $0 < \Phi(t), 0 < \Phi'(t),\ 0 < \Phi''(t)$, it is necessary that there exists some $T_{\max}$ such that

$$\Phi(t) \to \infty \,\text{as}\, t \to T_{\max}.$$

Let us check that $T_{\max} < \infty$. By virtue of Lemma 1.5 and (1.7), for a.e. $t \in (0, T)$

$$\begin{aligned}\|u(\cdot,t)\|_{2,\Omega} &\leq C\,\|u(\cdot,t)\|_{\sigma(\cdot,t),\Omega} \\ &\leq C \max\left\{\left(\int_{\Omega} |u|^{\sigma}\,dx\right)^{\frac{1}{\sigma^-}}, \left(\int_{\Omega} |u|^{\sigma}\,dx\right)^{\frac{1}{\sigma^+}}\right\},\end{aligned} \tag{11.40}$$

$$\begin{aligned}\|\nabla u(\cdot,t)\|_{2,\Omega} &\leq C\,\|\nabla u(\cdot,t)\|_{p(\cdot),\Omega} \\ &\leq C \max\left\{\left(\int_{\Omega} |\nabla u|^{p}\,dx\right)^{\frac{1}{p^-}}, \left(\int_{\Omega} |\nabla u|^{p}\,dx\right)^{\frac{1}{p^+}}\right\}.\end{aligned} \tag{11.41}$$

Due to (11.39) there is a finite constant $C$ such that

$$\|u_t\|_{2,\Omega}^2 + \int_{\Omega} |\nabla u(\cdot,t)|^{p(\cdot,t)}\,dx + \int_{\Omega} |u(\cdot,t)|^{\sigma(\cdot,t)}\,dx \leq C\Phi'',$$

which allows one to rewrite (11.40) and (11.41) in the form

$$\|u(\cdot,t)\|_{2,\Omega} \leq C \max\left\{(\Phi'')^{\frac{1}{\sigma^-}}, (\Phi'')^{\frac{1}{\sigma^+}}\right\},$$

$$\|\nabla u(\cdot,t)\|_{2,\Omega} \leq C \max\left\{(\Phi'')^{\frac{1}{p^-}}, (\Phi'')^{\frac{1}{p^+}}\right\}.$$

It follows that for every $\varepsilon \in [0, 1]$

$$\begin{aligned} 0 \leq \Phi' &= 2\int_\Omega u\, u_t dx + \frac{\varepsilon}{2}\int_\Omega \|\nabla u\|^2 dx \\ &\leq 2\,\|u(\cdot,t)\|_{2,\Omega}\,\|u_t(\cdot,t)\|_{2,\Omega} + \frac{1}{2}\int_\Omega |\nabla u|^2 dx \\ &\leq C\left(\max\left\{(\Phi'')^{\frac{1}{\sigma^-}+\frac{1}{2}}, (\Phi'')^{\frac{1}{\sigma^+}+\frac{1}{2}}\right\} + \max\left\{(\Phi'')^{\frac{2}{p^-}}, (\Phi'')^{\frac{2}{p^+}}\right\}\right). \end{aligned}$$

Since $0 < \Phi''(t)$, without loss of generality we may assume that $\Phi' \geq 1$. In this case

$$\begin{aligned} &C\,(\Phi')^\mu \leq \Phi'', \\ &\frac{1}{\mu} = \max\left\{\frac{1}{\sigma^-} + \frac{1}{2}, \frac{2}{p^-}\right\} < 1 \quad \text{if } \sigma^- > 2 \text{ and } p^- > 2. \end{aligned} \tag{11.42}$$

The constants $C$ that appear in the estimates between (11.40) and (11.42) depend only on $|\Omega|$, $p^\pm$ and $\sigma^\pm$. Integration of inequality (11.42) leads to the estimate

$$\Phi'(t) \geq \Phi'(0)\left(1 - \frac{t(\mu-1)}{C}\,(\Phi'(0))^{\mu-1}\right)^{-\frac{1}{\mu-1}} \to \infty \tag{11.43}$$

as

$$t \to T_{\max} = \frac{C}{\mu-1}\,(\Phi'(0))^{-\mu+1} < \infty. \tag{11.44}$$

Integrating inequality (11.43) we obtain

$$\begin{aligned} \Phi(t) &\geq \Phi'(0)\int_0^t \left(1 - \frac{s}{T_{\max}}\right)^{-\frac{1}{\mu-1}} ds + \Phi(0) \\ &= \Phi'(0) T_{\max}\frac{\mu-1}{2-\mu}\left[\left(1 - \frac{t}{T_{\max}}\right)^{-\frac{2-\mu}{\mu-1}} - 1\right] + \Phi(0) \end{aligned} \tag{11.45}$$

with $1 < \mu < 2$. It follows that $\Phi(t) \to \infty$ as $t \to T_{\max}$. □

Let us assume now that the exponents $p, \sigma$ are weakly dependent on $t$, that is, the constants $C_p, C_\sigma$ in (11.5) are small. The proof of the finite time blow up needs no changes, provided that

$$E(t) + \varepsilon \int_0^t \int_\Omega |\nabla u_t|^2 \, dxds \le 0, \qquad 0 \le t \le T_{\max} \tag{11.46}$$

with $T_{\max}$ already defined in Theorem 11.3. According to Lemma 11.2

$$E(t) + \varepsilon \int_0^t \int_\Omega |\nabla u_t|^2 \, dxds \le E(0) + T_{\max} \mathrm{e}(a_+ C_p + b_+ C_\sigma)|\Omega|,$$

and inequality (11.46) follows if we assume that

$$\delta = \max\left\{C_p, C_\sigma\right\} \le \frac{|E(0)|}{T_{\max} \mathrm{e}(a_+ + b_+)|\Omega|}, \qquad E(0) < 0. \tag{11.47}$$

These arguments are summarized in the following theorem.

**Theorem 11.4** *Let $u$ be a weak energy solution of problem* (11.1)–(11.3). *Assume that the conditions of* Lemma 11.2 *and* (11.47) *are fulfilled with $T_{\max}$ defined in* Theorem 11.3. *If*

$$E(0) < 0, \qquad 0 < (u_0, u_1)_{2,\Omega}, \qquad 2 \le p^- \le p^+ < \lambda < \sigma^-,$$

*then the solution $u(x,t)$ blows up in a finite time, that is, $\Phi(t)$ becomes unbounded within the interval* $(0, T_{\max})$.

**Corollary 11.2** *Under the conditions of* Theorem 11.4 *the blow-up moment $T$ can be estimated independently of $\varepsilon$.*

*Proof* The constant $C$ in formulas (11.42)–(11.44) doesn't depend on $\varepsilon$, while

$$\Phi'(0) = 2(u_0, u_1)_{2,\Omega} + \varepsilon \int_\Omega |\nabla u_0|^2 \, dx, \tag{11.48}$$

and

$$T_{\max} = \frac{C}{\mu - 1} \left(2(u_0, u_1)_{2,\Omega} + \varepsilon \int_\Omega |\nabla u_0|^2 \, dx\right)^{-\mu+1} \tag{11.49}$$

depend on $\varepsilon$. Using the independent of $\varepsilon$ estimates

$$0 < 2(u_0, u_1)_{2,\Omega} \le \Phi'(0) \le 2(u_0, u_1)_{2,\Omega} + \int_\Omega |\nabla u_0|^2 \, dx, \tag{11.50}$$

$$\frac{C}{\mu-1}\left(2(u_0,u_1)_{2,\Omega}+\int_\Omega |\nabla u_0|^2\,dx\right)^{-\mu+1} \\ \le T_{\max} \le T := \frac{2^{1-\mu}C}{\mu-1}(u_0,u_1)_{2,\Omega}^{1-\mu}, \qquad 1<\mu<2, \tag{11.51}$$

we rewrite inequality (11.45) in the form

$$\Phi(t) \ge \frac{2^{2-\mu}C}{2-\mu}(u_0,u_1)_{2,\Omega}^{2-\mu}\left(\left(1-\frac{t}{T}\right)^{-\frac{2-\mu}{\mu-1}}-1\right)+\|u_0\|_{2,\Omega}^2. \tag{11.52}$$

It follows that $\Phi(t)\to\infty$ as $t\to T$, where $T$ is defined in (11.51). □

**Corollary 11.3** *In the foregoing conditions,* $\|u(\cdot,t)\|_{\sigma(\cdot,t)}\to\infty$ *as* $t\to T-$.

*Proof* Using (11.14) and (11.17) we find that

$$\int_\Omega\left(\frac{1}{2}u_t^2+\frac{a}{p}|\nabla u|^p\right)dx+\varepsilon\int_0^t\int_\Omega|\nabla u_t|^2\,dxds \le \int_\Omega\frac{b}{\sigma}|u|^\sigma\,dx. \tag{11.53}$$

It follows that for $p\ge 2$, $\sigma\ge 2$ and $0\le\varepsilon\le 1$

$$\Phi(t) = 2\int_\Omega uu_t dx+\frac{\varepsilon}{2}\int_\Omega|\nabla u|^2\,dx \\ \le 2\|u\|_{2,\Omega}\|u_t\|_{2,\Omega}+\varepsilon\int_\Omega|\nabla u|^2\,dx = J_1+J_2.$$

By virtue of (11.53) and (1.6)

$$J_1 \le C\|u\|_{\sigma,\Omega}\left(\int_\Omega|u|^\sigma\,dx\right)^{\frac{1}{2}} \\ \le C\max\left\{\left(\int_\Omega|u|^\sigma\,dx\right)^{\frac{1}{\sigma^+}+\frac{1}{2}},\left(\int_\Omega|u|^\sigma\,dx\right)^{\frac{1}{\sigma^-}+\frac{1}{2}}\right\},$$

$$J_2 \le C\max\left\{\left(\int_\Omega|\nabla u|^p\,dx\right)^{\frac{2}{p^+}},\left(\int_\Omega|\nabla u|^p\,dx\right)^{\frac{2}{p^-}}\right\} \\ \le C\max\left\{\left(\int_\Omega|u|^\sigma\,dx\right)^{\frac{2}{p^+}},\left(\int_\Omega|u|^\sigma\,dx\right)^{\frac{2}{p^-}}\right\}$$

with independent of $\varepsilon$ constants $C$. Gathering these estimates we obtain:

$$\Phi(t) \le \Psi\left(\int_\Omega |u|^\sigma \, dx\right),$$
$$\Psi(s) = \max\left\{s^{\frac{2}{p^+}}, s^{\frac{2}{p^-}}, s^{\frac{1}{\sigma^-}+\frac{1}{2}}, s^{\frac{1}{\sigma^+}+\frac{1}{2}}\right\}, \quad s \ge 0.$$

Plugging this inequality into (11.52) we find that

$$\Psi\left(\int_\Omega |u|^\sigma \, dx\right) \ge \frac{2^{2-\mu}C}{2-\mu}(u_0, u_1)_{2,\Omega}^{2-\mu}\left(\left(1-\frac{t}{T}\right)^{-\frac{2-\mu}{\mu-1}} - 1\right) + \|u_0\|_{2,\Omega}^2 \quad (11.54)$$

with independent of $\varepsilon$ constants $C$, $\mu$ $T$. □

## 11.3 Young Measure Solutions

We consider the initial and boundary value problem

$$u_{tt} = \mathscr{L}u + \operatorname{div}(\varepsilon \nabla u_t) + f(x, t), \quad (x, t) \in Q_T, \quad (11.55)$$

$$\mathscr{L}u = \operatorname{div}\left(a(x, t)\, |\nabla u|^{p(x,t)-2}\, \nabla u\right) + b(x, t)\, |u|^{\sigma(x,t)-2}\, u,$$

$$u(x, 0) = u_0(x), \; u_t(x, 0) = u_1(x), \; x \in \Omega, \quad (11.56)$$

$$u|_{\Gamma_T} = 0, \; \Gamma_T = \partial\Omega \times (0, T) \quad (11.57)$$

with a constant parameter $\varepsilon > 0$ (which we will send to zero).

**Definition 11.2** (*YM-solutions*) Let $S(x, A) = |A|^{p(x,t)-2}A$. A pair $(u, \nu)$ is a Young measure solution of problem (11.55)–(11.57) with $\varepsilon = 0$ if for fixed $T > 0$

(i) $u \in W^{1,\infty}((0, T); L^2(\Omega)) \cap L^\infty((0, T); W_0^{1,p(\cdot)}(\Omega)) \cap L^\infty((0, T); L^{\sigma(\cdot)}(\Omega))$,

(ii) $\forall \phi \in C_0^\infty(\Omega)$

$$\begin{aligned} &\text{(a)} \quad \int_\Omega (u(x, t) - u_0(x))\phi(x)\, dx \to 0, \\ &\text{(b)} \quad \exists\, \{t_k\} : \quad \int_\Omega (u_t(x, t_k) - u_1(x))\phi(x)\, dx \to 0 \text{ as } t_k \to 0, \end{aligned} \quad (11.58)$$

(iii) $\forall \phi \in C^\infty((0, T); C_0^\infty(\Omega)), \phi(x, T) = 0$,

$$\int_{Q_T} \left(u_t\phi_t + a\langle \nu, S(x,\cdot)\rangle\nabla\phi - b\,u|^{\sigma(\cdot)-2}u\phi\right) dxdt$$
$$= \int_{\Omega} u_1\phi(\cdot,0)dx + \int_{Q_T} f\phi dx, \tag{11.59}$$

where $\nu = \nu_{x,t}$ is a probability measure for a.e. $x \in \Omega, t \in (0,T)$, and $\langle \nu, S(\cdot)\rangle$ is defined as a dual pairing of $S$ with the measure $\nu$, i.e.,

$$\langle \nu, S(\cdot)\rangle := \int S(A)d\nu(A).$$

Let $\varepsilon > 0$ be an arbitrary fixed number. If the data of problem (11.55)–(11.57) meet the conditions of Theorem 11.1, then there exists a global in time energy solution $u \in \mathbf{U}(Q_T)$. Under the conditions of Theorem 11.2 the energy solution of problem (11.55)–(11.57) exists locally in time. The initial condition (11.58)(a) immediately follows from (11.25): since $\|u_t\|_{2,Q_T} \leq K$ uniformly with respect to $\varepsilon$, then

$$\left|\int_{\Omega} \eta(x)(u(x,t) - u_0(x))\,dx\right| \leq \int_{Q_t} |\eta||v_t|\,dz \leq \sqrt{t}K\,\|\eta\|_{2,\Omega}.$$

The initial condition (11.58)(b) is fulfilled due to Lemma 11.10 and Remark 11.4. Under the conditions of Theorems 11.1 and 11.2 the energy solutions belong to $\mathbf{U}(Q_T) \cap C([0,T]; L^2(\Omega))$, but possess better regularity so that the conditions of item (i) of Definition 11.2 are also satisfied—see Remark 11.1.

## *11.3.1 Passage to the Limit as $\varepsilon \to 0$*

**Theorem 11.5** *Problem* (11.55)–(11.57) *with* $\varepsilon = 0$ *admits a YM-solution in the sense of* Definition 11.2. *The solution is global in time under the conditions of* Lemma 11.3 *or* Lemma 11.4, *and local in time under the conditions of* Lemma 11.5.

*Proof* Let us make the usual agreement: if a sequence contains a convergent subsequence, we assume that they coincide and use the same notation for both of them.

Let $\{u^\varepsilon\}$ be the family of solutions of problem (11.55)–(11.57) with $\varepsilon > 0$. We follow the proof in [210, Theorem 3.1]. The first step is to check that the Young measure $\nu_{x,t}$ generated by the sequence $\nabla u^\varepsilon$ is a probability measure for a.e. $x \in \Omega$, $t \in (0,T)$. This is true if, for instance,

$$\int_{\Omega} |\nabla u^\varepsilon|^q dx \leq C. \tag{11.60}$$

Estimate (11.60) holds for every $q < p_-$: by Young's inequality and (11.21) we have

$$\int_\Omega |\nabla u^\varepsilon|^q dx \le |\Omega| + \int_\Omega |\nabla u^\varepsilon|^{p(\cdot,t)} dx \le C.$$

The next step is to prove that the Young measure $\nu$ determines weak $L^1$ limits of the composite functions $S(x, t, \nabla u^\varepsilon)$, with $S \in C(\mathbb{R}^n \times \mathbb{R}^n)$, according to the rule

$$S(x, t, \nabla u^\varepsilon) \rightharpoonup \langle \nu, S(x, t, \cdot)\rangle \quad \text{in } L^1(\Omega). \tag{11.61}$$

Here we bear in mind the function $S(x, t, A) = |A|^{p(x,t)-2}A$, that appears in problem (11.55)–(11.57). However, property (11.61) holds only for those functions $S$ for which $S(x, t, \nabla u^\varepsilon)$ is weakly relatively compact in $L^1(\Omega)$, which makes it necessary to verify that it is true also for $S(x, t, \nabla u^\varepsilon) = |\nabla u^\varepsilon|^{p(x,t)-2}\nabla u^\varepsilon$. It is clear that $S(x, t, \nabla u^\varepsilon)$ will be weakly relatively compact in $L^1(\Omega)$ for each $t$ if we prove that

$$S(x, t, \nabla u^\varepsilon) \in L^\infty(0, T; L^{p'}(\Omega)).$$

By the energy estimate (11.21) we have

$$\begin{aligned}\sup_{t\in(0,T)} \int_\Omega |S(x, \nabla u^\varepsilon)|^{p'} dx &= \sup_{t\in(0,T)} \int_\Omega (|\nabla u^\varepsilon|^{p-1})^{p'} dx \\ &= \sup_{t\in(0,T)} \int_\Omega |\nabla u^\varepsilon|^p dx \le C.\end{aligned}$$

Thus, $S \in L^\infty(0, T; L^{p'(x,t)}(\Omega))$, $S$ is weakly relatively compact in $L^1(\Omega)$ for each $t$, and the convergence property (11.61) holds. Moreover, (11.21) and (11.25) allow one conclude that there exist a function $u$ and a subsequence $\{u^{\varepsilon_k}\}$ (we assume that it coincides with $\{u^\varepsilon\}$) such that for any $T < T_{\max}$

$$\begin{aligned} u^\varepsilon &\to u \text{ strongly in } L^{\sigma(\cdot,\cdot)-1}(Q_T), \\ u_t^\varepsilon &\rightharpoonup u_t \ *\text{-weak in } L^\infty(0, T; L^2(\Omega)), \\ \nabla u^\varepsilon &\rightharpoonup \nabla u \ *\text{-weak in } L^\infty(0, T; L^{p(\cdot,\cdot)}(\Omega)).\end{aligned}$$

There exist the limits as $\varepsilon \to 0$:

$$\varepsilon \left| \int_{Q_T} \nabla u_t^\varepsilon \nabla \phi \, dz \right| \le \sqrt{\varepsilon}\left(\varepsilon \|\nabla u^\varepsilon\|_{2,Q_T}^2\right)^{\frac{1}{2}} \|\nabla \phi\|_{2,Q_T} \le \sqrt{\varepsilon} C \to 0,$$

$$\int_{Q_T} b|u^\varepsilon|^{\sigma-2} u^\varepsilon \phi \, dz \to \int_{Q_T} b|u|^{\sigma-2} u \phi \, dz,$$

$$\int_{Q_T} a\left|\nabla u^{\varepsilon}\right|^{p-2}\nabla u^{\varepsilon}\nabla\phi\,dz \to \int_{Q_T} a\langle\nu, S(x,t,\cdot)\rangle\nabla\phi\,dz.$$

Plugging these limit relations into the integral identity (11.4) we arrive at the identity

$$\begin{aligned}\int_{Q_T}&\left(u_t\phi_t + a\langle\nu, S(x,t,\cdot)\rangle\nabla\phi - b|u|^{\sigma(\cdot)-2}u\phi\right)dz\\ &= \int_{\Omega} u_1\phi(\cdot,0)dx + \int_{Q_T} f\phi\,dx,\end{aligned} \tag{11.62}$$

which proves that a pair $(u,\nu)$ is a YM-solution of the problem. □

*Remark 11.5* In the proof of Theorem 11.5 condition (1.31) of logarithmic continuity of $p(x,t)$ was never used explicitly, whereas it is crucial for the proofs of the existence Theorems 11.1 and 11.2. This observation allows one to interpret Theorem 11.5 as the conditional result: the assertions of Theorem 11.5 hold true if for every $\varepsilon > 0$ problem (11.1)–(11.3) admits a solution $u^{\varepsilon}$ in the sense of Definition (11.1) and this solution satisfies estimates (11.21), (11.25).

### 11.3.2 Lower Estimates for $\|u^{\varepsilon}\|_{\sigma,\Omega}$ Blow up

We are now in position to show that for the sequence of Young measure solutions $\{u^{\varepsilon}\}$ of problem (11.55)–(11.57) the norms $\|u^{\varepsilon}\|_{\sigma,\Omega}$ can be estimated from below uniformly with respect to $\varepsilon$. The same estimate holds true for the limit solution with $\varepsilon = 0$.

**Lemma 11.12** *Let* $\{u^{\varepsilon}\}$ *be a sequence of solutions of problem* (11.55)–(11.57). *Let the conditions of the* Lemma 11.1 *be fulfilled and*

$$2 < p^- \le p^+ < \sigma^-, \qquad E(0) \le 0, \qquad (u_0, u_1)_{2,\Omega} > 0. \tag{11.63}$$

*Then*

$$\begin{aligned}\Psi&\left(\int_{\Omega}|u^{\varepsilon}|^{\sigma}\,dx\right)\\ &\ge \frac{2^{2-\mu}C}{2-\mu}(u_0,u_1)_{2,\Omega}^{2-\mu}\left(\left(1-\frac{t}{T}\right)^{-\frac{2-\mu}{\mu-1}} - 1\right) + \|u_0\|_{2,\Omega}^2 \to \infty \quad \text{as } t \to T\end{aligned} \tag{11.64}$$

*with*

$$\Psi(s) = \max\left\{s^{\frac{2}{p^+}}, s^{\frac{2}{p^-}}, s^{\frac{1}{\sigma^-}+\frac{1}{2}}, s^{\frac{1}{\sigma^+}+\frac{1}{2}}\right\}, \quad s \ge 0,$$

*T from* (11.51) *and independent of* $\varepsilon$ *constants* $C > 0$, $1 < \mu < 2$.

*Proof* Estimate (11.64) is an immediate byproduct of (11.54). □

In the conditions of Lemma 11.12 $p_t = \sigma_t = 0$. Let us now relax these hypotheses and assume that the exponents $p, \sigma$ are weakly dependent on $t$, that is, the constants $C_p$, $C_\sigma$ in conditions (11.5) are small. In this case the existence of blow up follows by the same arguments, provided that

$$E(t) + \varepsilon \int_0^t \int_\Omega |\nabla u_t|^2 \, dxds \leq 0 \quad \text{for } t \in [0, T] \tag{11.65}$$

with $T$ defined in (11.51). By Lemma 11.2

$$E(t) + \varepsilon \int_0^t \int_\Omega |\nabla u_t|^2 \, dxds \leq E(0) + T\left(a_+ \frac{C_p}{(p^-)^2} + b_+ \frac{C_\sigma}{(\sigma^-)^2}\right) |\Omega|.$$

Inequality (11.65) follows if we claim that $E(0) < 0$ and

$$\max\left\{C_p, C_\sigma\right\} \leq |E(0)| T \left(a_+ \frac{C_p}{(p^-)^2} + b_+ \frac{C_\sigma}{(\sigma^-)^2}\right) |\Omega|. \tag{11.66}$$

**Lemma 11.13** *Let the conditions of* Lemma 11.2 *be fulfilled. Assume that*

$$E(0) < 0, \qquad 0 < (u_0, u_1)_{2,\Omega}, \qquad 2 \leq p^- \leq p^+ < \lambda < \sigma^-,$$

*and that the data of problem* (11.55)–(11.57) *satisfy inequality* (11.66) *with T defined in* (11.51). *Then for the sequence* $\{u^\varepsilon\}$ *of solutions of problem* (11.55)–(11.57) *estimate* (11.64) *holds, which yields*

$$\int_\Omega |u^\varepsilon(\cdot, t)|^{\sigma(\cdot, t)} \, dx \to \infty \quad \text{as} \quad t \to T \text{ uniformly with respect to } \varepsilon.$$

# Chapter 12
# Semilinear Hyperbolic Equations

In this chapter we study the Dirichlet problem for the semilinear hyperbolic equation

$$\begin{aligned} u_{tt} &= \Delta u + b(x,t)|u|^{\sigma(x,t)-2}u + f(x,t) \quad \text{in } Q_T, \\ u(x,0) &= u_0(x), \quad u_t(x,0) = u_1(x) \text{ in } \Omega, \\ u &= 0 \text{ on } \Gamma_T \end{aligned}$$

with a bounded exponent $b(x,t)$ and a variable exponent $\sigma(x,t)$ satisfying the conditions

$$1 < \sigma^-(t) \le \sigma(x,t) \le \sigma^+(t) < \infty.$$

We address the following questions: (a) conditions of existence and uniqueness of weak and strong solutions and (b) conditions of nonexistence of global in time nonnegative strong solutions. The latter is proved by the Kaplan method of eigenfunctions: we show that every nonnegative strong solution become infinite in a finite time. The nonexistence result is extended then to the limit situations where $b(x,t)$ may vanish and $\sigma(x,t) \searrow 2$ as $t \to \infty$, thus transforming the equation into the linear wave equation. Another generalization consists in extending these conclusions to the nonnegative solutions of general semilinear equations

$$u_{tt} = Lu + f(x,t,u),$$

where $L$ is the linear elliptic operator

$$Lu = \sum_{i=1}^{n} D_i \left(a_{ij}(x)D_j u + a_i(x)u\right) + a_0(x)u$$



S. Antontsev and S. Shmarev, *Evolution PDEs with Nonstandard Growth Conditions*, Atlantis Studies in Differential Equations 4,
DOI 10.2991/978-94-6239-112-3_12

and the right-hand side is of the form

$$f(x,t,u) = \sum_{k=1}^{N} b_k(x,t)u^{\sigma_k(x,t)-1} + \sum_{i=N}^{K} c_i(x,t)\int_\Omega d_i(s,t)u^{\sigma_i(s,t)-1}\,ds$$

with nonnegative coefficients $b_k \geq 0$, $c_i \geq 0$, $d_i \geq 0$.

## 12.1 Weak and Strong Solutions

Let us consider the problem

$$\begin{cases} u_{tt} = \Delta u + b(x,t)|u|^{\sigma(x,t)-2}u + f(x,t) \quad \text{in } Q_T = \Omega \times (0,T), \\ u(x,0) = u_0(x), \\ u_t(x,0) = u_1(x) \quad \text{in } \Omega, \\ u = 0 \text{ on } \Gamma_T = \partial\Omega \times (0,T]. \end{cases} \tag{12.1}$$

It is assumed that

$$\begin{aligned} & b(x,t), \sigma(x,t) \in C^0(\overline{Q}_T), \\ & \text{there exist finite constants } b^{\pm}, b^+ > 0, \text{ and } \sigma^{\pm} \text{ such that} \\ & b^- \leq b(x,t) \leq b^+, \qquad 1 < \sigma^- \leq \sigma(x,t) \leq \sigma^+ \text{ in } \overline{Q}_T \end{aligned} \tag{12.2}$$

[we use the notation introduced in (1.1)].

**Definition 12.1** A function $u : Q_T \mapsto \mathbb{R}$ is called weak solution of problem (12.1) if:

(i) $|\nabla u|,\ u_t,\ |u|^{\sigma(x,t)-1} \in L^\infty\big(0,T;L^2(\Omega)\big)$,
(ii) $u(x,t) \to u_0(x)$ in $L^2(\Omega)$ as $t \to 0$,

$$\forall \phi \in C_0^\infty(\Omega)\ \exists\{t_k\} :\ t_k \to 0 \text{ and } \int_\Omega (u_t(x,t_k) - u_1(x))\phi(x)dx \to 0 \quad \text{as } k \to \infty,$$

(iii) $\forall \phi \in C^\infty\big(0,T;C_0^\infty(\Omega)\big)$, $\phi(x,T) = 0$,

$$\int_{Q_T}\left(-u_t\phi_t + \nabla u \cdot \nabla\phi - b\,|u|^{\sigma-2}u\phi\right)dz = \int_\Omega u_1(x)\phi(x,0)\,dx + \int_{Q_T} f\phi\,dz.$$

**Definition 12.2** A weak solution $u$ of problem (12.1) is called strong solution if

$$D^2_{x_i x_j} u, \nabla u_t \in L^\infty(0, T; L^2(\Omega)), \quad i, j = 1, \dots, n,$$
$$u \in L^\infty(0, T; L^\infty(\Omega)) \quad \text{if } n \le 4,$$
$$u \in L^\infty(0, T; L^{\frac{2n}{n-4}}(\Omega)) \quad \text{if } n > 4.$$

## 12.2 Local Existence Theorems

**Theorem 12.1** (Local in time existence) *Let conditions* (12.2) *be fulfilled and*

$$u_0 \in L^2(\Omega), \quad u_1 \in L^2(\Omega), \quad f \in L^2(0, T; L^2(\Omega)). \tag{12.3}$$

*Assume that*

$$\sigma^+ < \frac{2(n-1)}{n-2} \textit{ if } n > 2, \qquad \sigma^+ < \infty \textit{ if } n \le 2. \tag{12.4}$$

*Then problem* (12.1) *has at least one local in time weak solution in the sense of* Definition 12.1.

*Remark 12.1* In the case $\sigma^+ \le 2$ the solution exists in the cylinder $Q_T$ of arbitrary finite height $T$.

A weak solution of problem (12.1) is obtained as the limit of the sequence of Galerkin's approximations $u^{(m)} = \sum_{k=1}^m u_k(t)\psi_k(x)$, where $\{\psi_k\}$ are eigenfunctions of the Dirichlet problem for the Laplace operator in $\Omega$. For the data $u_1^{(m)}$, $u_0^{(m)}$, $f^{(m)}$ we take

$$u_1^{(m)} = \sum_{i=1}^m (u_1, \psi_i)_{2,\Omega}\psi_i \to u_1 \text{ in } L^2(\Omega),$$
$$u_0^{(m)} = \sum_{i=1}^m (u_0, \psi_i)_{2,\Omega}\psi_i \to u_1 \text{in } H_0^1(\Omega),$$
$$f^{(m)} = \sum_{i=1}^m (f, \psi_i)_{2,\Omega}\psi_i \to f \text{ in } L^2(0, T; L^2(\Omega)).$$

The coefficients $u_k(t)$ are defined from the system of ordinary differential equations

$$\left(u_{tt}^{(m)} - \Delta u^{(m)} - b|u^{(m)}|^{\sigma-2}u^{(m)} + f^{(m)}, \psi_k\right)_{2,\Omega} = 0, \qquad k = 1, \dots, m. \tag{12.5}$$

Equations (12.5) together with the initial conditions lead to the Cauchy problem for the system of second-order ordinary differential equations for the coefficients $u_k(t)$:

$$\begin{aligned} &u_k'' = \mathscr{F}_k(t, u_1, \ldots, u_m), \\ &u_k(0) = (u_0, \psi_k)_{2,\Omega}, \qquad u_k'(0) = (u_1, \psi_k)_{2,\Omega}, \quad k = 1, \ldots, m, \end{aligned} \tag{12.6}$$

where

$$\mathscr{F}_k = -\left(\nabla u^{(m)}, \nabla \psi_k\right)_{2,\Omega} + \left(b|u^{(m)}|^{\sigma-2}u^{(m)}, \psi_k\right)_{2,\Omega} + (f, \psi_k)_{2,\Omega}.$$

For every finite $m$ system (12.6) has a solution $u_k(t) \in C^2(0, T_m)$, $k = 1, \ldots, m$, on an interval $(0, T_m)$. Let us check that each of the approximate solutions $u^{(m)}$ can be continued to an independent of $m$ interval $(0, T)$. Multiplying each of equations (12.6) by $u_k'(t)$ and summing up we arrive at the following relation: (for the simplicity of notation we denote $u = u^{(m)}$)

$$\frac{1}{2}\frac{d}{dt}\left(\int_\Omega u_t^2\,dx + \int_\Omega |\nabla u|^2\,dx\right) = \int_\Omega b(x,t)|u|^{\sigma(x,t)-2}uu_t dx + \int_\Omega f^{(m)}u_t dx. \tag{12.7}$$

Let us introduce the function

$$X_m(t) = \|u_t(t)\|_{2,\Omega}^2 + \|\nabla u(t)\|_{2,\Omega}^2,$$

and consider separately the cases $\sigma^+ > 2$ and $\sigma^+ \le 2$. Accept the notation

$$X(0) = \|u_0\|_{2,\Omega}^2 + \|u_1\|_{2,\Omega}^2, \qquad \kappa = \frac{\sigma^+}{2} > 1.$$

**Lemma 12.1** *Let $\sigma^+ > 2$. Assume that inequalities* (12.4) *are fulfilled. Then there exists a moment $T_{\max}$, depending on $X(0)$, $\kappa$ and $\|f\|_{L^2(0,T;L^2(\Omega))}$, such that $X_m(t)$ and $\|u(t)\|_{2(\sigma^+-1),\Omega}$ are bounded uniformly in $m$ on every interval $[0, t] \subset [0, T_{\max})$.*

*Proof* It follows from (12.7) that

$$\begin{aligned} \frac{1}{2}X_m'(t) &\le b^+ \int_\Omega |u|^{\sigma(x,t)-1}|u_t|dx + \frac{1}{2}\left(\|f^{(m)}\|_{2,\Omega}^2 + \|u_t\|_{2,\Omega}^2\right) \\ &\le b^+ \int_\Omega |u|^{\sigma(x,t)-1}|u_t|dx + \left(\|f^{(m)}\|_{2,\Omega}^2 + X(t)\right) \\ &\le b^+ \int_\Omega |u|^{\sigma(x,t)-1}|u_t|dx + \left(1 + \|f^{(m)}\|_{2,\Omega}^2\right)(1 + X(t))^{\frac{\sigma^+}{2}}. \end{aligned}$$

To estimate the first term we use the inequalities of Hölder and Young,

$$\int_\Omega |u|^{\sigma(x,t)-1}|u_t|\,dx \le C\|u_t\|_{2,\Omega}\left(\int_\Omega |u|^{2(\sigma-1)}\,dx\right)^{\frac{1}{2}}$$
$$\le CX_m^{\frac{1}{2}}(t)\left(1+\int_\Omega |u|^{2(\sigma^+-1)}\,dx\right)^{\frac{1}{2}},$$

and then apply assumptions (12.4) and the embedding theorem in Sobolev spaces:

$$\int_\Omega |u|^{2(\sigma^+-1)}\,dx \le C\left(\int_\Omega |\nabla u|^2\,dx\right)^{\sigma^+-1} \le CX_m^{\frac{\sigma^+-1}{2}}(t) \le C(1+X_m(t))^{\frac{\sigma^+-1}{2}}.$$

Gathering these estimates we obtain the ordinary differential inequality for $X(t)$:

$$\begin{aligned} X_m'(t) &\le C(1+\|f^{(m)}\|_{2,\Omega}^2)\,(1+X_m(t))^\kappa \\ &\le C(1+\|f\|_{2,\Omega}^2)\,(1+X_m(t))^\kappa \end{aligned} \tag{12.8}$$

with an independent of $m$ constant $C$. Integrating this inequality and simplifying we arrive at the estimate

$$X_m(t) \le \frac{1+X_0}{\left(1-(\kappa-1)C(1+X_0)^{\kappa-1}\left(t+\int_0^t \|f\|_{2,\Omega}^2(\tau)\,d\tau\right)\right)^{\frac{1}{\kappa-1}}} - 1.$$

The right-hand side of the last inequality does not depend on $m$ and remains bounded on every interval $[0,t] \subset [0, T_{\max})$, where $T_{\max}$ is the least solution of the equation

$$t + \int_0^t \|f\|_{2,\Omega}^2(\tau)\,d\tau = \frac{1}{C(\kappa-1)(1+\|u_0\|_{2,\Omega}^2+\|u_1\|_{2,\Omega}^2)^{\kappa-1}}.$$

Since the left-hand side of this equation is monotone increasing in $t$, equals zero at $t=0$ and tends to infinity as $t\to\infty$, this equation has a unique solution $T_{\max}<\infty$. □

**Lemma 12.2** *If $\sigma^+ \le 2$, the functions $X_m(t)$ and $\|u(t)\|_{2,\Omega}$ are bounded uniformly with respect to $m$ on the interval $(0,T)$.*

*Proof* Let $\sigma(x,t) \le \sigma^+ \le 2$. Applying the inequalities of Young and Poincaré, we estimate

$$\begin{aligned}\int_\Omega |b||u|^{\sigma-1}|u_t|\,dx &\leq \frac{1}{2}\int_\Omega u_t^2\,dx + \frac{1}{2}(b^+)^2\int_\Omega |u|^{2(\sigma-1)}\,dx\\ &\leq \frac{1}{2}\int_\Omega u_t^2\,dx + \frac{C}{2}(b^+)^2\left(1+\|u(t)\|^2_{2,\Omega}\right)\\ &\leq \frac{1}{2}X_m(t) + C_1\left(1+\|\nabla u(t)\|^2_{2,\Omega}\right)\\ &= C_2 X_m(t) + C_3,\end{aligned}$$

$$\int_\Omega |f^{(m)}||u_t|\,dx \leq \frac{1}{2}X_m(t) + \|f^{(m)}(t)\|^2_{2,\Omega} \leq \frac{1}{2}X_m(t) + \|f(t)\|^2_{2,\Omega}.$$

By virtue of (12.7) $X_m(t)$ satisfies the differential inequality

$$\begin{cases} X_m'(t) \leq K\,X_m(t) + L\left(1+\|f(t)\|^2_{2,\Omega}\right),\\ X_m(0) \leq X(0)\end{cases}$$

with independent of $m$ constants $K$ and $L$. The conclusion follows now from the Gronwall inequality. □

From now on we assume that $T$ is an arbitrary finite number if $\sigma^+ \leq 2$ and $T < T_{\max}$ in the case $\sigma^+ > 2$. Let us fix some $m \in \mathbb{N}$ and take an arbitrary function

$$\phi \in \mathscr{P}_N = \left\{\phi(x,t) : \ \phi = \sum_{i=1}^N d_i(t)\psi_i(x),\ d_i(t) \in C^2[0,T],\ d_i(x,T) = 0\right\}$$

with $N \leq m$. Multiplying (12.5) by $d_k(t)$, integrating over the interval $(0,t)$ and summing up we arrive at the identity

$$\int_{Q_T}\left(u_{tt}^{(m)}\phi + \nabla u^{(m)}\nabla\phi - b|u^{(m)}|^{\sigma-2}u^{(m)}\phi - f\phi\right)dz = 0. \tag{12.9}$$

Integrating by parts in the first term of (12.9) we rewrite it in the form

$$\begin{aligned}\int_{Q_t}&\left(-u_t^{(m)}\phi_t + \nabla u^{(m)}\nabla\phi - b|u^{(m)}|^{\sigma-2}u^{(m)}\phi - f\phi\right)dz\\ &+\int_\Omega u_t^{(m)}(x,0)\phi(x,0)\,dx = 0.\end{aligned} \tag{12.10}$$

Due to the uniform estimates of Lemmas 12.1 and 12.2 there exists a function $u$ such that (up to a subsequence)

$$
\begin{aligned}
&u^{(m)} \rightharpoonup u \ *\text{-weakly in } L^{\infty}(0, T; L^{2(\sigma^{+}-1)}(\Omega)) \text{ and a.e. in } Q_T,\\
&u_t^{(m)} \rightharpoonup u_t \text{ in } L^2(Q_T) \text{ and } *\text{-weakly in } L^{\infty}(0, T; L^2(\Omega)),\\
&\nabla u^{(m)} \rightharpoonup \nabla u \text{ in } L^2(Q_T) \text{ and } *\text{-weakly in } L^{\infty}(0, T; L^2(\Omega)).
\end{aligned}
$$

Passing to the limit as $m \to \infty$ in (12.10) we find that for every fixed $N$ and $\phi_N \in \mathscr{P}_N$

$$
\begin{aligned}
&\int_{Q_T} \left(-u_t \phi_{N,t} + \nabla u \nabla \phi_N\right) dz - \int_{Q_T} \left(b\,|u|^{\sigma-2} u + f\right) \phi_N \, dz\\
&\quad + \int_{\Omega} u_t(x,0)\phi_N(x,0)\, dx = 0
\end{aligned}
$$

Since $C^{\infty}(0, T; C_0^{\infty}(\Omega)) \cap \mathscr{P}_N$ is dense in $\mathscr{P}_N$, this equality holds in the limit as $N \to \infty$ and $u$ satisfies condition (iii) of Definition 12.1. By Lemma 2.6 the inclusions $u_t$, $|\nabla u| \in L^{\infty}(0, T; L^2(\Omega))$ imply the inclusion $u \in C^0(0, T; L^2(\Omega))$. The fulfillment of the second initial condition follows from Lemma 11.10 with obvious simplifications.

*Remark 12.2* By virtue of (12.5) for every $\phi \in C^{\infty}(0, T; C_0^{\infty}(\Omega))$

$$
|(u_{tt}^{(m)}, \phi)_{2,\Omega}| \leq \|\nabla u^{(m)}\|_{2,\Omega}\|\nabla \phi\|_{2,\Omega} + b^{+}\|u^{(m)}\|_{\sigma,\Omega}\|\phi\|_{\sigma,\Omega} + \|f^{(m)}\|_{2,\Omega}\|\phi\|_{2,\Omega}.
$$

Applying the estimates of Lemmas 12.1, 12.2 and integrating in $t$ over the interval $(0, \tau)$ with $\tau < T_{\max}$ we find that for the constructed solution

$$
|(u_{tt}, \phi)_{2,Q_\tau}| \leq C\left(\|\nabla \phi\|_{2,Q_\tau} + \|\phi\|_{2,Q_\tau} + \|\phi\|_{\sigma,Q_\tau}\right).
$$

**Theorem 12.2** (Local in time existence of strong solutions) *Let under the conditions of* Theorem 12.1 $\partial\Omega \in C^{1,1}$ *and*

$$
\begin{aligned}
&\|\nabla b\|_{\infty,\Omega},\ \|\nabla \sigma\|_{\infty,\Omega},\ \|b\|_{\infty,\Omega} \leq K < \infty,\\
&f \in L^2(0, T; H_0^1(\Omega)), \quad D^2_{x_i x_j} u_0, \nabla u_1 \in L^2(\Omega), \quad i, j = 1, \ldots, n.
\end{aligned}
$$

*Assume that*

$$
2 \leq \sigma^{+} < \begin{cases} \infty & \text{if } n = 1, 2,\\ 2\dfrac{n-1}{n-2} & \text{if } n = 3, 4,\\ 2\dfrac{n-2}{n-4} & \text{if } n > 4. \end{cases} \tag{12.11}
$$

*Then the constructed weak solution of problem* (12.1) *possesses the properties*

$$\begin{aligned}
&D^2_{x_i x_j} u,\ \nabla u_t \in L^\infty(0, T; L^2(\Omega)), \quad i, j = 1, \ldots, n,\\
&u \in L^\infty(0, T; L^\infty(\Omega)) \quad \textit{if}\, n \le 4,\\
&u \in L^\infty(0, T; L^{\frac{2n}{n-4}}(\Omega)) \quad \textit{if}\, n > 4
\end{aligned} \tag{12.12}$$

*and is a strong solution of problem* (12.1).

*Proof* Inclusions (12.12) will follow if we prove that they hold for every term of the sequence $\{u^{(m)}\}$. For the sake of simplicity of notation, here we denote by $u$ the $m$th approximating function $u^{(m)}$. Multiplying (12.5) by $u_k'(t)\Delta\psi_k$, summing up in $k = 1, 2, \ldots, m$ and integrating over $\Omega$ we obtain the equality

$$\begin{aligned}
&\frac{1}{2}\frac{d}{dt}\left(\int_\Omega |\nabla u_t|^2\, dx + \int_\Omega |\Delta u|^2\, dx\right)\\
&\qquad = -\int_\Omega b|u|^{\sigma-2} u \Delta u_t\, dx + \int_\Omega \nabla f^{(m)} \cdot \nabla u_t\, dx.
\end{aligned} \tag{12.13}$$

Integrating by parts and using the representation

$$|u|^{\sigma-2} u = (\sigma - 1)\,\mathrm{sign}\, u \int_0^{|u|} s^{\sigma-2}\, ds,$$

it is straightforward to check that

$$\begin{aligned}
I &:= \int_\Omega b|u|^{\sigma-2} u \Delta u_t\, dx\\
&= -\int_\Omega \nabla\left(b|u|^{\sigma-2} u\right) \nabla u_t\, dx\\
&= -\int_\Omega \left(|u|^{\sigma-2} u \nabla b + b\left(|u|^{\sigma-2} u \ln|u| \nabla\sigma + (\sigma - 1)|u|^{\sigma-2} \nabla u\right)\right) \nabla u_t\, dx.
\end{aligned}$$

By Hölder's inequality

$$|I| \le C\left(\int_\Omega |\nabla u_t|^2\, dx\right)^{\frac{1}{2}} \left(\int_\Omega \left(u^{2(\sigma-1)} + u^{2(\sigma-1)} \ln^2|u| + u^{2(\sigma-2)} |\nabla u|^2\right) dx\right)^{\frac{1}{2}}$$

with a constant $C$ depending on $\|\nabla b\|_{\infty,\Omega}$, $\|\nabla\sigma\|_{\infty,\Omega}$, $\|b\|_{\infty,\Omega}$, $\|\sigma\|_{\infty,\Omega}$. Applying Hölder's and Young's inequalities we may estimate

$$\begin{aligned} I_1 &:= \left( \int_\Omega u^{2\sigma-4} |\nabla u|^2 \, dx \right)^{\frac{1}{2}} \\ &\leq \left( \int_\Omega |\nabla u|^{2q} \, dx \right)^{\frac{1}{2q}} \left( \int_\Omega u^{\frac{2(\sigma-2)}{q-1} q} \, dx \right)^{\frac{q-1}{2q}} \\ &\leq C \left( \int_\Omega |\nabla u|^{2q} \, dx \right)^{\frac{1}{2q}} \left( \int_\Omega u^{\frac{2(\sigma^+-2)}{q-1} q} \, dx + 1 \right)^{\frac{q-1}{2q}} \end{aligned}$$

with any $q > 1$. A suitable $q$ is still to be chosen. Let us fix some $t \in (0, T_{\max})$, define the function $\phi := \Delta u \in L^2(\Omega)$ and consider $u$ as the solution of the homogeneous Dirichlet problem for the Poisson equation $\Delta u = \phi$ in $\Omega$. It is a well-known result of the classical elliptic theory (see [187, Chap. 3] or [147, Chap. 9]) that for $\partial\Omega \in C^{1,1}$

$$\|u\|_{H^2(\Omega)\cap H_0^1(\Omega)} \leq C\|\phi\|_{2,\Omega} = C\|\Delta u\|_{2,\Omega}, \qquad C = C(\Omega, n).$$

Combining this estimate with the Sobolev embedding theorem we find that

$$\|\nabla u\|_{2q,\Omega} \leq C\|\Delta u\|_{2,\Omega} \quad \text{with} \quad \begin{cases} 2 < 2q \leq \dfrac{2n}{n-2} & \text{if } n \geq 3, \\ 2 < 2q < \infty & \text{if } n = 1, 2. \end{cases} \tag{12.14}$$

Let us choose $q$ such that the right-hand side of the estimate on $I_1$ is estimated through $\|\Delta u\|_{2,\Omega}$:

$$\begin{aligned} \left(1 + \int_\Omega u^{\frac{2(\sigma^+-2)}{q-1} q} \, dx\right)^{\frac{q-1}{2q}} &\leq C \left(1 + \left(\int_\Omega |\nabla u|^{2q} \, dx\right)^{\frac{\sigma^+-2}{q-1}}\right)^{\frac{q-1}{2q}} \\ &\leq C \left(1 + \int_\Omega |\Delta u|^2 \, dx\right)^{\frac{\sigma^+-2}{2}} \end{aligned} \tag{12.15}$$

provided that

$$\frac{\sigma^+ - 2}{q-1} \leq \frac{n}{n-2q} \text{ if } n > 2q, \qquad \sigma^+ < \infty \quad \text{if } n \leq 2q. \tag{12.16}$$

One may choose an exponent $q$ satisfying conditions (12.14) and (12.16) if $\sigma^+$ is subject to the following restrictions:

$$\sigma^+ < \infty \text{ for } n \leq 4, \qquad \sigma^+ < 2 + \frac{n+2}{n-4} = 3\frac{n-2}{n-4} \text{ for } n > 4.$$

Gathering the above estimates we obtain the inequality

$$I_1 \le C\left(1+\int_\Omega |\Delta u|^2 dx\right)^{\frac{\sigma^+-1}{2}}.$$

Applying Young's inequality and arguing by analogy we estimate

$$\begin{aligned} I_2 &:= \left(\int_\Omega u^{2(\sigma-1)}\,dx\right)^{\frac{1}{2}} \le \left(1+\int_\Omega u^{2(\sigma^+-1)}\,dx\right)^{\frac{1}{2}} \\ &\le C\left(1+\int_\Omega |\Delta u|^2\,dx\right)^{\frac{\sigma^+-1}{2}}, \end{aligned}$$

provided that

$$\sigma^+ < \infty \text{ if } n \le 4, \qquad \sigma^+ < 1+\frac{n}{n-4} = 2\frac{n-2}{n-4} \text{ if } n > 4. \tag{12.17}$$

To estimate the term

$$I_3 = \left(\int_\Omega u^{2(\sigma-1)}\ln^2|u|\,dx\right)^{\frac{1}{2}}$$

we notice that for every $\delta > 0$

$$|u|^{2(\sigma-1)}\ln^2|u| = |u|^{2(\sigma-1)+\delta}\left(|u|^{-\delta}\ln^2|u|\right) \le C(\delta)|u|^{2(\sigma^+-1)+\delta} \quad \text{if } |u| \ge 1,$$

and, since $\sigma^- > 1$, for every $0 < \delta < 2(\sigma^- - 1)$

$$|u|^{2(\sigma-1)}\ln^2|u| = |u|^{2(\sigma-1)-\delta}\left(|u|^{\delta}\ln^2|u|\right) \le C(\delta)|u|^{2(\sigma^--1)-\delta} \quad \text{if } |u| \le 1.$$

Choosing $0 < \delta < 2(\sigma^- - 1)$ and applying Young's inequality we find that

$$\begin{aligned} I_3 &\le C_1\left(\int_\Omega u^{2(\sigma^+-1)+\delta}\,dx\right)^{\frac{1}{2}} + C_2\left(\int_\Omega u^{2(\sigma^--1)-\delta}\,dx\right)^{\frac{1}{2}} \\ &\le C\left(1+\int_\Omega u^{2(\sigma^+-1)+\delta}\,dx\right)^{\frac{1}{2}}. \end{aligned}$$

If $\sigma^+$ satisfies (12.17), for the sufficiently small $\delta$ the last integral is estimated as $I_2$:

$$I_3 \le C\left(1+\int_\Omega |\Delta u|^2\,dx\right)^{\frac{\sigma^+-1}{2}+\frac{\delta}{4}} \quad \text{with some } \delta \in (0, 2(\sigma^- - 1)).$$

Finally,

$$\int_{\Omega} \nabla f^{(m)} \cdot \nabla u_t \, dx \le (1 + \||\nabla f^{(m)}|\|_{2,\Omega}^2)(1 + \||\nabla u_t|\|_{2,\Omega}^2). \tag{12.18}$$

Let us introduce the function

$$Y_m(t) = \int_{\Omega} |\nabla u_t|^2 \, dx + \int_{\Omega} |\Delta u|^2 \, dx.$$

Gathering the estimates on $I_1$, $I_2$, $I_3$ and (12.18), from (12.13) we obtain the differential inequality

$$\begin{cases} Y_m'(t) & \le C(1 + \||\nabla f|\|_{2,\Omega}^2)(1 + Y_m(t))^{\gamma} \\ Y_m(0) & = \|\Delta u_0^{(m)}\|_{2,\Omega}^2 + \|\nabla u_1^{(m)}\|_{2,\Omega}^2 \\ & \le \|\Delta u_0\|_{2,\Omega}^2 + \|\nabla u_1\|_{2,\Omega}^2 = Y(0) \end{cases} \tag{12.19}$$

with the exponent

$$\gamma = \max\left\{\frac{\sigma^+}{2} + \frac{\delta}{4}, \frac{\sigma^+ - 1}{2}\right\} = \kappa + \frac{\delta}{4} > 1.$$

Inequality (12.19) can be integrated exactly as (12.8). Since $\delta > 0$ can be taken arbitrarily small, $Y_m(t)$ are bounded uniformly in $m$ on every interval $[0, \tau] \subset [0, T_{\max})$. Reverting to inequality (12.15) and using the obtained estimate on $Y_m(t)$ we find that $\|u(t)\|_{p,\Omega}$ are bounded uniformly in $m$ for any $p < \infty$ if $n \le 4$ and for $p \le \dfrac{2n}{n-4}$ if $n > 4$. □

**Theorem 12.3** (Uniqueness of strong solutions) *Let us assume that*

$$2 \le \sigma(x,t) \le \sigma^+ < 2\frac{n-3}{n-4} \quad \text{if} \quad n > 4, \qquad 2 \le \sigma(x,t) \le \sigma^+ < \infty \quad \text{if} \quad n \le 4.$$

*Then problem* (12.1) *cannot have more than one strong solution.*

*Proof* Let $u$ and $v$ be two strong solutions of problem (12.1). Introduce the function $w = u - v$. By virtue of equation (12.1), the regularity conditions of Definition 12.2 and the assumptions on $\sigma^+$

$$w_{tt} = \Delta w + b\left(|u|^{\sigma-2}u - |v|^{\sigma-2}v\right) \in L^{\infty}(0, T; L^2(\Omega)). \tag{12.20}$$

Let $\zeta_{k,\tau}(t)$ be the cut-off function defined in (11.33). Take some $h > 0$ (small), multiply (12.20) by $(\zeta_{k,\tau+h}(t) - \zeta_{k,\tau}(t))w_t$ and integrate over $Q_T$. Passing to the limit as $k \to \infty$, we obtain: for a.e. $\tau, \tau + h \in (0, T)$

$$\frac{1}{2}\int_{\Omega}\left(w_t^2(t)+|\nabla w(t)|^2\right)dx\Big|_{t=\tau}^{t=\tau+h}=\int_{\tau}^{\tau+h}\int_{\Omega}b\left(|u|^{\sigma-2}u-|v|^{\sigma-2}v\right)w_t\,dz.$$

Dividing this equality by $h$, letting $h\to 0$ and applying the Lebesgue differentiation theorem we arrive at the equality

$$\frac{1}{2}\frac{d}{dt}\int_{\Omega}\left(w_t^2(t)+|\nabla w(t)|^2\right)dx=\int_{\Omega}b\left(|u|^{\sigma-2}u-|v|^{\sigma-2}v\right)w_t\,dx\quad\forall\,\text{a.e. } t\in(0,T).$$

Notice that

$$\begin{aligned}\left||u|^{\sigma-2}u-|v|^{\sigma-2}v\right|&=\left|\int_0^1\frac{d}{d\theta}\left(|\theta u+(1-\theta)v|^{\sigma-2}(\theta u+(1-\theta)v)\right)\right|d\theta\\&=(\sigma-1)\int_0^1|\theta u+(1-\theta)v|^{\sigma-2}\,d\theta|u-v|\\&\le(\sigma-1)\left(1+|u|+|v|\right)^{\sigma-2}|u-v|.\end{aligned}$$

Plugging this estimate into the previous equality we find that for a.e. $t\in(0,T)$

$$\begin{aligned}\frac{1}{2}\frac{d}{dt}\left(\int_{\Omega}(w_t^2+|\nabla w|^2\,dx\right)&\le b^+(\sigma^+-1)\int_{\Omega}(1+|u|+|v|)^{\sigma-2}|w||w_t|\,dx\\&\le C\left(\int_{\Omega}w_t^2dx\right)^{\frac{1}{2}}\left(\int_{\Omega}|w|^{\frac{2n}{n-2}}dx\right)^{\frac{n-2}{2n}}\left(1+\int_{\Omega}(1+|u|+|v|)^{n(\sigma^+-2)}dx\right)^{\frac{1}{n}}.\end{aligned}$$

The second factor on the right-hand side is bounded by virtue of the Sobolev embedding theorem, $\|u\|_{\frac{2n}{n-2},\Omega}\le C\|\nabla u\|_{2,\Omega}$. The third factor is bounded due to the regularity assumption in Definition 12.2 of the strong solution. In the case $n\le 4$ the estimate is obvious, in the case $n>4$ the claim $n(\sigma^+-2)<\dfrac{2n}{n-2}$ is equivalent to the inequality $\sigma^+<2\dfrac{n-3}{n-4}$, which holds by assumption. It follows that the function

$$Z_m(t)=\|w_t(t)\|_{2,\Omega}^2+\|\nabla w(t)\|_{2,\Omega}^2$$

satisfies the inequality

$$Z_m'(t)\le CZ_m(t),\qquad Z_m(0)=0$$

and the conclusion is a consequence of Gronwall's inequality. □

## 12.3 Nonexistence of Global Nonnegative Solutions

To prove nonexistence of global in time nonnegative solutions we apply the Kaplan method of eigenfunctions. Notice that the existence theorems proved in the previous section do not allow one to conclude that the solution of problem (12.1) has a definite sign, a fact that renders conditional the results below.

### *12.3.1 Ordinary Differential Inequalities*

We will need several technical results concerning the 2nd-order ordinary differential inequalities of the types

$$\mu''(t) \geq h(\mu) := -\lambda\mu + \alpha_0 \min\left\{\mu^{\sigma^- -1}, \mu^{\sigma^+ -1}\right\}, \quad \sigma^{\pm} = const, \tag{12.21}$$

$$\mu''(t) \geq h(t,\mu) := -\lambda\mu + \alpha_0 \min\left\{\mu^{\sigma^-(t)-1}, \mu^{\sigma^+(t)-1}\right\} \tag{12.22}$$

with positive constants $\lambda$ and $\alpha_0$.

**Lemma 12.3** *Let $\mu \in C^0[0,T)$ be a nonnegative function such that $\mu', \mu'' \in L^\infty(0,T)$. If $\mu(t)$ satisfies inequality* (12.21) *a.e. in $(0,T)$ and*

$$\begin{aligned} &\mu(0) = a_0 > 0, \quad \mu'(0) = a_1 > 0, \\ &h(a_0) \geq 0, \qquad h'(a_0) \geq 0, \\ &2 < \sigma^- \leq \sigma^+ < \infty, \end{aligned} \tag{12.23}$$

*then it is necessary that*

$$\sup\{\mu(t) | \, t \in (0,T)\} = +\infty.$$

*Proof* It is easy to see that $h(\mu)$ is a concave function:

$$h''(\mu) = \alpha_0 \min\left\{(\sigma^- -1)(\sigma^- -2)\mu^{\sigma^- -3}, (\sigma^+ -1)(\sigma^+ -2)\mu^{\sigma^+ -3}\right\} \geq 0.$$

Integrating this inequality and taking into account (12.23) we find that $h'(\mu) \geq h'(a_0) \geq 0$. Integration of the last inequality gives $h(\mu) \geq h(a_0) \geq 0$ for all $\mu \geq a_0$ and, correspondingly, $\mu''(t) \geq 0$. It follows that $\mu(t) \geq a_1 t + a_0$ and

$$h(\mu) \geq \frac{\alpha_0}{2}\mu^{\sigma^- -1} \text{ for } t \geq t_0 = \frac{1}{a_1}\max\left\{1, \left(\frac{2\lambda}{\alpha_0}\right)^{\frac{1}{\sigma^- -2}}\right\}.$$

Substituting this inequality into (12.21) we have

$$\mu''(t) \geq \frac{\alpha_0}{2}\mu^{\sigma^- -1} \text{ for } \quad t \geq t_0.$$

Now we multiply by $\mu'(t)$ and integrate over the interval $(0, t)$ to get

$$\mu'^2(t) \geq \frac{\alpha_0}{\sigma^-}\mu^{\sigma^-} + C, \qquad C = a_1^2 - \frac{\alpha_0}{\sigma^-}a_0^{\sigma^-}.$$

Since the integral

$$\int_{\mu(t_0)}^{\infty} \frac{ds}{\sqrt{\frac{\alpha_0}{\sigma^-}s^{\sigma^-} + C}}$$

is convergent, it follows from the previous inequality that

$$\int_{\mu(t_0)}^{\mu(t)} \frac{ds}{\sqrt{\frac{\alpha_0}{\sigma^-}s^{\sigma^-} + C}} \geq t - t_0$$

and it is necessary that there exists $T_{\max}$ such that $\mu(t) \to \infty$ as $t \to T_{\max}$. □

Let us assume now that

$$\begin{cases} 2 \leq \sigma^-(t) \leq \sigma^+(t) < \infty, \\ 0 \leq \sigma^-(t) - 2 = \varepsilon(t) \searrow 0 \text{ as } t \to \infty. \end{cases} \tag{12.24}$$

**Lemma 12.4** *Let $\mu \in C^0[0, T)$ be a nonnegative function such that $\mu', \mu'' \in L^\infty(0, T)$. Assume that $\mu(t)$ satisfies inequality* (12.21) *a.e. in* $(0, T)$ *with exponents $\sigma^\pm(t) \in C^0[0, T]$. If $\sigma^-(t)$ satisfies* (12.24) *and*

$$\begin{array}{ll} \mu(0) = a_0 \geq 1, & \mu'(0) = a_1 > 0, \\ h(t, a_0) \geq 0, & h'(t, a_0) \geq 0 \ \ \forall t \geq 0, \end{array}$$

$$\lambda - \alpha_0 = C_0 > 0, \qquad \int_{\mu_0}^{\infty} \left( \int_{\mu_0}^{\mu} s^{\varepsilon(\ln \frac{2s}{a_0})+1} \, ds \right)^{-\frac{1}{2}} d\mu < \infty \ \ \forall \mu_0 > 1,$$

*then it is necessary that*

$$\sup\{\mu(t) |\ t \in (0, T)\} = \infty.$$

*Proof* It is straightforward to check that the following relations hold:

$$\frac{\partial h(t, \mu)}{\partial \mu} = \begin{cases} -\lambda + \alpha_0 \left(\sigma^-(t) - 1\right) \mu^{\sigma^-(t)-2} \geq 0 & \text{if } \mu \geq 1, \\ -\lambda + \alpha_0 \left(\sigma^+(t) - 1\right) \mu^{\sigma^+(t)-2} \geq 0, & \text{if } \mu < 1, \end{cases}$$

$$\frac{\partial^2 h(t,\mu)}{\partial^2 \mu} = \begin{cases} \alpha_0\left(\sigma^-(t)-1\right)\left(\sigma^-(t)-2\right)\mu^{\sigma^-(t)-3} \geq 0 & \text{if } \mu \geq 1, \\ \alpha_0\left(\sigma^+(t)-1\right)\left(\sigma^+(t)-2\right)\mu^{\sigma^+(t)-3} \geq 0 & \text{if } \mu < 1, \end{cases}$$

$$\frac{\partial h(t,\mu)}{\partial \mu} = \int_{a_0}^{\mu} \frac{\partial^2 h(t,s)}{\partial^2 \mu} ds + \frac{\partial h(t,a_0)}{\partial \mu} \geq 0,$$

$$h(t,\mu) = \int_{a_0}^{\mu} \frac{\partial h(t,s)}{\partial \mu} ds + h(t,a_0) \geq 0.$$

Integration of the inequality $\mu''(t) \geq 0$ gives the estimate $\mu(t) \geq a_1 t + a_0 \geq 1$, whence for all $t \geq 0$

$$\begin{aligned} h(t,\mu) &= -\lambda\mu + \alpha_0 \mu^{\sigma^-(t)-1} = \mu^{\sigma^-(t)-1}\left(\alpha_0 - \lambda\mu^{2-\sigma^-(t)}\right) \\ &\geq \mu^{\sigma^-(t)-1}\left(\alpha_0 - \lambda\mu^{2-\sigma^-(t)}\right) \geq C_0 \mu^{\sigma^-(t)-1}. \end{aligned}$$

The function $\mu$ satisfies the conditions

$$\begin{cases} \mu''(t) \geq C_0 \mu^{\sigma^-(t)-1}, & t > 0, \\ \mu(0) = a_0 \geq 1, \quad \mu'(0) = a_1 > 0, \quad C_0 > 0. \end{cases} \tag{12.25}$$

Without loss of generality we may assume that $C_0 = 1$. Since $\mu(t) \geq 1$, it is necessary that $\mu(t)$ satisfies also the linear differential inequality

$$\begin{cases} \mu''(t) \geq \mu^{\sigma^-(t)-1} \geq \mu, \quad t \geq 0, \\ \mu(0) = a_0, \quad \mu'(0) = a_1. \end{cases}$$

Take the solution $v(t)$ of the problem

$$v'' - v = 0, \quad v(0) = a_0, \quad v'(0) = a_1 \tag{12.26}$$

and consider the function $\omega = \mu - v$, which satisfies the conditions

$$\omega'' \geq \omega, \quad \omega(0) = \omega'(0) = 0.$$

It is asserted that $\omega(t) \geq 0$. Let us argue by contradiction and assume that there exists $t_0 > 0$ such that $\omega(t) < 0$ and $\omega'(t) \leq 0$ on the interval $(0, t_0)$. Set $\Omega(t) = -\omega(t)$. This function satisfies the conditions $\Omega(t) > 0$, $\Omega'(t) \geq 0$ in $(0, t_0)$,

$$\Omega''(t) \leq \Omega(t) \text{ for } t \in (0, t_0), \quad \Omega(0) = \Omega'(0) = 0.$$

Multiplying the differential inequality for $\Omega(t)$ by $\Omega'(t)$ and integrating over the interval $(0,t)$, $t < t_0$, we find that $\Omega'(t) \leq \Omega(t)$ with $\Omega(t) > 0$. It follows that $\ln \dfrac{\Omega(t)}{\Omega(\tau)} \leq t - \tau$ for every $\tau \in (0,t)$ and, finally,

$$\Omega(t) \leq \Omega(\tau)e^{t-\tau} \to 0 \quad \text{as}\, \tau \to 0,$$

which contradicts the assumption $\Omega(t) = -\omega(t) > 0$ in $(0,t_0)$.

Comparison of $\mu(t)$ and the solution of the (12.26) $v(t)$ leads to the inequality

$$\mu(t) \geq v(t) = \frac{a_0 + a_1}{2}e^t + \frac{a_0 - a_1}{2}e^{-t} \geq \frac{a_0}{2}e^t = \gamma e^t, \quad \gamma = \frac{a_0}{2}$$

for all $t$ beginning with some $\tau > 0$. The last inequality is equivalent to the following one:

$$\ln \frac{\mu(t)}{\gamma} \geq t, \qquad t \geq \tau. \tag{12.27}$$

Since $\varepsilon(s)$ is a decreasing function by assumption (12.24), it follows from (12.27) that

$$\varepsilon\left(\ln \frac{\mu(t)}{\gamma}\right) \leq \varepsilon(t),$$

which transforms (12.25) into the autonomous inequality

$$\mu''(t) \geq \mu^{\varepsilon\left(\ln \frac{\mu}{\gamma}\right)+1}, \qquad t \geq \tau.$$

Multiplying it by $\mu'(t)$ and integrating we have

$$\frac{(\mu')^2(t)}{2} \geq \int_{\mu(\tau)}^{\mu(t)} s^{\varepsilon(\ln \frac{s}{\gamma})+1} ds + C, \quad C = \frac{(\mu')^2(\tau)}{2}.$$

Every function satisfying this inequality becomes infinite in a finite time because

$$t - \tau \leq \int_{\mu(\tau)}^{\mu(t)} \frac{dy}{\sqrt{\int_{\mu(\tau)}^{y} s^{\varepsilon(\ln \frac{s}{\gamma})+1} ds}} \leq L < \infty \tag{12.28}$$

with an independent of $t$ constant $L$. □

*Remark 12.3* Let us present an example of the function $\varepsilon(t) = \sigma^-(t) - 2 \geq 0$ which satisfies condition (12.28). Let us claim that for some $\delta > 0$

$$\sqrt{\int_{\mu_0}^{\mu} s^{\varepsilon(\ln \frac{s}{\gamma})+1} ds} \sim \mu \ln^{1+\delta} \mu \quad \text{as}\ \mu \to \infty. \tag{12.29}$$

By the L'Hôpitale rule, this is true if

$$\mu^{\varepsilon(\ln \frac{\mu}{\gamma})} \sim 2\ln^{2(1+\delta)} \mu + 2(1-\delta)\ln^{1+2\delta} \mu \quad \text{as } \mu \to \infty,$$

that is,

$$\ln\left(\ln^{2(1+\delta)} \mu\right) \sim \varepsilon\left(\ln \frac{\mu}{\gamma}\right) \ln \mu \quad \text{as } \mu \to \infty.$$

Setting $s = \ln \mu - \ln \gamma$, it is easy to transform the previous relation into the following one:

$$\varepsilon(s) \sim \frac{2(1+\delta)\ln(s+\ln\gamma)}{t+\ln\gamma} \quad \text{as } s \to \infty.$$

This gives the needed estimate on the behavior of $\sigma^-(t) = \varepsilon(t) + 2$ as $s \to \infty$.

### 12.3.2 Blow-up

Let us apply the auxiliary Lemmas 12.3 and 12.4 to derive conditions of nonexistence of nonnegative global in time solutions of problem (12.1). We use the Kaplan method of eigenfunctions. Let $\lambda > 0$ be the first eigenvalue and let $\phi \geq 0$ be the corresponding eigenfunction of the problem

$$-\Delta\phi = \lambda\phi \text{ in } \Omega, \qquad \phi = 0 \text{ on } \partial\Omega. \tag{12.30}$$

The eigenfunction is normalized by the condition $\|\phi\|_{1,\Omega} = 1$. Introduce the functions

$$\mu(t) = \int_\Omega u(\cdot, t)\phi(\cdot)\, dx,$$

$$\alpha(t) = \left(\int_\Omega \left(b(x,t)\phi^{2-\sigma(x,t)}(x)\right)^{\frac{1}{2-\sigma^-(t)}} dx\right)^{2-\sigma^-(t)} \geq \alpha_0 \tag{12.31}$$

with a positive constant $\alpha_0$.

**Theorem 12.4** *Let $u$ be a nonnegative strong solution of problem* (12.1)*. Assume that $f(x,t)$ and $b(x,t)$ are nonnegative and there exist constants $\sigma^\pm$ such that*

$$2 < \sigma^- \leq \sigma(x,t) \leq \sigma^+ < \infty.$$

*If condition* (12.31) *is fulfilled and*

$$\mu(0) = \int_\Omega u_0(x)\phi(x)dx = a_0 > 0,$$
$$\mu'(0) = \int_\Omega u_1(x)\phi(x)dx = a_1 > 0,$$
$$h(a_0) \geq 0, \ h'(a_0) \geq 0,$$

*then there exists a finite moment* $T_{\max}$ *such that*

$$\sup\{\|u(t)\|_{\infty,\Omega} \,|\, t \in (0, T_{\max})\} = +\infty. \tag{12.32}$$

*Proof* Under the regularity assumptions on the strong solution

$$\int_\Omega u_{tt}\phi\,dx = -\int_\Omega \nabla u\cdot\nabla\phi\,dx + \int_\Omega b(x,t)u^{\sigma(x,t)-1}\phi\,dx + \int_\Omega f\phi\,dx$$

for a.e. $t \in (0, T)$. This equality yields the inequality

$$\begin{aligned}\mu''(t) &= -\lambda\mu + \int_\Omega b\,u^{\sigma(x,t)-1}\phi\,dx + \int_\Omega f\phi\,dx \\ &\geq -\lambda\mu + \int_\Omega b\,u^{\sigma(x,t)-1}\phi\,dx.\end{aligned} \tag{12.33}$$

The last term on the right-hand side of (12.33) is estimated from below by the inverse Hölder inequality:

$$\int_\Omega |u||v|dx \geq \|u\|_{q,\Omega}\|v\|_{\frac{q}{q-1},\Omega} \tag{12.34}$$

with

$$q = \frac{1}{\sigma^-(t)-1} \in (0,1), \qquad w = u^{\sigma^-(t)-1}\phi^{\sigma^-(t)-1}, \qquad v = a\phi^{2-\sigma^-(t)}.$$

We have:

$$\begin{aligned}I &= \int_\Omega bu^{\sigma-1}\phi\,dx = \int_\Omega (u\phi)^{\sigma-1}b\phi^{2-\sigma}\,dx \\ &\geq \left(\int_\Omega (u\phi)^{\frac{\sigma-1}{\sigma^-(t)-1}}\,dx\right)^{\sigma^+-1}\left(\int_\Omega \left(b\phi^{2-\sigma}\right)^{\frac{1}{2-\sigma^-(t)}}\,dx\right)^{2-\sigma^-(t)} \\ &\geq \min\left\{\|u\phi\|^{\sigma^-(t)-1}_{\frac{\sigma-1}{\sigma^-(t)-1}}, \|u\phi\|^{\sigma^-(t)-1}_{\frac{\sigma-1}{\sigma^-(t)-1}}\right\}\left(\int_\Omega \left(b\phi^{2-\sigma}\right)^{\frac{1}{2-\sigma^-(t)}}\,dx\right)^{2-\sigma^-(t)}.\end{aligned} \tag{12.35}$$

Using the inequalities

$$\min\left\{\|u\phi\|^{\sigma^-(t)-1}_{\frac{\sigma-1}{\sigma^-(t)-1}}, \|u\phi\|^{\sigma^+(t)-1}_{\frac{\sigma-1}{\sigma^-(t)-1}}\right\} \le \left(\int_\Omega (u\phi)^{\frac{\sigma-1}{\sigma^-(t)-1}}\,dx\right)^{\sigma^-(t)-1},$$

$$\|u\phi\|_{1,\Omega} \le \|1\|_{\frac{\sigma-1}{\sigma-\sigma^-(t)},\Omega}\|u\phi\|_{\frac{\sigma-1}{\sigma^-(t)-1},\Omega} \le C\|u\phi\|_{\frac{\sigma-1}{\sigma^-(t)-1},\Omega}, \quad C = C(|\Omega|,\sigma^\pm)$$

we continue (12.35) as follows:

$$\begin{aligned} I &\ge \min\left\{\|u\phi\|_1^{\sigma^-(t)-1}, \|u\phi\|_1^{\sigma^+(t)-1}\right\}\alpha(t) \\ &\ge \alpha_0 \min\left\{\mu^{\sigma^+(t)-1}, \mu^{\sigma^+(t)-1}\right\}. \end{aligned} \tag{12.36}$$

Plugging (12.35) into (12.33) we arrive at the ordinary differential inequality

$$\mu''(t) \ge -\lambda\mu + \alpha(t)\min\left\{\mu^{\sigma^-(t)-1}, \mu^{\sigma^+(t)-1}\right\} := h(t,\mu) \tag{12.37}$$

with the function $h(t,\mu)$ which obeys the estimate

$$h(t,\mu) \ge h(\mu) := -\lambda\mu + \alpha_0 \min\left\{\mu^{\sigma^--1}, \mu^{\sigma^+-1}\right\}.$$

This estimate allows one to write (12.37) in the form (12.21):

$$\mu''(t) \ge -\lambda\mu + \alpha_0 \min\left\{\mu^{\sigma^--1}, \mu^{\sigma^+-1}\right\} = h(\mu)$$

with constant parameters $\lambda$, $\alpha_0$, $\sigma^-$, $\sigma^+$. The conclusion follows now from Lemma 12.3. □

**Theorem 12.5** *Let $u$ be a nonnegative strong solution to problem* (12.1) *with $f \ge 0$ and $b \ge 0$ in $Q_T$. Assume that*

$$2 \le \sigma^-(t) \le \sigma(x,t) \le \sigma^+(t) < \infty$$

*with $\sigma^\pm(t)$ satisfying* (12.24). *If condition* (12.31) *is fulfilled and if*

$$\begin{aligned} &\mu(0) = \int_\Omega u_0(x)\phi(x)dx = a_0 \ge 1, \\ &\mu'(0) = \int_\Omega u_1(x)\phi(x)dx = a_1 > 0, \\ &h(0,a_0) \ge \lambda - \alpha_0 = C_1 > 0, \\ &\int_{\mu_0}^\infty \left(\int_{\mu_0}^\mu s^{\varepsilon(\ln\frac{s}{\gamma})+1}\,ds\right)^{-\frac{1}{2}} d\mu < \infty \quad \forall \mu_0 > 1, \end{aligned}$$

*then there exists a finite moment* $T_{\max}$ *such that* $\sup\{\|u(t)\|_{\infty,\Omega} \mid t \in (0, T_{\max})\} = \infty$.

*Proof* Following the proof of Theorem 12.4 we find that the function $\mu(t)$ satisfies the ordinary differential inequality (12.22). The assertion of the theorem is an immediate byproduct of Lemma 12.4 □

*Remark 12.4* It is noteworthy that condition (12.31) may hold for nonnegative functions $b(x,t)$ which may vanish on a set of zero measure in $\Omega$. Let $M = \sup_{\Omega} \phi(x)$, where $\phi$ is the first eigenfunction of problem (12.30). Since $M < \infty$, condition (12.31) is fulfilled if we claim that

$$\int_{\Omega} \left(\frac{1}{b(x,t)}\right)^{\frac{1}{\sigma^{-}(t)-2}} dx \leq K$$

for all $t \in (0, T)$ with a finite constant $K$. In this case (12.31) reads

$$\alpha(t) = \left(\int_{\Omega} \left(b\phi^{2-\sigma}\right)^{\frac{1}{2-\sigma^{-}(t)}} dx\right)^{2-\sigma^{-}(t)} \geq \frac{1}{K \sup\limits_{Q_T} M^{\sigma-2}} = \alpha_0 > 0.$$

### 12.3.3 Regional Blow-up

Let $u$ be a nonnegative strong solution of problem (12.1) with $\sigma(x,t) \geq 2$. Let $D \subseteq \Omega$ be a simple-connected subdomain with $|D| > 0$ and $\partial D \in C^1$. Take the first weak eigenfunction $\phi(x)$ of the problem

$$-\Delta\phi = \lambda\phi \text{ in } D, \qquad \phi = 0 \text{ on } \partial D, \quad \|\phi\|_{1,D} = 1,$$

and define the function

$$\mu(t) = \int\limits_{D} u\phi dx.$$

Assume that $f$ and $b$ are nonnegative in $D \times (0, T)$. By virtue of equation (12.1)

$$\begin{aligned}\mu''(t) &= \int_{D} u_{tt}\phi\, dx = \int_{D} \phi\Delta u\, dx + \int\limits_{D} bu^{\sigma(x,t)-1}\phi\, dx + \int_{\Omega} f\phi\, dx \\ &\geq -\lambda\mu(t) - \int_{\partial D} \phi\,(\nabla u \cdot \mathbf{n})\, dS + \int\limits_{D} bu^{\sigma(x,t)-1}\phi dx, \qquad (12.38)\end{aligned}$$

where $\mathbf{n}$ is the outward normal to $\partial D$. Since $\phi = 0$ and for the strong solution $D_{x_i}u$ have traces on $\partial D$, it follows from (12.38) that

$$\mu''(t) \geq -\lambda\mu + \int_D b(x,t)u^{\sigma(x,t)-1}\phi\,dx.$$

Conditions of the finite time blow-up in the domain $D$ follow now by the same arguments that were used in the previous subsection.

## 12.4 Semilinear Equations of General Form

Let us consider a nonnegative solution $u(x,t)$ of the problem

$$\begin{cases} u_{tt} = Lu + f(x,t,u) & \text{in } Q_T, \\ u(x,0) = u_0(x) \text{ in } \Omega, \\ u_t(x,0) = u_1(x) \text{ in } \Omega, \\ u = 0 \text{ on } \Gamma_T, \end{cases} \tag{12.39}$$

where $L$ is the linear elliptic operator

$$Lu = D_i\left(a_{ij}(x)D_j u + a_i(x)u\right) + a_0(x)u,$$

and the right-hand side has the form

$$f(x,t,u) = \sum_{k=1}^{N} b_k(x,t)u^{\sigma_k(x,t)-1} + \sum_{i=N}^{K} c_i(x,t)\int_\Omega d_i(s,t)u^{\sigma_i(s,t)-1}ds$$

with $b_k \geq 0$, $c_i \geq 0$, $d_i \geq 0$. Let us denote by $L^*$ the conjugate operator

$$\begin{cases} L^*\phi \equiv -D_j\left(a_{ij}(x)D_i\phi\right) - a_i(x)D_{x_i}\phi + a_0(x)\phi, \\ (u, L^*\phi)_{2,\Omega} = (\phi, Lu)_{2,\Omega} \quad \forall u,\ \phi \in H_0^1(\Omega), \end{cases}$$

and assume that there exist a nonnegative function $\phi \geq 0$ and a positive number $\lambda > 0$ such that

$$L^*\phi = -\lambda\phi \text{ in } \Omega, \qquad \phi = 0 \quad \text{on} \quad \Gamma = \partial\Omega.$$

Formally multiplying (12.39) by $\phi$ and integrating over $\Omega$ we arrive at the relation

$$\mu''(t) = \int_D u_{tt}\phi dx = -\lambda\mu + I_1 + I_2, \tag{12.40}$$

where

$$I_1 = \int_\Omega \left( \sum_{k=1}^{N} b_k(x,t) u^{\sigma_k(x,t)-1} \right) \phi \, dx,$$

$$I_2 = \int_\Omega \left( \sum_{i=N}^{K} c_i(x,t) \int_\Omega d_i(s,t) u^{\sigma_i(s,t)-1} ds \right) \phi \, dx.$$

$I_1$, $I_2$ are estimated from below in the following way:

$$\begin{aligned} I_1 &\geq \left( \sum_{k=1}^{N} \int_\Omega \left( b_k \phi^{2-\sigma_k} \right)^{\frac{1}{2-\sigma_k^-}} dx \right)^{2-\sigma_k^-} \min \left\{ \mu^{\sigma_k^-(t)-1}, \mu^{\sigma_k^+(t)-1} \right\} \\ &= \alpha_k(t) \min \left\{ \mu^{\sigma_k^-(t)-1}, \mu^{\sigma_k^+(t)-1} \right\} \\ &= \alpha_k(t) \min \left\{ \mu^{\sigma_k^-(t)-1}, \mu^{\sigma_k^+(t)-1} \right\}, \qquad k = 1, \ldots, N, \end{aligned}$$

$$\begin{aligned} I_2 \geq & \sum_{i=N+1}^{K} \int_\Omega c_i(x,t)\phi(x,t) \min \left\{ \mu^{\sigma_k^-(t)-1}, \mu^{\sigma_k^+(t)-1} \right\} \\ & \times \left( \int_\Omega d_i^{\frac{1}{2-\sigma_i^-(t)}}(s,t) \phi^{\frac{\sigma_i-1}{\sigma_i-2}}(s,t) ds \right)^{2-\sigma_i^-} dx \\ = & \sum_{i=N+1}^{Q} \alpha_i(t) \min \left\{ \mu^{\sigma_k^-(t)-1}, \mu^{\sigma_k^+(t)-1} \right\}, \qquad k = N+1, \ldots, K. \end{aligned}$$

These formulas lead to the nonlinear ordinary differential inequalities of the types (12.21) and (12.22):

$$\begin{aligned} \mu''(t) &\geq -\lambda\mu + \sum_{i=1}^{K} \alpha_i(t) \min \left\{ \mu^{\sigma_k^-(t)-1}, \mu^{\sigma_k^+(t)-1} \right\} \\ &\geq -\lambda\mu + \alpha_0 \min \left\{ \mu^{\sigma^-(t)-1}, \mu^{\sigma^+(t)-1} \right\}, \end{aligned}$$

where

$$\alpha_0 = \min_t \min_k \alpha_k, \qquad \sigma^-(t) = \min_k \sigma_k^-(t), \qquad \sigma^+(t) = \max_k \sigma_k^+(t).$$

The blow-up property of the functions satisfying these inequalities is studied in Lemmas 12.3 and 12.4.

*Remark 12.5* The above arguments are applicable to equations

$$u_{tt} = Lu + f(x, t, u)$$

with an abstract operator $L$. It is sufficient to claim that $L$ is an autonomous operator, i.e., the coefficients of $L$ are independent of $t$, and that there exist a nonnegative function $\phi$ and a positive number $\lambda$ such that $L\phi = \lambda\phi$ and

$$(Lu, \phi)_{2,\Omega} = \lambda(u, \phi)_{2,\Omega}.$$

# References

1. R. Aboulaich, D. Meskine, and A. Souissi, *New diffusion models in image processing*, Comput. Math. Appl., 56 (2008), pp. 874–882.
2. E. Acerbi and G. Mingione, *Regularity results for electrorheological fluids: the stationary case*, C. R. Math. Acad. Sci. Paris, 334 (2002), pp. 817–822.
3. E. Acerbi and G. Mingione, *Regularity results for stationary electro-rheological fluids*, Arch. Ration. Mech. Anal., 164 (2002), pp. 213–259.
4. E. Acerbi, G. Mingione, and G. A. Seregin, *Regularity results for parabolic systems related to a class of non-Newtonian fluids*, Ann. Inst. H. Poincaré Anal. Non Linéaire, 21 (2004), pp. 25–60.
5. R. A. Adams, *Sobolev spaces*, Academic Press [A subsidiary of Harcourt Brace Jovanovich, Publishers], New York-London, 1975. Pure and Applied Mathematics, Vol. 65.
6. G. Akagi, *Doubly nonlinear parabolic equations involving variable exponents*, Discrete Contin. Dyn. Syst. Ser. S, 7 (2014), pp. 1–16.
7. G. Akagi and K. Matsuura, *Well-posedness and large-time behaviors of solutions for a parabolic equation involving $p(x)$-Laplacian*, Discrete Contin. Dyn. Syst., (2011), pp. 22–31.
8. G. Akagi and K. Matsuura, *Nonlinear diffusion equations driven by the $p(\cdot)$-Laplacian*, NoDEA Nonlinear Differential Equations Appl., 20 (2013), pp. 37–64.
9. G. Akagi and G. Schimperna, *Subdifferential calculus and doubly nonlinear evolution equations in $L^p$-spaces with variable exponents*, J. Funct. Anal., 267 (2014), pp. 173–213.
10. Y. Alkhutov, *The Harnack inequality and the Hölder property of solutions of nonlinear elliptic equations with a nonstandard growth condition*, Differ. Uravn., 33 (1997), pp. 1651–1660, 1726.
11. Y. Alkhutov, S. Antontsev, and V. Zhikov, *Parabolic equations with variable order of nonlinearity*, in Collection of works of the Mathematics Institute of the National Academy of Science of Ukraine, vol. 6 (1), 2009, pp. 23–50.
12. Y. Alkhutov and V. Zhikov, *Existence theorems and qualitative properties of solutions of parabolic equations with a variable order of nonlinearity*, Dokl. Akad. Nauk, 430 (2010), pp. 295–299.
13. Y. Alkhutov and V. Zhikov, *Existence theorems for solutions of parabolic equations with a variable order of nonlinearity*, Tr. Mat. Inst. Steklova, 270 (2010), pp. 21–32.
14. Y. Alkhutov and V. Zhikov, *Existence and uniqueness theorems for solutions of parabolic equations with a variable nonlinearity exponent*, Mat. Sb., 205 (2014), pp. 3–14.



S. Antontsev and S. Shmarev, *Evolution PDEs with Nonstandard Growth Conditions*, Atlantis Studies in Differential Equations 4,
DOI 10.2991/978-94-6239-112-3

15. H. W. Alt and E. di Benedetto, *Nonsteady flow of water and oil through inhomogeneous porous media*, Ann. Scuola Norm. Sup. Pisa Cl. Sci. (4), 12 (1985), pp. 335–392.
16. H. W. Alt and S. Luckhaus, *Quasilinear elliptic-parabolic differential equations*, Math. Z., 183 (1983), pp. 311–341.
17. P. Amorim and S. Antontsev, *Young measure solutions for the wave equation with $p(x,t)$-Laplacian: Existence and blow-up*, Nonlinear Anal., 92 (2013), pp. 153–167.
18. D. Andrade, M. A. Jorge Silva, and T. F. Ma, *Exponential stability for a plate equation with $p$-Laplacian and memory terms*, Math. Methods Appl. Sci., 35 (2012), pp. 417–426.
19. B. Andreianov, M. Bendahmane, and S. Ouaro, *Structural stability for variable exponent elliptic problems, I: the $p(x)$-Laplacian kind problems*, Nonlinear Anal., 73 (2010), pp. 2–24.
20. B. Andreianov, M. Bendahmane, and S. Ouaro, *Structural stability for variable exponent elliptic problems. II. The $p(u)$-Laplacian and coupled problems*, Nonlinear Anal., 72 (2010), pp. 4649–4660.
21. F. Andreu, V. Caselles, and J. M. Mazón, *Limited flux diffusion equations*, Bol. Soc. Esp. Mat. Apl. SeMA, (2006), pp. 113–136.
22. F. Andreu, V. Caselles, and J. M. Mazón, *Diffusion equations with finite speed of propagation*, in Functional analysis and evolution equations, Birkhäuser, Basel, 2008, pp. 17–34.
23. F. Andreu-Vaillo, V. Caselles, and J. M. Mazón, *Parabolic quasilinear equations minimizing linear growth functionals*, vol. 223 of Progress in Mathematics, Birkhäuser Verlag, Basel, 2004.
24. S. Antontsev, *Localization of solutions of degenerate equations of continuum mechanics*, Akad. Nauk SSSR Sibirsk. Otdel. Inst. Gidrodinamiki, Novosibirsk, 1986. (in Russian; "Lokalizatsiya resheniĭ vyrozhdayushchikhsya uravneniĭ mekhaniki sploshnoĭ sredy").
25. S. Antontsev, *Wave equation with $p(x,t)$-Laplacian and damping term: Existence and blow-up*, in Abstracts, Nonlinear Models in Partial Differential Equations, An international congress on occasion of J.I. Diaz's 60th birthday, Toledo, Spain, June 14-17, 2011, 2011, p. 8.
26. S. Antontsev, *Wave equation with $p(x,t)$-Laplacian and damping term: Existence and blow-up*, Differ. Equ. Appl., 3 (2011), pp. 503–525.
27. S. Antontsev, *Wave equation with $p(x,t)$-Laplacian and damping term: Blow-up of solutions*, C.R. Mecanique, 339 (2011), pp. 751–755.
28. S. Antontsev and M. Chipot, *Anisotropic equations: uniqueness and existence results*, Differential Integral Equations, 21 (2008), pp. 401–419.
29. S. Antontsev, M. Chipot, and S. Shmarev, *Uniqueness and comparison theorems for solutions of doubly nonlinear parabolic equations with nonstandard growth conditions*, Commun. Pure Appl. Anal., 12 (2013), pp. 1527–1546.
30. S. Antontsev, M. Chipot, and Y. Xie, *Uniqueness results for equations of the $p(x)$-Laplacian type*, Adv. Math. Sci. Appl., 17 (2007), pp. 287–304.
31. S. Antontsev and H. B. de Oliveira, *The Oberbeck-Boussinesq problem modified by a thermo-absorption term*, J. Math. Anal. Appl., 379 (2011), pp. 802–817.
32. S. Antontsev and H. B. de Oliveira, *Asymptotic behavior of trembling fluids*, Nonlinear Anal. Real World Appl., 19 (2014), pp. 54–66.
33. S. Antontsev, J. I. Díaz, and A. Domanskiĭ, *Stability and stabilization of generalized solutions of degenerate problems of two-phase filtration*, Dokl. Akad. Nauk, 325 (1992), pp. 1151–1155. (English translation in "Soviet Phys. Dokl.", v.37, n.8, (1993), 411–413.
34. S. Antontsev, J. I. Díaz, and A. Domanskiĭ, *Continuous dependence and stabilization of solutions of the degenerate system in two-phase filtration*, Dinamika Sploshnoi Sredy, Fluids dynamics with free boundary, 107 (1993), pp. 11–25.
35. S. Antontsev, J. I. Díaz, and S. Shmarev, *The support shrinking properties for solutions of quasilinear parabolic equations with strong absorption terms*, Ann. Fac. Sci. Toulouse Math. (6), 4 (1995), pp. 5–30.
36. S. Antontsev, J. I. Díaz, and S. Shmarev, *Energy Methods for Free Boundary Problems: Applications to Non-linear PDEs and Fluid Mechanics*, Bikhäuser, Boston, 2002. Progress in Nonlinear Differential Equations and Their Applications, Vol. 48.

37. S. Antontsev, A. Domanskii, and V. Penkovskii, *Filtration in by-well zone of the formation and the well productivity stimulation problems*, Lavrentyev Institute of Hydrodynamics, Siberian Branch of the USSR Academy of Science, Novosibirsk, 1989. (in Russian).
38. S. Antontsev and J. Ferreira, *Existence, uniqueness and blowup for hyperbolic equations with nonstandard growth conditions*, Nonlinear Analysis Series A: Theory, Methods, Applications, 94 (2013), pp. 62–77.
39. S. Antontsev and J. Ferreira, *On a viscoelastic plate equation with strong damping and $\overrightarrow{p}(x,t)-$Laplacian. Existence and uniqueness*, Differential Integral Equations, 27 (2014), pp. 1147–1170.
40. S. Antontsev, F. Miranda, and L. Santos, *A class of electromagnetic p-curl systems: blow-up and finite time extinction*, Nonlinear Anal., 75 (2012), pp. 3916–3929.
41. S. Antontsev and J. F. Rodrigues, *On stationary thermo-rheological viscous flows*, Ann. Univ. Ferrara, Sez., VII. Sci. Mat., 52 (2006), pp. 19–36.
42. S. Antontsev and S. Shmarev, *A model porous medium equation with variable exponent of nonlinearity: existence, uniqueness and localization properties of solutions*, Nonlinear Anal., 60 (2005), pp. 515–545
43. S. Antontsev and S. Shmarev, *On the localization of solutions of elliptic equations with nonhomogeneous anisotropic degeneration*, Sibirsk. Mat. Zh., 46 (2005), pp. 963–984.
44. S. Antontsev and S. Shmarev, *Elliptic equations and systems with nonstandard growth conditions: existence, uniqueness and localization properties of solutions*, Nonlinear Anal., 65 (2006), pp. 728–761.
45. S. Antontsev and S. Shmarev, *Elliptic equations with anisotropic nonlinearity and nonstandard growth conditions*, vol. 3 of Handbook of Differential Equations, Stationary Partial Differential Equations, Elsevier B.V., 2006, ch. 1, pp. 1–100.
46. S. Antontsev and S. Shmarev, *Existence and uniqueness of solutions of degenerate parabolic equations with variable exponents of nonlinearity*, Fundam. Prikl. Mat., 12 (2006), pp. 3–19.
47. S. Antontsev and S. Shmarev, *Parabolic equations with anisotropic nonstandard growth conditions*, in Free boundary problems, vol. 154 of Internat. Ser. Numer. Math., Birkhäuser, Basel, 2007, pp. 33–44.
48. S. Antontsev and S. Shmarev, *Extinction of solutions of parabolic equations with variable anisotropic nonlinearities*, in Proceedings of the Steklov Institute of Mathematics, vol. 261, 2008, pp. 11–21.
49. S. Antontsev and S. Shmarev, *Anisotropic parabolic equations with variable nonlinearity*, Publ. Sec. Mat. Univ. Autònoma Barcelona, (2009), pp. 355–399.
50. S. Antontsev and S. Shmarev, *Directional localization of solutions to elliptic equations with nonstandard anisotropic growth conditions*, in More Progresses in Analysis, Proceedings of the 5th International ISAAC Congress, 25-30 July 2005, University of Catania, Italy. Editors H.G. W Begehr, F. Nicolosi, World Scientific Publishing Co. Pte. Ltd, 2009, pp. 681–690.
51. S. Antontsev and S. Shmarev, *Localization of solutions of anisotropic parabolic equations*, Nonlinear Anal., 71 (2009), pp. e725–e737.
52. S. Antontsev and S. Shmarev, *Blow-up of solutions to parabolic equations with non-standard growth conditions*, J. Comp. Applied Mathematics, 234 (2010), pp. 2633–2645.
53. S. Antontsev and S. Shmarev, *On the blow-up of solutions to anisotropic parabolic equations with variable nonlinearity*, Tr. Mat. Inst. Steklova, 270 (2010), pp. 33–48.
54. S. Antontsev and S. Shmarev, *Vanishing solutions of anisotropic parabolic equations with variable nonlinearity*, J. Math. Anal. Appl., 361 (2010), pp. 371–391.
55. S. Antontsev and S. Shmarev, *Elliptic equations with triple variable nonlinearity*, Complex Var. Elliptic Equ., 56 (2011), pp. 573–597.
56. S. Antontsev and S. Shmarev, *Parabolic equations with double variable nonlinearities*, Mathematics and Computers in Simulation, 81 (2011), pp. 2018–2032.
57. S. Antontsev and S. Shmarev, *Doubly degenerate parabolic equations with variable nonlinearity I: Existence of bounded strong solutions*, Advances in Differential Equations, 17 (2012), pp. 1181–1212.

58. S. Antontsev and S. Shmarev, *Energy solutions of evolution equations with nonstandard growth conditions*, Monografías de la Real Academia de Ciencias de Zaragoza, 38 (2012), pp. 85–111.
59. S. Antontsev and S. Shmarev, *Existence and uniqueness for double nonliner parabolic equations with non standard growth conditions*, Differential equations and applications, DEA, 4 (2012), pp. 67–94.
60. S. Antontsev and S. Shmarev, *Doubly degenerate parabolic equations with variable nonlinearity II: Blow-up and extinction in a finite time*, Nonlinear Anal., 95 (2014), pp. 483–498.
61. S. Antontsev and V. Zhikov, *Higher integrability for parabolic equations of $p(x,t)$-Laplacian type*, Adv. Differential Equations, 10 (2005), pp. 1053–1080.
62. S. N. Antontsev and A. V. Domanskiĭ, *Uniqueness of generalized solutions of degenerate problems of two-phase filtration*, Chisl. Metody Mekh. Sploshn. Sredy, 15 (1984), pp. 15–27. (in Russian).
63. S. N. Antontsev and G. Gagneux, *Petits paramètres et passages à la limite dans les problèmes de filtration diphasique*, Progress in partial differential equations. (Pont-*à*- Mousson 1997, Pitman Res. Notes in Mathematics Series, Longman, Harlow), 1 (1998).
64. D. G. Aronson, *The porous medium equation*, in Nonlinear diffusion problems (Montecatini Terme, 1985), vol. 1224 of Lecture Notes in Math., Springer, Berlin, 1986, pp. 1–46.
65. G. Autuori, P. Pucci, and M. C. Salvatori, *Asymptotic stability for anisotropic Kirchhoff systems*, J. Math. Anal. Appl., 352 (2009), pp. 149–165.
66. G. Autuori, P. Pucci, and M. C. Salvatori, *Global nonexistence for nonlinear Kirchhoff systems*, Arch.Rational Mech.Anal., 196 (2011), pp. 489–516.
67. K. Baghaei, M. B. Ghaemi, and M. Hesaaraki, *Lower bounds for the blow-up time in a semilinear parabolic problem involving a variable source*, Appl. Math. Lett., 27 (2014), pp. 49–52.
68. X. Bai and S. Zheng, *A semilinear parabolic system with coupling variable exponents*, Ann. Mat. Pura Appl. (4), 190 (2011), pp. 525–537.
69. G. I. Barenblatt, *On self-similar motions of a compressible fluid in a porous medium*, Akad. Nauk SSSR. Prikl. Mat. Meh., 16 (1952), pp. 679–698.
70. G. I. Barenblatt, *On some unsteady motions of a liquid and gas in a porous medium*, Akad. Nauk SSSR. Prikl. Mat. Meh., 16 (1952), pp. 67–78.
71. J. W. Barrett and W. B. Liu, *Finite element approximation of the parabolic p-Laplacian*, SIAM J. Numer. Anal., 31 (1994), pp. 413–428.
72. A. Benaissa and S. Mokeddem, *Decay estimates for the wave equation of p-Laplacian type with dissipation of m-Laplacian type*, Math. Methods Appl. Sci., 30 (2007), pp. 237–247.
73. M. Bendahmane and K. H. Karlsen, *Nonlinear anisotropic elliptic and parabolic equations in $\mathbb{R}^N$ with advection and lower order terms and locally integrable data*, Potential Anal., 22 (2005), pp. 207–227.
74. M. Bendahmane and K. H. Karlsen, *Anisotropic doubly nonlinear degenerate parabolic equations*, in Numerical mathematics and advanced applications, Springer, Berlin, 2006, pp. 381–386.
75. M. Bendahmane and K. H. Karlsen, *Renormalized solutions of an anisotropic reaction-diffusion-advection system with $L^1$ data*, Commun. Pure Appl. Anal., 5 (2006), pp. 733–762.
76. M. Bendahmane, K. H. Karlsen, and M. Saad, *Nonlinear anisotropic elliptic and parabolic equations with variable exponents and $L^1$ data*, Commun. Pure Appl. Anal., 12 (2013), pp. 1201–1220.
77. M. Bendahmane, P. Wittbold, and A. Zimmermann, *Renormalized solutions for a nonlinear parabolic equation with variable exponents and $L^1$-data*, J. Differential Equations, 249 (2010), pp. 1483–1515.
78. P. Bénilan, L. Boccardo, and M. A. Herrero, *On the limit of solutions of $u_t = \Delta u^m$ as $m \to \infty$*, Rend. Sem. Mat. Univ. Politec. Torino, (1989), pp. 1–13 (1991). Some topics in nonlinear PDEs (Turin, 1989).
79. F. Bernis, *Existence results for doubly nonlinear higher order parabolic equations on unbounded domains*, Math. Ann., 279 (1988), pp. 373–394.

80. P. Blanchard and E. Brüning, *Variational methods in mathematical physics. A unified approach*, Texts and Monographs in Physics, Springer-Verlag, Berlin, 1992.
81. L. Boccardo, T. Gallouët, and J. L. Vazquez, *Solutions of nonlinear parabolic equations without growth restrictions on the data*, Electron. J. Differential Equations, (2001), pp. No. 60, 20.
82. V. Bögelein and F. Duzaar, *Higher integrability for parabolic systems with non-standard growth and degenerate diffusions*, Publ. Mat., 55 (2011), pp. 201–250.
83. V. Bögelein and F. Duzaar, *Hölder estimates for parabolic $p(x,t)$-Laplacian systems*, Math. Ann., 354 (2012), pp. 907–938.
84. M. Bokalo, *Almost periodic solutions of anisotropic elliptic-parabolic equations with variable exponents of nonlinearity*, Electron. J. Differential Equations, (2014), pp. No. 169, 13.
85. M. M. Bokalo, O. M. Buhrii, and R. A. Mashiyev, *Unique solvability of initial-boundary-value problems for anisotropic elliptic-parabolic equations with variable exponents of nonlinearity*, J. Nonlinear Evol. Equ. Appl., (2013), pp. 67–87.
86. T. M. Bokalo, *Initial-boundary problems for systems of a high order doubly nonlinear parabolic equations with variable exponent of nonlinearity*, Mat. Stud., 38 (2012), pp. 68–79.
87. T. M. Bokalo and O. M. Buhrii, *Doubly nonlinear parabolic equations with variable exponents of nonlinearity*, Ukrainian Math. J., 63 (2011), pp. 709–728.
88. M.-M. Boureanu, *Infinitely many solutions for a class of degenerate anisotropic elliptic problems with variable exponent*, Taiwanese J. Math., 15 (2011), pp. 2291–2310.
89. M.-M. Boureanu and F. Preda, *Infinitely many solutions for elliptic problems with variable exponent and nonlinear boundary conditions*, NoDEA Nonlinear Differential Equations Appl., 19 (2012), pp. 235–251.
90. M.-M. Boureanu, P. Pucci, and V. D. Rădulescu, *Multiplicity of solutions for a class of anisotropic elliptic equations with variable exponent*, Complex Var. Elliptic Equ., 56 (2011), pp. 755–767.
91. M.-M. Boureanu and V. D. Rădulescu, *Anisotropic Neumann problems in Sobolev spaces with variable exponent*, Nonlinear Anal., 75 (2012), pp. 4471–4482.
92. O. M. Buhrii and R. A. Mashiyev, *Uniqueness of solutions of the parabolic variational inequality with variable exponent of nonlinearity*, Nonlinear Anal., 70 (2009), pp. 2325–2331.
93. N. Calvo, J. I. Díaz, J. Durany, E. Schiavi, and C. Vázquez, *On a doubly nonlinear parabolic obstacle problem modelling ice sheet dynamics*, SIAM J. Appl. Math., 63 (2002), pp. 683–707 (electronic).
94. C. Carstensen and M. O. Rieger, *Young-measure approximations for elastodynamics with non-monotone stress-strain relations*, M2AN Math. Model. Numer. Anal., 38 (2004), pp. 397–418.
95. C. Chen, *Global existence and $L^\infty$ estimates of solution for doubly nonlinear parabolic equation*, J. Math. Anal. Appl., 244 (2000), pp. 133–146.
96. M. Y. Chen, *Blow-up for doubly degenerate nonlinear parabolic equations*, Acta Math. Sin. (Engl. Ser.), 24 (2008), pp. 1525–1532.
97. Y. Chen, S. Levine, and M. Rao, *Variable exponent, linear growth functionals in image restoration*, SIAM J. Appl. Math, 66 (2006), pp. 1383–1406.
98. M. Chipot, *Elliptic equations: an introductory course*, Birkhäuser Advanced Texts: Basler Lehrbücher. [Birkhäuser Advanced Texts: Basel Textbooks], Birkhäuser Verlag, Basel, 2009.
99. M. Chipot and M. Michaille, *Uniqueness results and monotonicity properties for strongly nonlinear elliptic variational inequalities*, Ann. Scuola Norm. Sup. Pisa Cl. Sci., (4)16 (1989), pp. 137–166.
100. M. Chipot and M. Michaille, *Uniqueness results and monotonicity properties for the solutions of some variational inequalities. existence of a free boundary free boundary problems*, in Free boundary problems:theory and applications, Vol. I, (Irsee,1987), Pitman Res. Notes Math. Ser.185, Logman Sci. Tech., Horlow, 1990, pp. 271–276.
101. M. Chipot and J.-F. Rodrigues, *Comparison and stability of solutions to a class of quasilinear parabolic problems*, Proc. Roy. Soc. Edinburgh Sect. A, 110 (1988), pp. 275–285.
102. M. Chipot and T. Sideris, *An upper bound for the waiting time for nonlinear degenerate parabolic equations*, Trans. Amer. Math. Soc., 288 (1985), pp. 423–427.

103. P. Cianci, A. V. Martynenko, and A. F. Tedeev, *The blow-up phenomenon for degenerate parabolic equations with variable coefficients and nonlinear source*, Nonlinear Anal., 73 (2010), pp. 2310–2323.
104. J. Clements, *Existence theorems for a quasilinear evolution equation*, SIAM J. Appl. Math., 26 (1974), pp. 745–752.
105. D. V. Cruz-Uribe and A. Fiorenza, *Variable Lebesgue spaces*, Applied and Numerical Harmonic Analysis, Birkhäuser/Springer, Heidelberg, 2013. Foundations and harmonic analysis.
106. G. D'Aguì and A. Sciammetta, *Infinitely many solutions to elliptic problems with variable exponent and nonhomogeneous Neumann conditions*, Nonlinear Anal., 75 (2012), pp. 5612–5619.
107. S. P. Degtyarev and A. F. Tedeev, *Estimates for the solution of the Cauchy problem with increasing initial data for a parabolic equation with anisotropic degeneration and double nonlinearity*, Dokl. Akad. Nauk, 417 (2007), pp. 156–159.
108. S. P. Degtyarev and A. F. Tedeev, $L_1 - L_\infty$ *estimates for the solution of the Cauchy problem for an anisotropic degenerate parabolic equation with double nonlinearity and growing initial data*, Mat. Sb., 198 (2007), pp. 45–66.
109. J. Díaz and J. Padial, *Uniqueness and existence of a solution in* $BV_t(q)$ *space to a doubly nonlinear parabolic problem*, Publ. Mat., 40 (1996), pp. 527–560.
110. J. Díaz and F. Thélin, *On a nonlinear parabolic problem arising in some models related to turbulent flows*, SIAM J. Math. Anal, 25 (1994), pp. 1085–1111.
111. J. I. Díaz and G. Galiano, *On the boussinesq system with non linear thermal diffusion*, Nonlinear Anal., Theory, Meth. & App., 30 (1997), pp. 3255–3263.
112. J. I. Díaz and G. Galiano, *Existence and uniqueness of solutions of the Boussinesq system with nonlinear thermal diffusion*, Topol. Methods Nonlinear Anal., 11 (1998), pp. 59–82.
113. J. I. Díaz and L. Véron, *Local vanishing properties of solutions of elliptic and parabolic quasilinear equations*, Trans. Amer. Math. Soc., 290 (1985), pp. 787–814.
114. E. DiBenedetto, *Degenerate parabolic equations*, Universitext, Springer-Verlag, New York, 1993.
115. L. Diening, P. Harjulehto, P. Hästö, and M. Růžička, *Lebesgue and Sobolev Spaces with Variable Exponents*, Springer, Berlin, 2011. Series: Lecture Notes in Mathematics, Vol. 2017, 1st Edition.
116. L. Diening, P. Nägele, and M. Růžička, *Monotone operator theory for unsteady problems in variable exponent spaces*, Complex Var. Elliptic Equ., 57 (2012), pp. 1209–1231.
117. L. Diening, M. Růžička, and J. Wolf, *Existence of weak solutions for unsteady motions of generalized Newtonian fluids*, Ann. Sc. Norm. Super. Pisa Cl. Sci. (5), 9 (2010), pp. 1–46.
118. M. Dreher, *The wave equation for the p-Laplacian*, Hokkaido Math. J., 36 (2007), pp. 21–52.
119. J. A. Dubinskiĭ, *Weak convergence for nonlinear elliptic and parabolic equations*, Mat. Sb. (N.S.), 67 (109) (1965), pp. 609–642.
120. Y. Dubinskii, *Nonlinear elliptic and parabolic equations*, in Itogi nauki i tekhniki. Ser. Sovremennye problemy matematiki, vol. 9, VINITI, Moscow, 1976, pp. 5–126.
121. J. C. M. Duque, R. M. P. Almeida, and S. N. Antontsev, *Convergence of the Finite Element Method for the Porous Media Equation with Variable Exponent*, SIAM J. Numer. Anal., 51 (2013), pp. 3483–3504.
122. J. C. M. Duque, R. M. P. Almeida, and S. N. Antontsev, *Numerical study of the porous medium equation with absorption, variable exponents of nonlinearity and free boundary*, Appl. Math. Comput., 235 (2014), pp. 137–147.
123. A. El Hamidi and J. Vétois, *Sharp Sobolev asymptotics for critical anisotropic equations*, Arch. Ration. Mech. Anal., 192 (2009), pp. 1–36.
124. C. M. Elliott, M. A. Herrero, J. R. King, and J. R. Ockendon, *The mesa problem: diffusion patterns for* $u_t = \nabla \cdot (u^m \nabla u)$ *as* $m \to +\infty$, IMA J. Appl. Math., 37 (1986), pp. 147–154.
125. F. Ettwein and M. Růžička, *Existence of strong solutions for electrorheological fluids in two dimensions: steady Dirichlet problem*, Springer-Verlag, Berlin, 2003. Geometric Analysis and Nonlinear Partial Differential Equations, Edit. by S. Hildebrant and H. Karcher.

126. X. Fan, *Anisotropic variable exponent Sobolev spaces and* $\mathbf{p(x)}$*-Laplacian equations*, Complex Var. Elliptic Equ., 56 (2011), pp. 623–642.
127. X. Fan, J. Shen, and D. Zhao, *Sobolev embeddings theorems for spaces* $W^{k,p(x)}$, J. Math. Anal. Appl., 262, no 2 (2001), pp. 749–760.
128. X. Fan and D. Zhao, *On the spaces* $L^{p(x)}(\Omega)$ *and* $W^{m,p(x)}(\Omega)$, J. Math. Anal. Appl., 263 (2001), pp. 424–446.
129. J. Fernández Bonder, N. Saintier, and A. Silva, *On the Sobolev embedding theorem for variable exponent spaces in the critical range*, J. Differential Equations, 253 (2012), pp. 1604–1620.
130. R. Ferreira, A. de Pablo, M. Pérez-LLanos, and J. D. Rossi, *Critical exponents for a semilinear parabolic equation with variable reaction*, Proc. Roy. Soc. Edinburgh Sect. A, 142 (2012), pp. 1027–1042.
131. I. Fragalà, F. Gazzola, and B. Kawohl, *Existence and nonexistence results for anisotropic quasilinear elliptic equations*, Ann. Inst. H. Poincaré Anal. Non Linéaire, 21 (2004), pp. 715–734.
132. G. Fragnelli, *Positive periodic solutions for a system of anisotropic parabolic equations*, J. Math. Anal. Appl., 367 (2010), pp. 204–228.
133. A. Friedman, *Partial differential equations*, Holt, Rinehart and Winston, Inc., New York, 1969.
134. Y. Fu and N. Pan, *Existence of solutions for nonlinear parabolic problem with* $p(x)$*-growth*, J. Math. Anal. Appl., 362 (2010), pp. 313–326.
135. Y. Fu and N. Pan, *Local boundedness of weak solutions for nonlinear parabolic problem with* $p(x)$*-growth*, J. Inequal. Appl., (2010), pp. Art. ID 163269, 16.
136. V. A. Galaktionov and S. I. Pohozaev, *Blow-up and critical exponents for nonlinear hyperbolic equations*, Nonlinear Analysis, 53 (2003), pp. 453–466.
137. V. A. Galaktionov and J. L. Vázquez, *The problem of blow-up in nonlinear parabolic equations*, Discrete Contin. Dyn. Syst., 8 (2002), pp. 399–433. Current developments in partial differential equations (Temuco, 1999).
138. V. A. Galaktionov and J. L. Vázquez, *A stability technique for evolution partial differential equations*, Progress in Nonlinear Differential Equations and their Applications, 56, Birkhäuser Boston, Inc., Boston, MA, 2004. A dynamical systems approach.
139. V. A. Galaktionov and J. L. Vázquez, *A stability technique for evolution partial differential equations*, Progress in Nonlinear Differential Equations and their Applications, 56, Birkhäuser Boston Inc., (2004).
140. G. Galiano, *Spatial and time localization of solutions of the Boussinesq system with nonlinear thermal diffusion*, Nonlinear Anal., 42 (2000), pp. 423–438.
141. H. Gao and T. F. Ma, *Global solutions for a nonlinear wave equation with p-Laplacian operator*, EJQTDE., (1999), pp. 1–13.
142. W. Gao and B. Guo, *Existence and localization of weak solutions of nonlinear parabolic equations with variable exponent of nonlinearity*, Ann. Mat. Pura Appl. (4), 191 (2012), pp. 551–562.
143. Y. Gao and W. Gao, *Extinction and asymptotic behavior of solutions for nonlinear parabolic equations with variable exponent of nonlinearity*, Bound. Value Probl., (2013), pp. 2013:164, 10.
144. Y.-Z. Gao and W.-J. Gao, *Existence and blow-up of solutions for a nonlinear parabolic system with variable exponents*, Commun. Math. Res., 29 (2013), pp. 61–67.
145. V. Georgiev and G. Todorova, *Existence of a solution of the wave equation with nonlinear damping and source terms*, J. Differential Equations, 109 (1994), pp. 295–308.
146. J. Giacomoni and G. Vallet, *Some results about an anisotropic* $p(x)$*-Laplace-Barenblatt equation*, Adv. Nonlinear Anal., 1 (2012), pp. 277–298.
147. D. Gilbarg and N. S. Trudinger, *Elliptic partial differential equations of second order*, Classics in Mathematics, Springer-Verlag, Berlin, 2001. Reprint of the 1998 edition.
148. B. H. Gilding and R. Kersner, *The characterization of reaction-convection-diffusion processes by travelling waves*, J. Differential Equations, 124 (1996), pp. 27–79.

149. B. H. Gilding and R. Kersner, *Travelling waves in nonlinear diffusion-convection reaction*, Progress in Nonlinear Differential Equations and their Applications, 60, Birkhäuser Verlag, Basel, 2004.
150. B. Guo and W. Gao, *Study of weak solutions for a fourth-order parabolic equation with variable exponent of nonlinearity*, Z. Angew. Math. Phys., 62 (2011), pp. 909–926.
151. B. Guo and W. Gao, *Study of weak solutions for parabolic equations with nonstandard growth conditions*, J. Math. Anal. Appl., 374 (2011), pp. 374–384.
152. B. Guo and W. Gao, *Existence and asymptotic behavior of solutions for nonlinear parabolic equations with variable exponent of nonlinearity*, Acta Math. Sci. Ser. B Engl. Ed., 32 (2012), pp. 1053–1062.
153. Z. Guo, Q. Liu, J. Sun, and B. Wu, *Reaction-diffusion systems with $p(x)$-growth for image denoising*, Nonlinear Anal. Real World Appl., 12 (2011), pp. 2904–2918.
154. J. Haehnle and A. Prohl, *Approximation of nonlinear wave equations with nonstandard anisotropic growth conditions*, Math. Comp., 79 (2010), pp. 189–208.
155. M. R. Hajiaboli, *An anisotropic fourth-order diffusion filter for image noise removal*, Int. J. Comput. Vis., 92 (2011), pp. 177–191.
156. T. Halsey, *Electrorheological fluids*, Science, 258 (1992), pp. 761–766.
157. P. Harjulehto, P. Hästö, Ú. V. Lê, and M. Nuortio, *Overview of differential equations with non-standard growth*, Nonlinear Anal., 72 (2010), pp. 4551–4574.
158. J. Haškovec and C. Schmeiser, *A note on the anisotropic generalizations of the Sobolev and Morrey embedding theorems*, Monatsh. Math., 158 (2009), pp. 71–79.
159. E. Henriques, *Regularity for the porous medium equation with variable exponent: the singular case*, J. Differential Equations, 244 (2008), pp. 2578–2601.
160. E. Henriques, *Concerning the regularity of the anisotropic porous medium equation*, J. Math. Anal. Appl., 377 (2011), pp. 710–731.
161. E. Henriques and J. M. Urbano, *Intrinsic scaling for PDE's with an exponential nonlinearity*, Indiana Univ. Math. J., 55 (2006), pp. 1701–1721.
162. S.-Y. Hsu, *Removable singularities and non-uniqueness of solutions of a singular diffusion equation*, Math. Ann., 325 (2003), pp. 665–693.
163. K. M. Hui, *Singular limit of solutions of the equation $u_t = \Delta(u^m/m)$ as $m \to 0$*, Pacific J. Math., 187 (1999), pp. 297–316.
164. K. M. Hui, *Asymptotic behaviour of solutions of $u_t = \Delta \log u$ in a bounded domain*, Differential Integral Equations, 14 (2001), pp. 175–188.
165. K. Ishige, *On the existence of solutions of the Cauchy problem for a doubly nonlinear parabolic equation*, SIAM J. Math. Anal., 27 (1996), pp. 1235–1260.
166. A. I. Ivanov and J. F. Rodrigues, *Existence and uniqueness of a weak solution to the initial mixed boundary-value problem for quasilinear elliptic-parabolic equations*, Journal of Mathematical Sciences, Vol. 109, No. 5 (2002), pp. 1851–1866.
167. J. Jena, *Lie group of transformations to study anisotropic diffusion of scalar images*, Lobachevskii J. Math., 33 (2012), pp. 232–238.
168. H. Jian and B. Song, *Solutions of the anisotropic porous medium equation in $\mathbb{R}^n$ under an $L^1$-initial value*, Nonlinear Anal., 64 (2006), pp. 2098–2111.
169. J. Kačur, *On a solution of degenerate elliptic-parabolic systems in Orlicz-Sobolev spaces. I*, Math. Z., 203 (1990), pp. 153–171.
170. J. Kačur, *On a solution of degenerate elliptic-parabolic systems in Orlicz-Sobolev spaces. II*, Math. Z., 203 (1990), pp. 569–579.
171. A. S. Kalashnikov, *Some problems of the qualitative theory of second-order nonlinear degenerate parabolic equations*, Uspekhi Mat. Nauk, 42 (1987), pp. 135–176, 287.
172. A. S. Kalashnikov, *Some nonlinear mathematical models in the theory of heat conduction that contain a small parameter in the exponents*, Differentsial'nye Uravneniya, 30 (1994), pp. 1039–1049, 1102–1103.
173. A. S. Kalashnikov, *On some problems in the nonlinear theory of heat conduction with data containing a small parameter in the exponents*, Zh. Vychisl. Mat. i Mat. Fiz., 35 (1995), pp. 1077–1094.

174. A. S. Kalashnikov, *On some nonlinear problems in mathematical physics with exponents that are close to critical*, Tr. Semin. im. I. G. Petrovskogo, (1996), pp. 73–98, 346.
175. S. Kamin and J. L. Vázquez, *Fundamental solutions and asymptotic behaviour for the p-Laplacian equation*, Rev. Mat. Iberoamericana, 4 (1988), pp. 339–354.
176. S. Kamin and J. L. Vázquez, *Asymptotic behaviour of solutions of the porous medium equation with changing sign*, SIAM J. Math. Anal., 22 (1991), pp. 34–45.
177. S. Kaplan, *On the growth of the solutions of quasilinear parabolic equations*, Comm. Pure Appl. Math., 16 (1963), pp. 305–330.
178. T. Kato, *Blow up of solutions of some nonlinear hyperbolic equations*, Manusscripta Math., 28 (1980), pp. 235–268.
179. T. S. Khin and N. Su, *Propagation property for anisotropic nonlinear diffusion equation with convection*, J. Math. Anal. Appl., 354 (2009), pp. 220–228.
180. T. S. Khin and N. Su, *Propagation property for nonlinear parabolic equations of p-Laplacian-type*, Int. J. Math. Anal. (Ruse), 3 (2009), pp. 591–602.
181. P. E. Kloeden and J. Simsen, *Pullback attractors for non-autonomous evolution equations with spatially variable exponents*, Commun. Pure Appl. Anal., 13 (2014), pp. 2543–2557.
182. M. O. Korpusov, *Solution blowup for the heat equation with double nonlinearity*, Theoret. and Math. Phys., 172 (2012), pp. 1173–1176. Translation of Teoret. Mat. Fiz. 172 (2012), no. 3, 339–343.
183. M. O. Korpusov, *On the blow-up of solutions of a class of parabolic equations with double nonlinearity*, Mat. Sb., 204 (2013), pp. 19–42.
184. O. Kováčik and J. Rákosník, *On spaces $L^{p(x)}$ and $W^{k,p(x)}$*, Czechoslovak Math. J., 41(116) (1991), pp. 592–618.
185. O. A. Ladyženskaja, V. A. Solonnikov, and N. N. Ural'ceva, *Linear and quasilinear equations of parabolic type*, Translated from the Russian by S. Smith. Translations of Mathematical Monographs, Vol. 23, American Mathematical Society, Providence, R.I., 1968.
186. O. Ladyzhenskaya, *Attractors for semigroups and evolution equations*, Lezioni Lincee. [Lincei Lectures], Cambridge University Press, Cambridge, 1991.
187. O. A. Ladyzhenskaya and N. N. Ural'tseva, *Linear and quasilinear elliptic equations*, Translated from the Russian by Scripta Technica, Inc. Translation editor: Leon Ehrenpreis, Academic Press, New York, 1968.
188. G. I. Laptev, *Solvability of second-order quasilinear parabolic equations with double degeneration*, Sibirsk. Mat. Zh., 38 (1997), pp. 1335–1355, iii.
189. G. I. Laptev, *Weak solutions of second-order quasilinear parabolic equations with double non-linearity*, Sbornik: Mathematics, 188 (1997), p. 1343.
190. H. Levine and G. Todorova, *Blow-up of solutions of the cauchy problem for a wave equations with nonlinear damping and source terms and positive initial energy*, in Proceedings of the American Mathematical Society, vol. 129, 2000, pp. 793–805.
191. S. Levine, Y. Chen, and J. Stanich, *Image restoration via nonstandard diffusion*, Technical Report ♯ 04-01, Dept. of Mathematics and Computer Science, Duquesne University, 2004.
192. F. Li, *Anisotropic parabolic equations with measure data. II*, Math. Nachr., 279 (2006), pp. 1585–1596.
193. F. Li and B. Liu, *Asymptotic analysis for blow-up solutions in parabolic equations involving variable exponents*, Appl. Anal., 92 (2013), pp. 651–664.
194. F. Li and H. Zhao, *Anisotropic parabolic equations with measure data*, J. Partial Differential Equations, 14 (2001), pp. 21–30.
195. S. Lian, W. Gao, C. Cao, and H. Yuan, *Study of the solutions to a model porous medium equation with variable exponent of nonlinearity*, Journal of Mathematical Analysis and Applications, 342 (2008), pp. 27 – 38.
196. S. Lian, W. Gao, H. Yuan, and C. Cao, *Existence of solutions to an initial Dirichlet problem of evolutional $p(x)$-Laplace equations*, Ann. Inst. H. Poincaré Anal. Non Linéaire, 29 (2012), pp. 377–399.
197. J.-L. Lions, *Quelques méthodes de résolution des problèmes aux limites non linéaires*, Dunod, 1969.

198. B. Liu and F. Li, *Blowup solutions and their blowup rates for parabolic equations with non-standard growth conditions*, Electron. J. Qual. Theory Differ. Equ., (2011), pp. No. 77, 11.
199. B. Liu and F. Li, *Non-simultaneous blowup in heat equations with nonstandard growth conditions*, J. Differential Equations, 252 (2012), pp. 4481–4502.
200. C. Liu, J. Gao, and S. Lian, *Existence of solutions for the evolution $p(x)$-Laplacian equation not in divergence form*, J. Appl. Math., (2012), pp. Art. ID 835495, 21.
201. J. Málek, J. Nečas, M. Rokyta, and M. Růžička, *Weak and measure-valued solutions to evolutionary PDEs*, Chapman Hall, London, 1996.
202. R. A. Mashiyev and O. M. Buhrii, *Existence of solutions of the parabolic variational inequality with variable exponent of nonlinearity*, J. Math. Anal. Appl., 377 (2011), pp. 450–463.
203. V. S. Melnik and J. Valero, *On attractors of multivalued semi-flows and differential inclusions*, Set-Valued Anal., 6 (1998), pp. 83–111.
204. S. A. Messaoudi, *On the decay of solutions for a class of quasilinear hyperbolic equations with non-linear damping and source terms*, Mathematical Methods in the Applied Sciences, (2003).
205. S. A. Messaoudi and B. Said Houari, *Global non-existence of solutions of a class of wave equations with non-linear damping and source terms*, Math. Methods Appl. Sci., 27 (2004), pp. 1687–1696.
206. M. Mihăilescu, G. Moroşanu, and V. Rădulescu, *Eigenvalue problems in anisotropic Orlicz-Sobolev spaces*, C. R. Math. Acad. Sci. Paris, 347 (2009), pp. 521–526.
207. M. Mihăilescu, G. Moroşanu, and V. Rădulescu, *Eigenvalue problems for anisotropic elliptic equations: an Orlicz-Sobolev space setting*, Nonlinear Anal., 73 (2010), pp. 3239–3253.
208. M. Mihăilescu and C. Varga, *Multiplicity results for some elliptic problems with nonlinear boundary conditions involving variable exponents*, Comput. Math. Appl., 62 (2011), pp. 3464–3471.
209. È. Mitidieri and S. I. Pokhozhaev, *A priori estimates and the absence of solutions of nonlinear partial differential equations and inequalities*, Tr. Mat. Inst. Steklova, 234 (2001), pp. 1–384.
210. S. Müller, *Variational models for microstructure and phase transitions*, Calculus of variations and geometric evolution problems (Cetraro, 1996), Lecture Notes in Math., 1713, Springer, Berlin, (1999), pp. 85–210.
211. J. Musielak, *Orlicz spaces and modular spaces*, vol. 1034 of Lecture Notes in Mathematics, Springer-Verlag, Berlin, 1983.
212. H. Nakano, *Modulared Semi-Ordered Linear Spaces*, Maruzen Co. Ltd., Tokyo, 1950.
213. H. Nakano, *Modulared linear spaces*, J. Fac. Sci. Univ. Tokyo. Sect. I., 6 (1951), pp. 85–131.
214. W. Niu, *Long-time behavior for a nonlinear parabolic problem with variable exponents*, J. Math. Anal. Appl., 393 (2012), pp. 56–65.
215. W. Orlicz, *Über konjugierte exponentenfolgen*, Studia Math, 3 (1931), pp. 200–211.
216. S. E. Pastukhova, *Compensated compactness principle and solvability of generalized Navier-Stokes equations*, J. Math. Sci. (N. Y.), 173 (2011), pp. 769–802. Problems in mathematical analysis. No. 55.
217. P. Perona and J. Malik, *Scale-space and edge detection using anisotropic diffusion*, IEEE Transactions on Pattern Analysis and Machine Intelligence, 12 (1990), pp. 629–639.
218. J. P. Pinasco, *Blow-up for parabolic and hyperbolic problems with variable exponents*, Nonlinear Anal., 71 (2009), pp. 1094–1099.
219. M. M. Porzio, *$L^\infty$-regularity for degenerate and singular anisotropic parabolic equations*, Boll. Un. Mat. Ital. A (7), 11 (1997), pp. 697–707.
220. M. M. Porzio, *Existence, uniqueness and behavior of solutions for a class of nonlinear parabolic problems*, Nonlinear Anal., 74 (2011), pp. 5359–5382.
221. P. Sun, M. Liu, and C. Cao, *Extinction in finite time of solutions to the nonlinear diffusion equations involving $p(x,t)$-laplacian operator*, Boundary Value Problems, (2013), p. 2013:259.
222. P. Quittner and P. Souplet, *Superlinear parabolic problems*, Birkhäuser Advanced Texts: Basler Lehrbücher. [Birkhäuser Advanced Texts: Basel Textbooks], Birkhäuser Verlag, Basel, 2007. Blow-up, global existence and steady states.

223. K. Rajagopal and M. Růžička, *Mathematical modelling of electro-rheological fluids*, Cont. Mech. Therm., 13 (2001), pp. 59–78.
224. J. Rákosník, *Some remarks to anisotropic Sobolev spaces. I*, Beiträge Anal., (1979), pp. 55–68.
225. J. Rákosník, *Some remarks to anisotropic Sobolev spaces. II*, Beiträge Anal., (1980), pp. 127–140 (1981).
226. J.-M. Rakotoson, *T-sets and relaxed solutions for parabolic equations*, J. Differential Equations, 111 (1994), pp. 458–471.
227. M. Rammaha, D. Toundykov, and Z. Wilstein, *Global existence and decay of energy for a nonlinear wave equation with p-Laplacian damping*, Discrete Contin. Dyn. Syst., 32 (2012), pp. 4361–4390.
228. M. Růžička, *Electrorheological fluids: modeling and mathematical theory*, Springer, Berlin, 2000. Lecture Notes in Mathematics, 1748.
229. M. O. Rieger, *Young measure solutions for nonconvex elastodynamics*, SIAM J. Math. Anal., 34 (2003), pp. 1380–1398 (electronic).
230. A. Rodríguez and J. L. Vázquez, *Non-uniqueness of solutions of nonlinear heat equations of fast diffusion type*, Ann. Inst. H. Poincaré Anal. Non Linéaire, 12 (1995), pp. 173–200.
231. A. Rodriguez, J. L. Vazquez, and J. R. Esteban, *The maximal solution of the logarithmic fast diffusion equation in two space dimensions*, Adv. Differential Equations, 2 (1997), pp. 867–894.
232. E. S. Sabinina, *On a class of non-linear degenerate parabolic equations*, Dokl. Akad. Nauk SSSR, 143 (1962), pp. 794–797.
233. A. A. Samarskii, V. A. Galaktionov, S. P. Kurdyumov, and A. P. Mikhailov, *Blow-up in quasilinear parabolic equations*, Walter de Gruyter & Co., Berlin, 1995. Translated from the 1987 Russian original by Michael Grinfeld and revised by the authors.
234. M. Sango, *On a doubly degenerate quasilinear anisotropic parabolic equation*, Analysis (Munich), 23 (2003), pp. 249–260.
235. J. Simon, *Compact sets in the space $L^p(0,T;B)$*, Ann. Mat. Pura Appl. (4), 146 (1987), pp. 65–96.
236. J. Simsen, *A global attractor for a $p(x)$-Laplacian parabolic problem*, Nonlinear Anal., 73 (2010), pp. 3278–3283.
237. J. Simsen, *A global attractor for a $p(x)$-Laplacian inclusion*, C. R. Math. Acad. Sci. Paris, 351 (2013), pp. 87–90.
238. J. Simsen and M. S. Simsen, *On $p(x)$-Laplacian parabolic problems*, Nonlinear Stud., 18 (2011), pp. 393–403.
239. J. Simsen and M. S. Simsen, *PDE and ODE limit problems for $p(x)$-Laplacian parabolic equations*, J. Math. Anal. Appl., 383 (2011), pp. 71–81.
240. J. Simsen and M. S. Simsen, *Existence and upper semicontinuity of global attractors for $p(x)$-Laplacian systems*, J. Math. Anal. Appl., 388 (2012), pp. 23–38.
241. J. Simsen, M. S. Simsen, and M. R. T. Primo, *Continuity of the flows and upper semicontinuity of global attractors for $p_s(x)$-Laplacian parabolic problems*, J. Math. Anal. Appl., 398 (2013), pp. 138–150.
242. J. Simsen, M. Stefanello Simsen, and F. B. Rocha, *Existence of solutions for some classes of parabolic problems involving variable exponents*, Nonlinear studies, 21 (2014), pp. 113–128.
243. S. L. Sobolev, *Some applications of functional analysis in mathematical physics*, vol. 90 of Translations of Mathematical Monographs, American Mathematical Society, Providence, RI, 1991. Translated from the third Russian edition by Harold H. McFaden, With comments by V. P. Palamodov.
244. K. N. Soltanov, *Some nonlinear equations of the nonstable filtration type and embedding theorems*, Nonlinear Anal., 65 (2006), pp. 2103–2134.
245. B. Song, *Anisotropic diffusions with singular advections and absorptions. part 1. existence*, Applied Mathematics Letters, 14 (2001), pp. 811–816.
246. B. Song, *Anisotropic diffusions with singular advections and absorptions. part 2. uniqueness*, Applied Mathematics Letters, 14 (2001), pp. 817–823.

247. W. J. Song, B. Guo, and W. J. Gao, *Blow-up and global boundedness for a weakly coupled parabolic system with variable exponents*, Acta Math. Sci. Ser. A Chin. Ed., 33 (2013), pp. 285–291.
248. W. J. Song, B. Guo, and W. J. Gao, *Blow-up and global boundedness for a weakly coupled parabolic system with variable exponents*, Acta Math. Sci. Ser. A Chin. Ed., 33 (2013), pp. 285–291.
249. N. M. Stavrakakis and A. N. Stylianou, *Global attractor for some wave equations of $p$- and $p(x)$-Laplacian type*, Differential Integral Equations, 24 (2011), pp. 159–176.
250. A. F. Tedeev, *The interface blow-up phenomenon and local estimates for doubly degenerate parabolic equations*, Appl. Anal., 86 (2007), pp. 755–782.
251. M. Troisi, *Teoremi di inclusione per spazi di Sobolev non isotropi*, Ricerche Mat., 18 (1969), pp. 3–24.
252. J. L. Vázquez, *Asymptotic beahviour for the porous medium equation posed in the whole space*, J. Evol. Equ., 3 (2003), pp. 67–118. Dedicated to Philippe Bénilan.
253. J. L. Vázquez, *Darcy's law and the theory of shrinking solutions of fast diffusion equations*, SIAM J. Math. Anal., 35 (2003), pp. 1005–1028 (electronic).
254. J. L. Vázquez, *The porous medium equation*, Oxford Mathematical Monographs, The Clarendon Press Oxford University Press, Oxford, 2007. Mathematical theory.
255. J. Vétois, *The blow-up of critical anisotropic equations with critical directions*, NoDEA Nonlinear Differential Equations Appl., 18 (2011), pp. 173–197.
256. J. Vétois, *Strong maximum principles for anisotropic elliptic and parabolic equations*, Adv. Nonlinear Stud., 12 (2012), pp. 101–114.
257. H. Wang and Y. He, *On blow-up of solutions for a semilinear parabolic equation involving variable source and positive initial energy*, Appl. Math. Lett., 26 (2013), pp. 1008–1012.
258. Y. Wang, *Intrinsic Harnack inequalities for parabolic equations with variable exponents*, Nonlinear Anal., 83 (2013), pp. 12–30.
259. Y. Wang, W. Ren, and H. Wang, *Anisotropic second and fourth order diffusion models based on Convolutional Virtual Electric Field for image denoising*, Comput. Math. Appl., 66 (2013), pp. 1729–1742.
260. J. Weickert, *Anisotropic diffusion in image processing*, European Consortium for Mathematics in Industry, B. G. Teubner, Stuttgart, 1998.
261. Z. Wilstein, *Global well-posedness for a nonlinear wave equation with $p$-Laplacian damping*, Dissertation, University of Nebraska-Lincoln, http://digitalcommons.unl.edu/mathstudent/24, (2011), pp. 1–116.
262. J. Wolf, *Existence of weak solutions to the equations of non-stationary motion of non-Newtonian fluids with shear rate dependent viscosity*, J. Math. Fluid Mech., 9 (2007), pp. 104–138.
263. X. Wu, B. Guo, and W. Gao, *Blow-up of solutions for a semilinear parabolic equation involving variable source and positive initial energy*, Appl. Math. Lett., 26 (2013), pp. 539–543.
264. M. Xu and Y. Z. Chen, *Hölder continuity of weak solutions for parabolic equations with nonstandard growth conditions*, Acta Math. Sin. (Engl. Ser.), 22 (2006), pp. 793–806.
265. Z. Yang and G. Chen, *Global existence of solutions for quasi-linear wave equations with viscous damping*, J. Math. Anal. Appl., 285 (2003), pp. 604–618.
266. F. Yao, *Hölder regularity of the gradient for the non-homogeneous parabolic $p(x, t)$-laplacian equations.*, Math. Meth. Appl. Sci., (2013). doi:10.1002/mma.2953.
267. J. Yin, J. Li, and C. Jin, *Non-extinction and critical exponent for a polytropic filtration equation*, Nonlinear Anal., 71 (2009), pp. 347–357.
268. M. Yoshihiro, O. Takao, S. Tetsu, and S. Naoki, *Compact embeddings for Sobolev spaces of variable exponents and existence of solutions for nonlinear elliptic problems involving the $p(x)$-laplacian and its critical exponent*, Annales Academiæ Scientiarum Fennicæ Mathematica, 35 (2010), pp. 115–130.
269. M. Yu and X. Lian, *Boundedness of solutions of parabolic equations with anisotropic growth conditions*, Canad. J. Math., 49 (1997), pp. 798–809.

270. Y. B. Zel'dovič and A. S. Kompaneec, *On the theory of propagation of heat with the heat conductivity depending upon the temperature*, in Collection in honor of the seventieth birthday of academician A. F. Ioffe, Izdat. Akad. Nauk SSSR, Moscow, 1950, pp. 61–71.
271. C. Zhang and S. Zhou, *Renormalized and entropy solutions for nonlinear parabolic equations with variable exponents and $L^1$ data*, J. Differential Equations, 248 (2010), pp. 1376–1400.
272. J. N. Zhao, *Existence and nonexistence of solutions for $u_t = \mathrm{div}(|\nabla u|^{p-2}\nabla u) + f(\nabla u, u, x, t)$*, J. Math. Anal. Appl., 172 (1993), pp. 130–146.
273. Y. Zhijian, *Global existence, asymptotic behavior and blowup of solutions for a class of nonlinear wave equations with dissipative term*, J. Differential Equations, 187 (2003), pp. 520–540.
274. Y. Zhijian, *Initial boundary value problem for a class of non-linear strongly damped wave equations*, Math. Methods Appl. Sci., 26 (2003), pp. 1047–1066.
275. V. V. Zhikov, *On the density of smooth functions in Sobolev-Orlicz spaces*, Zap. Nauchn. Sem. S.-Peterburg. Otdel. Mat. Inst. Steklov. (POMI), 310 (2004), pp. 67–81, 226.
276. V. V. Zhikov, *Solvability of the three-dimensional thermistor problem*, Tr. Mat. Inst. Steklova, 261 (2008), pp. 101–114.
277. V. V. Zhikov, *On an approach to the solvability of generalized Navier-Stokes equations*, Funktsional. Anal. i Prilozhen., 43 (2009), pp 33–53.
278. V. V. Zhikov and S. E. Pastukhova, *On the solvability of the Navier-Stokes system for a nonhomogeneous non-Newtonian fluid*, Dokl. Akad. Nauk, 426 (2009), pp. 592–596.
279. V. V. Zhikov and S. E. Pastukhova, *Global solvability of Navier-Stokes equations for a non-homogeneous non-Newtonian fluid*, in Nonlinear partial differential equations and related topics, vol. 229 of Amer. Math. Soc. Transl. Ser. 2, Amer. Math. Soc., Providence, RI, 2010, pp. 233–252.
280. V. V. Zhikov and S. E. Pastukhova, *On the property of higher integrability for parabolic systems of variable order of nonlinearity*, Mat. Zametki, 87 (2010), pp. 179–200.

# Index


**B**
Blow-up, 241, 322
  doubly nonlinear parabolic equation, 323
  eventually linear equation, 250
  higher-order equations, 262
  hyperbolic equation with damping, 359
  nonlocal equations, 252
  $p(x)$-Laplacian with sources, 255
  regional, 251
  regional, hyperbolic equation, 388
  semilinear equation, 242
  vanishing sources, 247
  Young measure solution, 367

**C**
Carathéodory function, 4

**D**
Darcy's law, 100
Differential inequality
  1st order, homogeneous, 90
  1st order, nonhomogeneous, 92, 196, 198, 204
  2nd order, 259, 381
Diffusion, 175
  fast, 185
  slow, 185
Directional localization, 198
  nonpropagation, 204
  stationary solutions, 208
  unbounded solutions, 216
Doubly nonlinear equation
  energy strong solution, 322
  strong solution, 278, 293

**E**
Electromagnetism, 263
Electro-rheological fluids, 175
Embedding theorems
  anisotropic spaces, 34, 327
Energy functional, 348
Equation of state, 100
Evolution $p(x,t)$-Laplacian
  anisotropic, 107
  Barenblatt-type equation, 174
  Cauchy problem, 144, 215
  discontinuous $p(x,t)$, 178
  energy solution, 110
  entropy solution, 177
  higher-order, 174
  $H$ solution, 179
  local weak solution, 182, 188
  Neumann problem, 173
  renormalized solution, 177
  singular perturbation, 141
  stationary solutions, 161
  systems, 173, 183
  systems, 172, 173
  $W$ solution, 179
Extinction in a finite time, 81, 185, 219, 335
  at a prescribed moment, 336
  critical growth, 339
  equations of general form, 238




S. Antontsev and S. Shmarev, *Evolution PDEs with Nonstandard Growth Conditions*, Atlantis Studies in Differential Equations 4,
DOI 10.2991/978-94-6239-112-3

eventually linear equations, 339
limit cases, 97, 220, 236, 337

F
Fick's law, 185
Finite speed of propagation, 81, 185
Friedrichs's mollifier, 19
Function spaces
$H^k(\Omega)$, $H_0^k(\Omega)$, 3
$L^M(\Omega)$, 7
$L^\infty(\Omega)$, 2
$L^p(0,T;L^q(\Omega))$, 3
$L^p(0,T;W_0^{1,p}(\Omega))$, 3
$V$, 146
$W$, 145
$W'$, 145
$W_0^{1,p(\cdot)}(\Omega)$, 18
$W^{k,q}(\Omega)$, 2
$\mathbf{U}'(Q)$, 27
$\mathbf{U}(Q)$, 27
$\mathbf{V}_+(\Omega)$, 23
$\mathbf{V}_\sigma(\Omega)$, 161
$\mathbf{V}_t(\Omega)$, 22, 292
$\mathbf{W}(Q)$, 22, 292
$\mathbf{W}_+(Q)$, 141
$\mathbf{W}'(Q)$, 22, 292
$\mathscr{V}(Q)$, 278
$L^{p(\cdot)}(\Omega)$, 4

G
Global attractor, 240
Gronwall's inequality, 3

H
Hölder's inequality, 3, 9
inverse, 244, 386
Higher-order equations, 263
Hyperbolic equation
blow-up, 385
nonlocal, 344, 389
$p(x)$-Laplacian, 344
strong solution, 371
semilinear, weak solution, 370
system, 344
weak energy. solution, 345
Young measure solutions, 364

I
Ideal barotropic gas, 100
Image processing, 175
Interpolation inequalities, 40
$L^{p(\cdot)}(Q)$, 42
trace interpolation, 41

K
Kaplan's method of eigenfunctions, 242, 385

L
Large time behavior, 340
Leray-Lions operator, 141, 180, 240
Logarithmic continuity
$C_{\log}(\Omega)$, 18
$C_{\log}(Q_T)$, 24
Luxemburg's norm, 5

M
Mass conservation law, 100
Modular
$A_{p(\cdot)}$, 4 , 23
Monotonicity, 120

N
Navier-Stokes equations, 175, 264
Non-newtonian fluids, 175, 264
Notation
$Q$, 1
$Q_T$, 2
$\Gamma$, 1
$\Gamma_T$, 2
$\Omega$, 1
$\phi^\pm$, 2
$\phi^\pm(t)$, 2

O
Orlicz's norm, 12

P
$p(x,t)$-curl systems, 263
Poincaré's inequality, 18
Porous Medium Equation, 45
anisontopic, weak solution, 70
anisotropic, 69, 79
anisotropic, Cauchy problem, 79
local weak solution, 83
signed, 45

stationary solutions, 70
variable exponent, 46
weak solution, 46, 61
with lower-Order terms, 60
Pullback attractor, 240

**S**
Schauder Fixed Point Theorem, 47
Steklov's means, 146

**V**
Viscoelastic plate, 263

**W**
Waiting time, 81, 197
infinite, 205

**Y**
Young's inequality, 3, 245

MIX
Papier aus verantwortungsvollen Quellen
Paper from responsible sources
FSC® C105338

If you have any concerns about our products,
you can contact us on
**ProductSafety@springernature.com**

In case Publisher is established outside the EU,
the EU authorized representative is:
**Springer Nature Customer Service Center GmbH**
**Europaplatz 3, 69115 Heidelberg, Germany**

Printed by Libri Plureos GmbH
in Hamburg, Germany